# Advances in Heart Valve Biomechanics

Michael S. Sacks • Jun Liao
Editors

# Advances in Heart Valve Biomechanics

Valvular Physiology, Mechanobiology, and Bioengineering

*Editors*
Michael S. Sacks
The Oden Institute and the
Department of Biomedical Engineering
The University of Texas at Austin
Austin, TX, USA

Jun Liao
The Department of Bioengineering
The University of Texas at Arlington
Arlington, TX, USA

ISBN 978-3-030-01991-4     ISBN 978-3-030-01993-8 (eBook)
https://doi.org/10.1007/978-3-030-01993-8

© Springer Nature Switzerland AG 2018
This work is subject to copyright. All rights are reserved by the Publisher, whether the whole or part of the material is concerned, specifically the rights of translation, reprinting, reuse of illustrations, recitation, broadcasting, reproduction on microfilms or in any other physical way, and transmission or information storage and retrieval, electronic adaptation, computer software, or by similar or dissimilar methodology now known or hereafter developed.
The use of general descriptive names, registered names, trademarks, service marks, etc. in this publication does not imply, even in the absence of a specific statement, that such names are exempt from the relevant protective laws and regulations and therefore free for general use.
The publisher, the authors, and the editors are safe to assume that the advice and information in this book are believed to be true and accurate at the date of publication. Neither the publisher nor the authors or the editors give a warranty, express or implied, with respect to the material contained herein or for any errors or omissions that may have been made. The publisher remains neutral with regard to jurisdictional claims in published maps and institutional affiliations.

This Springer imprint is published by the registered company Springer Nature Switzerland AG
The registered company address is: Gewerbestrasse 11, 6330 Cham, Switzerland

# Preface

Heart valves (HVs) are cardiac structures that ensure unidirectional blood flow during the cardiac cycle. However, this description does not adequately describe their biomechanical function, which is multimodal, and their loading cycle is repeated every second. While they are primarily passive structures driven by forces exerted by the surrounding blood and heart, this description also does not adequately describe their elegant and complex biomechanical performance. The semilunar valves (pulmonary (PV) and aortic (AV)) prevent retrograde flow back into the ventricles during diastole, while the atrioventricular valves (tricuspid (TV) and mitral (MV)) prohibit reverse flow from the ventricle to the atrium during systole. They must replicate this feat with each heartbeat; over a single lifetime, HVs will open and close at least $3 \times 10^9$ times.

As in many physiological systems, one can approach heart valve biomechanics from a multi-length scale approach, since mechanical stimuli occur and have biological impact at the organ, tissue, and cellular scales. For example, valve interstitial cells (VICs) are known to respond to local tissue stress by altering cellular stiffness through valvular remodeling and valvular pathologies. On the other hand, the fact that AV diseases preferentially occur in the aortic side (fibrosa) of the valvular leaflets, where they are exposed to unstable flow conditions, highlights the importance of shear in AV biology and pathobiology. Another important point is that valvular extracellular matrix (ECM) is composed of dense network of collagen, elastin, and GAGs and is thus functionally and mechanically very different from other cardiovascular structures (e.g., blood vessels, myocardium). In fact, valvular ECM is structurally and behaves mechanically much more like the dense planar connective tissues of the musculoskeletal system. They are unique, however, in that they must function within a blood-contacting environment and are thus coated with endothelial cell monolayer. Moreover, there is evidence that endothelium/interstitial cell communication may play an important role in valvular ECM homeostasis. Yet, despite its clinical importance, the unique and demanding valvular biological/biomechanical environment is relatively unexplored, with most research focusing on valvular prosthetic design and development. In addition to deepening our understanding of physiological and biological function, simulation and imaging technologies have reached a level of sophistication that major contributions to medicine and pathophysiology are now possible. Heart valves represent one of the most demanding areas in cardiovascular simulation due to issues such as complex 3D surface geometry, rapid motion, hemodynamic force levels, complex tissue mechanical behavior, and strong mechanical interactions with the surrounding cardiac and vascular structures. This book focuses on heart valve functional biomechanics. Specifically, we refer to the unique aspects of valvular function, valvular mechanobiology, mechanical behavior at various hierarchical levels, tissue remodeling in repair/regeneration, hemodynamics, assessment technologies, and simulation technologies that achieve this remarkable feat.

The editors would like to thank all the chapter contributors, whose hard work, invaluable support, and patience made this book possible.

Austin, TX, USA  Michael S. Sacks
Arlington, TX, USA  Jun Liao

# Abbreviations

| | |
|---|---|
| 5-HT/5-HTR | Serotonin/serotonin receptor SALS |
| AC | Against curvature |
| AFM | Atomic force microscopy |
| ALP | Alkaline phosphatase |
| APC | Adenomatous polyposis coli |
| AV | Aortic valve |
| AV | Atrioventricular |
| AVC | Atrioventricular canal |
| aVIC | Activated valve interstitial cell |
| AVIC | Aortic valve interstitial cell |
| BAV | Bicuspid aortic valve |
| BMP | Bone morphogenic protein |
| BMP/BMPR | Bone morphogenic protein/bone morphogenic protein receptor |
| CAVD | Calcific aortic valve disease |
| CFA | Collagen fiber architecture |
| CHD | Congenital heart disease |
| CRGDS | Adhesive peptide sequence (Cys-Arg-Gly-asp-Ser) |
| dpc | Days post conception |
| E | Embryonic day |
| ECM | Extracellular matrix |
| $E_{\text{eff}}$ | Effective stiffness |
| EGF | Epidermal growth factor |
| EMT | Endothelial-to-mesenchymal transformation |
| EndoMT | Endothelial-mesenchymal transformation |
| ERK | Extracellular signal-regulated kinase |
| ET-1 | Endothelin 1 |
| FBN1 | Fibrillin-1 |
| FE | Finite element |
| FGF | Fibroblast growth factor |
| FIB-SEM | Focused ion beam—scanning electron microscope |

| | |
|---|---|
| HSP47 | Heat shock protein 47 |
| IP3 | Inositol triphosphate |
| LPS | Lipopolysaccharide |
| LRP | Low-density lipoprotein receptor-related protein |
| MA | Micropipette aspiration |
| MEK | Mitogen-activated protein kinase |
| MMP | Matrix metalloproteinase |
| MV | Mitral valve |
| MVAL | Mitral valve anterior leaflet |
| MVIC | Mitral valve interstitial cell |
| MVP | Mitral valve prolapse |
| NAR | Nuclear Apsect ratio |
| NFAT | Nuclear factor of activated T-cells |
| NFATc1 | Nuclear factor of activated T-cells |
| NF-κB | Nuclear factor kappa-light-chain-enhancer of activated B cells |
| NICD | Notch intracellular domain |
| NO | Nitric oxide |
| NOI | Normalized orientation index |
| NOS3 | Nitric oxide synthase 3 |
| OFT | Outflow tract |
| pAVIC | Porcine aortic valve interstitial cell |
| PDGFR | Platelet derived growth factor receptor |
| PECAM | Platelet/endothelial cell adhesion molecule |
| PEG | Poly (ethylene glycol) |
| PKC | Protein kinase C |
| PLC | Phospholipase C |
| PV | Pulmonary valve |
| PVIC | Pulmonary valve interstitial cell |
| qVIC | Quiescent valve interstitial cell |
| RNA | Ribonucleic acid |
| RVE | Representative volume element |
| SALS | Small angle light scattering |
| SAXS | Small angle X-ray scattering |
| Shp2 | Src homology phosphatase 2 |
| SL | Semilunar |
| SMA | Smooth muscle actin |
| SVK | Saint-Venant Kirchhoff |
| TEHV | Tissue engineered heart valve |
| TEM | Transmission Electron microscopy |
| TGF-β | Transforming growth factor-β |
| TGF-β1 | Transforming growth factor beta 1 |
| TIMPs | Tissue inhibitor of metalloproteinases |
| TLR4 | Toll-like receptor 4 |
| TV | Tricuspid valve |

| | |
|---|---|
| TVIC | Tricuspid valve interstitial cell |
| TVP | Transvalvular pressure |
| US | United States |
| VEC | Valve endothelial cell |
| VEC | Valve interstitial cell |
| VE-cadherin | Vascular endothelial-cadherin |
| VEGF | Vascular endothelial growth factor |
| VEGF/VEGFR | Vascular endothelial growth factor/vascular endothelial growth factor receptor |
| VIC | Valve endothelial cell |
| VIC | Valve interstitial cell |
| WC | With curvature |
| Wnt | Wingless-related integration site |
| α-SMA | α-Smooth muscle actin |

# Contents

**Part I Native Heart Valve Function**

**Biological Mechanics of the Heart Valve Interstitial Cell**............ 3
Alex Khang, Rachel M. Buchanan, Salma Ayoub, Bruno V. Rego,
Chung-Hao Lee, and Michael S. Sacks
1 Introduction.................................................. 4
2 Major Questions and Challenges............................... 5
3 Advances in Investigating VIC Mechanobiology................. 6
   3.1 Isolated Cell Studies.................................... 6
   3.2 In Situ Tissue-Level Evaluation of VIC Contraction
       Behaviors............................................ 14
   3.3 Interlayer Micromechanics of the AV Leaflet............. 16
   3.4 Down-Scale Model of the VIC Within ECM............... 18
   3.5 Use of 3D Hydrogel for Mechanobiological Studies........ 21
   3.6 Uniaxial Planar Stretch Bioreactors..................... 23
   3.7 Model-Driven Experimental Design..................... 28
4 Future Directions............................................. 31
References...................................................... 33

**Endothelial Mechanotransduction**............................... 37
James N. Warnock
1 Introduction.................................................. 37
   1.1 Mechanical Environment of the Aortic Valve Endothelium.... 38
   1.2 Mechanisms of Mechanosensation....................... 39
   1.3 Pathological Implications of Mechanotransduction......... 40
   1.4 In Vitro and Ex Vivo Methods for Mechanotransduction
       Experiments.......................................... 41
2 Shear Stress.................................................. 43
   2.1 Aortic Valve Endothelial Cells Under Physiologic
       Flow Conditions...................................... 43
   2.2 Role of Shear Stress in Aortic Valve Pathology............ 44

| | | | |
|---|---|---|---|
| 3 | Cyclic Strain | | 46 |
| | 3.1 | Endothelial Layer Integrity Under Elevated Cyclic Strain | 47 |
| | 3.2 | Pro-inflammatory Response of Endothelial Cells to Cyclic Strain | 49 |
| 4 | Summary | | 50 |
| References | | | 51 |

**The Role of Proteoglycans and Glycosaminoglycans in Heart Valve Biomechanics** .................................................. 59
Varun K. Krishnamurthy and K. Jane Grande-Allen

| | | |
|---|---|---|
| 1 | Introduction and Chapter Overview | 59 |
| 2 | Proteoglycans and Glycosaminoglycans in Valves: Composition, Distribution, and Interactions | 60 |
| 3 | Glycosaminoglycan Stability Affects Tissue Biomechanics in Bioprostheses | 62 |
| 4 | Theoretical and Experimental Determination of Glycosaminoglycan Biomechanics in Native Valve Tissue | 64 |
| 5 | Bioreactor Studies to Evaluate Valve Tissue Biomechanics and ECM Remodeling | 66 |
| 6 | Correlating Glycosaminoglycan Levels with Tissue Biomechanics (Normal vs. Disease) | 69 |
| 7 | Effect of Cell Stimulation on Proteoglycan and Glycosaminoglycan Response | 72 |
| 8 | Summary | 74 |
| References | | 75 |

**On the Unique Functional Elasticity and Collagen Fiber Kinematics of Heart Valve Leaflets** ..................................... 81
Jun Liao and Michael S. Sacks

| | | | |
|---|---|---|---|
| 1 | Introduction | | 82 |
| | 1.1 | Function of Heart Valve Leaflets | 82 |
| | 1.2 | Composition and Ultrastructure and Heart Valve Leaflets | 82 |
| | 1.3 | Valvular Diseases and Viscoelastic Properties of Leaflet Tissues | 84 |
| 2 | Conventional Concepts on Viscoelasticity of Soft Tissues | | 85 |
| 3 | The Unique Viscoelastic Properties of Heart Valve Tissues: Normal Stress Relaxation but Negligible Creep | | 87 |
| 4 | Collagen Fibril Kinematics in Valve Leaflet Stretch, Stress Relaxation, and Creep | | 89 |
| | 4.1 | Hierarchical Structure of Collagen in Heart Valve Leaflets | 89 |
| | 4.2 | SAXS, Biaxial Device for SAXS Beamline, and Experimental Protocols | 92 |
| | 4.3 | Collagen Fibril Kinematics in MV Leaflets | 96 |
| 5 | Final Remarks | | 99 |
| References | | | 100 |

**Tricuspid Valve Biomechanics: A Brief Review** .................... 105
William D. Meador, Mrudang Mathur, and Manuel K. Rausch
1 Introduction ............................................. 105
2 Tricuspid Leaflets ........................................ 106
  2.1 Morphology and Nomenclature ....................... 106
  2.2 In Vivo Studies ..................................... 106
  2.3 In Vitro Studies .................................... 107
  2.4 In Silico Studies ................................... 108
3 Tricuspid Annulus ....................................... 108
  3.1 Morphology and Nomenclature ....................... 108
  3.2 In Vivo Studies ..................................... 109
  3.3 In Vitro Studies .................................... 109
  3.4 In Silico Studies ................................... 110
4 Tricuspid Chordae Tendineae ............................. 110
  4.1 Morphology and Nomenclature ....................... 110
  4.2 In Vivo Studies ..................................... 111
  4.3 In Vitro Studies .................................... 111
  4.4 In Silico Studies ................................... 111
5 Most Recent Studies ..................................... 112
6 Conclusion .............................................. 112
References ................................................. 112

**Measurement Technologies for Heart Valve Function** .............. 115
Morten O. Jensen, Andrew W. Siefert, Ikechukwu Okafor,
and Ajit P. Yoganathan
1 Introduction ............................................. 116
2 Experimental Models .................................... 116
  2.1 Aortic and Pulmonary Pulse Duplicators ............. 116
  2.2 Right and Left Heart Bench Models ................. 116
  2.3 Steady Backpressure Model ......................... 119
  2.4 Restored Contractility Model ....................... 119
  2.5 Large Animal Models ............................... 119
  2.6 Patients ............................................ 120
3 Assessing Valve Geometry, Dynamics, and Tissue Deformation ...... 120
  3.1 Echocardiography ................................... 120
  3.2 Magnetic Resonance Imaging ........................ 121
  3.3 Cardiac Computed Tomography and Micro Computed
      Tomography ........................................ 123
  3.4 Sonomicrometry .................................... 123
  3.5 Biplane Videofluoroscopy ........................... 124
  3.6 Stereophotogrammetry .............................. 124
  3.7 High-Speed Photography and Videography ........... 125
  3.8 Trigonometry ....................................... 125

| | | | |
|---|---|---|---|
| 4 | Assessing Flow Characteristics | | 126 |
| | 4.1 | Bulk Flow Characteristics | 126 |
| | 4.2 | Doppler Echocardiography | 128 |
| | 4.3 | Detailed Flow Field Characterization | 130 |
| 5 | Technologies for Quantifying Atrioventricular Valve and Repair Device Mechanics | | 134 |
| | 5.1 | Assessing Subvalvular Mechanics of the Atrioventricular Valves | 135 |
| | 5.2 | Assessing Annular-Based Mechanics | 137 |
| | 5.3 | Heart Valve Force Transducers with External Anchoring | 140 |
| 6 | Additional Modalities for Assessing Prosthetic Heart Valves | | 141 |
| 7 | General Discussion and Final Remarks | | 141 |
| References | | | 142 |

## Part II  Heart Valve Disease and Treatment

**Calcific Aortic Valve Disease: Pathobiology, Basic Mechanisms, and Clinical Strategies** .................................................... 153
Payal Vyas, Joshua D. Hutcheson, and Elena Aikawa

| | | | |
|---|---|---|---|
| 1 | Introduction | | 153 |
| 2 | Normal Aortic Valve Physiology and Function | | 155 |
| | 2.1 | Anatomy of Normal Aortic Valve | 155 |
| | 2.2 | Function of Normal Aortic Valve | 157 |
| | 2.3 | Aortic Valve Microstructure | 157 |
| | 2.4 | Aortic Valve Cell Biology | 159 |
| 3 | Calcific Aortic Valve Disease Pathobiology | | 161 |
| 4 | Mechanisms of Calcific Aortic Valve Disease | | 163 |
| | 4.1 | Cellular Mechanisms | 163 |
| | 4.2 | Molecular Mechanisms | 167 |
| 5 | Diagnostic and Therapeutic Approaches | | 169 |
| | 5.1 | Aortic Valve Replacement | 169 |
| | 5.2 | Potential Drug Targets and Clinical Trials | 170 |
| | 5.3 | Clinical Imaging and Identification | 171 |
| 6 | Future Perspectives | | 171 |
| References | | | 172 |

**Remodelling Potential of the Mitral Heart Valve Leaflet** .............. 181
Bruno V. Rego, Sarah M. Wells, Chung-Hao Lee, and Michael S. Sacks

| | | | |
|---|---|---|---|
| 1 | Introduction | | 182 |
| 2 | Methods | | 184 |
| | 2.1 | Mechanical Database and Processing | 184 |
| | 2.2 | Accounting for Changes in Leaflet Dimensions | 185 |
| | 2.3 | Constitutive Model | 187 |

|   | 2.4 | Parameter Estimation | 189 |
|---|---|---|---|
|   | 2.5 | Interpretation of Parameter Values | 191 |
|   | 2.6 | MVIC Geometry | 192 |
| 3 | Results | | 193 |
|   | 3.1 | Leaflet Dimensions | 193 |
|   | 3.2 | Collagen and Elastin Fractions | 194 |
|   | 3.3 | Elastin Moduli | 195 |
|   | 3.4 | Collagen Fibre Architecture | 195 |
|   | 3.5 | Structural Effects of Growth and Remodelling | 196 |
|   | 3.6 | Quantitative MVIC Geometry | 197 |
| 4 | Discussion | | 199 |
|   | 4.1 | General Overview | 199 |
|   | 4.2 | Delayed Restoration of Homeostasis | 201 |
|   | 4.3 | Generalization and Special Considerations | 202 |
|   | 4.4 | Limitations | 203 |
|   | 4.5 | Conclusions | 203 |
| References | | | 204 |

**Molecular and Cellular Developments in Heart Valve Development and Disease** .......... 207
Lindsey J. Anstine, Anthony S. Baker, and Joy Lincoln

| 1 | Introduction | | 208 |
|---|---|---|---|
| 2 | Structure-Function Relationships of Mature Heart Valves | | 208 |
|   | 2.1 | Extracellular Matrix | 209 |
|   | 2.2 | Valve Cell Populations | 210 |
| 3 | Heart Valve Development | | 211 |
|   | 3.1 | Embryonic Precursor Cells | 211 |
|   | 3.2 | Molecular Regulators of Endocardial Cushion Formation | 211 |
|   | 3.3 | The Contribution of Non-endothelial Cell Lineages to the Developing Heart Valve Structures | 217 |
|   | 3.4 | Molecular Regulation of Embryonic Heart Valve Remodeling | 218 |
| 4 | Heart Valve Disease | | 219 |
|   | 4.1 | Bicuspid Aortic Valve | 220 |
|   | 4.2 | Mitral Valve Prolapse | 222 |
|   | 4.3 | Pathobiology of Valve Interstitial Cells | 225 |
|   | 4.4 | Pathobiology of Valve Endothelial Cells | 226 |
| 5 | Current Treatments and Clinical Perspectives | | 229 |
| References | | | 229 |

**Mechanical Mediation of Signaling Pathways in Heart Valve Development and Disease** .......... 241
Ishita Tandon, Ngoc Thien Lam, and Kartik Balachandran

| 1 | Introduction | 241 |
|---|---|---|
| 2 | The Mechanical Environment of the Developing and Adult Heart Valve | 243 |

|     |     | 2.1 | The Developing Heart Valve | 243 |
| --- | --- | --- | --- | --- |
|     |     | 2.2 | The Mature Heart Valve | 244 |
|     |     | 2.3 | Signaling Pathways Involved During Valve Morphogenesis | 246 |
|     |     | 2.4 | Signaling Pathways During Valve Disease | 251 |
|     | 3 | Mechanobiology of Valvulogenesis and Disease | | 254 |
|     | 4 | Conclusion | | 258 |
| References | | | | 258 |

**Tissue Engineered Heart Valves** ............................................. 263
Jay M. Reimer and Robert T. Tranquillo

| 1 | Background | | 263 |
| --- | --- | --- | --- |
|   | 1.1 | Existing Therapies | 265 |
| 2 | Tissue Engineered Heart Valves | | 267 |
|   | 2.1 | Cell Sources | 267 |
|   | 2.2 | Scaffold Materials | 269 |
|   | 2.3 | Stimulation Paradigms | 271 |
| 3 | Valve Design | | 273 |
| 4 | In Vitro Functional Testing | | 275 |
| 5 | In Vivo Testing | | 276 |
|   | 5.1 | Animal Models for TEHVs | 276 |
|   | 5.2 | Preclinical Testing Results with TEHVs | 277 |
| 6 | Remaining Challenges | | 280 |
| 7 | Summary | | 281 |
| References | | | 281 |

**Decellularization in Heart Valve Tissue Engineering** ............... 289
Katherine M. Copeland, Bo Wang, Xiaodan Shi, Dan T. Simionescu,
Yi Hong, Pietro Bajona, Michael S. Sacks, and Jun Liao

| 1 | Introduction | | 290 |
| --- | --- | --- | --- |
| 2 | Fundamentals of Tissue-Engineered Heart Valves | | 292 |
| 3 | Components and Ultrastructure of Aortic Valves | | 293 |
|   | 3.1 | Structure of Aortic Valve | 293 |
|   | 3.2 | Cellular Components of Aortic Valve | 293 |
| 4 | Decellularization and Applications in TEHV | | 294 |
|   | 4.1 | Concept of Decellularization and Acellular HV Scaffolds | 294 |
|   | 4.2 | Decellularization Techniques Available for HV Decellularization | 295 |
| 5 | Effects of Decellularization Protocols on Leaflet Biomechanics and a Study on Fatigue Mechanism | | 296 |
| 6 | Cell Sources for TEHV | | 302 |
| 7 | Bioreactors for TEHV Development | | 304 |
| 8 | TEHV via Decellularization Approach: Large Animal Studies and Clinical Trials | | 306 |

| | | |
|---|---|---|
| 9 | Current Challenges in TEHV via Decellularization | 307 |
| 10 | Final Remarks | 310 |
| References | | 313 |

**Novel Bioreactors for Mechanistic Studies of Engineered Heart Valves** .................................................. 319
Kristin Comella and Sharan Ramaswamy

| | | | |
|---|---|---|---|
| 1 | Introduction | | 320 |
| 2 | Bioreactors | | 320 |
| | 2.1 | Design | 320 |
| | 2.2 | Mechanical Environment | 322 |
| 3 | Scaffolds | | 325 |
| | 3.1 | Synthetic | 326 |
| | 3.2 | Natural | 326 |
| 4 | Cell Types | | 327 |
| | 4.1 | Mesenchymal Stem Cells (MSCs) | 327 |
| | 4.2 | Fibroblasts/Smooth Muscle Cells/Valvular Interstitial Cells | 328 |
| | 4.3 | Endothelial Progenitor Cells | 328 |
| | 4.4 | Embryonic/IPS/Amniotic/Cord | 328 |
| 5 | Preclinical/Clinic Studies | | 330 |
| 6 | Conclusions | | 331 |
| References | | | 332 |

**Bioprosthetic Heart Valves: From a Biomaterials Perspective** ......... 337
Naren Vyavahare and Hobey Tam

| | | |
|---|---|---|
| 1 | Introduction | 338 |
| 2 | Growing the Heart Valve Market Space | 338 |
| 3 | Current Indications for Use | 339 |
| 4 | Current Treatments for Valvular Disease | 341 |
| 5 | Current Gold Standards in Heart Valve Replacements | 344 |
| 6 | Mechanical Heart Valves (MHVs) | 344 |
| 7 | Bioprosthetic Heart Valves Overview (BHVs) | 346 |
| 8 | BHV Base Materials: Bovine Pericardium (BP) and Porcine Aortic Heart Valve Leaflets (PAVs) | 347 |
| 9 | BHV Drawbacks | 352 |
| 10 | Industry Practice of BHV Fabrication | 356 |
| 11 | Underlying Factors in Failure Modes of BHVs | 358 |
| 12 | Mechanisms of Calcification | 358 |
| 13 | Mechanisms of Structural Degradation | 360 |
| 14 | Barriers to Innovation | 365 |
| 15 | Add on Methods to Curb Shortcomings of Glutaraldehyde Crosslinking in BHVs | 366 |
| 16 | Alternative Crosslinking Chemistries to Innovate a Novel Biomaterial for BHV Fabrication | 367 |

| | | |
|---|---|---|
| 17 | Combined Approaches to Curb Structural Degradation and Calcification | 370 |
| 18 | Serving a New Demographic | 372 |
| 19 | Contrasting Patient Demographics; Contrasting Design Requirements for Heart Valves | 373 |
| 20 | Designing for Emerging Markets and Developed Markets | 374 |
| 21 | Summary and Conclusion | 374 |
| References | | 375 |

## Part III  Computational Approaches for Heart Valve Function

**Computational Modeling of Heart Valves: Understanding and Predicting Disease** ........................................ 385
Ahmed A. Bakhaty, Ali Madani, and Mohammad R. K. Mofrad

| | | |
|---|---|---|
| 1 | Introduction to Computational Modeling of Heart Valves | 386 |
| | 1.1 Computational Mechanics | 387 |
| | 1.2 Tissue Modeling | 389 |
| | 1.3 Computational Fluid Dynamics (CFD) and Fluid Structure Interaction (FSI) | 390 |
| | 1.4 Molecular Dynamics | 392 |
| | 1.5 Multiscale Methods | 393 |
| 2 | Aortic Valve Models | 394 |
| | 2.1 Aortic Valve Computational Models | 394 |
| | 2.2 Modeling of Calcific Aortic Stenosis | 396 |
| | 2.3 Bicuspid Aortic Valve | 396 |
| | 2.4 Aortic Valve Surgical Modeling | 397 |
| 3 | Mitral Valve Models | 398 |
| | 3.1 Mitral Valve Computational Models | 398 |
| | 3.2 Mitral Valve Repair | 399 |
| 4 | Pulmonary and Tricuspid Valve Models | 399 |
| | 4.1 Pulmonary Valve | 399 |
| | 4.2 Tricuspid Valve | 401 |
| 5 | Artificial Valves | 401 |
| | 5.1 Artificial Valve Modeling | 401 |
| | 5.2 Mechanical Valves | 403 |
| | 5.3 Bioprosthetic Valves | 404 |
| | 5.4 Tissue-Engineered Valves | 404 |
| 6 | Future Directions | 405 |
| | 6.1 Summary: A Discussion on the Validity of Models | 405 |
| | 6.2 Heart Systems | 405 |
| | 6.3 Multiscale Considerations | 406 |
| | 6.4 Patient-Specific Modeling | 406 |
| | 6.5 Intraoperative Modeling | 406 |
| | 6.6 Non-Physics-Based Computation | 406 |
| References | | 407 |

**Biomechanics and Modeling of Tissue-Engineered Heart Valves** . . . . . . 413
T. Ristori, A. J. van Kelle, F. P. T. Baaijens, and S. Loerakker
1   Introduction . . . . . . . . . . . . . . . . . . . . . . . . . . . . . . . . . . . . . . . . . . 414
    1.1   In Vitro Heart Valve Tissue Engineering . . . . . . . . . . . . . . . . . . . 414
    1.2   In Situ Heart Valve Tissue Engineering . . . . . . . . . . . . . . . . . . . 415
    1.3   Limitations and Challenges . . . . . . . . . . . . . . . . . . . . . . . . . . . . 415
    1.4   Outline . . . . . . . . . . . . . . . . . . . . . . . . . . . . . . . . . . . . . . . . . . 416
2   Biomechanical Properties of Native and Tissue-Engineered
    Heart Valves . . . . . . . . . . . . . . . . . . . . . . . . . . . . . . . . . . . . . . . . . . 416
    2.1   Heart Valve Structure, Function, and Biomechanics . . . . . . . . . . 416
    2.2   TEHV Biomechanical Properties . . . . . . . . . . . . . . . . . . . . . . . . 418
    2.3   Computational Simulations of TEHVs . . . . . . . . . . . . . . . . . . . . 418
3   Mathematical Models of Collagen Remodeling . . . . . . . . . . . . . . . . . 419
    3.1   Collagen Architecture in Native Aortic Heart Valves . . . . . . . . . 419
    3.2   Collagen Remodeling in Response to Mechanical Stimuli . . . . . . 420
    3.3   Early Computational Models . . . . . . . . . . . . . . . . . . . . . . . . . . . 420
    3.4   Realignment with Principal Loading Directions . . . . . . . . . . . . . 421
    3.5   Realignment in Between Principal Loading Directions . . . . . . . . 422
    3.6   Inclusion of Fiber Dispersity . . . . . . . . . . . . . . . . . . . . . . . . . . . 422
    3.7   Main Limitations . . . . . . . . . . . . . . . . . . . . . . . . . . . . . . . . . . . 423
4   Modeling Stress Fiber Remodeling . . . . . . . . . . . . . . . . . . . . . . . . . . 424
    4.1   Actin Stress Fibers . . . . . . . . . . . . . . . . . . . . . . . . . . . . . . . . . . 424
    4.2   SF Remodeling . . . . . . . . . . . . . . . . . . . . . . . . . . . . . . . . . . . . 424
    4.3   Computational Models for SF Remodeling . . . . . . . . . . . . . . . . . 425
    4.4   Stress Homeostasis . . . . . . . . . . . . . . . . . . . . . . . . . . . . . . . . . . 425
    4.5   Stress and Strain Homeostasis . . . . . . . . . . . . . . . . . . . . . . . . . . 427
    4.6   SF Assembly Dependent on Strain and Strain Rate . . . . . . . . . . . 428
    4.7   Model Based on Thermodynamics . . . . . . . . . . . . . . . . . . . . . . . 429
    4.8   Main Challenges . . . . . . . . . . . . . . . . . . . . . . . . . . . . . . . . . . . . 429
5   Prestress and Cell-Mediated Collagen Remodeling . . . . . . . . . . . . . . . 430
    5.1   Cell-Mediated Collagen Turnover . . . . . . . . . . . . . . . . . . . . . . . 430
    5.2   Cell-Mediated Prestress . . . . . . . . . . . . . . . . . . . . . . . . . . . . . . 431
    5.3   Phenomenological Models of Cell-Mediated
          Collagen Remodeling . . . . . . . . . . . . . . . . . . . . . . . . . . . . . . . . 431
    5.4   Biologically Motivated Models for Cell-Mediated Collagen
          Remodeling . . . . . . . . . . . . . . . . . . . . . . . . . . . . . . . . . . . . . . . 432
    5.5   Inclusion of SF Remodeling . . . . . . . . . . . . . . . . . . . . . . . . . . . 433
    5.6   Remodeling of TEHVs . . . . . . . . . . . . . . . . . . . . . . . . . . . . . . . 434
    5.7   Future Scope . . . . . . . . . . . . . . . . . . . . . . . . . . . . . . . . . . . . . . 434
6   Future Directions . . . . . . . . . . . . . . . . . . . . . . . . . . . . . . . . . . . . . . 434
    6.1   Growth . . . . . . . . . . . . . . . . . . . . . . . . . . . . . . . . . . . . . . . . . . 435
    6.2   Agent-Based Models . . . . . . . . . . . . . . . . . . . . . . . . . . . . . . . . 436
    6.3   How Far Should We Go? . . . . . . . . . . . . . . . . . . . . . . . . . . . . . 436
    6.4   Implications for TEHVs . . . . . . . . . . . . . . . . . . . . . . . . . . . . . . 437
References . . . . . . . . . . . . . . . . . . . . . . . . . . . . . . . . . . . . . . . . . . . . . . . 437

**Fluid–Structure Interaction Analysis of Bioprosthetic Heart Valves: the Application of a Computationally-Efficient Tissue Constitutive Model** .......................................... 447
Rana Zakerzadeh, Michael C. H. Wu, Will Zhang, Ming-Chen Hsu, and Michael S. Sacks
1 Introduction ........................................... 448
2 Modeling Framework .................................... 450
   2.1 Thin Shell Formulations for the BHV Leaflets .............. 450
   2.2 Leaflet Tissue Material Model .......................... 452
   2.3 Implementation Verification of the Material Model .......... 457
3 Numerical Simulations ................................... 458
   3.1 Setup of the Immersogeometric FSI Simulation ............. 459
   3.2 Different Levels and Directions of Anisotropy Using Effective Model ......................................... 460
   3.3 A Qualitative Study of the Crosslinking Effects on FSI Simulation of BHV ................................... 462
4 Discussion ............................................. 465
5 Conclusions ............................................ 467
References ................................................ 467

**Towards Patient-Specific Mitral Valve Surgical Simulations** ......... 471
Amir H. Khalighi, Bruno V. Rego, Andrew Drach, Robert C. Gorman, Joseph H. Gorman, and Michael S. Sacks
1 Introduction ........................................... 472
2 Image-Guided 3D MV Analysis ............................ 473
3 A Comprehensive Pipeline for Multi-Resolution Modeling of the Mitral Valve ...................................... 474
4 Exploitation of Image-Based Biomechanical Modeling ........... 477
   4.1 rt-3DE In Vivo Data Acquisition and Segmentation .......... 478
   4.2 In Vivo Leaflet Geometric Model Development Pipeline ...... 478
   4.3 Development of the MV FE Model ...................... 479
   4.4 Development of a Functionally Equivalent Subvalvular Apparatus ................................ 480
5 Future Directions ....................................... 482
References ................................................ 483

# Contributors

**Elena Aikawa** Department of Cardiovascular Medicine, Brigham and Women's Hospital, Harvard Medical School, Boston, MA, USA

**Lindsey J. Anstine** Center for Cardiovascular Research and The Heart Center at Nationwide Children's Hospital Research Institute, Columbus, OH, USA

**Salma Ayoub** James T. Willerson Center for Cardiovascular Modeling and Simulation, The Oden Institute and the Department of Biomedical Engineering, The University of Texas at Austin, Austin, TX, USA

**F. P. T. Baaijens** Department of Biomedical Engineering, Eindhoven University of Technology, Eindhoven, The Netherlands

Institute for Complex Molecular Systems, Eindhoven University of Technology, Eindhoven, The Netherlands

**Pietro Bajona** Department of Cardiovascular and Thoracic Surgery, University of Texas Southwestern Medical Center, Dallas, TX, USA

**Anthony S. Baker** Health Sciences Library Medical Visuals, The Ohio State University, Columbus, OH, USA

**Ahmed A. Bakhaty** Department of Civil and Environmental Engineering, University of California, Berkeley, CA, USA

Department of Electrical Engineering and Computer Science, University of California, Berkeley, CA, USA

Department of Mathematics, University of California, Berkeley, CA, USA

**Kartik Balachandran** Department of Biomedical Engineering, University of Arkansas, Fayetteville, AR, USA

**Rachel M. Buchanan** James T. Willerson Center for Cardiovascular Modeling and Simulation, The Oden Institute and the Department of Biomedical Engineering, The University of Texas at Austin, Austin, TX, USA

**Kristin Comella** Tissue Engineered Mechanics, Imaging and Materials Laboratory (TEMIM Lab), Department of Biomedical Engineering, College of Engineering and Computing, Florida International University, Miami, FL, USA

**Katherine M. Copeland** Department of Bioengineering, University of Texas at Arlington, Arlington, TX, USA

**Andrew Drach** James T. Willerson Center for Cardiovascular Modeling and Simulation, Institute for Computational Engineering and Sciences, Department of Biomedical Engineering, The University of Texas at Austin, Austin, TX, USA

**Joseph H. Gorman** Gorman Cardiovascular Research Group, Smilow Center for Translational Research, Perelman School of Medicine, University of Pennsylvania, Philadelphia, PA, USA

**Robert C. Gorman** Gorman Cardiovascular Research Group, Smilow Center for Translational Research, Perelman School of Medicine, University of Pennsylvania, Philadelphia, PA, USA

**K. Jane Grande-Allen** Department of Bioengineering, Rice University, Houston, TX, USA

**Yi Hong** Department of Bioengineering, University of Texas at Arlington, Arlington, TX, USA

**Ming-Chen Hsu** Department of Mechanical Engineering, Iowa State University, Ames, IA, USA

**Joshua D. Hutcheson** Department of Cardiovascular Medicine, Brigham and Women's Hospital, Harvard Medical School, Boston, MA, USA

**Morten O. Jensen** Department of Biomedical Engineering, University of Arkansas, Fayetteville, AR, USA

**Amir H. Khalighi** James T. Willerson Center for Cardiovascular Modeling and Simulation, Institute for Computational Engineering and Sciences, Department of Biomedical Engineering, The University of Texas at Austin, Austin, TX, USA

**Alex Khang** James T. Willerson Center for Cardiovascular Modeling and Simulation, The Oden Institute and the Department of Biomedical Engineering, The University of Texas at Austin, Austin, TX, USA

**Varun K. Krishnamurthy** Department of Bioengineering, Rice University, Houston, TX, USA

**Ngoc Thien Lam** Department of Biomedical Engineering, University of Arkansas, Fayetteville, AR, USA

**Chung-Hao Lee** School of Aerospace and Mechanical Engineering, The University of Oklahoma, Norman, OK, USA

James T. Willerson Center for Cardiovascular Modeling and Simulation, Institute for Computational Engineering and Sciences, Department of Biomedical Engineering, The University of Texas at Austin, Austin, TX, USA

**Jun Liao** Tissue Biomechanics and Bioengineering Laboratory, The Department of Bioengineering, The University of Texas at Arlington, Arlington, TX, USA

**Joy Lincoln** Center for Cardiovascular Research and The Heart Center at Nationwide Children's Hospital Research Institute, Columbus, OH, USA

Department of Pediatrics, The Ohio State University, Columbus, OH, USA

**S. Loerakker** Department of Biomedical Engineering, Eindhoven University of Technology, Eindhoven, The Netherlands

Institute for Complex Molecular Systems, Eindhoven University of Technology, Eindhoven, The Netherlands

**Ali Madani** Department of Applied Science and Technology and Mechanical Engineering, University of California, Berkeley, CA, USA

**Mrudang Mathur** Department of Mechanical Engineering, University of Texas at Austin, Austin, TX, USA

**William D. Meador** Department of Biomedical Engineering, University of Texas at Austin, Austin, TX, USA

**Mohammad R. K. Mofrad** Department of Mechanical and Bioengineering, University of California, Berkeley, CA, USA

**Ikechukwu Okafor** School of Chemical and Biomolecular Engineering, Georgia Institute of Technology, Atlanta, GA, USA

**Sharan Ramaswamy** Tissue Engineered Mechanics, Imaging and Materials Laboratory (TEMIM Lab), Department of Biomedical Engineering, College of Engineering and Computing, Florida International University, Miami, FL, USA

**Manuel K. Rausch** Department of Aerospace Engineering and Engineering Mechanics, University of Texas at Austin, Austin, TX, USA

**Bruno V. Rego** James T. Willerson Center for Cardiovascular Modeling and Simulation, Institute for Computational Engineering and Sciences, Department of Biomedical Engineering, The University of Texas at Austin, Austin, TX, USA

**Jay M. Reimer** Department of Biomedical Engineering, University of Minnesota, Minneapolis, MN, USA

**T. Ristori** Department of Biomedical Engineering, Eindhoven University of Technology, Eindhoven, The Netherlands

Institute for Complex Molecular Systems, Eindhoven University of Technology, Eindhoven, The Netherlands

**Michael S. Sacks** The Oden Institute and the Department of Biomedical Engineering, The University of Texas at Austin, Austin, TX, USA

**Xiaodan Shi** Department of Bioengineering, University of Texas at Arlington, Arlington, TX, USA

**Andrew W. Siefert** Cardiac Implants LLC, Tarrytown, NY, USA

**Dan T. Simionescu** Department of Bioengineering, Clemson University, Clemson, SC, USA

**Hobey Tam** Clemson University, Clemson, SC, USA

**Ishita Tandon** Department of Biomedical Engineering, University of Arkansas, Fayetteville, AR, USA

**Robert T. Tranquillo** Department of Biomedical Engineering, University of Minnesota, Minneapolis, MN, USA

**A. J. van Kelle** Department of Biomedical Engineering, Eindhoven University of Technology, Eindhoven, The Netherlands

Institute for Complex Molecular Systems, Eindhoven University of Technology, Eindhoven, The Netherlands

**Payal Vyas** Department of Cardiovascular Medicine, Brigham and Women's Hospital, Harvard Medical School, Boston, MA, USA

**Naren Vyavahare** Clemson University, Clemson, SC, USA

**Bo Wang** College of Science, Mathematics and Technology, Alabama State University, Montgomery, AL, USA

**James N. Warnock** School of Chemical, Materials and Biomedical Engineering, University of Georgia, Athens, GA, USA

**Sarah M. Wells** School of Biomedical Engineering, Dalhousie University, Halifax, NS, Canada

**Michael C. H. Wu** Department of Mechanical Engineering, Iowa State University, Ames, IA, USA

**Ajit P. Yoganathan** School of Chemical and Biomolecular Engineering, Georgia Institute of Technology, Atlanta, GA, USA

Wallace H. Coulter Department of Biomedical Engineering, Georgia Institute of Technology, Atlanta, GA, USA

**Rana Zakerzadeh** James T. Willerson Center for Cardiovascular Modeling and Simulation, The Oden Institute and the Department of Biomedical Engineering, The University of Texas at Austin, Austin, TX, USA

**Will Zhang** James T. Willerson Center for Cardiovascular Modeling and Simulation, The Oden Institute and the Department of Biomedical Engineering, The University of Texas at Austin, Austin, TX, USA

# Part I
# Native Heart Valve Function

# Biological Mechanics of the Heart Valve Interstitial Cell

Alex Khang, Rachel M. Buchanan, Salma Ayoub, Bruno V. Rego, Chung-Hao Lee, and Michael S. Sacks

**Abstract** Heart valves are composed of multilayered tissues that contain a population of vascular endothelial cells (VEC) on the blood contacting surfaces and valve interstitial cells (VIC) in the bulk tissue mass that maintain homeostasis and respond to injury. The mechanosensitive nature of VICs facilitates the regulation of growth and remodeling of heart valve leaflets throughout different stages of life. However, pathological phenomenon such as mitral valve regurgitation and calcific aortic valve disease lead to pathological micromechanical environments. Such scenarios highlight the importance of studying the mechanobiology of VICs to better understand their mechanical and biosynthetic behavior. In the present chapter, we review use of novel experimental-computational techniques to link VIC biosynthetic response to changes in in vivo deformation in health and disease. In addition, we discuss the development of tissue-level models that shed light on the biomechanical state of VICs in situ. To conclude, we outline future directions for heart valve mechanobiology including model-driven experiments and highlight the need for high-fidelity, multi-scale models to link the cell-, tissue-, and organ-level events of heart valve growth and remodeling.

---

A. Khang · R. M. Buchanan · S. Ayoub
James T. Willerson Center for Cardiovascular Modeling and Simulation, The Oden Institute and the Department of Biomedical Engineering, The University of Texas at Austin, Austin, TX, USA

B. V. Rego
James T. Willerson Center for Cardiovascular Modeling and Simulation, Institute for Computational Engineering and Sciences, Department of Biomedical Engineering, The University of Texas at Austin, Austin, TX, USA

C.-H. Lee
School of Aerospace and Mechanical Engineering, The University of Oklahoma, Norman, OK, USA

M. S. Sacks (✉)
The Oden Institute and the Department of Biomedical Engineering, The University of Texas at Austin, Austin, TX, USA
e-mail: msacks@ices.utexas.edu

**Keywords** Multiscale modeling · Heart valve interstitial cell mechanobiology · Model-driven experiments · 3D hydrogel culture · Remodeling · Cell mechanics · Finite element method · Cellular contraction · Myofibroblast

# 1 Introduction

Proper blood flow through the heart is made possible through healthy and functioning heart valves. The heart has four valves: the mitral (MV), aortic (AV), pulmonary (PV), and tricuspid valve (TV). The MV and TV are atrioventricular valves and control blood flow from the atria to the ventricles, whereas the PV and AV are semilunar valves and regulate blood flow through the pulmonary artery and aorta, respectively. The MV and AV are on the left side of the heart and help in transporting oxygenated blood from the lungs to the rest of the body. On the right side of the heart, the TV and PV guide the flow of deoxygenated blood from the body to the lungs. Heart valves perform within a highly mechanically demanding environment and undergo tension, flexure, and shear stress throughout the cardiac cycle [1]. At the left side of the heart, the AV and MV experience greater transvalvular pressures (80 and 120 mmHg, respectively) than the PV and TV do at the right side of the heart (10 and 25 mmHg, respectively) [2]. Previous work has shown that heart valves experience mean peak strain rates of 300–400% $s^{-1}$ in the radial direction and 100–130% $s^{-1}$ in the circumferential direction [3–6]. The complex mechanical demands of the heart system highlight the need for repair and remodeling mechanisms of valvular tissues.

Heart valves are known to remodel and repair throughout life [7]. All heart valves contain leaflets composed of multilayered, high-functioning tissues designed to withstand the above driving stresses. For example, the AV has three distinct layers: the collagen-rich fibrosa, the spongiosa which contains a high concentration of proteoglycans, and the ventricularis that features dense collagen and elastin fiber networks [7, 8]. In contrast, the MV has four distinct layers in the central leaflet region: the spongiosa, fibrosa, ventricularis, and a fourth layer called the atrialis that faces towards the atria. In semilunar heart valves, the fibrosa layer faces towards the aorta while the ventricularis faces towards the ventricle and the spongiosa is located in between both. In atrioventricular heart valves the ventricularis also faces the ventricle and is adjacent to the fibrosa. The spongiosa layer lies between the fibrosa and atrialis. Recent evidence has shown that the spongiosa is essentially a specialized extension of the fibrosa and ventricularis layers, both in terms of mechanical functionality and extracellular matrix (ECM) components [9, 10]. The unique organization of valve leaflet layers and underlying ECM components allow for the simultaneous and proper function of semilunar and atrioventricular heart valves, accounting for the differences in location within the heart and orientation to the direction of blood flow. Previous studies have quantified in detail the transmural structural heterogeneity in heart valve leaflet tissues, and have elucidated the role that this unique and finely regulated ECM architecture plays in maintaining a homeostatic distribution of stress in vivo [11, 12]. These regulatory functions occur at the hands of the underlying VICs embedded throughout all layers. VICs are mechanocytes and are sensitive to their mechanical environment. They display multiple, reversible phenotypes and remain quiescent under normal conditions and behave similarly to fibroblasts. In response to pathological alterations in force and hemodynamics, VICs display a myofibroblast phenotype. Valvular damage causes VICs to upregulate ECM biosynthesis and become contractile through expression of alpha-smooth muscle actin ($\alpha$-SMA). Moreover, further investigations have shown that the geometry, deformation, and biosynthetic behavior of VICs depend highly on local mechanical properties of the surrounding ECM [12–15].

Recent work has reinforced the hypothesis that VIC deformation is a major driver in regulating heart valve repair and remodeling mechanisms [16–18]. VIC response to deformation may be motivated by biological pressure to return to a homeostatic state through remodeling of the leaflet ECM [12, 18]. However, little is known about the underlying mechanisms that make this possible. For example, it is unclear whether the intrinsic stiffness of VICs or the stiffness of the environment plays a role in this process. In addition, more work is necessary to link cellular deformations to biosynthetic activity and relate these observations to in vivo function. Multi-scale approaches that utilize both experimental and computational techniques are thus required to study the complex biomechanical states of VICs [19].

## 2 Major Questions and Challenges

It has long been hypothesized that deformation plays a critical role in regulating the biosynthetic activity of mechanocytes and therefore governs catabolic mechanisms. In the case of heart valves, cyclic loading causes the deformation of VICs, which in turn provides the mechano-regulation necessary for maintaining homeostasis. Alterations in force and hemodynamics can cause VICs to display an activated, myofibroblast phenotype and increase matrix remodeling activity. If activation persists, drastic changes in valve leaflet structure and function can occur, furthering the onset of disease.

In certain disease states such as myxomatous mitral valve degeneration and calcific aortic valve disease (CAVD), the pathological tissue structure plays a large role in the mechano-regulation of VIC biosynthetic activity. Myxomatous mitral valve degeneration can occur due to alterations in mechanical stress and genetic abnormalities [20]. Some hallmarks of this disease include an increase in glyosaminoglycans within the spongiosa layer, degradation of collagen, and fragmentation of elastin networks. The net effect of the ECM remodeling causes the leaflet tissue to thicken and weaken. This in turn induces VICs to become activated and increase expression of proteolytic enzymes. In addition, VICs begin to proliferate within the spongiosa layer. These symptoms can result in mitral valve regurgitation due to "floppy" mitral valve leaflets. On the other hand, CAVD is caused by calcium accumulation onto aortic valve leaflets. As the disease progresses, the valve leaflets thicken, and calcium nodules begin to form. This occurs due to VICs undergoing phenotypic transitions to an osteoblastic phenotype leading to the secretion of bone-like ECM [21]. Late stages of CAVD can cause the aortic valve to be less efficient, resulting in aortic stenosis, severe obstruction to cardiac outflow from the left ventricle.

Myxomatous mitral valve degeneration and CAVD highlight the role of VIC-ECM interactions in valvular disease. VIC deformation and thus biosynthetic regulation is dependent upon these interactions. Due to these reasons, a need exists to study VIC-ECM coupling to better understand valve disease in the hope of engineering effective therapeutics. A need also exists to elucidate the intricate relationship between VICs and VECs. VIC phenotype is highly dependent upon chemokines and cytokines excreted by VECs. Factors like endothelin-1 and transforming growth factor beta (TGF-β1) can directly affect the contractility and activation levels of VICs, respectively. Future efforts should also be directed towards investigating the synergistic effects of mechanical stimulation, VIC-ECM coupling, and VIC-VEC interactions on repair and remodeling mechanisms.

Studying the behavior of VICs under physiologically relevant conditions is of the utmost importance for developing accurate and trustworthy models. Popular methods among researchers to study VIC biology and mechanics include two-dimensional cultures and assays. Although these studies are insightful, they fall short of mimicking the complexity of the native VIC microenvironment and fail to elicit accurate behavior of VICs. To circumvent this issue, some researchers have resorted to using valve leaflet explants for mechanobiological studies and utilizing soft hydrogel matrices to seed and study VICs in 3D. Such environments allow for the examination of cellular deformation, specifically nuclear deformation, in response to mechanical loading. In recent years, nuclear aspect ratio (NAR), the dimensionless ratio between the nuclear major and minor axes, has been used as a cell-scale metric for correlating mechanical stimulation with biological function [13, 18, 22, 23]. Analysis of NAR has been tremendously valuable in studying the underlying mechanobiology responsible for valve tissue remodeling and repair. However, rigorous elucidation of the role that cellular deformation plays in homeostatic regulation of normal and diseased valves remains a challenge. In addition, more work is needed in studying the role of pathological tissue structure in local VIC mechano-regulation. In this review, contemporary approaches for the study of VIC mechanobiology and the associated cellular phenomena are summarized.

## 3 Advances in Investigating VIC Mechanobiology

### 3.1 Isolated Cell Studies

#### 3.1.1 Overview

The mechanosensitive nature of VICs allows for the maintenance and repair of valvular tissue. Ex situ observations of valve leaflets from the four heart valves reveal that leaflets belonging to the left side of the heart (AV and MV) are far stiffer than those from the right side of the heart (PV and TV) [24–26]. This is most likely due to the stark difference in transvalvular pressure (TVP) between the two sides. It is hypothesized that since the left side of the heart experiences higher TVP, larger stresses are imposed on the underlying VICs, thus inducing them to form more developed cytoskeletal networks and increase collagen production. To test this hypothesis, techniques like micropipette aspiration (MA) and atomic force microscopy (AFM) are used to measure the stiffness of VICs isolated from leaflets of all four valves [24, 25, 27]. Protein assays are performed to compliment the single-cell mechanical tests and provide insight into how VIC biosynthetic behavior varies between valves. These studies are a crucial first step in studying the biophysical state of VICs and how it is altered with respect to the mechanical demands of the cellular milieu.

### 3.1.2 Micropipette Aspiration Studies of VIC Biomechanical Behaviors

VICs isolated from all four heart valves were cultured and used for MA stiffness measurements and protein quantification assays. For MA measurements, VICs were suspended in culture media and tested with micropipettes with inner diameters ranging from 6 to 9 μm (Fig. 1a). The pressure through the micropipette was adjustable and used to achieve consistent testing of VICs using the following protocol: (1) an initial tare pressure (~50 Pa for 60 s) was applied to capture cells and form a seal in between the cell and micropipette, (2) the pressure was increased to ~250 Pa and held consistent for 120 s, and (3) the pressure was then increased to ~500 Pa and held for the last 120 s of the test. Extended experimental details can be found here [27]. The actual pressure applied at each phase was recorded and images throughout the test were captured with a camera coupled with a bright-field microscope. From the images, the aspiration length of the cell was measured. VIC effective stiffness ($E_{eff}$) was determined through use of a half-space model (punch model) that models the cell as an isotropic, elastic, incompressible half-space material [28] and was determined from the following equation:

$$E_{eff} = \Phi(\eta)[(3r)/(2\pi)](\Delta P/L) \tag{1}$$

where $\Phi(\eta)$ is a dimensionless parameter determined from the ratio of the micropipette inner radius to the wall thickness with a set value of 2.1, $r$ is the radius of the micropipette inner radius, and $\Delta P/L$ is the change in applied pressure divided by the aspiration length. Here, $E_{eff}$ is used as a convenient parameter to compare the intrinsic stiffness of the different VIC types.

It was observed experimentally that pulmonary and tricuspid VICs (PVIC and TVIC, respectively) displayed a larger aspiration length in response to the applied pressure compared to mitral (MVICs) and aortic VICs (AVICs). This translates to AVICs and MVICs being significantly stiffer than PVICs and TVICs (Fig. 1b). VICs from the same side of the heart (i.e., AVICs and MVICs) did not vary significantly in $E_{eff}$ however. It was found that the $E_{eff}$ of the various VIC types was linearly related

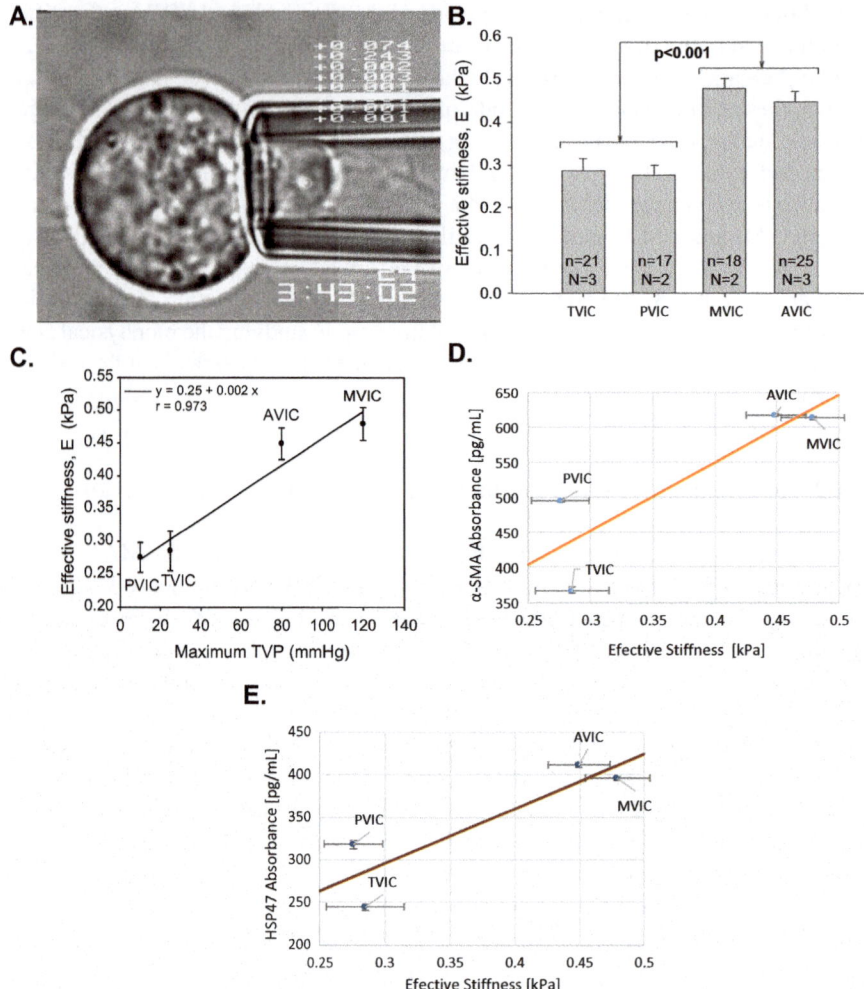

**Fig. 1** (a) Mechanical evaluation of a VIC via micropipette aspiration. (b) The effective stiffness of the various VIC types evaluated from micropipette aspiration. (c) VIC effective stiffness with respect to the maximum transvalvular pressure (TVP) of the different heart valves. (d) α-SMA and (e) HSP47 absorbance determined via ELISA plotted against the effective stiffness of the various VIC types. Modified from [27]

to the maximum TVP felt by each valve suggesting that VIC stiffness is, at the least, correlated to the stress present within the microenvironment (Fig. 1c).

Elisa protein assays were performed on both in vitro and in situ VICs specifically to assess the levels of smooth muscle actin (SMA) and heat shock protein 47 (HSP47) among the different valve types. SMA is present within VICs displaying an activated phenotype and is hypothesized to alter the cellular stiffness and play a role in collagen biosynthesis, thus making it a protein of interest. Production of type I collagen has been shown to be dependent upon HSP47 and therefore HSP47 serves as a surrogate for the level of collagen biosynthesis [29]. Analysis of the in vitro Elisa results shows higher levels of SMA and HSP47 among valve cells from the left side of the heart (Fig. 1d, e respectively). In situ results display a similar trend as well. When taken together, these results suggest that VICs residing within the MV and AV adapt to the higher levels of stress through regulation of cytoskeletal organization and higher collagen production rates.

### 3.1.3 Microindentation Studies

AFM is another method used to measure the stiffness of VICs through deforming the cells with a cantilever [30]. In short, the test samples are prepared by seeding VICs in a monolayer onto collagen coated glass coverslips.

AFM measurements were performed using the "tapping mode" in which the cantilever makes ~70 indentations in a rectilinear grid sampling pattern for every VIC (Fig. 2a). Stiffness ($E$) is calculated from this method using the following equation:

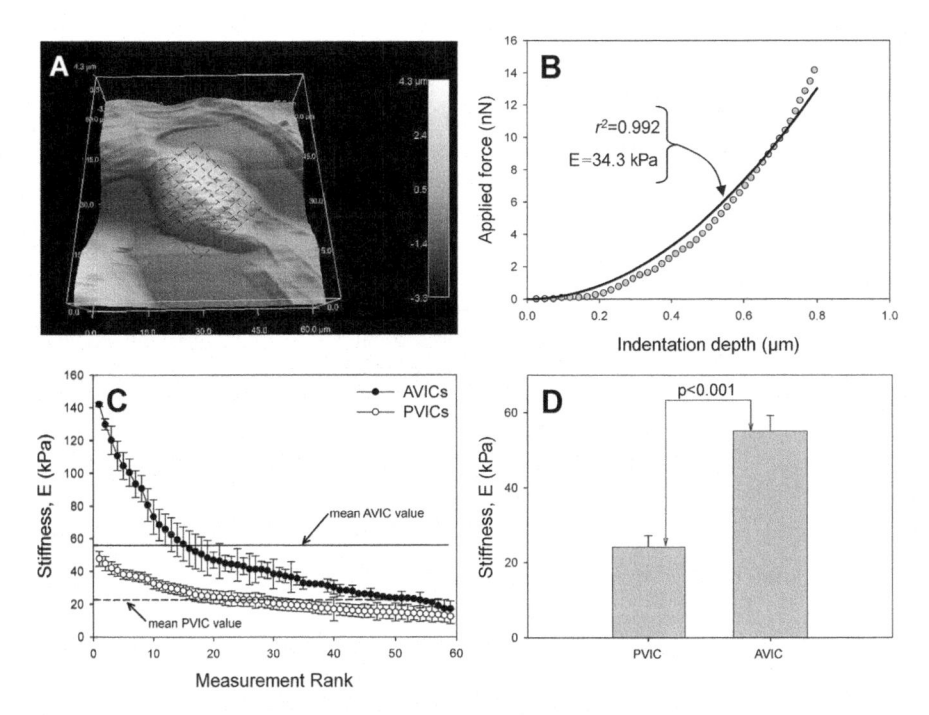

**Fig. 2** (a) Height map developed from atomic force microcopy measurements of a VIC. (b) Representative applied force vs. indentation depth generated from atomic force microscopy measurements. The experimental data is fit to Eq. (4) and $E$ is backed out from the resulting fit. (c) AVIC and PVIC stiffness computed from atomic force microscopy measurements ranked in order of magnitude. (d) Average AVIC and PVIC stiffness. From [30]

$$E = \frac{F(1-v^2)}{\pi \cdot \Phi(\delta)} \quad (2)$$

where $v$ is the Poisson's ratio (0.5 due to incompressibility assumption), $F$ is the force applied on the VIC by the AFM tip, and $\Phi(\delta)$ is a function of the conical tip probe geometry defined as [31]:

$$\Phi(\delta) = \delta^2 \left[ \frac{2 \cdot \tan(\alpha)}{\pi^2} \right] \quad (3)$$

where $\alpha$ represents the probe opening angle (35°) and $\delta$ denotes the indentation depth of the probe. Incorporation of Eq. (3) simplifies Eq. (2) to the following:

$$E = \frac{F}{0.594 \cdot \delta^2} \quad (4)$$

$E$ was then determined from fitting the experimental data to Eq. (4) (Fig. 2b). Similar to the previously mentioned MA studies, $E$ serves as a parameter for comparison between the various VIC types.

The $E$ values obtained from ~70 AFM measurements (Fig. 2c) were averaged for each VIC and it was shown that AVICs were approximately twice as stiff as PVICs (Fig. 2d), consistent with the observations made with MA. However, VIC stiffness reported from the AFM measurements were approximately 100 times larger than those obtained from MA and is consistent with studies performed on other cell types [32–35]. The drastic differences may be owing to the assumptions of the models used to describe VIC mechanical behavior and the difference in deformation methods. AFM utilizes a localized force to make measurements whereas MA tests a larger region. Intuitively, a localized measurement would be largely affected by the stiffness of nearby cellular components like local stress fibers and the nucleus whereas measurements made in larger regions would consider cell membrane properties and the local cytoskeleton network. Although deviations are present in the absolute stiffness values from both techniques, the general trend observed among different types of VICs is consistent.

### 3.1.4 Isolated VIC Mechanical Models

The disparity between stiffness measurements from MA and AFM may arise not only from the difference in deformation modes but also the activation state of the VICs and the corresponding levels of α-SMA expression during the experiment. During MA, VICs are presumably in an inactivated state due to the lack of adhesion and during AFM experiments they are assumed to be highly activated resulting from adhesion on a 2D substrate. The disparity between stiffness measurements arising from different experimental techniques and VIC activation states suggests that

## A. Micropipette Aspiration

| $\mu^{sf}$=390 Pa | $\mu^{cyto}$=5.0 Pa |
|---|---|
| $f$ | $\mu^{nuc}$ |

⬇ $\mu^{sf}, \mu^{cyto}$

## B. AFM on Peripheral Region

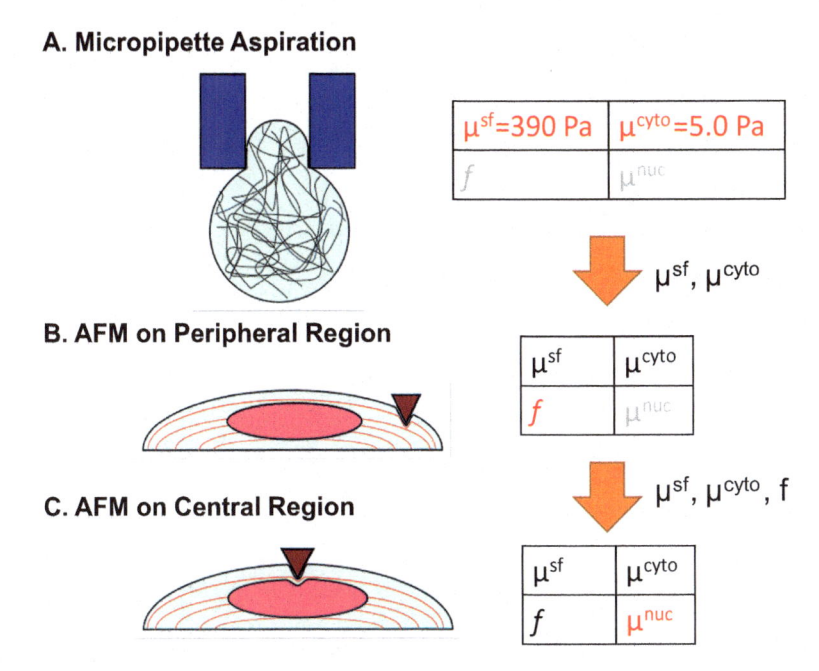

| $\mu^{sf}$ | $\mu^{cyto}$ |
|---|---|
| $f$ | $\mu^{nuc}$ |

⬇ $\mu^{sf}, \mu^{cyto}, f$

## C. AFM on Central Region

| $\mu^{sf}$ | $\mu^{cyto}$ |
|---|---|
| $f$ | $\mu^{nuc}$ |

**Fig. 3** (a) Data from micropipette aspiration experiments are used to calibrate the shear modulus of the stress fibers ($\mu^{sf}$) and the cytoplasm ($\mu^{cyto}$). (b) Using the previously determined parameters and atomic force microscopy data from peripheral regions of VICs, the contraction force ($f$) is computed. (c) The shear modulus of the nucleus ($\mu^{nuc}$) is calculated last, incorporating all the previously determined parameters and atomic force microscopy measurements made over the nucleus. From [24]

analysis of the effective stiffness of VICs alone is insufficient and a need exists for more rigorous methods to better understand VIC mechanobiology. To accomplish this, isolated VIC mechanical models were developed which incorporated the orientation of the cytoskeletal network, the relative expression of α-SMA stress fibers, and the mechanical properties of the nucleus [24, 25].

The model was designed to estimate the effective mechanical behaviors of the major VIC subcellular components by integrating the MA and AFM experimental data from both AVICs and PVICs (Fig. 3). Analysis of the MA data required consideration of the cellular components assumed to most likely contribute during the experiments, namely the passive stress fibers and the cytoplasm. Modeling the respective contributions of these subcellular components using a shear modulus allows for a single parameter to represent their mechanical state, simplifying determination of differences between VIC types. The first step in developing this model is taken from the MA data analysis [27], where the total Cauchy stress $\boldsymbol{T}$ was considered the sum of the passive cytoskeletal and passive (i.e., inactive) stress fibers

$$\boldsymbol{T} = \boldsymbol{T}_{cyto} + \boldsymbol{T}_{sf} \tag{5}$$

Next, from the AFM experiments, contributions from the cytoplasm, passive stress fibers, stress fiber active stress ($f$), and the nucleus were considered in the total Cauchy stress using

$$T = T_{\text{cyto}} + T_{\text{sf}} + T_f + T_{\text{nuc}} \quad (6)$$

For parameter optimization, the MA data was first used to back out the shear modulus of the stress fibers ($\mu^{\text{sf}}$) and the cytoplasm ($\mu^{\text{cyto}}$) (Fig. 3a). Briefly, the cytoskeleton was modeled as an isotropic hyperelastic material using a nearly incompressible neo-Hookean model. The passive component of individual α-SMA fibers was modeled as a transversely isotropic material and both these components were incorporated into the expanded total Cauchy stress equation used for MA analysis

$$T = 2\frac{1}{J}F\frac{\partial}{\partial C}\left[\frac{\mu^{\text{cyto}}}{2}(\bar{I}_1 - 3) + \frac{1}{2}K(\ln J)^2\right]F^{\text{T}}$$
$$+ \frac{1}{4\pi}\int_0^{2\pi}\int_0^{\pi}\left[H(I_4 - 1)2\frac{I_4}{J}\frac{\partial}{\partial I_4}\left(\frac{\mu^{\text{sf}}\bar{\varphi}^{\text{sf}}}{2}(I_4-1)^2\right)m\otimes m\right]\sin\theta d\theta d\phi \quad (7)$$

The parameters $\mu^{\text{cyto}}$ and $\mu^{\text{sf}}$ were calibrated through minimization of the least square error between the simulated and experimental aspiration lengths while taking into consideration the applied pressure and normalized α-SMA expression levels ($\bar{\varphi}^{\text{sf}}$).

After proper calibration of these parameters, AFM data obtained on peripheral regions of VICs was used to back out the contractile force ($f$) of the stress fibers (Fig. 3b) using the following expanded total Cauchy stress equation for AFM analysis

$$T = 2\frac{1}{J}F\frac{\partial}{\partial C}\left[\frac{\mu^{\text{cyto}}}{2}(\bar{I}_1 - 3) + \frac{1}{2}K(\ln J)^2\right]F^{\text{T}}$$
$$+ \int_0^{2\pi}\int_0^{\pi}\left[\Gamma_t(m)H(I_4 - 1)2\frac{I_4}{J}\frac{\partial}{\partial I_4}\left(\frac{\mu^{\text{sf}}\bar{\varphi}^{\text{sf}}}{2}(I_4-1)^2\right)m\otimes m\right]\sin\theta d\theta d\phi \quad (8)$$
$$+ \int_0^{2\pi}\int_0^{\pi}\left[\Gamma_t(m)f\cdot\bar{\varphi}^{\text{sf}}\frac{I_4}{J}m\otimes m\right]\sin\theta d\theta d\phi$$

After finding the parameter $f$, AFM data on the central region (above the nucleus) was used to calibrate the respective shear modulus of the nuclei ($\mu^{\text{nuc}}$) (Fig. 3c) by replacing the first term of Eq. (8) with

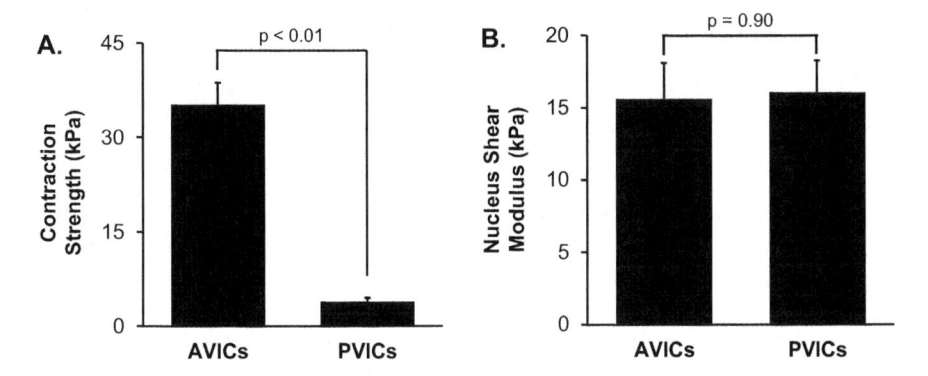

Fig. 4 (a) The contraction strength of AVICs and PVICs backed out from the isolated VIC mechanical models. The model predicts that AVICs display a larger contraction strength than PVICs. (b) There was no significant difference between the nucleus shear modulus of AVICs and PVICs as determined from the model. From [7]

$$T = 2\frac{1}{J}F\frac{\partial}{\partial C}\left[\frac{\mu^{\text{nuc}}}{2}(\bar{I}_1 - 3) + \frac{1}{2}K(\ln J)^2\right]F^{\text{T}} \tag{9}$$

and minimizing the least square error between simulated and experimental indentation depth vs. force data. Detailed information on the formulation of the model can be found in [24].

Simulations demonstrated that AVICs displayed a ~10 times stronger contractile force than PVICs, highlighting the model's ability to account for the higher levels of α-SMA expression within AVICs (Fig. 4a). Similarities in nuclear stiffness between AVICs and PVICs were also captured by the model (Fig. 4b). This suggests that any major difference in stiffness between the VIC types is most likely due to the stiffness of α-SMA and not the nucleus. Modifications to the model were made to consider f-actin and α-SMA expression levels along with stress fiber orientation, strain rate effects, and the relation between stress fiber length and tension [25]. The model outputted AVIC contraction strength under various activation scenarios and it was found that contraction strength was greatest when the cells were treated with activation inducing transforming growth factor beta 1 (TGF-β1) and then with 90 mM of potassium chloride (KCL). The adjusted model was also able to showcase that inhibiting stress fiber formation through use of Cytochalasin D was effective in reducing cellular contraction strength.

These studies demonstrate that isolated cell models are beneficial tools for discovery and offer an explanation for the difference in stiffness levels observed between VIC types and also between testing methods. These models mark a crucial first step in capturing the mechanical behavior of VICs. In addition, they have the potential to be incorporated into multi-scale models and be adapted to represent the mechanics of VIC populations. Future work will focus on coupling VIC mechanical models with agent-based models to thoroughly capture the biomechanical and biosynthetic cellular events occurring within valve leaflet tissues.

## 3.2 In Situ Tissue-Level Evaluation of VIC Contraction Behaviors

### 3.2.1 Overview

To elicit accurate biological behavior, physiologically relevant forms of deformation are desirable in mechanobiological studies. Planar membrane tension and flexural deformation testing are two highly suitable methods that aim to mimic the natural strain experienced by heart valve leaflets. While biaxial tensile testing has proven highly effective for characterizing the mechanical properties of the native leaflet and individual tissue layers [10, 11], flexure-based approaches are advantageous in two regards: (1) it is extremely sensitive at low stresses and strains and (2) it can reflect differences in ECM components and architecture throughout the different layers of the valve leaflets. In addition, beam-bending models, such as the Euler-Bernoulli relation [36], can be utilized to determine an instantaneous effective modulus $E_{eff}$ of the valve leaflet tissue to gauge VIC contraction behavior and biophysical state from tissue-level studies.

### 3.2.2 VIC-ECM Coupling and Bidirectional Valve Leaflet Bending Response

Flexural deformation tests have been used to show significant increases in $E_{eff}$ arising from cellular contraction (90 mM KCL). This form of testing has also been used to capture a decrease in $E_{eff}$ with the loss of the cellular basal tonus (Fig. 5) [37].

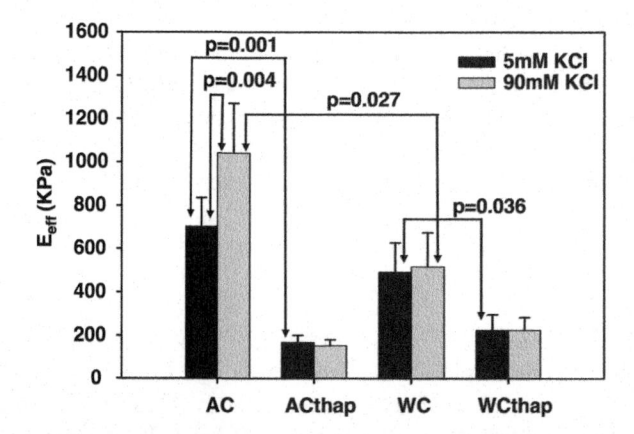

**Fig. 5** Tissue effective stiffness of normal and contraction inhibited (thap) aortic valve leaflet test specimens bent with (WC) and against (AC) curvature after treatment with 5 mM KCL (normal) and 90 mM KCL (hypertensive). The tissue effective stiffness was computed from the derivative of the moment vs. curvature plots. The effective stiffness increased significantly in the against curvature direction after treatment with 90 mM KCL. From [37]

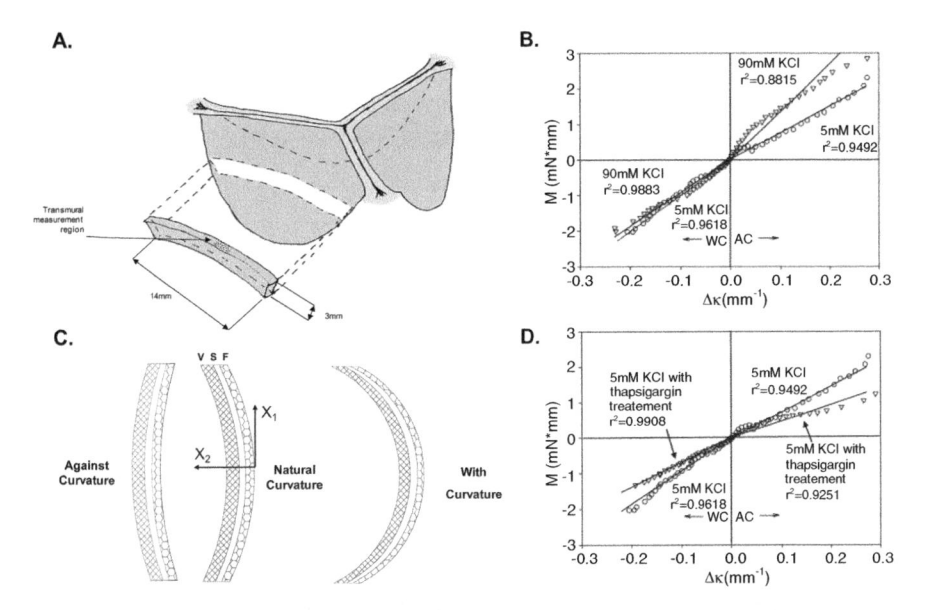

**Fig. 6** (a) Testing specimens are excised from aortic valve leaflets. (b) Representative moment vs. curvature plot of an aortic valve leaflet specimen under normal (5 mM KCL) and hypertensive (90 mM KCL) conditions. (c) Broad overview of mechanical testing experimental protocol subjecting leaflet specimens to bending deformations with and against curvature. The aortic valve layers (*V* ventricularis, *S* spongiosa, *F* fibrosa) were noted to ensure proper orientation of the valve specimen while testing. (d) Representative moment vs. curvature plot of aortic valve leaflet specimen treated with a stress fiber inhibiting solution (thapsigargin) under normal (5 mM KCL) and hypertensive (90 mM KCL) conditions. Modified from [37]

In short, this is done through tracking dots placed along the transmural cross section of the test specimen excised from AV leaflets (Fig. 6a) to keep track of its change in curvature ($\Delta\kappa$) from an initial, reference configuration. The moment ($M$) for each $\Delta\kappa$ is calculated throughout the test and $M$ vs. $\Delta\kappa$ plots are generated. From these plots, the tissue effective stiffness $E_{\text{eff}}$ is determined through use of the Euler-Bernoulli relation:

$$\frac{M}{I} = E_{\text{eff}}\Delta\kappa \tag{10}$$

where $I$ is the second moment of inertia and a function of the sample dimensions $\left(I = \frac{t^3 w}{12}\right)$. For more experimental details, the reader is referred to the following references [37–39]. The moment-curvature plots also revealed the direction-dependent bending response of AV leaflets (Fig. 6b, d). Valve leaflet tissues were bent "with" and "against" the natural leaflet curvature (WC and AC, respectively, Fig. 6c) and it was observed that the tissue modulus was greatest when bent in the AC direction. This is so due to the highly organized and collagen-rich fibrosa layer being under tension, causing a rise in $E_{\text{eff}}$. When bent in the WC direction, the ventricularis is under tension and as a result the tissue effective modulus was much lower owing to the collagen and compliant elastin fibers residing within that layer. The bending response of AV leaflets was found to be linear in both bending directions (Fig. 6b, d), which is unlike that of the nonlinear planar tensile tissue response [40, 41].

## 3.3 Interlayer Micromechanics of the AV Leaflet

Subsequent flexure studies have focused on examining the intricate interlayer micromechanics of the AV leaflet [9]. The behavior of the layers with respect to one another was of special interest to delineate the way valve leaflets deform, either as a bonded unit or separately. To better understand this phenomenon, analysis of the transmural strain of AV leaflets was conducted. Microscopic India ink tracking dots were airbrushed onto the transmural wall of AV leaflet testing specimens. Opposite of this side were macro-level tracking dots used for assessment of the bending test and development of moment-curvature plots. Both sets of tracking dots were tracked throughout the flexural deformation test at $\Delta\kappa$ ranges of 0 (reference image), 0.1, 0.2, and 0.3 mm$^{-1}$. The tissue-level deformation gradient $\mathbf{F}$ was computed from the reference and deformed coordinates of the microscopic markers and decomposed into stretch and rotation tensors to remove rigid body deformations, leaving only axial stretch information. The total axial stretch ($\Lambda_1$) was assessed across the normalized transmural cross section of the test specimens with $\Lambda_1 > 1$ representing tension, $\Lambda_1 < 1$ representing compression, and $\Lambda_1 = 0$ representing the neutral axis in which no deformations occur.

By plotting $\Lambda_1$ against the normalized leaflet thickness, a linear trend was observed (Fig. 7a, b) for both with and against curvature tests. This observation confirms that the spongiosa does not allow for shearing to occur between the fibrosa and ventricularis under physiologically relevant deformations, a phenomenon that some believed to occur before [42–44]. If the spongiosa did allow for shearing between the layers, each layer would exhibit its own neutral axis when undergoing flexure. This was discovered through performing interlayer bonding simulations of the AV leaflet. In short, an AV leaflet finite element model that considered the relative thickness of each layer (fibrosa-45%, spongiosa-30%, ventricularis-25%) as determined from histological measurements was developed in COMSOL Multiphysics v4.3 (COMSOL, Burlington, MA). A parametric study was performed on the shear modulus of the spongiosa layer ($\mu_s$). $\mu_s$ was set at 1 Pa, 0.1 kPa, 1 kPa, 10 kPa, and 45 kPa and it was determined that for the spongiosa to have a significant difference in axial stretch from the fibrosa and ventricularis, $\mu_s$ would have to be less than 1 kPa (Fig. 7c). No discernable difference was found experimentally and instead evidence suggests that AV leaflets function as a single bonded unit. Furthermore, analysis of histological sections revealed gradual transmural changes in ECM components throughout the AV leaflet layers which further supports the conclusion

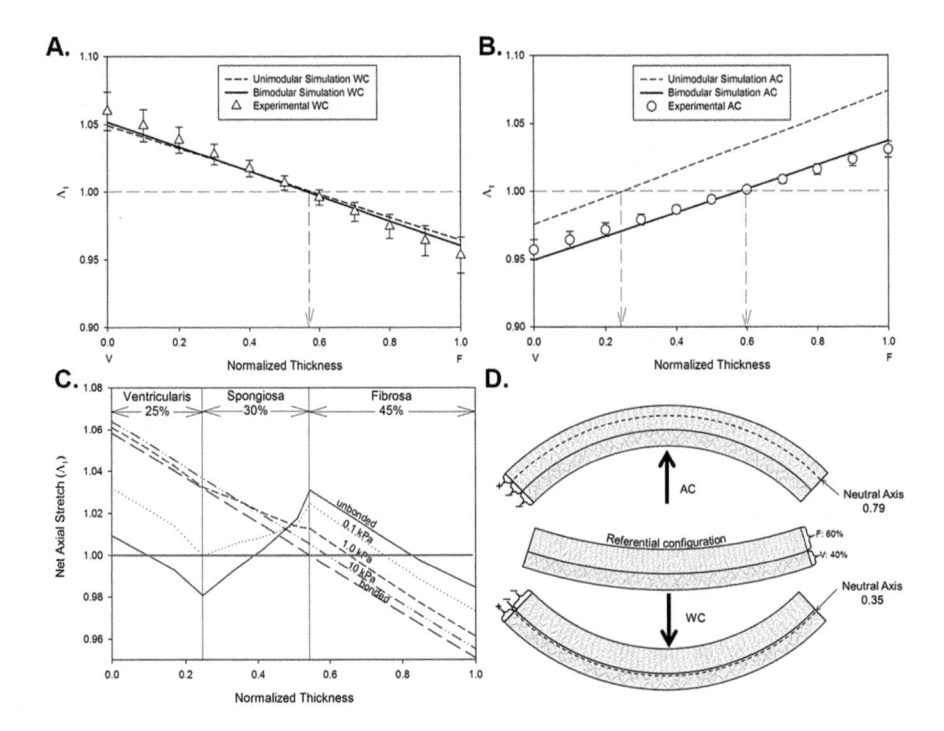

**Fig. 7** Axial stretch plotted against the normalized leaflet thickness in the (**a**) with curvature and (**b**) against curvature directions. Note that a unimodular material model is insufficient in capturing the bending response in both directions. Thus, a bimodular material model is needed. (**c**) Results from the simulated interlayer bonding parametric study. For measurable differences to occur between the transmural strain of the heart valve layers, notably the ventricularis and the fibrosa, the spongiosa must have a shear modulus of less than 1 kPa. (**d**) The location of the neutral axis in both the against curvature and with curvature bending directions. Modified from [9]

that AV leaflets behave as a functionally graded, bonded unit containing contiguous features [10].

It was also discovered that a unimodular material model was insufficient in capturing the bidirectional bending behavior of AV leaflets. Instead a bimodular material model is needed to accomplish this, which further highlights the complex, heterogeneous ECM components and architecture across valve tissue layers. In addition, it suggests that the spongiosa can be viewed as a contiguous extension of both the fibrosa and ventricularis in that it doesn't play a major mechanical role in bending deformations. The final bimodular incompressible neo-Hookean material model used is as follows

$$W^{\pm} = \frac{\mu_L^{\pm}}{2}(I_1 - 3) - p(I_3 - 1) \tag{11}$$

where $\mu_L^\pm$ represents four total moduli. $\mu_F^\pm$ and $\mu_V^\pm$ are the fibrosa and ventricularis shear moduli, respectively, in tension (+) and compression (−). Equation (11) successfully captured the bidirectional bending behavior of the aortic valve leaflet and was able to estimate layer- and deformation-specific shear moduli. This bimodular model would later be adjusted and incorporated into a down-scale model of the AV leaflet in the next section.

Computation of the axial stretch during AV bending tests revealed that the neutral axis (NA) is located at different locations depending on whether the test is performed with or against curvature (Fig. 7d). In the WC direction, the NA shifts towards the ventricularis, ~0.35 of the normalized thickness. This implies that the entire fibrosa, located above the NA, is in compression. In the AC direction, the NA shifts towards the fibrosa, ~0.79 of the normalized thickness. This suggests that the fibrosa is far stiffer while under tension than in compression. The following results are due to the distinct ECM components within each layer that allow the leaflets to be less stiff when the valve opens (WC) and stiffer when the valve closes (AC) to ensure proper blood flow through the heart.

## 3.4 Down-Scale Model of the VIC Within ECM

Elucidating the biomechanical state of the VIC in its natural environment is desirable to shed light on how it affects the valve leaflet tissue. However, it is extremely difficult to observe how VICs are responding mechanically in situ. To circumvent this issue, a finite element (FE) model of the AV leaflet under bending deformations was developed in Abaqus version 6.14 (Dassault Systemes, Johnston, RI, USA) to estimate layer-specific VIC contractility, connectivity, and stiffness [26]. The FE model was designed to consider microstructural details like AVIC size, shape, distribution, and orientation as determined from histological data.

The model features a macro and micro (down-scale) component (Fig. 8). Representative volume elements (RVE) were optimized statistically for each layer to represent the native 3D structure. A brief overview of the formulation of the FE model follows.

The macro component is formulated to estimate layer- and contractile state-dependent mechanics of the AV leaflet under bending deformations. The AV leaflet specimen is simulated as a bi-layered and bimodular neo-Hookean isotropic nearly incompressible material

$$W^\pm = \frac{{}_{\text{state}}\mu_{\text{Macro}}^{\text{layer}\pm}}{2}(I_1 - 3) - p(I_3 - 1) \tag{12}$$

where ${}_{\text{state}}\mu_{\text{Macro}}^{\text{layer}\pm}$ is the shear modulus of a specific AV leaflet layer under compression (−) and tension (+), $I_1$ and $I_3$ are the first and third invariants of the left Cauchy-Green deformation tensor $\mathbf{C}=\mathbf{F}^T\mathbf{F}$, respectively, and $p$ is the Lagrange multiplier

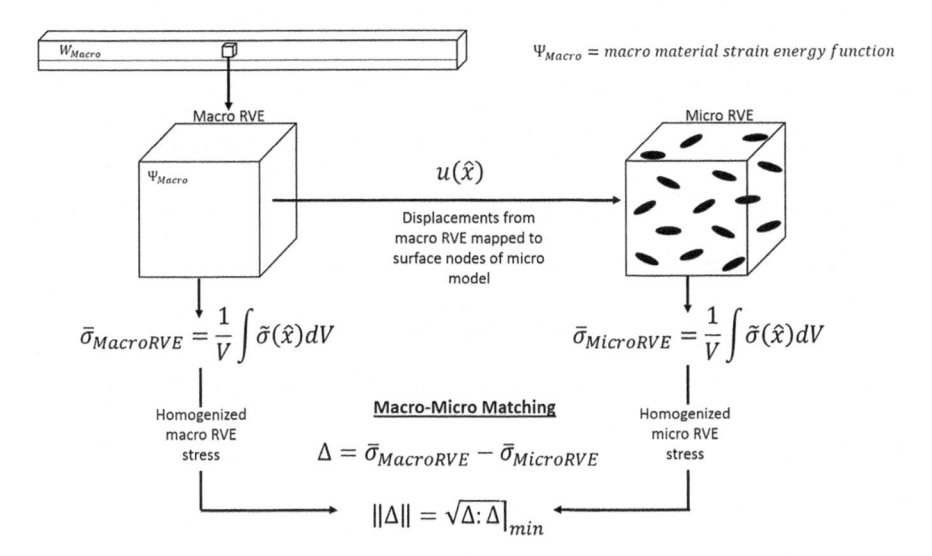

**Fig. 8** Overview of macro-micro finite element methodology. Macro-level simulations are performed and the resulting displacements ($u(\hat{x})$) from the macro RVE are mapped to the micro RVE as boundary conditions. The average stress across the macro ($\bar{\sigma}_{\text{MacroRVE}}$) and micro ($\bar{\sigma}_{\text{MicroRVE}}$) RVEs are then compared and finite element simulations are performed until the difference between both values are minimized. From [26]

used to enforce incompressibility. Here, the subscript "state" denotes the contractile state of the AVICs (normal, hypertensive, inactivated) during testing and is used for the comparison of shear moduli between the contractile states. From the macro model, $_{\text{state}}\mu_{\text{Macro}}^{\text{layer}\pm}$ is determined from matching experimental moment-curvature data obtained from flexure tests to Eq. (12). The simulated displacements of the macro RVE from the center of the specimen (the point of greatest moment and curvature) are then mapped to the surface nodes of the micro model (Fig. 8).

Within the framework of the micro model, AVIC stiffness, contractility, and connectivity are built in as variable parameters (Fig. 9). AVICs were modeled using neo-Hookean, ellipsoidal inclusions within the micro RVE. The contractile capability of AVICs was incorporated through use of isometric thermal expansion to mimic cellular contraction by inducing isothermal deformations of the ellipsoidal inclusions [45]. Consistent with the macro model, a bi-layered and bimodular neo-Hookean isotropic nearly incompressible material model was used in the micro component as well

$$W^{\pm} = \frac{\mu_{\text{AVIC}}^{\text{layer}}}{2}(I_1 - 3) + \frac{1}{D_1}(J^{\text{el}} - 1)^2 \tag{13}$$

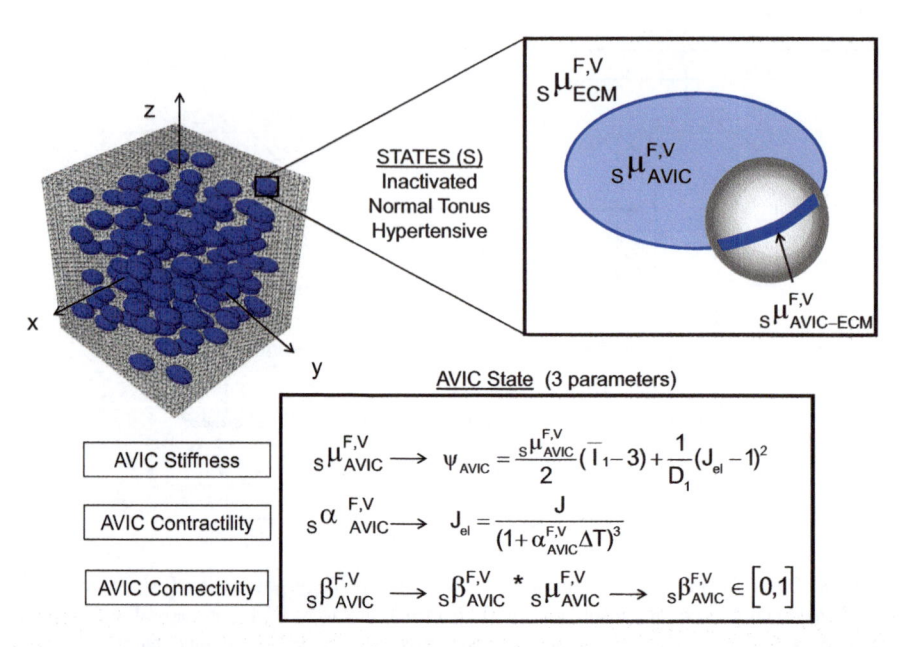

**Fig. 9** The micro RVE with ellipsoidal VIC inclusions used to back out AVIC stiffness, contractility, and connectivity. From [26]

where $\mu_{AVIC}^{layer}$ is the layer-specific shear modulus of AVIC inclusions, $D_1$ is constant defined by $\frac{2}{\kappa}$ where $\kappa$ is the bulk modulus of the material and $J^{el}$ is the elastic volume ratio computed through

$$J^{el} = \frac{J}{(1+\varepsilon^{th})^3} \quad (14)$$

where $\varepsilon^{th}$ is the thermal expansion strain and a function of temperature ($T$) and the thermal expansion coefficient ($\alpha$)

$$\varepsilon^{th} = \alpha \Delta T \quad (15)$$

In the simulation, $\alpha$ is correlated with AVIC contraction and is parametrically altered while $\Delta T$ is kept constant.

AVIC connectivity to the ECM is represented with a thin interface boundary around each AVIC inclusion (Fig. 10). The modulus of this interface is equivalent to $\mu_{AVIC}^{layer}$ and is parametrically altered by multiplying it with the variable $\beta_{AVIC}^{F,V\pm}$ ($\beta_{AVIC}^{F,V\pm} \in [0, 1]$) with $\beta = 1$ denoting complete binding of the AVIC inclusions to the ECM and $\beta = 0$ denoting no binding. The average stress across the macro

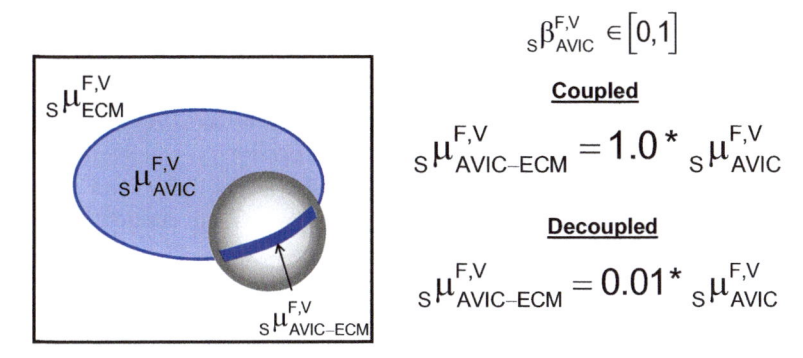

# Biological Mechanics of the Heart Valve Interstitial Cell

$$_S\beta_{AVIC}^{F,V} \in [0,1]$$

**Coupled**

$$_S\mu_{AVIC-ECM}^{F,V} = 1.0 *\ _S\mu_{AVIC}^{F,V}$$

**Decoupled**

$$_S\mu_{AVIC-ECM}^{F,V} = 0.01 *\ _S\mu_{AVIC}^{F,V}$$

**Fig. 10** The interfacial layer between the VIC inclusions and the simulated ECM. The modulus of the interfacial layer is multiplied by the term $\beta$ ($\beta \in [0,1]$) to simulate various levels of VIC connectivity throughout the different layers of the AV leaflet. From [26]

and micro RVE are calculated and a finite element simulation is performed by adjusting AVIC stiffness ($\mu_{AVIC}^{layer}$), contractility ($\alpha$), and connectivity ($\beta_{AVIC}^{F,V\pm}$) until the average stress across both RVEs match. Once an acceptable match is achieved, the estimated parameters are reported. For more detailed information regarding the formulation of the model, readers are directed to the following source [26].

These FE models were highly beneficial tools to estimate mechanical parameters that are impractical or even impossible to measure experimentally. Such tools are especially useful in studies of native tissues with poor optical properties that disallow the direct assessment of VIC behavior. In addition, modeling also provides techniques to assess the layer-specific response of VICs.

## 3.5 Use of 3D Hydrogel for Mechanobiological Studies

Native valve leaflet tissue studies are highly desirable because they elicit the most accurate behavior of VICs. Although extremely insightful, native tissue environments are limited mainly due to (1) their inability to be tuned to answer-specific mechanobiological questions and (2) the difficulty involved with detailed imaging of cellular and subcellular components. Thus, a need exists for a 3D platform that mimics the native micromechanical environment while featuring favorable optical properties for visualization. One avenue that has shown promise in this regard are peptide-modified poly(ethylene glycol) (PEG) hydrogels [46]. These matrices consist of norbornene functionalized PEG molecules that bind covalently with peptides that contain cysteine residues via thiol:ene reactions. Incorporation of the adhesive peptide sequence CRGDS (Cys-Arg-Gly-Asp-Ser) and matrix metalloproteinase (MMP) degradable peptide crosslinkers allow for cellular binding, growth, and microenvironmental remodeling. Through use of the following equations

$$G = \frac{\rho_{xL} RT}{Q^{1/3}} \tag{16}$$

where $\rho_{xL}$ is the crosslinking density, $R$ is the ideal gas constant, $T$ is the temperature, and $Q$ is the volumetric swelling ratio, the shear modulus ($G$) can be estimated and fine-tuned by adjusting the amount of MMP degradable crosslinkers and essentially the thiol:ene ratio of the hydrogel chemistry [47]. From $G$, the elastic modulus ($E$) can be computed from the equation:

$$E = 2G(1 + v) \tag{17}$$

where $v$ is the Poisson's ratio of the hydrogel material, which is assumed to be 0.5 (incompressible). In addition to tunable mechanics, the PEG hydrogel matrices also offer modulation of adhesion levels through adjustment of the concentration of CRGDS peptide sequences. Coupled with tunable mechanics, the hydrogel system can be utilized to mimic healthy and diseased environments.

Preliminary work has been done to quantify the contractility of AVICs within 10 kPa hydrogels through use of end-loading, flexural deformation testing. The experimental testing protocol was adapted from [37]. The VIC embedded hydrogels were treated with a normal (5 mM KCL) solution and subjected to flex-testing. The testing solution was then switched for a hypertensive solution (90 mM KCL) to elicit VIC active contraction and the specimens were tested again. Finally, contractility was halted by essentially killing the cells with 70% methanol before testing the gels a final time.

The moment–curvature plots (Fig. 11a) produced from these tests display higher moment values resulting from the hypercontractile state and lower moment values in the inactive state, compared to the normal condition. Initial observations revealed that the hydrogel material displayed a nonlinear moment–curvature response. Thus,

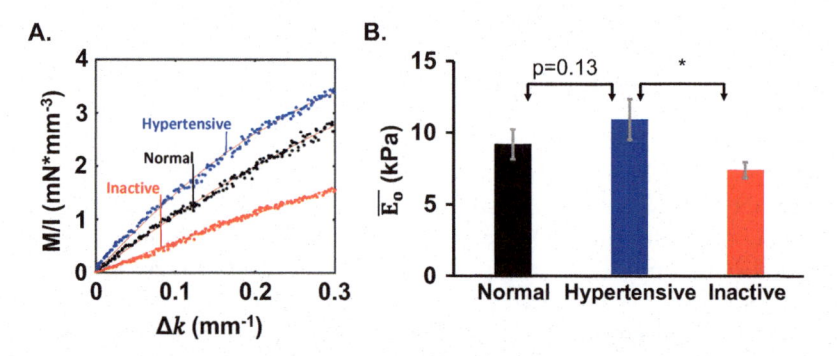

**Fig. 11** (a) Moment vs. curvature plots for VIC-hydrogels under hypertensive, normal, and inactive conditions. (b) The average initial effective stiffness for VIC-hydrogels under hypertensive, normal, and inactive conditions obtained from the initial slope of nonlinear moment vs. curvature plots

the moment curvature data was modeled using a second order polynomial of the form

$$\frac{M}{I} = a\Delta\kappa^2 + b\Delta\kappa + c \tag{18}$$

This analysis technique was borrowed from [38]. Through use of the Euler-Bernoulli relation Eq. (10) and the derivative of the second order polynomial (Eq. (18)), a linear relation between the hydrogel $E_{\text{eff}}$ and $\Delta\kappa$ was established. $E_{\text{eff}}$ at small strains ($\Delta\kappa = 0$ mm$^{-1}$) was used as a parameter for comparison between the contractile states and it was observed that the hydrogel was stiffer when the embedded VICs were in a hypertensive state and less stiff when they were inactivated, compared to the control condition (Fig. 11b). These preliminary results reflect the appreciable cell-material interactions and show great promise in using tunable hydrogel environments for future mechanobiological studies. In addition, hydrogel environments are an intriguing tool for the development of down-scale FEM models similar to the one mentioned in the prior section to estimate the biomechanical state of VICs. Another avenue of interest involves taking advantage of the optical clarity of the hydrogels to observe cellular and subcellular events and capturing them within agent-based models.

## 3.6 Uniaxial Planar Stretch Bioreactors

### 3.6.1 Overview

Planar tensile bioreactors allow for highly controlled, mechanical conditioning of valve leaflet tissues [18, 48–51]. Such systems allow for the emulation of hypo-physiological, normal, and hyper-physiological strain levels to study how tissue-level deformations drive the biosynthetic response of VICs. To maximize relevance of these studies, a macro-micro finite element model is employed to interpret in vitro findings and relate them to in vivo functional states. In the following section, we summarize a novel experimental-computational approach to link VIC biosynthetic response to cellular deformation and explain the implications this may have on surgically repaired MV.

### 3.6.2 Bioreactor Design

MV anterior leaflets (MVAL) were excised from porcine hearts and rectangular specimens were dissected from the clear zone measuring 11.5 mm in the circumferential direction and 7.5 mm in the radial direction (Fig. 12a). A tissue strip bioreactor was engineered based off previous designs [48–51]. The system featured an environmental specimen chamber that housed the tissue specimens (Fig. 12b). Metallic

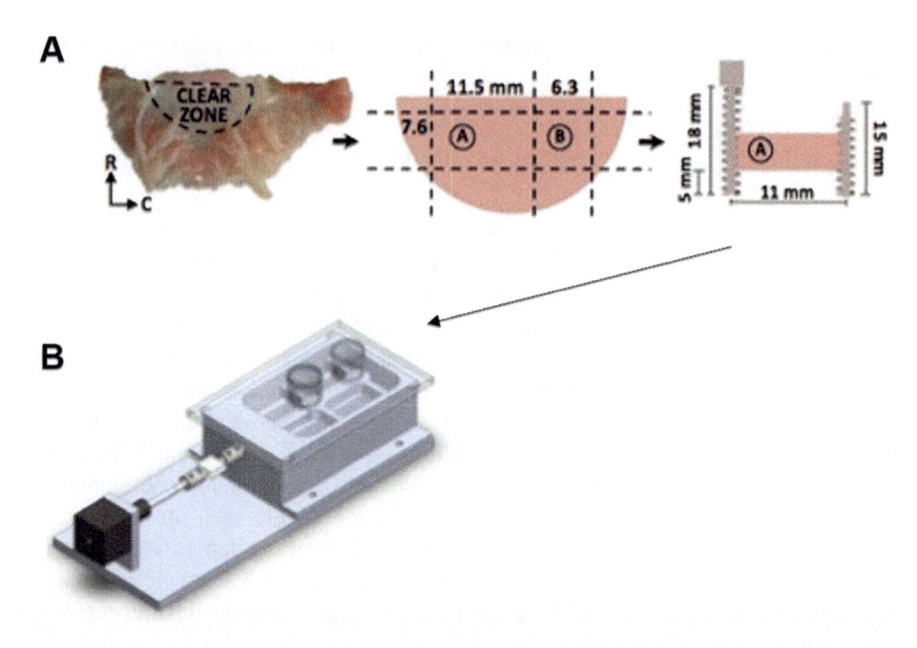

**Fig. 12** (**a**) A mitral valve anterior leaflet excised into two distinct sections used for (**a**) mechanical conditioning (**b**) protein quantification. (**b**) A uniaxial tissue strip tensile bioreactor used to impose controlled deformations of test specimens. From [18]

springs were used to pierce the MV leaflet specimens along the radial width on each side to establish five points of contact with proximity to one another. The spring was used to attach one side of the specimen to a stationary post within the environmental chamber and the other side to a metal rod connected to a linear actuator used to impose deformations. Cyclic stretch tests were performed for 48 h before downstream assessment of the test specimens.

### 3.6.3 Collagen Fiber Architecture Alters Nuclear and Cytoplasmic Geometry

Small angle light scattering (SALS) was used to assess the collagen fiber architecture (CFA) of the test samples and return a normalized orientation index (NOI) with an NOI value of 100% representing uniformly aligned collagen networks and 0% denoting a completely random fiber orientation. Extended experimental details can be found here [52].

CFA orientation appeared to increase drastically at 30% strain (Fig. 13a, c). Transmission electron microscopy (TEM) revealed that nuclear aspect ratio (ratio between major and minor axis, NAR) and cytoplasmic aspect ratio increased in a similar fashion at 30% strain (Fig. 13b, d). It was also observed that the cytoplasmic aspect ratio closely resembled the NAR at lower strains (0–20%) but decouples from

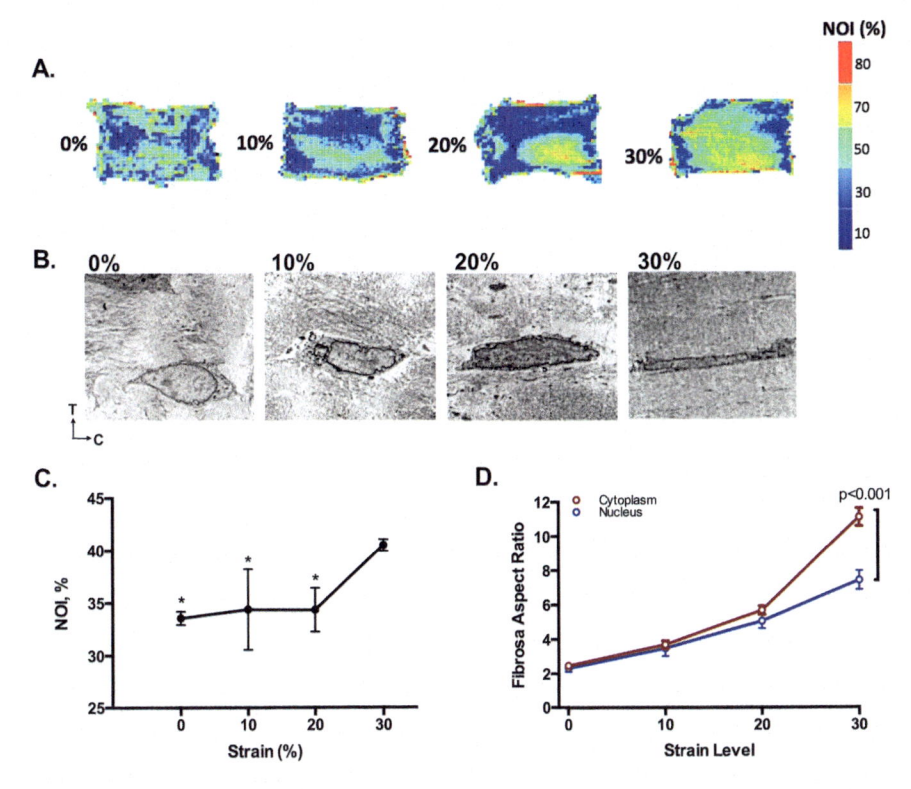

**Fig. 13** (a) Normalized orientation index of the collagen fiber architecture with respect to increasing strain levels. (b) Transmission electron micrographs of MVICs displaying nuclear and cytoplasmic deformation due to increasing strains. (c) Quantification of the normalized orientation index shows a drastic increase in the 30% strain level compared to the lower strain levels. (d) The nuclear and cytoplasmic aspect ratio of MVICs with respect to strain level. Note how the cytoplasmic aspect ratio decouples from the nuclear aspect ratio at 30% strain. From [18]

the NAR at hyper-physiological strain levels (30%, Fig. 13d). This phenomenon provides evidence that MVICs may behave differently under nonphysiological strain levels.

### 3.6.4 Large Strains Cause MVIC Activation

RT-qPCR revealed a linear trend between $\alpha$-SMA and Type I Collagen gene expression in relation to increasing strain levels suggesting MVICs become activated in response to large strains (Fig. 14a, b). Consistent with these observations, colorimetric assays displayed a significant increase in collagen production within the 30% strain group after 48 h compared to measurements made postmortem (Fig. 14d). An increase in Sulfated GAG content was also observed in this group

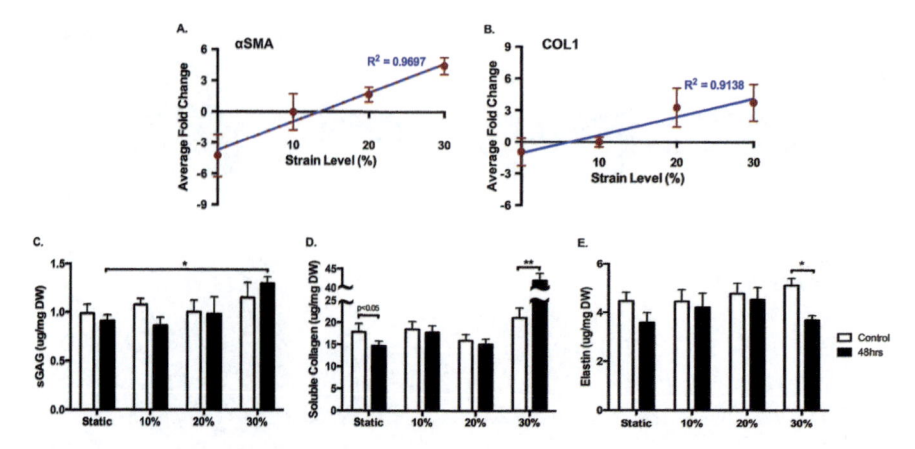

**Fig. 14** The average fold change of (**a**) α-SMA and (**b**) type I collagen normalized to the 10% strain as determined from RT-qPCR. Colorimetric assay results at different strain levels for (**c**) sulfated GAGs, (**d**) soluble collagen, and (**e**) elastin content of both post mortem samples and samples subjected to cyclic strain for 48 h. From [18]

(Fig. 14c). Interestingly, Elastin content decreased at higher strains possibly due to an increase in proteolytic activity of MMPs, TIMPs, and cathepsins (Fig. 14e) [49]. When taken together, these results provide evidence to support the phenotypic switch of MVICs from a normal, quiescent state to an activated state exhibiting increased ECM remodeling activity. These results provide insight to how MVICs may respond to diseased induced stress alterations in vivo.

### 3.6.5 Clinical Relevance: Relating In Vitro Observations to In Vivo Function

The deformations of normal and surgically repaired MVAL have been measured previously through use of sonomicrometry array techniques (Fig. 15a, b) [53]. This information was incorporated into a previously developed MVIC microenvironment model [13]. In short, an RVE was used to model the fibrosa layer of the MVAL including MVICs as ellipsoidal inclusions (Fig. 15c). Cell dimensions, density, and orientations were determined previously via histology and were reflected within the model [13, 54]. ECM was modeled using a simplified structural constitutive model and the cellular inclusions were modeled using a modified Saint-Venant Kirchhoff (SVK) material model [13]. FE simulations were performed inside ABAQUS 6.13 framework (Simulia, Dassault Systèmes, Providence, RI, USA) by assigning time-dependent, tissue deformations from sonomicrometry arrays as boundary conditions on the RVE. The predicted deformation field was used to compute the in vivo

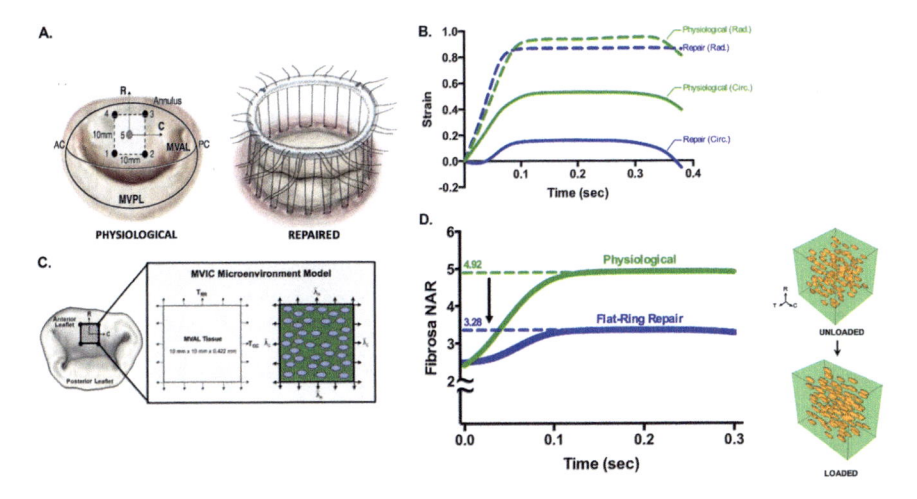

**Fig. 15** (a) Schematic of normal and surgically repaired mitral valves with five sonocrystals placed on the anterior leaflets. (b) Circumferential (solid line) and radial (dashed line) strain for both physiological (green) and repair (blue) mitral valves. Surgical repair leads to a decrease in circumferential strain. (c) Schematic of the RVE used to model the anterior leaflet. The tissue-level strains from (b) are applied as boundary conditions. Finite element simulations are performed to predict deformation fields and essentially to estimate the nuclear aspect ratio of the ellipsoidal MVIC inclusions. (d) The estimated NAR of MVICs within the fibrosa layer from the finite element simulations. The model estimates that a decrease in NAR occurs due to surgical repair. From [18]

deformation of the MVIC inclusions through analysis of the NAR. It was found that the NAR of MVICs at peak end-systole is 4.92 in normal MVs (Fig. 15d). The model predicted a decrease in NAR to 3.28 in surgically repaired valves due to the decrease in circumferential deformation (Fig. 15d) [53]. The results suggest that nuclear deformation is closely linked to the circumferential deformation of MVALs. More recently, the same simulation framework was applied in a novel study of MV leaflet remodeling during pregnancy [12]. Remarkably, changes in NAR were found to correlate closely with geometric changes to the valve apparatus that occur in early pregnancy. This initial perturbation of VIC geometry was then found to trigger a complex layer-specific cascade of fiber remodeling mechanisms in the ECM, which work in concert to restore VIC homeostasis by the end of gestation.

Through combining the experimental data and the simulated NARs, a link between in vivo cell deformations and biosynthetic response is made (Fig. 16). The NAR is used as a dimensionless parameter to loosely bracket ranges of biological behavior within hypo-physiological (<3.30), normal (3.30–5.0), and hyper-physiological (>5.0) states. From the analysis of the bracketed ranges, it is speculated that deformation scenarios leading to hypo- and hyper-physiological NAR values may cause a change in VIC phenotype and essentially alter biosynthetic response in attempt to return to homeostasis.

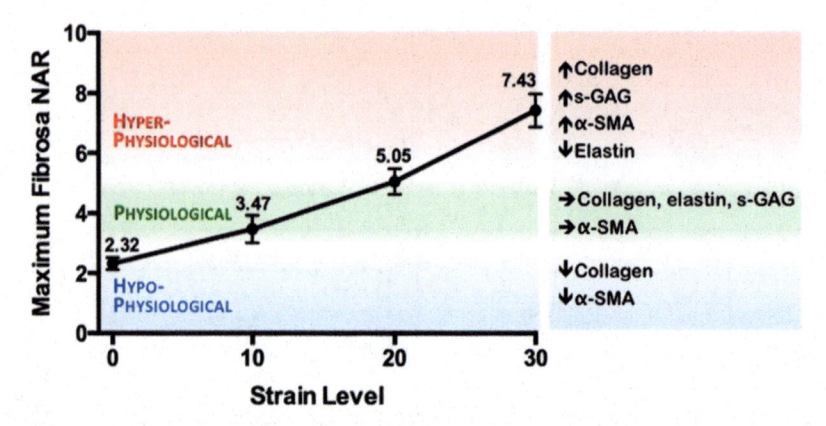

Fig. 16 MVIC biosynthetic behavior is bracketed into hyper-physiological, physiological, and hypo-physiological ranges based off the simulated NARs for the normal and surgically repaired case. NAR below 3.28 is deemed as hypo-physiological. An NAR between 3.28 and 4.92 is considered normal and above 4.92 is hyper-physiological. From [18]

## 3.7 Model-Driven Experimental Design

### 3.7.1 FE Models Relating Tissue-Level and MVIC Nuclear Deformations

It has been shown experimentally that increased cellular deformation correlates with abnormal biosynthetic activity. To delineate the role that tissue-level deformations have on cellular biomechanical states and geometry, experimental and computational techniques were utilized [13].

MVALs were loaded under equi-biaxial tension and the resulting layer-specific deformations of MVICs were measured using multi-photon microscopy (Fig. 17a). A FE MVIC microenvironment model was developed to describe the relation between planar tension and layer-specific MVIC nuclear deformation (Fig. 17b) and it was observed that the stiffness of MVICs varied little between the different MVAL layers (Fig. 17c–f). This suggests that differences in MVIC NAR across MVAL layers are mainly due to the deformation of the surrounding ECM and not the intrinsic stiffness of the cells. Documenting the nuclear deformation of MVICs in response to in vitro equi-biaxial loading was a crucial first step towards analysis of cellular function within surgically repaired MVs.

Additional MV models focus on estimating the in vivo stress and strain through sonomicrometry arrays [19].

Through tracking of sonocrystals placed on Dorsett sheep MVALs, the deformation between several kinematic states of normal and surgically repaired MVs was measured making it possible to map in vitro deformations back to the in vivo state (Fig. 18a–c). In addition, in vivo stresses were characterized through use of MVAL kinematic data and inverse modeling techniques. The in vivo stresses where estimated to be ~360 kPa in the circumferential direction and ~450 kPa in the radial

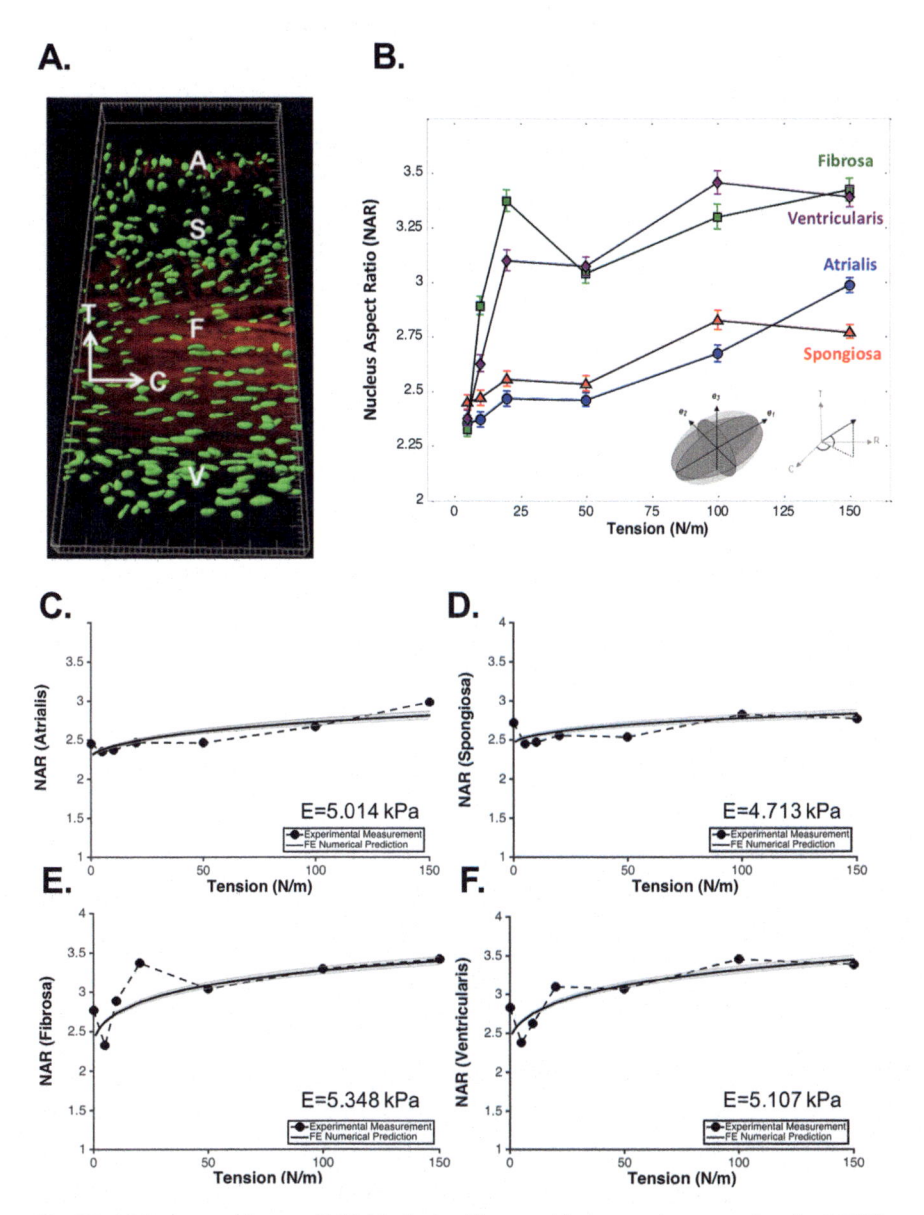

Biological Mechanics of the Heart Valve Interstitial Cell

**Fig. 17** (a) Transmural image of MVAL obtained from multi-photon microscopy showing MVICs across all layers. (b) Layer-specific nuclear aspect ratio measurements made at different equi-biaxial membrane tensions. (c–f) Experimentally determined nuclear aspect ratio is fit to the FE model and layer-specific stiffness of MVICs is backed out. Modified from [13]

direction when not considering the effects of pre-strain. With pre-strain, the stresses were estimated to increase to ~510 kPa in the circumferential direction and ~740 kPa in the radial direction.

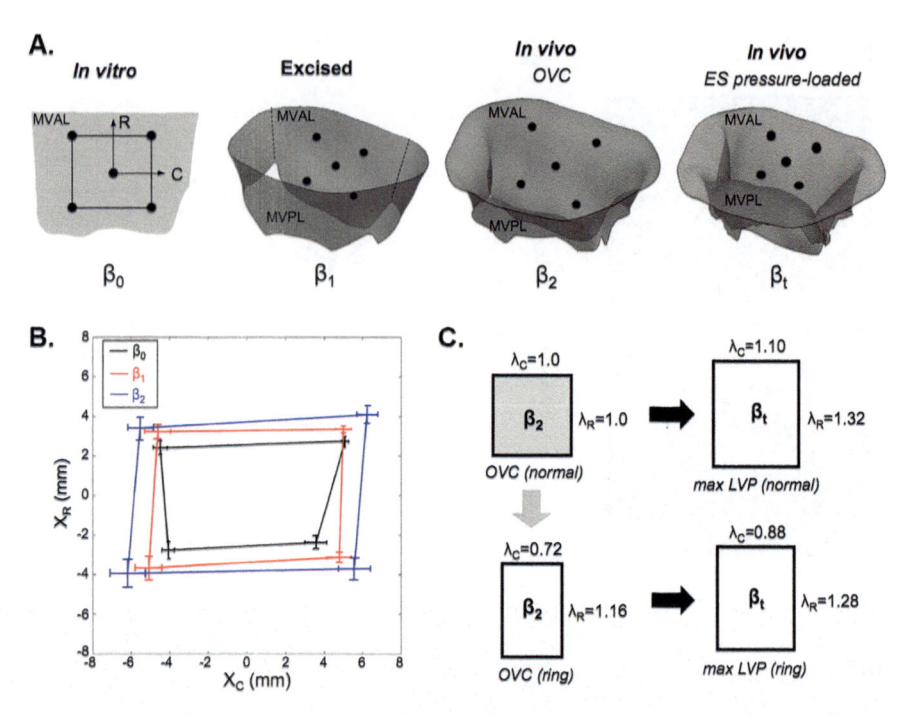

Fig. 18 (a) Schematic of the different MVAL kinematic states. Sonocrystals are tracked and their displacements are recorded across the in vitro configuration ($\beta_0$), the excised configuration ($\beta_1$), the in vivo configuration just before the onset of ventricular contraction (OVC) where the MV leaflets are coapted ($\beta_2$), and the in vivo configuration at the end-systolic (ES) time point where the MV is under transvalvular pressure ($\beta_t$). (b) The population averaged location of the sonocrystals with respect to the radial ($X_R$) and circumferential ($X_C$) directions for key kinematic states. (c) The circumferential ($\lambda_C$) and radial ($\lambda_R$) deformation of the center region of MVALs due to surgical ring repair. From [19]

Through coupling both previously discussed MV models by applying the in vivo tissue deformations as boundary conditions within the MVIC microenvironment model, in vivo layer-specific NAR was estimated for a representative cardiac cycle (Fig. 19a). Using kinematic data gathered on surgically repaired MVs from [53], NAR estimates were also made for surgically repaired MVs (Fig. 19c). The layer-specific NAR rate of change is also reported (Fig. 19b, d) and it was observed that the fibrosa displayed the highest rate of change within normal valves. Interestingly, among repaired valves the atrialis displayed the highest NAR values and rate of change. This is hypothesized to be a result of leaflet contraction in the circumferential direction and expansion in the radial direction after flat ring repair. This notable deviation from normal valve geometry most likely causes altered distribution of pressure throughout the anterior leaflet, thus resulting in altered, layer-specific MVIC deformations. The ability to estimate in vivo stresses and deformations for normal and surgically repaired MVALs highlights the benefits of taking an experimental-computational approach to studying VIC mechanobiology.

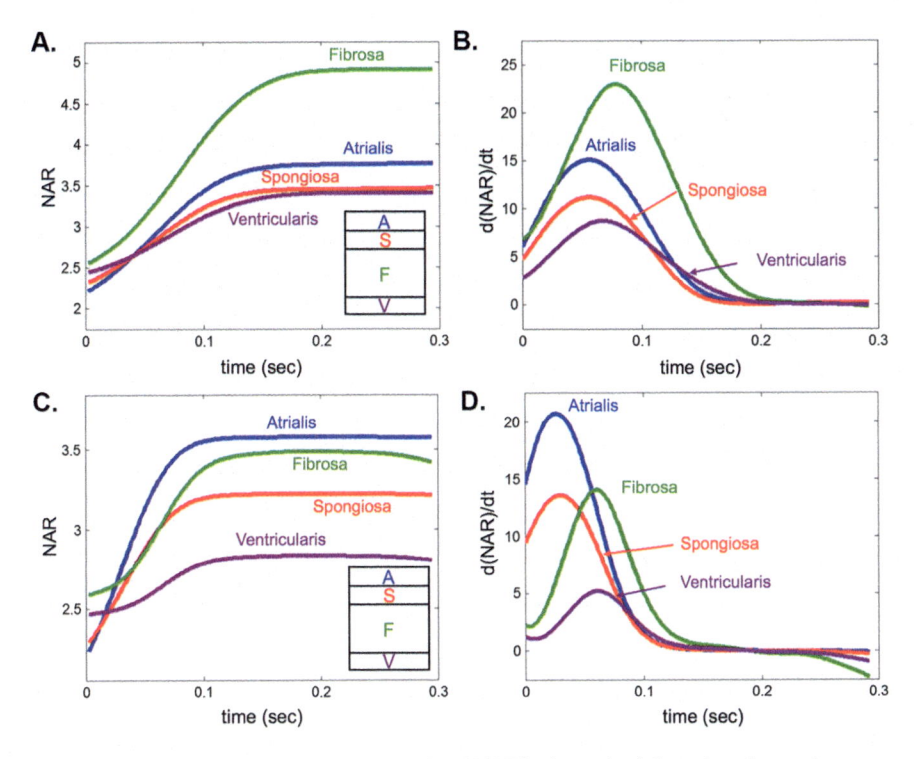

Biological Mechanics of the Heart Valve Interstitial Cell 31

**Fig. 19** The layer-specific nuclear aspect ratio of MVICs determined from the microenvironment model for the (**a**) normal and (**c**) surgical repair scenarios. The NAR rate of change for (**b**) normal and (**d**) surgical repair highlights the differences in layer-specific cell deformations as a result of altered tissue-level deformations. From [19]

## 4 Future Directions

A need exists for a full multi-scale approach to study and capture the complexity of the heart system. Combining experimental and computational methods is key to the refinement of our current knowledge base on VIC mechanobiology. Linking the cell-, tissue-, and organ-level response of the heart system not only widens our understanding of the heart's physiology, but also offers a unique opportunity to simulate diseased states. Recent computational modeling studies have elucidated the importance of including microstructurally informed tissue properties in predictive simulation frameworks, as both CFA and macro-scale geometric properties vary considerably over local tissue regions in heart valves [55–59]. In light of recent findings quantifying the link between local ECM structural properties and VIC mechanical and biosynthetic behavior [18], it has become increasingly clear that a full understanding of valvular function, maintenance, and adaptation must be rooted in a detailed knowledge of the interrelationships between organ-level physiology and tissue/cell coupling.

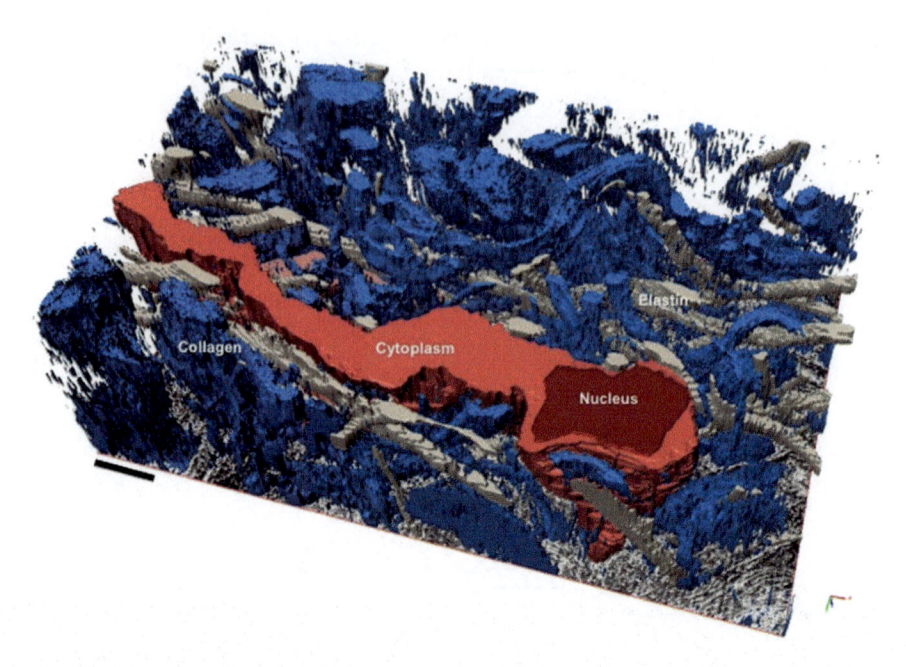

**Fig. 20** 3D reconstruction of the VIC microenvironment from FIB-SEM images. Note how the VIC is embedded within networks of collagen and elastin fibers

Currently, a promising avenue for investigating complex valvular tissue systems is through the development of agent-based models that can combine the VIC mechanical environment, biomechanical states, and their resulting biosynthetic activity within a cohesive modeling pipeline. For this, high detailed cellular-level observations are crucial towards the formulation of VIC models.

State-of-the-art imaging technologies like the Focused Ion Beam Scanning Electron Microscope (FIB-SEM) have produced high-quality images of the native valve leaflet structure and help shed light on the intricate VIC-ECM interactions (Fig. 20). This preliminary work reflects the complex VIC microenvironment and highlights considerations to be made when formulating agent-based models.

Linking tissue-level deformations to changes in MVIC NAR in the in vivo state marks a huge success in gauging the biophysical state of these cells. To take these studies further, quantification of MVIC biosynthetic response within the model-prescribed physiological stresses is currently underway. With use of biaxial bioreactor tissue strip systems, MVALs will undergo mechanical conditioning and the ECM components will be quantified for the normal and valve repair cases. From these experiments, we hope to validate the NAR values estimated from the model and document the biosynthetic behavior of MVICs at the predicted stress levels. This model-driven study serves as an example of how experimental-computational techniques can broaden our understanding of the underlying mechanobiology within living systems.

To increase predictive capabilities of VIC models, parameters correlating physical cellular characteristics and downstream expression of genes are needed. One avenue that researchers have used to do this is through analysis of VIC nuclear geometry, specifically NAR [16–18]. Preliminary correlations have been made to associate NAR with specific stress environments and biosynthesis rates. However, nuclear deformations may play a larger role in controlling intracellular pathways and regulating the behavior of VICs. Future efforts should focus on closely examining the cellular and subcellular events of VIC biological function and relating it back to measurable parameters (i.e., NAR). Advances in hydrogel chemistry with favorable optical properties and tunable characteristics show great promise in this endeavor. Hydrogel environments that offer tunable mechanics and peptide modifications present a unique opportunity to emulate healthy and diseased states. In addition, the ability to gather mechanical and optical data from these matrices make them extremely attractive as a tool for future 3D mechanobiological studies.

**Acknowledgements** This work was supported by the National Institutes of Health (NIH) Grants R01HL119297. CHL was in part supported by start-up funds from the School of Aerospace and Mechanical Engineering (AME) at the University of Oklahoma, and the American Heart Association Scientist Development Grant Award (16SDG27760143).

# References

1. Merryman WD, Lukoff HD, Long RA, Engelmayr GC Jr, Hopkins RA, Sacks MS. Synergistic effects of cyclic tension and transforming growth factor-beta1 on the aortic valve myofibroblast. Cardiovasc Pathol. 2007;16(5):268–76.
2. Guyton AC. Textbook of medical physiology. 5th ed. Philadelphia: W.B. Saunders Company; 1976.
3. He Z, Ritchie J, Grashow JS, Sacks MS, Yoganathan AP. In vitro dynamic strain behavior of the mitral valve posterior leaflet. J Biomech Eng. 2005;127(3):504–11.
4. He Z, Sacks MS, Baijens L, Wanant S, Shah P, Yoganathan AP. Effects of papillary muscle position on in-vitro dynamic strain on the porcine mitral valve. J Heart Valve Dis. 2003;12(4):488–94.
5. Sacks MS, Enomoto Y, Graybill JR, Merryman WD, Zeeshan A, Yoganathan AP, et al. In-vivo dynamic deformation of the mitral valve anterior leaflet. Ann Thorac Surg. 2006;82(4):1369–77.
6. Sacks MS, He Z, Baijens L, Wanant S, Shah P, Sugimoto H, et al. Surface strains in the anterior leaflet of the functioning mitral valve. Ann Biomed Eng. 2002;30(10):1281–90.
7. Ayoub S, Ferrari G, Gorman RC, Gorman JH, Schoen FJ, Sacks MS. Heart valve biomechanics and underlying mechanobiology. Compr Physiol. 2016;6(4):1743–80.
8. Sacks MS, Yoganathan AP. Heart valve function: a biomechanical perspective. Philos Trans R Soc Lond Ser B Biol Sci. 2007;362(1484):1369–91.
9. Buchanan RM, Sacks MS. Interlayer micromechanics of the aortic heart valve leaflet. Biomech Model Mechanobiol. 2014;13(4):813–26.
10. Rego BV, Sacks MS. A functionally graded material model for the transmural stress distribution of the aortic valve leaflet. J Biomech. 2017;54:88–95.
11. Stella JA, Sacks MS. On the biaxial mechanical properties of the layers of the aortic valve leaflet. J Biomech Eng. 2007;129(5):757–66.

12. Rego BV, Wells SM, Lee CH, Sacks MS. Mitral valve leaflet remodelling during pregnancy: insights into cell-mediated recovery of tissue homeostasis. J R Soc Interface. 2016;13(125):20160709.
13. Lee CH, Carruthers CA, Ayoub S, Gorman RC, Gorman JH 3rd, Sacks MS. Quantification and simulation of layer-specific mitral valve interstitial cells deformation under physiological loading. J Theor Biol. 2015;373:26–39.
14. Pierlot CM, Lee JM, Amini R, Sacks MS, Wells SM. Pregnancy-induced remodeling of collagen architecture and content in the mitral valve. Ann Biomed Eng. 2014;42(10):2058–71.
15. Pierlot CM, Moeller AD, Lee JM, Wells SM. Pregnancy-induced remodeling of heart valves. Am J Physiol Heart Circ Physiol. 2015;309(9):H1565–78.
16. Lam NT, Muldoon TJ, Quinn KP, Rajaram N, Balachandran K. Valve interstitial cell contractile strength and metabolic state are dependent on its shape. Integr Biol. 2016;8(10):1079–89.
17. Tandon I, Razavi A, Ravishankar P, Walker A, Sturdivant NM, Lam NT, et al. Valve interstitial cell shape modulates cell contractility independent of cell phenotype. J Biomech. 2016;49(14):3289–97.
18. Ayoub S, Lee C-H, Driesbaugh KH, Anselmo W, Hughes CT, Ferrari G, et al. Regulation of valve interstitial cell homeostasis by mechanical deformation: implications for heart valve disease and surgical repair. J R Soc Interface. 2017;14:20170580.
19. Lee CH, Zhang W, Feaver K, Gorman RC, Gorman JH 3rd, Sacks MS. On the in vivo function of the mitral heart valve leaflet: insights into tissue-interstitial cell biomechanical coupling. Biomech Model Mechanobiol. 2017;16:1613.
20. Rabkin E, Aikawa M, Stone JR, Fukumoto Y, Libby P, Schoen FJ. Activated interstitial myofibroblasts express catabolic enzymes and mediate matrix remodeling in myxomatous heart valves. Circulation. 2001;104(21):2525–32.
21. Rajamannan NM, Evans FJ, Aikawa E, Grande-Allen KJ, Demer LL, Heistad DD, et al. Calcific aortic valve disease: not simply a degenerative process: a review and agenda for research from the National Heart and Lung and Blood Institute Aortic Stenosis Working Group. Executive summary: calcific aortic valve disease-2011 update. Circulation. 2011;124(16):1783–91.
22. Sacks MS, Merryman WD, Schmidt DE. On the biomechanics of heart valve function. J Biomech. 2009;42(12):1804–24.
23. Carruthers CA, Good B, D'Amore A, Liao J, Amini R, Watkins SC, et al., editors. Alterations in the microstructure of the anterior mitral valve leaflet under physiological stress. In: ASME 2012 summer bioengineering conference. American Society of Mechanical Engineers; 2012.
24. Sakamoto Y, Buchanan RM, Sacks MS. On intrinsic stress fiber contractile forces in semilunar heart valve interstitial cells using a continuum mixture model. J Mech Behav Biomed Mater. 2016;54:244–58.
25. Sakamoto Y, Buchanan RM, Sanchez-Adams J, Guilak F, Sacks MS. On the functional role of valve interstitial cell stress fibers: a continuum modeling approach. J Biomech Eng. 2017;139(2):021007.
26. Buchanan RM. An integrated computational-experimental approach for the in situ estimation of valve interstitial cell biomechanical state. Austin: The University of Texas at Austin; 2016.
27. Merryman WD, Youn I, Lukoff HD, Krueger PM, Guilak F, Hopkins RA, et al. Correlation between heart valve interstitial cell stiffness and transvalvular pressure: implications for collagen biosynthesis. Am J Physiol Heart Circ Physiol. 2006;290(1):H224–31.
28. Theret DP, Levesque MJ, Sato M, Nerem RM, Wheeler LT. The application of a homogeneous half-space model in the analysis of endothelial cell micropipette measurements. J Biomech Eng. 1988;110(3):190–9.
29. Rocnik EF, van der Veer E, Cao H, Hegele RA, Pickering JG. Functional linkage between the endoplasmic reticulum protein Hsp47 and procollagen expression in human vascular smooth muscle cells. J Biol Chem. 2002;277(41):38571–8.
30. Merryman WD, Liao J, Parekh A, Candiello JE, Lin H, Sacks MS. Differences in tissue-remodeling potential of aortic and pulmonary heart valve interstitial cells. Tissue Eng. 2007;13(9):2281–9.

31. Costa KD, Yin FC. Analysis of indentation: implications for measuring mechanical properties with atomic force microscopy. J Biomech Eng. 1999;121(5):462–71.
32. Mathur AB, Collinsworth AM, Reichert WM, Kraus WE, Truskey GA. Endothelial, cardiac muscle and skeletal muscle exhibit different viscous and elastic properties as determined by atomic force microscopy. J Biomech. 2001;34(12):1545–53.
33. Sato M, Theret DP, Wheeler LT, Ohshima N, Nerem RM. Application of the micropipette technique to the measurement of cultured porcine aortic endothelial cell viscoelastic properties. J Biomech Eng. 1990;112(3):263–8.
34. Guilak F, Ting-Beall HP, Baer AE, Trickey WR, Erickson GR, Setton LA. Viscoelastic properties of intervertebral disc cells. Identification of two biomechanically distinct cell populations. Spine. 1999;24(23):2475–83.
35. Na S, Sun Z, Meininger GA, Humphrey JD. On atomic force microscopy and the constitutive behavior of living cells. Biomech Model Mechanobiol. 2004;3(2):75–84.
36. Frisch-Fay R. Flexible bars. Washington, DC: Butterworths; 1962. 220 p.
37. Merryman WD, Huang HY, Schoen FJ, Sacks MS. The effects of cellular contraction on aortic valve leaflet flexural stiffness. J Biomech. 2006;39(1):88–96.
38. Mirnajafi A, Raymer J, Scott MJ, Sacks MS. The effects of collagen fiber orientation on the flexural properties of pericardial heterograft biomaterials. Biomaterials. 2005;26(7):795–804.
39. Mirnajafi A, Raymer JM, McClure LR, Sacks MS. The flexural rigidity of the aortic valve leaflet in the commissural region. J Biomech. 2006;39(16):2966–73.
40. Billiar KL, Sacks MS. Biaxial mechanical properties of the native and glutaraldehyde-treated aortic valve cusp: Part II—A structural constitutive model. J Biomech Eng. 2000;122 (4):327–35.
41. Billiar KL, Sacks MS. Biaxial mechanical properties of the natural and glutaraldehyde treated aortic valve cusp—Part I: Experimental results. J Biomech Eng. 2000;122(1):23–30.
42. Mohri H, Reichenback D, Merendino K. Biology of homologous and heterologous aortic valves. In: Ionescu M, Ross D, Wooler G, editors. Biological tissue in heart valve replacement. London: Butterworths; 1972. p. 137.
43. Vesely I, Boughner D. Analysis of the bending behaviour of porcine xenograft leaflets and of natural aortic valve material: bending stiffness, neutral axis and shear measurements. J Biomech. 1989;22(6/7):655–71.
44. Song T, Vesely I, Boughner D. Effects of dynamic fixation on shear behavior of porcine xenograft valves. Biomaterials. 1990;11:191–6.
45. Lu SCH, Pister KS. Decomposition of deformation and representation of the free energy function for isotropic thermoelastic solids. Int J Solids Struct. 1975;11(7–8):927–34.
46. Benton JA, Fairbanks BD, Anseth KS. Characterization of valvular interstitial cell function in three dimensional matrix metalloproteinase degradable PEG hydrogels. Biomaterials. 2009;30 (34):6593–603.
47. Byrant SJ, Anseth KS. Photopolymerization of hydrogel scaffolds. In: Ma PX, Elisseeff J, editors. Scaffolding in tissue engineering. New York: CRC Press; 2005. p. 71–90.
48. Balachandran K, Alford PW, Wylie-Sears J, Goss JA, Grosberg A, Bischoff J, et al. Cyclic strain induces dual-mode endothelial-mesenchymal transformation of the cardiac valve. Proc Natl Acad Sci U S A. 2011;108(50):19943–8.
49. Balachandran K, Hussain S, Yap CH, Padala M, Chester AH, Yoganathan AP. Elevated cyclic stretch and serotonin result in altered aortic valve remodeling via a mechanosensitive 5-HT(2A) receptor-dependent pathway. Cardiovasc Pathol. 2012;21(3):206–13.
50. Balachandran K, Konduri S, Sucosky P, Jo H, Yoganathan A. An ex vivo study of the biological properties of porcine aortic valves in response to circumferential cyclic stretch. Ann Biomed Eng. 2006;34(11):1655–65.
51. Balachandran K, Sucosky P, Jo H, Yoganathan AP. Elevated cyclic stretch alters matrix remodeling in aortic valve cusps: implications for degenerative aortic valve disease. Am J Physiol Heart Circ Physiol. 2009;296(3):H756–64.

52. Sacks MS, Smith DB, Hiester ED. A small angle light scattering device for planar connective tissue microstructural analysis. Ann Biomed Eng. 1997;25(4):678–89.
53. Amini R, Eckert CE, Koomalsingh K, McGarvey J, Minakawa M, Gorman JH, et al. On the in vivo deformation of the mitral valve anterior leaflet: effects of annular geometry and referential configuration. Ann Biomed Eng. 2012;40(7):1455–67.
54. Carruthers CA, Alfieri CM, Joyce EM, Watkins SC, Yutzey KE, Sacks MS. Gene expression and collagen Fiber micromechanical interactions of the semilunar heart valve interstitial cell. Cell Mol Bioeng. 2012;5(3):254–65.
55. Lee CH, Rabbah JP, Yoganathan AP, Gorman RC, Gorman JH 3rd, Sacks MS. On the effects of leaflet microstructure and constitutive model on the closing behavior of the mitral valve. Biomech Model Mechanobiol. 2015;14(6):1281–302.
56. Khalighi AH, Drach A, Bloodworth CH, Pierce EL, Yoganathan AP, Gorman RC, et al. Mitral valve chordae tendineae: topological and geometrical characterization. Ann Biomed Eng. 2017;45(2):378–93.
57. Khalighi AH, Drach A, Gorman RC, Gorman JH 3rd, Sacks MS. Multi-resolution geometric modeling of the mitral heart valve leaflets. Biomech Model Mechanobiol. 2018;17(2):351–66.
58. Drach A, Khalighi AH, Sacks MS. A comprehensive pipeline for multi-resolution modeling of the mitral valve: validation, computational efficiency, and predictive capability. Int J Numer Methods Biomed Eng. 2018;34(2)
59. Sacks MS, Khalighi A, Rego B, Ayoub S, Drach A. On the need for multi-scale geometric modelling of the mitral heart valve. Healthc Technol Lett. 2017;4(5):150.

# Endothelial Mechanotransduction

**James N. Warnock**

**Abstract** The aortic valve resides in a dynamic mechanical environment, with constant exposure to fluid shear stresses and cyclic strain. These physical forces directly impact the valve endothelium and these cells will regulate the biological response of the valve. Under normal physiological conditions, the endothelium provides a monolayer that protects the valve tissue. However, under abnormal conditions, a cascade of events can lead to a loss of integrity in the endothelium, expression of pro-inflammatory molecules that recruit monocytes to the tissue and paracrine signaling that could cause degradation of the valve tissue and extracellular matrix. Understanding the molecular events caused by changes in the mechanical environment is paramount in the identification of biomolecular markers for disease diagnosis and in the development of novel therapeutic strategies that could alleviate the need for surgical intervention.

**Keywords** Shear stress · Cyclic strain · Mechanobiology · Fibrosa · Ventricularis · Aortic stenosis · Aortic valve disease

## 1 Introduction

Mechanotransduction refers to the process through which cells sense and respond to mechanical forces and convert them to biochemical signals that elicit a specific biological response [1]. Mechanotransduction is particularly pertinent for the heart valves, which experience a plethora of mechanical stimuli during each cardiac cycle, including compression, tension, flexure and laminar and disturbed shear stress. Of the four heart valves, the aortic valve exists in the harshest physical environment. Within the aortic valve, two cell populations exist, and those populations experience

J. N. Warnock (✉)
School of Chemical, Materials and Biomedical Engineering, University of Georgia, Athens, GA, USA
e-mail: james.warnock@uga.edu

different mechanical forces and respond accordingly. The endothelial cells form a monolayer, covering the surface of the valve leaflet, and are exposed to shear stress but may also experience strain as the leaflet surface extends at different points in the cardiac cycle. A mixed population of myofibroblast, fibroblast, and smooth muscle cells occupies the valve interstitium. These cells react to compressive forces during diastole and may also experience strain.

Mechanical forces are important for maintaining valve homeostasis and their absence results in altered cell phenotype, increased apoptosis, and increased expression of pro-inflammatory genes [2, 3]. Conversely, excessive levels of force can lead to pathological changes within the valve [4–9]. Additionally, mechanical forces have been shown to play a role in heart valve morphogenesis and development [10, 11].

## 1.1 Mechanical Environment of the Aortic Valve Endothelium

During the cardiac cycle, the aortic valve endothelium will experience shear stresses. These shear stresses differ in magnitude and complexity between the ventricular and the aortic facing fibrosa surfaces of the valve. Flow across the ventricular surface is laminar in nature, whereas eddy structures form in the sinus regions with flow into the coronary arteries, resulting in disturbed flow on the aortic side of the valve [12, 13]. In a normal healthy adult under resting conditions, cardiac output is 5 L/min, resulting in a peak flow rate of ~20–25 L/min. Studies have been performed using laser doppler velocimetry (LDV) to calculate the wall shear stress produced at these flow rates. Measurements on the ventricular surface have been performed using a trileaflet polymetric valve [14, 15]. Under steady fluid flow, shear stress increases with flow rate with a maximum value of 79 dyne/cm$^2$ at 22.5 L/min. Shear stress is greatest when the leaflet is at 30° and decreases at 0° and 60° [14]. Under pulsatile conditions, which have more physiological relevance, peak shear stresses occur approximately 30–55 ms before peak volumetric flow rate. Valve diameter also has an influence on peak shear stress; valves with a smaller inner annulus diameter (24 mm) have a peak shear stress of 64 dyne/cm$^2$ compared to valves with a larger inner annulus diameter (24.8 mm) that have a peak shear stress of 71 dyne/cm$^2$ [15].

Studies to determine the shear stress on the valve fibrosa have utilized glutaraldehyde-preserved native porcine aortic valves. Unlike the ventricular surface, the aortic side experiences shear stress during both systole and diastole. The higher flow rates in the valve sinuses during systole cause shear stress to increase to 21.3 dyne/cm$^2$ at the mid-point. Diastolic shear stress is highest during early diastole and peaks at 3.8 dyne/cm$^2$ before gradually decreasing to zero [16].

In addition to shear stress, the endothelium will undergo anisotropic biaxial strain through the cardiac cycle. During diastole, the pressure difference between the aorta and left ventricle is 80 mmHg but can increase to 120 mmHg in hypertensive patients. The compressive force acting on the valve leaflet produces tensile strain, which has been estimated to be 11% in the circumferential direction and 13–25% in the radial direction, although this can increase to >15 and 31% in the circumferential and radial directions, respectively, under disease conditions [17].

## 1.2 Mechanisms of Mechanosensation

Many of the components of the plasma membrane have been identified as being involved in endothelial mechanotransduction including integrins [18–23], ion channels ($K^+$ channels and stretch activated $Ca^{2+}$ channels) [24–30], heterotrimeric G-proteins [31–34], and cell surface proteoglycans [21]. However, studies suggest it is unlikely the plasma membrane is the primary sensor of shear stress as the glycocalyx attenuates shear stress near the membrane surface [35, 36]. The glycocalyx is composed of glycolipids containing oligosaccharides and sialic acids and proteoglycans with glycosaminoglycan (GAG) side chains. Approximately 50–90% of the GAGs are heparin sulfate with the remaining GAGs being chondroitin sulfate and hyaluronic acid. The arrangement of oligosaccharides can be extremely diverse and although most chains contain less than 15 sugar residues they are often branched and can form covalent bonds. The glycocalyx can be observed microscopically and has the appearance of a dense, brush-like structure on the surface of the cell. For recent reviews of the glycocalyx structure and function in mechanotransduction, see [36–40]. Microparticle image velocimetry has been used to demonstrate that bulk fluid flow is excluded from the glycocalyx [41]. This further supports the hypothesis that the plasma membrane is not a primary sensor of shear stress. Instead, the glycocalyx may amplify the force via cell membrane focal adhesions or glycolipid linkages.

Cell signaling in response to shear stress is not limited to the luminal surface of the endothelial cell. Surface forces are transmitted through the body of the cell by the cytoskeletal actin filaments, intermediate filaments, and microtubules [42]. The endothelial cell cytoskeleton exists in a state of intracellular tension or prestress arising from the association of cytoskeletal elements with external forces developing from adjacent cells and attachment to the extracellular matrix. When the cell undergoes physical deformation, caused by changes in the external mechanical environment, the cytoskeletal elements will reorganize to maintain the balance between internal and external forces. The principles of tensional integrity, also known as tensegrity, have been described in detail elsewhere [43, 44]. Forces are transmitted to the intercellular junctions or to the transmembrane integrins that bind to the extracellular matrix. This has been demonstrated by studies that show phosphorylation of cell junction molecules and the involvement of VE-cadherin, VEGFRs, and platelet-endothelium cell adhesion molecule-1 (PECAM-1) [45–47]. The decentralized model of force transmission in cells has been described in detail by Helmke and Davies [48]. The spatiotemporal response of cells to mechanical deformation is shown in Fig. 1.

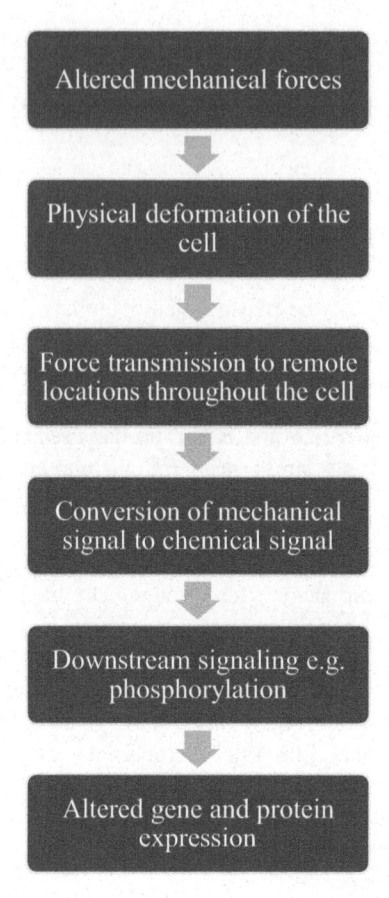

**Fig. 1** Changes to the hemodynamic environment cause the mechanical forces acting on the endothelial cells to be altered. This can lead to physical deformation of the cell; in the case of altered shear stress the deformation will occur at the luminal surface. For altered cyclic strain, the alteration may cause deformation of cell shape and impact cell-cell junctions or cell-matrix adhesions. Cells transmit the forces from the local point of deformation to remote points throughout the cell via the cytoskeleton. Consequently, mechanical forces can impact the cell at locations that are distinct from the origin. The mechanical forces are converted to chemical signals, which result in acute downstream signaling typically through phosphorylation of various intracellular molecules. Signaling can initiate changes in gene and protein expression

## 1.3 Pathological Implications of Mechanotransduction

The primary motivation for studying endothelial mechanotransduction is to understand the underlying mechanisms of aortic valve disease (AVD) and to identify potential diagnostic biomarkers or therapeutic strategies. A number of clinical risk factors are associated with AVD, including increased age, smoking, hypertension, and elevated lipoprotein (a) and low-density lipoprotein (LDL) cholesterol levels

[49]. These factors cause endothelial dysfunction and activation, characterized by expression of adhesion molecules such as VCAM-1, ICAM-1, and E-selectin. Concurrent expression of cytokines, such as MCP-1, results in recruitment of monocyte-macrophage cells to the valve. The activated macrophages produce multiple cytokines (including RANKL, IL-1β, and TNF-α) and express enzymes generating oxidants ($O_2^-$) that promote LDL oxidation [50]. These cytokines, along with increased NO produced by ECs, activate the NFκB, BMP-2, and Runx2 signaling pathways in VICs, causing an increase in alkaline phosphatase, osteocalcin, osteopontin, and transdifferentiation into an osteoblast-like phenotype [51]. IL-1β produced by leukocytes is also able to up-regulate expression of MMP-1 in VICs [52]. Additionally, MMP-2, -3, and -9 and TIMP-1 and TIMP-2 are involved in the pathogenesis of AVD [53–55]. The remodeling attributed to these molecules, along with increased synthesis but disorganized deposition of new collagen fibers, results in ECM and collagen disarray [56]. Accumulation of proteoglycans, lipids, minerals and possibly cellular proliferation also contribute to thickening of the valve [57, 58]. Although it is known that the ECM undergoes pathological remodeling, little is known about the earliest initiation events in AVD and the molecular triggers still need to be identified. Mechanical factors are strong initiators of the inflammatory cascade in vascular endothelium; it is therefore probable that valvular endothelial cells can also be stimulated by abnormal hemodynamics and tissue mechanics [59]. Although AVD shares many hallmarks of atherosclerosis, and some investigators have proposed that AVD is atherosclerosis of the aortic valve [60, 61], the two diseases are not synonymous [62, 63].

## 1.4 In Vitro and Ex Vivo Methods for Mechanotransduction Experiments

Prior to the 1970s, when endothelial cell culture was first introduced, studies on the effects of hemodynamic forces on endothelial cells were limited to ex vivo blood vessel organ culture, where the integrity of the endothelium was rapidly lost, or observational studies of fixed blood vessels [42, 64]. In recent years, numerous bioreactor systems have been developed ranging from simple systems designed to expose cell cultures to a single mechanical force, to elaborate systems capable of simultaneously exposing intact heart valves to multiple mechanical forces, to systems allowing for the real-time observation of mechanobiological changes. For a comprehensive review on methods used to study mechanically induced signaling in endothelial cells, see Reinhart-King et al. [65].

Fundamental studies have looked at the effects of shear stress on cultured endothelial cells. Isolated endothelial cells are cultured on glass or tissue culture plastic slides and placed in a parallel plate flow chamber. Although some variations exist, the basic apparatus consists of a polycarbonate chamber containing the glass slide, a silastic gasket, and a medium reservoir [66, 67]. Flow through the parallel plate chamber can be either through hydrostatic pressure or through the use of a peristaltic pump. Polyester spacers have been used to vary the height of the chamber. In doing so, a channel with well-defined geometry is created. Consequently, the Navier–Stokes equation of fluid flow with constant density is used to determine the wall shear stress ($\tau$) relative to flow rate and channel geometry by the formula:

$$\tau = \frac{6Q}{\mu b h^2}$$

where $Q$ is the flow rate, $\mu$ is the fluid viscosity (0.012 poise), and $b$ and $h$ are channel width and height, respectively [68]. As studies have revealed the significance of laminar vs. disturbed flow and distinct differences in the endothelium in different flow environments [69, 70], parallel plate systems have been developed to induce oscillatory flow by the inclusion of an oscillator [71, 72].

Plate and cone systems have also been developed to study endothelial cell responses to fluid flow [73–75]. Cells can be grown in a 35 mm or 100 mm culture dish that is customized for use in the flow system. The cone is made from Teflon and has a 0.5° angle. A magnetic stir bar is located inside the cone so that it can be stirred with a magnetic stirrer and the entire system can be placed in an incubator to maintain temperature. The shear stress is calculated from the rotational speed of the cone using the equation:

$$\tau = \frac{\omega \mu}{\theta}$$

where $\omega$ is the rotational speed, $\mu$ is the medium viscosity, and $\theta$ is the angle of the cone [65]. As with the parallel plate system, the cone and plate system can be used to exert laminar shear stress or oscillatory shear stress on cells [76, 77].

Endothelial biology is regulated not only by the mechanical environment but also by the underlying extracellular matrix and the VICs. Consequently, the aforementioned fluid flow systems have been adapted to enable them to culture intact valve tissue [78–81]. Studies have also been performed with tissue engineered coculture systems. In contrast to using native valve tissue, VICs are immobilized in a type I collagen hydrogen and endothelial cells are then grown on the luminal surface [82].

Several commercial devices are available for the cyclic stretching of cells, including the Flexcell Tension Systems (Flexercell International Corporation) [9, 83, 84] and the STREX mechanical cell strain instruments (B-Bridge International, Inc.) [85]. These systems, along with those developed by individual researchers, require cells to attach to a flexible membrane that can be stretched while ensuring that cells are stretched to the same extent as the membrane. To meet this objective, thin silicone membranes are used, often coated with ECM proteins such as collagen or fibronectin [65].

As with systems used for studying shear stress, systems for studying strain have been developed to use ex vivo tissue samples. The simplest of these systems mounts strips of aortic valve tissue in parallel chambers and applies a uniaxial strain. The chambers contain culture medium and can be placed in an incubator to maintain cell viability within the tissue in excess of 48 h. Strain is applied to the tissue through the use of a computer-controlled actuator [5, 86–88]. Adaptations of this design have permitted anisotropic biaxial stretching of tissue and live *en face* imaging of the endothelium [8]. More sophisticated versions of this system have been subsequently developed that introduces additional mechanical forces [89, 90]. This has been motivated by the understanding that the aortic valve does not experience a single mechanical force in isolation but a combination of forces simultaneously and that they could have synergistic or antagonistic effects on mechanotransduction. Possibly the most complex systems are those that have cultured the entire aortic valve and root, providing a physical environment that most closely mimics in vivo conditions [91, 92].

## 2 Shear Stress

### 2.1 Aortic Valve Endothelial Cells Under Physiologic Flow Conditions

Until relatively recently, it had been assumed that aortic valve endothelial cells (EC) were phenotypically similar to vascular endothelial cells. In the absence of fluid flow, EC form cobblestone morphology; however, when flow is applied, vascular EC align in the direction of flow. Conversely, valvular EC align perpendicular to flow [68]. This is consistent with the in vivo scenario where endothelial cells align circumferentially on the valve surface, which is also perpendicular to flow [93]. This alignment difference is regulated by changes in: (1) integrin clustering ($\alpha$ 1 integrin); (2) plaque formation (vinculin); (3) integrin-mediated signaling (focal adhesion kinase; FAK); and (4) cytoskeletal (f-actin) arrangement. Furthermore, this process is disrupted in both cells in vitro by blocking Rho kinase (with Y-27632) and calpain (calpain inhibitor-1). PI 3-kinase, on the other hand, regulates arterial but not valvular endothelial cell alignment [94].

In a comparison of transcriptional profiles between valvular EC exposed to unidirectional flow and static conditions, cDNA mircoarrays revealed that shear stress inhibits oxidative and pro-inflammatory gene expression. Additionally, shear stress down-regulates genes involved in the initiation of calcification of underlying interstitial cells, such as BMP-4 and osteopontin [70].

In addition to differences in gene and protein expression profiles between vascular and valvular ECs, differences also exist between ECs lining the fibrosa and ventricularis of the valve. The endothelium on the fibrosa (aortic side) is more susceptible to developing calcific lesions through increased expression of genes related to skeletal development and vascular calcification. Additionally, ECs on the fibrosa have significantly lower expression of genes known to inhibit vascular calcification including osteoprotegerin (OPG; tumor necrosis factor receptor superfamily; member 11b), C-type natriuretic peptide (CNP), and parathyroid hormone (PTH), each of which has been shown to inhibit cardiovascular calcification [95]. However, under normal physiological conditions, the fibrosa ECs do not differentially express genes associated with adhesion molecules E-selectin, vascular cell adhesion molecule -1 (VCAM-1), or intracellular adhesion molecule 1 (ICAM-1). The transcripts for galectin-1 and CNP, both of which are up-regulated by pro-inflammatory mediators [96, 97], are underexpressed in the fibrosa ECs [95]. Furthermore, EC nitric oxide (NO) is protective against AVD [98], but is reduced in calcified valves [99–101]. EC-derived NO signaling is side specific and regulated by shear stress. This is seen through the increase in cGMP production by EC when exposed to fluid shear stress. A unidirectional, pulsatile flow pattern that mimics the ventricularis hemodynamics produces significantly more cGMP than oscillatory shear stress, indicative of the fibrosa-sided shear stress [102]. These results suggest a role for hemodynamics in the development of aortic stenosis and calcification.

## 2.2 Role of Shear Stress in Aortic Valve Pathology

While physiological shear stresses maintain homeostasis of endothelial cells [78, 81, 103], abnormalities in shear stresses lead to endothelial activation and inflammation. Changes to the hemodynamic environment can be caused by aging, a known risk factor for AVD, whereby progressive stiffening of the leaflet ECM causes an increase in the shear stress magnitude [104]. Bicuspid valves also experience increases in shear stress magnitude and frequency, and are known to be particularly susceptible to calcification [105]. An increase in shear stress magnitude or frequency results in elevated levels of MMP-2, MMP-9, cathepsin L, and cathepsin S in the subendothelial space [106]. These proteolytic enzymes are involved in ECM degradation and have been detected in stenotic aortic valves [107, 108]; however, it is not clear if the expression of these enzymes following abnormal shear stress is involved in pathological changes to the leaflet or if it increases the turnover of collagen as a means of maintaining the integrity of leaflet biomechanics. Other studies have shown that increases in cyclic pressure magnitude and frequency increase biosynthetic activity in valve leaflets, with an increase in collagen and sGAG synthesis [109]. The elevated levels of MMP-9 are mediated by TGF-$\beta$1 signaling from endothelial cells under supra-physiologic hemodynamic stresses [103].

Increased shear stress magnitude, but not increased shear stress frequency, increases endothelial activation as evidenced by ICAM-1 and VCAM-1 expression and paracrine signaling, as seen by TGF-$\beta$1 and BMP-4 expression [103, 106]. Altered shear stress will increase the expression of TGF-$\beta$1, which can migrate to the subendothelial space or remain within the cell. Immunohistochemistry studies show that TGF-$\beta$1 is contained within the ECM of calcified valve leaflets at levels significantly higher than normal leaflets. These elevated levels of TGF-$\beta$1 cause interstitial cell calcification via apoptosis [110] as well as the aforementioned increases in MMP-2, and -9 and cathepsin L and S. Intracellular BMP-4 up-regulates the expression of VCAM-1 and ICAM-1. This phenomenon is only observed in the fibrosa endothelium and is absent in the ventricular endothelium [79]. The relationship between intracellular TGF-β1 and BMP-4 is still in question. When the fibrosa is exposed to elevated shear stress but no shear stress is imposed on the ventricularis, BMP-4 expression is inhibited by blocking TGF-β1 expression; however, when shear stresses are imposed simultaneously on both sides of the valve, BMP-4 expression is independent of TGF-β1 [103, 106].

Another potential mechanism for the development of degenerative valve disease is through endothelial to mesenchymal transformation (EndMT) [111]. The process of EndMT is an initiating event in embryonic valvulogenesis [112, 113]. Embryonic EC lose cell-cell contacts and separate from the monolayer. The cells invade and migrate into the interstitial space while simultaneously increasing expression of mesenchymal markers that include alpha smooth muscle actin (α-SMA) and loosing expression of endothelial markers, such as CD31/PECAM-1 [114]. In adult physiology, EndMT has been suggested as a potential mechanism of disease and is driven by inflammatory signaling. The cytokines IL-6 and TNFα, which have both been associated with diseased aortic valves [115–117], cause a dose-dependent loss in cell-cell contacts, decreases in PECAM-1 and VE-cadherin expression, increased α-SMA expression and cell invasion in adult ECs cultured in 3D collagen gels in vitro. Both Il-6 and TNF-α stimulate nuclear translocation of nuclear factor κB (NFκB), and the overexpression of TNF-α and NFκB lead to increased expression of pro-EndMT gene expression. Further analysis shows that NFκB activation and translocation occurs through the Akt signaling pathway but not the MEK1 or STAT3 pathways [118]. Oscillatory shear stress of all magnitudes can induce expression of pro-EndMT genes in addition to expression of inflammatory genes NFκB TGF-β1, and ICAM-1. In contrast, high steady shear stress is protective against inflammation and EndMT [111]. The combined effects of elevated shear stress is summarized in Fig. 2.

A third way that shear stress contributes to aortic valve stenosis is through the regulation of micro-RNAs (miRNA). miRNAs are short RNA sequences of approximately 22 nucleotides that bind to protein encoding mRNA. Repression of the mRNA targets leads to decreased translational efficiency and/or decreased mRNA levels [119]. Stress-induced up-regulation of miRNAs can lead to the down-regulation of a set of targeted mRNAs, while down-regulation of miRNAs can result in up-regulation of target mRNAs because of the loss of tonic inhibitory control of the miRNA on its target mRNA [120]. Bicuspid valves collected from patients requiring aortic valve replacement showed significant decreases in MiR-26a, MiR30b, and MiR-195 expression. Further, when valve interstitial cells are treated with miR-26a or miR-30b mimics in vitro, expression levels of pro-calcification mRNAs decrease [121]. Microarray analysis has been used to identify shear and side-specific mRNA and miRNA from cultured human aortic valve ECs. The data show that 30 shear-dependent and three side-dependent miRNAs are present in aortic valve ECs. Functional analysis of the miRNAs and shear-sensitive mRNAs

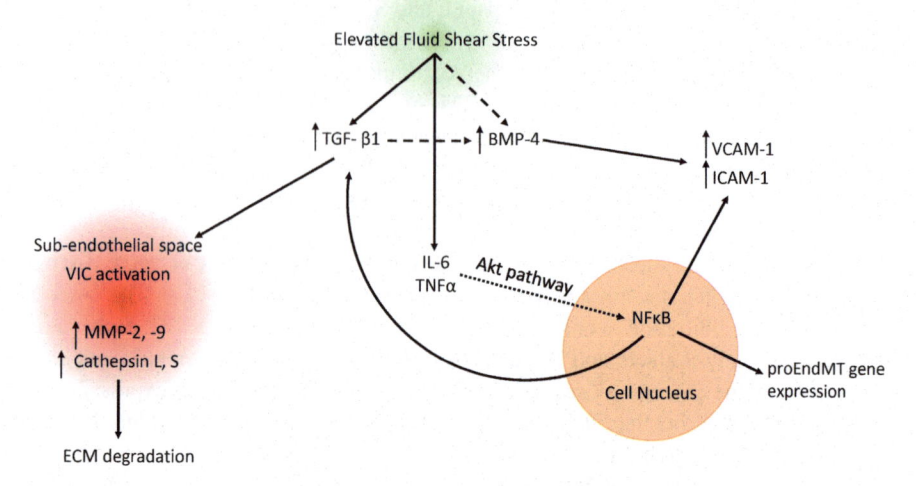

**Fig. 2** The combined effects of elevated shear stress on endothelial cells. IL-6 and TNFα expression is increased, causing NFκB to translocate to the cell nucleus via the Akt signaling pathway. This results in increased expression of ICAM-1, proEndMT genes, and TGF-β1. TGF-β1 can also be directly increased by elevated shear stress. This cytokine can migrate to the subendothelial space causing activation of the valve interstitial cells and increasing the expression of the matrix degrading enzymes MMP-2 and -9 and cathepsins L and S. Intracellular TGF-β1 may cause an increase in BMP-4, although BMP-4 may be directly increased in response to elevated shear stress. BMP-4 is directly responsible for an increase in the pro-inflammatory molecules VCAM-1 and ICAM-1

has predicted a number of genes regulated by changes in miRNA expression; these genes are linked to a number of cellular functions associated with AVD, including apoptosis, inflammation, calcification, cell cycle, cell proliferation, and cell movement [122].

## 3 Cyclic Strain

A strong correlation exists between regions of calcification and high mechanical strain in the aortic valve. In operatively excised stenotic or regurgitant valves, approximately 87% have calcification along the line of leaflet coaptation or radially from the area of leaflet attachment [123]. The traditional "wear and tear" hypothesis was that elevated mechanical forces, coupled with aging, caused disruption of the endothelium or damage to the interstitial matrix leading to lesion formation [124, 125]. Recent studies demonstrate convincingly that aortic sclerosis is an active cell-mediated process involving chronic inflammation [126, 127] and active calcification [50, 124, 128, 129]. Consistent with this new understanding of the pathogenesis of valve disease is increasing evidence that mechanical forces may contribute to valve disease by actively regulating valve cell biology. Valvular endothelial cells and valvular interstitial cells are exquisitely sensitive to mechanical forces, often responding with unique behaviors that distinguish them from seemingly similar cells [94].

## 3.1 Endothelial Layer Integrity Under Elevated Cyclic Strain

Endothelial Cells employ an adhesion protein motif that ensures stable yet flexible connections to both the ECM and neighboring cells. Patterns of focal adhesion can indicate cell functionality, as they are instrumental in signaling, migration, and adhesion to the ECM. Changes in the expression of cell-cell or cell-matrix adhesion molecules compromise the integrity of the endothelial layer and can increase the susceptibility of the tissue to monocyte migration and diapedesis.

Real-time, live *en face* imaging of aortic valve leaflets in an ex vivo stretch device shows that endothelial viability and integrity are maintained under normal, physiological conditions where the tissue is stretched 10% in the circumferential direction and 35% in the radial direction. Viability is maintained when strain is increased to represent hypertensive conditions, with 20% circumferential strain and 40% radial strain [8]. ECs are somewhat resilient to cyclic stretch in this respect as other studies show that interstitial cells have significant increases in apoptosis and proliferation at 15 and 20% circumferential strain within 48 h [130]. Although viability is maintained, the endothelial integrity begins to deteriorate under elevated strain (see Fig. 3). Analysis of cell-cell and cell-matrix adhesion proteins reveals changes in expression that could be responsible for the loss of integrity and could potentially lead to the development of a pro-inflammatory environment. Consistent with studies on the effect of shear stress on ECS, a differential response is observed between the fibrosa and ventricularis cell populations.

Platelet endothelial cell adhesion molecule 1 (PECAM-1), also known as CD31, is abundantly expressed in ECs. It becomes concentrated in regions of cell-cell contact and plays a role in forming and maintaining the contact inhibited state of ECs [131]. Fibrosa endothelial cells (FEC) show a significant increase in PECAM-1 gene expression at 20% cyclic strain, relative to static and 10% strain conditions [132] and greater expression of PECAM-1 adhesion molecules [83]. PECAM-1 senses mechanical stimuli and helps assist an inflammatory response by mediating transmigration of monocytes [133, 134]. PECAM-1 has also been implicated in the activation of eNOS [135]. In calcific aortic stenosis, oxidative stress is the result of enzymatic uncoupling of eNOS, causing increased production of superoxide [136]. Superoxide is greatly increased near calcified regions in human stenotic valves, while superoxide dismutase is greatly decreased [137]. This suggests that an increase in PECAM-1 gene and protein expression in FEC could render the fibrosa susceptible to monocyte migration and oxidative stress.

Elevated cyclic strain (20%) also causes FECs but not ventricularis endothelial cells (VEC) to increase $\beta_1$-integrin expression. For valvular studies, $\beta_1$-integrin

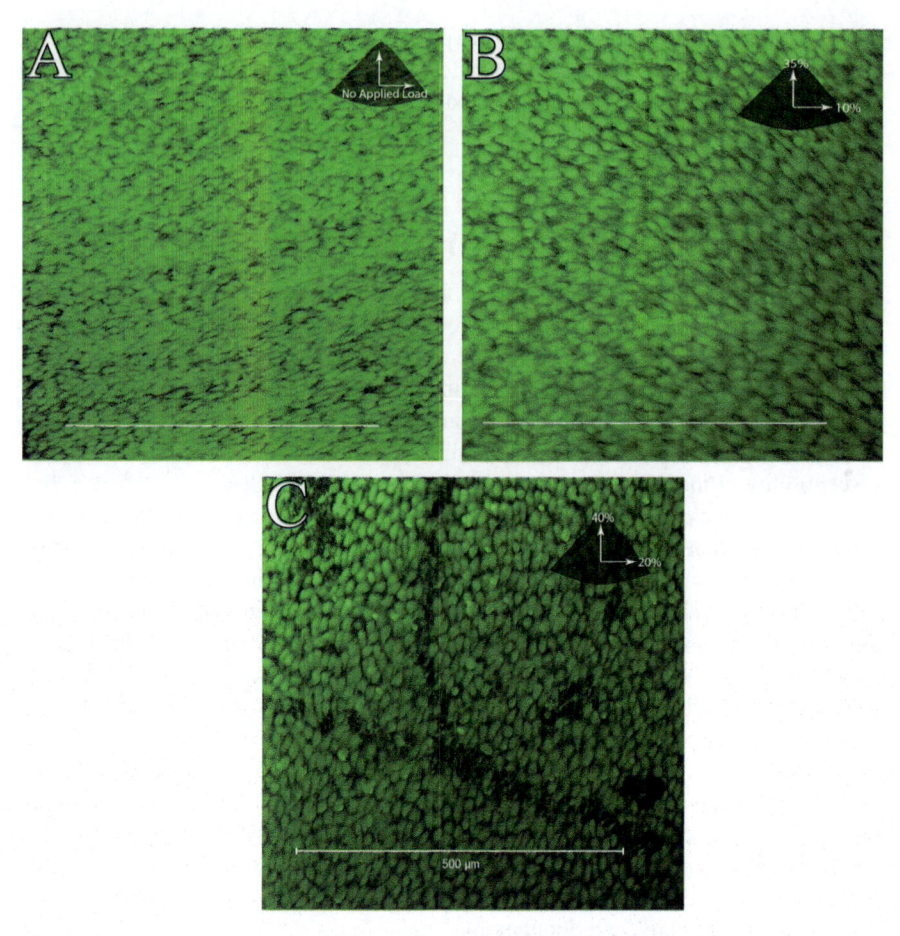

**Fig. 3** Real-time live cell imaging of the aortic valve endothelium visualized with calcein on Zeiss LSM 510 under (**a**) no applied tension, (**b**) physiological levels of biaxial stretch, and (**c**) hypertensive simulating biaxial stretch levels. Scale bars represent 500 μm. Reprinted from: Metzler SA, Digesu CS, Howard JI, Filip To SD, Warnock JN. Live en face imaging of aortic valve leaflets under mechanical stress. Biomech Model Mechanobiol. 2012;11(3–4):355–61 with permission from Springer

behavior is often investigated due to its predominant interaction with collagen. The differential expression of $\beta_1$-integrin between FEC and VEC at 20% cyclic strain could be related to the composition of the underlying ECM present in the aortic valve. The fibrosa is primarily composed of circumferentially oriented collagen, whereas the ventricularis is primarily composed of radially aligned elastin [138]. Therefore, it would be expected that FEC would express $\beta_1$-integrin at a higher level than VEC. Additionally, when FEC are strained, they may increase expression of $\beta_1$-integrin to increase adhesion to the collagen ECM. Conversely, VEC would not increase expression of $\beta_1$-integrin as this would not increase adhesion to the underlying ECM in the ventricularis.

The intracellular, cytoplasmic domain of $\beta_1$-integrin binds to integrin-linked kinase (ILK). ILK lies upstream of many intracellular signaling pathways and plays a pivotal role in cell-matrix interactions, ECM-mediated signaling, as well as other cellular functions [139–141]. Recent studies show that ILK stabilizes eNOS with Heat shock protein (Hsp) 90 and protects eNOS from enzymatic uncoupling. Consequently, physiological production of nitric oxide is achieved rather than oxidative stress caused by production of superoxide, thus preventing endothelial dysfunction [142]. Increased expression of $\beta_1$-integrin, and presumably an increase in ILK, would not only increase cell-matrix adhesion but could also have a protective role through stabilization of eNOS and prevention of oxidative stress that would promote a pro-inflammatory state.

The VEC respond differently to elevated cyclic strain. Whereas no significant changes are seen in PECAM-1 or $\beta_1$-integrin expression at 20% cyclic strain, significant increases are seen in Vinculin and VE-Cadherin expression. Vinculin is a crucial accessory molecule involved in force transduction and mediating cellular response. Vinculin binds integrins to the F-actin network at the intracellular face of the plasma membrane, and subsequently mediates cytoskeletal mechanics [143]. The observation of increased vinculin production with increased cell stress is attributed to an increase and strengthening of cell adhesions in order to help the tissue withstand strains. In the absence of strain, vinculin production decreases because without the presence of strain they are not stimulated to produce vinculin. Increased cell adhesions protect the ventricularis from monocyte invasion.

Cell-cell dynamics in ECs are regulated in part by Cadherin 5, type 2, or VE-Cadherin (vascular epithelium). VE-Cadherin is responsible for cell/cell adhesions and cell motility. More importantly, it is responsible for control of intercellular junctions and their integrity and is required to maintain a restrictive endothelial barrier [144]. The state of VE-Cadherin outside the cell is directly proportional to the permeability of the endothelium. VE-Cadherin expression significantly increases in VEC at 20% strain, indicating an increase of cell/cell adhesions. In contrast, fibrosa cells do not show any change in VE-Cadherin expression with respect to strain. This indicates that under pathological strain conditions, increased VE-Cadherin expression impedes monocyte transmigration on the ventricularis surface by reinforcing the EC monolayer, whereas the fibrosa does not exhibit increased resistance to monocyte infiltration.

## 3.2 *Pro-inflammatory Response of Endothelial Cells to Cyclic Strain*

Endothelial injury leads to an inflammatory response and a cascade of cell-mediated inflammatory events [145]. The activated endothelial layer generates specific surface molecules involved in monocyte and T-cell migration, adhesion and congregation. The pro-inflammatory proteins include intercellular adhesion molecules (ICAM), vascular cell adhesion molecules (VCAM), and endothelial leukocyte adhesion molecules (E-selectin or ELAM). These proteins act concurrently with chemotactic molecules, such as monocyte chemoattractant protein 1 (MCP-1), expressed by the endothelium and smooth muscle cells, to attract monocytes into the tissue.

When ECs are subjected to 20% strain for 24 hours, a significant increase in ICAM-1, VCAM-1, and E-selectin is observed, compared to 5 or 10% strain [9]. This demonstrates that a pro-inflammatory environment may develop in vivo when the endothelium is exposed to supraphysiological conditions, as is the case in hypertension. Aortic valve leaflets exposed to elevated have increased levels of BMP-2 and -4. The proteins are strongly expressed on the fibrosa side, yet absent on the ventricularis side, and show strong co-localization with von Willebrand factor, demonstrating that endothelial cells express the BMPs [146]. Studies have not been performed to determine if the strain-induced expression of BMP-2 or -4 is responsible for the expression of pro-inflammatory molecules, as is the case for elevated fluid shear stress. However, studies do show that elevated strain coupled with a pro-osteogenic humoral environment results in ex vivo calcification of aortic valves in a BMP-dependent manner. Inhibition of BMP by the antagonist noggin blocks calcification events including ALP activity, increased calcium content and Runx2 and osteocalcin expression in the valve cusp [146].

## 4 Summary

Mechanical forces play an important role in heart valve homeostasis and disease. The mechanical environment has been well characterized and studies utilizing various bioreactor systems have been able to show how deviations from physiological levels of shear stress and cyclic strain lead to changes in cell signaling causing a biological change within the endothelium. These changes lead to a pro-inflammatory environment whereby the tissue is susceptible to monocyte infiltration. Additionally, the endothelium increases MMP expression within the valve interstitium through paracrine signaling, causing remodeling of the leaflet, disorganized collagen deposition, and changes to the leaflet biomechanics.

Our knowledge of heart valve endothelial mechanobiology has advanced tremendously during the previous 15 years. It is imperative that as this body of knowledge continues to grow and the information is used to improve patient care. One area that this understanding can be applied is in the identification of molecular biomarkers for early disease detection and to monitor progression and severity. Valve dysfunction is currently diagnosed by detection of a heart murmur and subsequent ultrasound echocardiography. However, because aortic stenosis is asymptomatic, the disease is rarely discovered until it has progressed to an advanced stage [147]. Early diagnosis would allow for better disease management and the opportunity for interventional therapy and life-style changes to slow disease progression.

Additionally, the results from these studies should be used to develop novel therapies for the treatment of aortic valve disease. Two-thirds of valve disorders occur in the aortic position with ≈50% resulting in cardiothoracic surgery. Reasons for not undergoing aortic valve replacement include high perioperative risk, age, lack of symptoms, and patient/family refusal [148]. There are no alternative therapies to surgery, although several clinical studies have examined the use of lipid-lowering drugs (statins) and ACE inhibitors. Unfortunately, lipid-lowering therapy does not reduce aortic stenosis progression [149] and ACE inhibitors have had mixed efficacy [150, 151]. As our understanding of the molecular mechanisms of aortic valve disease increases, there is greater potential to develop therapies that could alleviate the need for valve surgery.

# References

1. Mofrad MR, Kamm RD. Cellular mechanotransduction: diverse perspectives from molecules to tissues. Cambridge: Cambridge University Press; 2009.
2. Konduri S, Xing Y, Warnock JN, He Z, Yoganathan AP. Normal physiological conditions maintain the biological characteristics of porcine aortic heart valves: an ex vivo organ culture study. Ann Biomed Eng. 2005;33(9):1158–66.
3. Smith KE, Metzler SA, Warnock JN. Cyclic strain inhibits acute pro-inflammatory gene expression in aortic valve interstitial cells. Biomech Model Mechanobiol. 2010;9(1):117–25.
4. Back M, Gasser TC, Michel JB, Caligiuri G. Biomechanical factors in the biology of aortic wall and aortic valve diseases. Cardiovasc Res. 2013;99(2):232–41.
5. Balachandran K, Konduri S, Sucosky P, Jo H, Yoganathan AP. An ex vivo study of the biological properties of porcine aortic valves in response to circumferential cyclic stretch. Ann Biomed Eng. 2006;34(11):1655–65.
6. Fisher CI, Chen J, Merryman WD. Calcific nodule morphogenesis by heart valve interstitial cells is strain dependent. Biomech Model Mechanobiol. 2013;12(1):5–17.
7. Merryman WD, Schoen FJ. Mechanisms of calcification in aortic valve disease: role of mechanokinetics and mechanodynamics. Curr Cardiol Rep. 2013;15(5):355.
8. Metzler SA, Digesu CS, Howard JI, Filip To SD, Warnock JN. Live en face imaging of aortic valve leaflets under mechanical stress. Biomech Model Mechanobiol. 2012;11(3–4):355–61.
9. Metzler SA, Pregonero CA, Butcher JT, Burgess SC, Warnock JN. Cyclic strain regulates pro-inflammatory protein expression in porcine aortic valve endothelial cells. J Heart Valve Dis. 2008;17(5):571–7; discussion 8.
10. Heckel E, Boselli F, Roth S, Krudewig A, Belting HG, Charvin G, et al. Oscillatory flow modulates mechanosensitive klf2a expression through trpv4 and trpp2 during heart valve development. Curr Biol. 2015;25(10):1354–61.
11. Tan H, Biechler S, Junor L, Yost MJ, Dean D, Li J, et al. Fluid flow forces and rhoA regulate fibrous development of the atrioventricular valves. Dev Biol. 2013;374(2):345–56.
12. Nicosia MA, Cochran RP, Einstein DR, Rutland CJ, Kunzelman KS. A coupled fluid-structure finite element model of the aortic valve and root. J Heart Valve Dis. 2003;12(6):781–9.
13. Ge L, Sotiropoulos F. Direction and magnitude of blood flow shear stresses on the leaflets of aortic valves: is there a link with valve calcification? J Biomech Eng. 2010;132(1):014505.
14. Weston MW, LaBorde DV, Yoganathan AP. Estimation of the shear stress on the surface of an aortic valve leaflet. Ann Biomed Eng. 1999;27(4):572–9.
15. Yap CH, Saikrishnan N, Yoganathan AP. Experimental measurement of dynamic fluid shear stress on the ventricular surface of the aortic valve leaflet. Biomech Model Mechanobiol. 2012;11(1–2):231–44.
16. Yap CH, Saikrishnan N, Tamilselvan G, Yoganathan AP. Experimental measurement of dynamic fluid shear stress on the aortic surface of the aortic valve leaflet. Biomech Model Mechanobiol. 2012;11(1–2):171–82.

17. Yap CH, Kim HS, Balachandran K, Weiler M, Haj-Ali R, Yoganathan AP. Dynamic deformation characteristics of porcine aortic valve leaflet under normal and hypertensive conditions. Am J Physiol Heart Circ Physiol. 2010;298(2):H395–405.
18. Loufrani L, Retailleau K, Bocquet A, Dumont O, Danker K, Louis H, et al. Key role of alpha (1)beta(1)-integrin in the activation of PI3-kinase-Akt by flow (shear stress) in resistance arteries. Am J Physiol Heart Circ Physiol. 2008;294(4):H1906–13.
19. Jalali S, del Pozo MA, Chen K, Miao H, Li Y, Schwartz MA, et al. Integrin-mediated mechanotransduction requires its dynamic interaction with specific extracellular matrix (ECM) ligands. Proc Natl Acad Sci U S A. 2001;98(3):1042–6.
20. Chen KD, Li YS, Kim M, Li S, Yuan S, Chien S, et al. Mechanotransduction in response to shear stress. Roles of receptor tyrosine kinases, integrins, and Shc. J Biol Chem. 1999;274 (26):18393–400.
21. Muller JM, Chilian WM, Davis MJ. Integrin signaling transduces shear stress-dependent vasodilation of coronary arterioles. Circ Res. 1997;80(3):320–6.
22. Tzima E, del Pozo MA, Shattil SJ, Chien S, Schwartz MA. Activation of integrins in endothelial cells by fluid shear stress mediates Rho-dependent cytoskeletal alignment. EMBO J. 2001;20(17):4639–47.
23. Wang Y, Miao H, Li S, Chen KD, Li YS, Yuan S, et al. Interplay between integrins and FLK-1 in shear stress-induced signaling. Am J Physiol Cell Physiol. 2002;283(5):C1540–7.
24. Brakemeier S, Kersten A, Eichler I, Grgic I, Zakrzewicz A, Hopp H, et al. Shear stress-induced up-regulation of the intermediate-conductance $Ca(2+)$-activated $K(+)$ channel in human endothelium. Cardiovasc Res. 2003;60(3):488–96.
25. Lieu DK, Pappone PA, Barakat AI. Differential membrane potential and ion current responses to different types of shear stress in vascular endothelial cells. Am J Physiol Cell Physiol. 2004;286(6):C1367–75.
26. Olesen SP, Clapham DE, Davies PF. Haemodynamic shear stress activates a $K+$ current in vascular endothelial cells. Nature. 1988;331(6152):168–70.
27. Naruse K, Sai X, Yokoyama N, Sokabe M. Uni-axial cyclic stretch induces c-src activation and translocation in human endothelial cells via SA channel activation. FEBS Lett. 1998;441 (1):111–5.
28. Naruse K, Sokabe M. Involvement of stretch-activated ion channels in $Ca2+$ mobilization to mechanical stretch in endothelial cells. Am J Phys. 1993;264(4 Pt 1):C1037–44.
29. Naruse K, Yamada T, Sokabe M. Involvement of SA channels in orienting response of cultured endothelial cells to cyclic stretch. Am J Phys. 1998;274(5 Pt 2):H1532–8.
30. Tatsukawa Y, Kiyosue T, Arita M. Mechanical stretch increases intracellular calcium concentration in cultured ventricular cells from neonatal rats. Heart Vessel. 1997;12(3):128–35.
31. Jin ZG, Wong C, Wu J, Berk BC. Flow shear stress stimulates Gab1 tyrosine phosphorylation to mediate protein kinase B and endothelial nitric-oxide synthase activation in endothelial cells. J Biol Chem. 2005;280(13):12305–9.
32. Otte LA, Bell KS, Loufrani L, Yeh JC, Melchior B, Dao DN, et al. Rapid changes in shear stress induce dissociation of a G alpha(q/11)-platelet endothelial cell adhesion molecule-1 complex. J Physiol. 2009;587.(Pt 10:2365–73.
33. Wang S, Iring A, Strilic B, Albarran Juarez J, Kaur H, Troidl K, et al. P2Y(2) and Gq/G(1) (1) control blood pressure by mediating endothelial mechanotransduction. J Clin Invest. 2015;125(8):3077–86.
34. Zeng H, Zhao D, Yang S, Datta K, Mukhopadhyay D. Heterotrimeric G alpha q/G alpha 11 proteins function upstream of vascular endothelial growth factor (VEGF) receptor-2 (KDR) phosphorylation in vascular permeability factor/VEGF signaling. J Biol Chem. 2003;278 (23):20738–45.
35. Secomb TW, Hsu R, Pries AR. Effect of the endothelial surface layer on transmission of fluid shear stress to endothelial cells. Biorheology. 2001;38(2–3):143–50.
36. Weinbaum S, Zhang X, Han Y, Vink H, Cowin SC. Mechanotransduction and flow across the endothelial glycocalyx. Proc Natl Acad Sci U S A. 2003;100(13):7988–95.

37. Reitsma S, Slaaf DW, Vink H, van Zandvoort MA, oude Egbrink MG. The endothelial glycocalyx: composition, functions, and visualization. Pflugers Arch. 2007;454(3):345–59.
38. Salmon AH, Satchell SC. Endothelial glycocalyx dysfunction in disease: albuminuria and increased microvascular permeability. J Pathol. 2012;226(4):562–74.
39. Tarbell JM, Ebong EE. The endothelial glycocalyx: a mechano-sensor and -transducer. Sci Signal. 2008;1(40):pt8.
40. Tarbell JM, Pahakis MY. Mechanotransduction and the glycocalyx. J Intern Med. 2006;259 (4):339–50.
41. Smith ML, Long DS, Damiano ER, Ley K. Near-wall micro-PIV reveals a hydrodynamically relevant endothelial surface layer in venules in vivo. Biophys J. 2003;85(1):637–45.
42. Davies P, Helmke B. Endothelial mechanotransduction. In: Mofrad MR, Kamm RD, editors. Cellular mechanotransduction. New York: Cambridge University Press; 2009.
43. Ingber DE. Cellular mechanotransduction: putting all the pieces together again. FASEB J. 2006;20(7):811–27.
44. Ingber DE, Wang N, Stamenovic D. Tensegrity, cellular biophysics, and the mechanics of living systems. Rep Prog Phys. 2014;77(4):046603.
45. Conway DE, Breckenridge MT, Hinde E, Gratton E, Chen CS, Schwartz MA. Fluid shear stress on endothelial cells modulates mechanical tension across VE-cadherin and PECAM-1. Curr Biol. 2013;23(11):1024–30.
46. Coon BG, Baeyens N, Han J, Budatha M, Ross TD, Fang JS, et al. Intramembrane binding of VE-cadherin to VEGFR2 and VEGFR3 assembles the endothelial mechanosensory complex. J Cell Biol. 2015;208(7):975–86.
47. Tzima E, Irani-Tehrani M, Kiosses WB, Dejana E, Schultz DA, Engelhardt B, et al. A mechanosensory complex that mediates the endothelial cell response to fluid shear stress. Nature. 2005;437(7057):426–31.
48. Helmke BP, Davies PF. The cytoskeleton under external fluid mechanical forces: hemodynamic forces acting on the endothelium. Ann Biomed Eng. 2002;30(3):284–96.
49. Stewart BF, Siscovick D, Lind BK, Gardin JM, Gottdiener JS, Smith VE, et al. Clinical factors associated with calcific aortic valve disease. Cardiovascular Health Study. J Am Coll Cardiol. 1997;29(3):630–4.
50. O'Brien KD. Pathogenesis of calcific aortic valve disease: a disease process comes of age (and a good deal more). Arterioscler Thromb Vasc Biol. 2006;26(8):1721–8.
51. Elmariah S, Mohler ER 3rd. The pathogenesis and treatment of the valvulopathy of aortic stenosis: beyond the SEAS. Curr Cardiol Rep. 2010;12(2):125–32.
52. Kaden JJ, Dempfle CE, Grobholz R, Tran HT, Kilic R, Sarikoc A, et al. Interleukin-1 beta promotes matrix metalloproteinase expression and cell proliferation in calcific aortic valve stenosis. Atherosclerosis. 2003;170(2):205–11.
53. Edep ME, Shirani J, Wolf P, Brown DL. Matrix metalloproteinase expression in nonrheumatic aortic stenosis. Cardiovasc Pathol. 2000;9(5):281–6.
54. Kaden JJ, Vocke DC, Fischer CS, Grobholz R, Brueckmann M, Vahl CF, et al. Expression and activity of matrix metalloproteinase-2 in calcific aortic stenosis. Z Kardiol. 2004;93 (2):124–30.
55. Soini Y, Satta J, Maatta M, Autio-Harmainen H. Expression of MMP2, MMP9, MT1-MMP, TIMP1, and TIMP2 mRNA in valvular lesions of the heart. J Pathol. 2001;194(2):225–31.
56. Chen JH, Simmons CA. Cell-matrix interactions in the pathobiology of calcific aortic valve disease: critical roles for matricellular, matricrine, and matrix mechanics cues. Circ Res. 2011;108(12):1510–24.
57. Hinton RB Jr, Lincoln J, Deutsch GH, Osinska H, Manning PB, Benson DW, et al. Extracellular matrix remodeling and organization in developing and diseased aortic valves. Circ Res. 2006;98(11):1431–8.
58. Otto CM, Kuusisto J, Reichenbach DD, Gown AM, O'Brien KD. Characterization of the early lesion of 'degenerative' valvular aortic stenosis. Histological and immunohistochemical studies. Circulation. 1994;90(2):844–53.

59. Butcher JT, Nerem RM. Valvular endothelial cells and the mechanoregulation of valvular pathology. Philos Trans R Soc B Biol Sci. 2007;362(1484):1445–57.
60. Agmon Y, Khandheria BK, Meissner I, Sicks JR, O'Fallon WM, Wiebers DO, et al. Aortic valve sclerosis and aortic atherosclerosis: different manifestations of the same disease? Insights from a population-based study. J Am Coll Cardiol. 2001;38(3):827–34.
61. Mohler ER 3rd. Mechanisms of aortic valve calcification. Am J Cardiol. 2004;94 (11):1396–402. A6
62. Otto CM, O'Brien KD. Why is there discordance between calcific aortic stenosis and coronary artery disease? Heart. 2001;85(6):601–2.
63. Butcher JT, Mahler GJ, Hockaday LA. Aortic valve disease and treatment: the need for naturally engineered solutions. Adv Drug Deliv Rev. 2011;63(4–5):242–68.
64. Davies P, Reidy M, Goode T, Bowyer D. Scanning electron microscopy in the evaluation of endothelial integrity of the fatty lesion in atherosclerosis. Atherosclerosis. 1976;25(1):125–30.
65. Reinhart-King CA, Fujiwara K, Berk BC. Physiologic stress-mediated signaling in the endothelium. Methods Enzymol. 2008;443:25–44.
66. Frangos J, Eskin S, McIntire L, Ives C. Flow effects on prostacyclin production by cultured human endothelial cells. Science. 1985;227(4693):1477–9.
67. Levesque MJ, Nerem RM. The elongation and orientation of cultured endothelial cells in response to shear stress. J Biomech Eng. 1985;107(4):341–7.
68. Butcher JT, Penrod AM, Garcia AJ, Nerem RM. Unique morphology and focal adhesion development of valvular endothelial cells in static and fluid flow environments. Arterioscler Thromb Vasc Biol. 2004;24(8):1429–34.
69. Davies PF, Passerini AG, Simmons CA. Aortic valve: turning over a new leaf(let) in endothelial phenotypic heterogeneity. Arterioscler Thromb Vasc Biol. 2004;24(8):1331–3.
70. Butcher JT, Tressel S, Johnson T, Turner D, Sorescu G, Jo H, et al. Transcriptional profiles of valvular and vascular endothelial cells reveal phenotypic differences: influence of shear stress. Arterioscler Thromb Vasc Biol. 2006;26(1):69–77.
71. Chen XL, Grey JY, Thomas S, Qiu FH, Medford RM, Wasserman MA, et al. Sphingosine kinase-1 mediates TNF-alpha-induced MCP-1 gene expression in endothelial cells: upregulation by oscillatory flow. Am J Physiol Heart Circ Physiol. 2004;287(4):H1452–8.
72. Guo D, Chien S, Shyy JY. Regulation of endothelial cell cycle by laminar versus oscillatory flow: distinct modes of interactions of AMP-activated protein kinase and Akt pathways. Circ Res. 2007;100(4):564–71.
73. Dewey CF Jr, Bussolari SR, Gimbrone MA Jr, Davies PF. The dynamic response of vascular endothelial cells to fluid shear stress. J Biomech Eng. 1981;103(3):177–85.
74. Remuzzi A, Dewey CF Jr, Davies PF, Gimbrone MA Jr. Orientation of endothelial cells in shear fields in vitro. Biorheology. 1984;21(4):617–30.
75. Go Y-M, Boo YC, Park H, Maland MC, Patel R, Pritchard KA, et al. Protein kinase B/Akt activates c-Jun NH2-terminal kinase by increasing NO production in response to shear stress. J Appl Physiol. 2001;91(4):1574–81.
76. Sorescu GP, Song H, Tressel SL, Hwang J, Dikalov S, Smith DA, et al. Bone morphogenic protein 4 produced in endothelial cells by oscillatory shear stress induces monocyte adhesion by stimulating reactive oxygen species production from a nox1-based NADPH oxidase. Circ Res. 2004;95(8):773–9.
77. Sorescu GP, Sykes M, Weiss D, Platt MO, Saha A, Hwang J, et al. Bone morphogenic protein 4 produced in endothelial cells by oscillatory shear stress stimulates an inflammatory response. J Biol Chem. 2003;278(33):31128–35.
78. Platt MO, Xing Y, Jo H, Yoganathan AP. Cyclic pressure and shear stress regulate matrix metalloproteinases and cathepsin activity in porcine aortic valves. J Heart Valve Dis. 2006;15 (5):622–9.
79. Sucosky P, Balachandran K, Elhammali A, Jo H, Yoganathan AP. Altered shear stress stimulates upregulation of endothelial VCAM-1 and ICAM-1 in a BMP-4- and TGF-beta1-dependent pathway. Arterioscler Thromb Vasc Biol. 2009;29(2):254–60.

80. Sucosky P, Padala M, Elhammali A, Balachandran K, Jo H, Yoganathan AP. Design of an ex vivo culture system to investigate the effects of shear stress on cardiovascular tissue. J Biomech Eng. 2008;130(3):035001.
81. Weston MW, Yoganathan AP. Biosynthetic activity in heart valve leaflets in response to in vitro flow environments. Ann Biomed Eng. 2001;29(9):752–63.
82. Butcher JT, Nerem RM. Valvular endothelial cells regulate the phenotype of interstitial cells in co-culture: effects of steady shear stress. Tissue Eng. 2006;12(4):905–15.
83. McIntosh CT, Warnock JN. Side-specific characterization of aortic valve endothelial cell adhesion molecules under cyclic strain. J Heart Valve Dis. 2013;22(5):631–9.
84. Carrion K, Dyo J, Patel V, Sasik R, Mohamed SA, Hardiman G, et al. The long non-coding HOTAIR is modulated by cyclic stretch and WNT/β-CATENIN in human aortic valve cells and is a novel repressor of calcification genes. PLoS One. 2014;9(5):e96577.
85. Takeda H, Komori K, Nishikimi N, Nimura Y, Sokabe M, Naruse K. Bi-phasic activation of eNOS in response to uni-axial cyclic stretch is mediated by differential mechanisms in BAECs. Life Sci. 2006;79(3):233–9.
86. Engelmayr GC Jr, Hildebrand DK, Sutherland FW, Mayer JE Jr, Sacks MS. A novel bioreactor for the dynamic flexural stimulation of tissue engineered heart valve biomaterials. Biomaterials. 2003;24(14):2523–32.
87. Engelmayr GC Jr, Rabkin E, Sutherland FW, Schoen FJ, Mayer JE Jr, Sacks MS. The independent role of cyclic flexure in the early in vitro development of an engineered heart valve tissue. Biomaterials. 2005;26(2):175–87.
88. Merryman WD, Lukoff HD, Long RA, Engelmayr GC Jr, Hopkins RA, Sacks MS. Synergistic effects of cyclic tension and transforming growth factor-beta1 on the aortic valve myofibroblast. Cardiovasc Pathol. 2007;16(5):268–76.
89. Engelmayr GC Jr, Soletti L, Vigmostad SC, Budilarto SG, Federspiel WJ, Chandran KB, et al. A novel flex-stretch-flow bioreactor for the study of engineered heart valve tissue mechanobiology. Ann Biomed Eng. 2008;36(5):700–12.
90. Masoumi N, Howell MC, Johnson KL, Niesslein MJ, Gerber G, Engelmayr GC Jr. Design and testing of a cyclic stretch and flexure bioreactor for evaluating engineered heart valve tissues based on poly(glycerol sebacate) scaffolds. Proc Inst Mech Eng H J Eng Med. 2014;228(6):576–86.
91. Warnock JN, Konduri S, He Z, Yoganathan AP. Design of a sterile organ culture system for the ex vivo study of aortic heart valves. J Biomech Eng. 2005;127(5):857–61.
92. Durst CA, Jane Grande-Allen K. Design and physical characterization of a synchronous multivalve aortic valve culture system. Ann Biomed Eng. 2010;38(2):319–25.
93. Deck JD. Endothelial cell orientation on aortic valve leaflets. Cardiovasc Res. 1986;20(10):760–7.
94. Butcher JT, Simmons CA, Warnock JN. Mechanobiology of the aortic heart valve. J Heart Valve Dis. 2008;17(1):62–73.
95. Simmons CA, Grant GR, Manduchi E, Davies PF. Spatial heterogeneity of endothelial phenotypes correlates with side-specific vulnerability to calcification in normal porcine aortic valves. Circ Res. 2005;96(7):792–9.
96. Suga S, Itoh H, Komatsu Y, Ogawa Y, Hama N, Yoshimasa T, et al. Cytokine-induced C-type natriuretic peptide (CNP) secretion from vascular endothelial cells--evidence for CNP as a novel autocrine/paracrine regulator from endothelial cells. Endocrinology. 1993;133(6):3038–41.
97. La M, Cao TV, Cerchiaro G, Chilton K, Hirabayashi J, Kasai K, et al. A novel biological activity for galectin-1: inhibition of leukocyte-endothelial cell interactions in experimental inflammation. Am J Pathol. 2003;163(4):1505–15.
98. Farrar EJ, Huntley GD, Butcher J. Endothelial-derived oxidative stress drives myofibroblastic activation and calcification of the aortic valve. PLoS One. 2015;10(4):e0123257.
99. Bosse K, Hans CP, Zhao N, Koenig SN, Huang N, Guggilam A, et al. Endothelial nitric oxide signaling regulates Notch1 in aortic valve disease. J Mol Cell Cardiol. 2013;60:27–35.

100. El-Hamamsy I, Balachandran K, Yacoub MH, Stevens LM, Sarathchandra P, Taylor PM, et al. Endothelium-dependent regulation of the mechanical properties of aortic valve cusps. J Am Coll Cardiol. 2009;53(16):1448–55.
101. Rajamannan NM, Subramaniam M, Stock SR, Stone NJ, Springett M, Ignatiev KI, et al. Atorvastatin inhibits calcification and enhances nitric oxide synthase production in the hypercholesterolaemic aortic valve. Heart. 2005;91(6):806–10.
102. Richards J, El-Hamamsy I, Chen S, Sarang Z, Sarathchandra P, Yacoub MH, et al. Side-specific endothelial-dependent regulation of aortic valve calcification: interplay of hemodynamics and nitric oxide signaling. Am J Pathol. 2013;182(5):1922–31.
103. Sun L, Sucosky P. Bone morphogenetic protein-4 and transforming growth factor-beta1 mechanisms in acute valvular response to supra-physiologic hemodynamic stresses. World J Cardiol. 2015;7(6):331–43.
104. Weinberg EJ, Schoen FJ, Mofrad MR. A computational model of aging and calcification in the aortic heart valve. PLoS One. 2009;4(6):e5960.
105. Chandra S, Rajamannan NM, Sucosky P. Computational assessment of bicuspid aortic valve wall-shear stress: implications for calcific aortic valve disease. Biomech Model Mechanobiol. 2012;11(7):1085–96.
106. Sun L, Rajamannan NM, Sucosky P. Defining the role of fluid shear stress in the expression of early signaling markers for calcific aortic valve disease. PLoS One. 2013;8(12):e84433.
107. Aikawa E, Nahrendorf M, Sosnovik D, Lok VM, Jaffer FA, Aikawa M, et al. Multimodality molecular imaging identifies proteolytic and osteogenic activities in early aortic valve disease. Circulation. 2007;115(3):377–86.
108. Helske S, Syvaranta S, Lindstedt KA, Lappalainen J, Oorni K, Mayranpaa MI, et al. Increased expression of elastolytic cathepsins S, K, and V and their inhibitor cystatin C in stenotic aortic valves. Arterioscler Thromb Vasc Biol. 2006;26(8):1791–8.
109. Xing Y, Warnock JN, He Z, Hilbert SL, Yoganathan AP. Cyclic pressure affects the biological properties of porcine aortic valve leaflets in a magnitude and frequency dependent manner. Ann Biomed Eng. 2004;32(11):1461–70.
110. Jian B, Narula N, Li QY, Mohler ER 3rd, Levy RJ. Progression of aortic valve stenosis: TGF-beta1 is present in calcified aortic valve cusps and promotes aortic valve interstitial cell calcification via apoptosis. Ann Thorac Surg. 2003;75(2):457–65; discussion 65–6.
111. Mahler GJ, Frendl CM, Cao Q, Butcher JT. Effects of shear stress pattern and magnitude on mesenchymal transformation and invasion of aortic valve endothelial cells. Biotechnol Bioeng. 2014;111(11):2326–37.
112. Markwald RR, Fitzharris TP, Manasek FJ. Structural development of endocardial cushions. Am J Anat. 1977;148(1):85–119.
113. Markwald RR, Fitzharris TP, Smith WN. Structural analysis of endocardial cytodifferentiation. Dev Biol. 1975;42(1):160–80.
114. Nakajima Y, Mironov V, Yamagishi T, Nakamura H, Markwald RR. Expression of smooth muscle alpha-actin in mesenchymal cells during formation of avian endocardial cushion tissue: a role for transforming growth factor beta3. Dev Dyn. 1997;209(3):296–309.
115. Cote N, Mahmut A, Fournier D, Boulanger MC, Couture C, Despres JP, et al. Angiotensin receptor blockers are associated with reduced fibrosis and interleukin-6 expression in calcific aortic valve disease. Pathobiology. 2014;81(1):15–24.
116. Kaden JJ, Dempfle CE, Grobholz R, Fischer CS, Vocke DC, Kilic R, et al. Inflammatory regulation of extracellular matrix remodeling in calcific aortic valve stenosis. Cardiovasc Pathol. 2005;14(2):80–7.
117. Kaden JJ, Kilic R, Sarikoc A, Hagl S, Lang S, Hoffmann U, et al. Tumor necrosis factor alpha promotes an osteoblast-like phenotype in human aortic valve myofibroblasts: a potential regulatory mechanism of valvular calcification. Int J Mol Med. 2005;16(5):869–72.
118. Mahler GJ, Farrar EJ, Butcher JT. Inflammatory cytokines promote mesenchymal transformation in embryonic and adult valve endothelial cells. Arterioscler Thromb Vasc Biol. 2013;33(1):121–30.

119. Guo H, Ingolia NT, Weissman JS, Bartel DP. Mammalian microRNAs predominantly act to decrease target mRNA levels. Nature. 2010;466(7308):835–40.
120. van Rooij E, Marshall WS, Olson EN. Toward microRNA-based therapeutics for heart disease—the sense in antisense. Circ Res. 2008;103(9):919–28.
121. Nigam V, Sievers HH, Jensen BC, Sier HA, Simpson PC, Srivastava D, et al. Altered microRNAs in bicuspid aortic valve: a comparison between stenotic and insufficient valves. J Heart Valve Dis. 2010;19(4):459–65.
122. Holliday CJ, Ankeny RF, Jo H, Nerem RM. Discovery of shear- and side-specific mRNAs and miRNAs in human aortic valvular endothelial cells. Am J Physiol Heart Circ Physiol. 2011;301(3):H856–67.
123. Thubrikar MJ, Aouad J, Nolan SP. Patterns of calcific deposits in operatively excised stenotic or purely regurgitant aortic valves and their relation to mechanical stress. Am J Cardiol. 1986;58(3):304–8.
124. Freeman RV, Otto CM. Spectrum of calcific aortic valve disease: pathogenesis, disease progression, and treatment strategies. Circulation. 2005;111(24):3316–26.
125. Robicsek F, Thubrikar MJ, Fokin AA. Cause of degenerative disease of the trileaflet aortic valve: review of subject and presentation of a new theory. Ann Thorac Surg. 2002;73 (4):1346–54.
126. Olsson M, Thyberg J, Nilsson J. Presence of oxidized low density lipoprotein in nonrheumatic stenotic aortic valves. Arterioscler Thromb Vasc Biol. 1999;19(5):1218–22.
127. O'Brien KD, Reichenbach DD, Marcovina SM, Kuusisto J, Alpers CE, Otto CM. Apolipoproteins B, (a), and E accumulate in the morphologically early lesion of 'degenerative' valvular aortic stenosis. Arterioscler Thromb Vasc Biol. 1996;16(4):523–32.
128. Rajamannan NM, Subramaniam M, Rickard D, Stock SR, Donovan J, Springett M, et al. Human aortic valve calcification is associated with an osteoblast phenotype. Circulation. 2003;107(17):2181–4.
129. Rajamannan NM, Subramaniam M, Springett M, Sebo TC, Niekrasz M, McConnell JP, et al. Atorvastatin inhibits hypercholesterolemia-induced cellular proliferation and bone matrix production in the rabbit aortic valve. Circulation. 2002;105(22):2660–5.
130. Balachandran K, Sucosky P, Jo H, Yoganathan AP. Elevated cyclic stretch alters matrix remodeling in aortic valve cusps: implications for degenerative aortic valve disease. Am J Physiol Heart Circ Physiol. 2009;296(3):H756–64.
131. Fujiwara K. Platelet endothelial cell adhesion molecule-1 and mechanotransduction in vascular endothelial cells. J Intern Med. 2006;259(4):373–80.
132. Metzler SA. The aortic valve endothelial cell: a multi-scale study of strain mechanobiology. Mississippi State: Mississippi State University; 2010.
133. Nakada MT, Amin K, Christofidou-Solomidou M, O'Brien CD, Sun J, Gurubhagavatula I, et al. Antibodies against the first Ig-like domain of human platelet endothelial cell adhesion molecule-1 (PECAM-1) that inhibit PECAM-1-dependent homophilic adhesion block in vivo neutrophil recruitment. J Immunol. 2000;164(1):452–62.
134. Vaporciyan AA, DeLisser HM, Yan HC, Mendiguren II, Thom SR, Jones ML, et al. Involvement of platelet-endothelial cell adhesion molecule-1 in neutrophil recruitment in vivo. Science. 1993;262(5139):1580–2.
135. Dusserre N, L'Heureux N, Bell KS, Stevens HY, Yeh J, Otte LA, et al. PECAM-1 interacts with nitric oxide synthase in human endothelial cells: implication for flow-induced nitric oxide synthase activation. Arterioscler Thromb Vasc Biol. 2004;24(10):1796–802.
136. Heistad DD, Wakisaka Y, Miller J, Chu Y, Pena-Silva R. Novel aspects of oxidative stress in cardiovascular diseases. Circ J. 2009;73(2):201–7.
137. Miller JD, Chu Y, Brooks RM, Richenbacher WE, Pena-Silva R, Heistad DD. Dysregulation of antioxidant mechanisms contributes to increased oxidative stress in calcific aortic valvular stenosis in humans. J Am Coll Cardiol. 2008;52(10):843–50.
138. El-Hamamsy I, Chester AH, Yacoub M. Cellular regulation of the structure and function of aortic valves. J Adv Res. 2010;1(1):5–12.

139. Chiswell BP, Zhang R, Murphy JW, Boggon TJ, Calderwood DA. The structural basis of integrin-linked kinase-PINCH interactions. Proc Natl Acad Sci U S A. 2008;105(52):20677–82.
140. Lal H, Verma SK, Foster DM, Golden HB, Reneau JC, Watson LE, et al. Integrins and proximal signaling mechanisms in cardiovascular disease. Front Biosci. 2009;14:2307–34.
141. Legate KR, Montanez E, Kudlacek O, Fassler R. ILK, PINCH and parvin: the tIPP of integrin signalling. Nat Rev Mol Cell Biol. 2006;7(1):20–31.
142. Herranz B, Marquez S, Guijarro B, Aracil E, Aicart-Ramos C, Rodriguez-Crespo I, et al. Integrin-linked kinase regulates vasomotor function by preventing endothelial nitric oxide synthase uncoupling: role in atherosclerosis. Circ Res. 2012;110(3):439–49.
143. Critchley DR. Genetic, biochemical and structural approaches to talin function. Biochem Soc Trans. 2005;33.(Pt 6:1308–12.
144. Vestweber D. VE-cadherin: the major endothelial adhesion molecule controlling cellular junctions and blood vessel formation. Arterioscler Thromb Vasc Biol. 2008;28(2):223–32.
145. Ross R. Atherosclerosis is an inflammatory disease. Am Heart J. 1999;138(5 Pt 2):S419–20.
146. Balachandran K, Sucosky P, Jo H, Yoganathan AP. Elevated cyclic stretch induces aortic valve calcification in a bone morphogenic protein-dependent manner. Am J Pathol. 2010;177(1):49–57.
147. Rosenhek R, Binder T, Porenta G, Lang I, Christ G, Schemper M, et al. Predictors of outcome in severe, asymptomatic aortic stenosis. N Engl J Med. 2000;343(9):611–7.
148. Bach DS. Prevalence and characteristics of unoperated patients with severe aortic stenosis. J Heart Valve Dis. 2011;20(3):284–91.
149. Teo KK, Corsi DJ, Tam JW, Dumesnil JG, Chan KL. Lipid lowering on progression of mild to moderate aortic stenosis: meta-analysis of the randomized placebo-controlled clinical trials on 2344 patients. Can J Cardiol. 2011;27(6):800–8.
150. Rosenhek R, Rader F, Loho N, Gabriel H, Heger M, Klaar U, et al. Statins but not angiotensin-converting enzyme inhibitors delay progression of aortic stenosis. Circulation. 2004;110(10):1291–5.
151. O'Brien KD, Zhao XQ, Shavelle DM, Caulfield MT, Letterer RA, Kapadia SR, et al. Hemodynamic effects of the angiotensin-converting enzyme inhibitor, ramipril, in patients with mild to moderate aortic stenosis and preserved left ventricular function. J Investig Med. 2004;52(3):185–91.

# The Role of Proteoglycans and Glycosaminoglycans in Heart Valve Biomechanics

Varun K. Krishnamurthy and K. Jane Grande-Allen

**Abstract** Proteoglycans (PGs) and glycosaminoglycans (GAGs) have long been recognized as constituents of heart valves but their precise functions have been mysterious and underrepresented. The heterogeneity, dynamic processing and viscoelastic nature of PGs and GAGs, and their ability to exist independently or in association with other extracellular matrix (ECM) components, contribute to the overall complexity of valve ECM (structure), and impact local tissue biomechanics (function). This chapter will discuss various approaches for elucidating the biomechanical properties of valvular PGs and GAGs, will relate valve tissue hemodynamic alterations and valve cell biomechanical stimuli to differences in PG/GAG expression (and misexpression), and will address directions for future studies.

**Keywords** Aortic and mitral valves · Cell and organ culture · Bioreactor · Animal models · Valve disease · Tissue engineering · Bioprostheses

## 1 Introduction and Chapter Overview

Proteoglycans (PGs) and glycosaminoglycans (GAGs) have long been recognized as constituents of heart valves but their precise functions have been mysterious and underrepresented. The heterogeneity, dynamic processing and viscoelastic nature of PGs and GAGs, and their ability to exist independently or in association with other extracellular matrix (ECM) components, contribute to the overall complexity of valve ECM (structure), and impact local tissue biomechanics (function). This chapter will discuss various approaches for elucidating the biomechanical properties of valvular PGs and GAGs, will relate valve tissue hemodynamic alterations and valve cell biomechanical stimuli to differences in PG/GAG expression (and misexpression), and will address directions for future studies.

V. K. Krishnamurthy · K. Jane Grande-Allen (✉)
Department of Bioengineering, Rice University, Houston, TX, USA
e-mail: grande@rice.edu

## 2 Proteoglycans and Glycosaminoglycans in Valves: Composition, Distribution, and Interactions

Heart valves consist of a multilayered structure with distinct fibrosa, spongiosa, and ventricularis (or atrialis) layers. While the dense, longitudinally oriented collagen fibers in the fibrosa confer durability and the radially oriented elastic fibers in the ventricularis impart extensibility, the PG-rich central spongiosa layer provides cohesiveness and lubrication between the two outer layers and serves to decrease shear stress associated with leaflet flexure [1, 2]. The primary valvular PGs and GAG are versican, aggrecan, perlecan, decorin, biglycan (PGs), and hyaluronan (GAG) [3–5]. All these components are synthesized by valve interstitial cells (VICs), the predominant cell type within valves [6]. With the exception of hyaluronan, GAGs exist in vivo as structural components of PGs; a PG consists of a protein core with at least one GAG chain attached (Fig. 1).

Structurally, GAGs are linear polysaccharides with repeating disaccharide units, consisting of an amino sugar (N-acetylglucosamine, glucosamine that is variously N-substituted, or N-acetylgalactosamine) and a uronic acid (glucuronic acid or iduronic acid) or galactose (Fig. 1). Remarkably, there exists a high degree of inherent heterogeneity within the GAG structure, with respect to the patterns of chain length, sulfation, and epimerization (collectively termed "fine structure") that make the investigation of GAG structure-function relationships very challenging. While versican is a large PG with long, multiple GAG chains, the small leucine-rich PGs (SLRPs) decorin and biglycan are composed of only one (decorin) or two (biglycan) GAG chains [7, 8]. Further, while hyaluronan, the major GAG in valves, is unsulfated, the sulfated GAGs such as glucosaminoglycans attached to heparan sulfate (HS) PG cores in perlecan, and the galactosaminoglycans attached to chondroitin sulfate (CS) and dermatan sulfate (DS) PG cores in versican, decorin and biglycan also predominate in valves. In addition, epimerization of glucuronic acid to iduronic acid occurs primarily in DS and HS, whereas sulfation occurs at the 4-O and 6-O sites in N-acetylgalactosamine residues or on the 2-C site in uronic acid residues of CS and DS entities, in different permutations [9]. Overall, there is enormous intricacy and diversity in GAG microstructure; however, it is noteworthy that the exact positions of where precise sulfation and epimerization patterns occur in which GAG chain still remain unclear.

Many of the PG and GAG studies in heart valves are centered around aortic and mitral valves since most prevalent valve diseases are of aortic or mitral origin [10]. In these valves, studies have shown that versican and hyaluronan are most strongly expressed in the spongiosa layer [11]. Versican, with its multiple 6-sulfated CS GAG chains, often associates with hyaluronan resulting in large aggregates. These aggregates have large hydrodynamic volumes and provide valves with hydration and lubrication while facilitating responsiveness to variable compressive loads, making them ideal for withstanding high transvalvular pressures [12–14]. During myxomatous mitral valve disease, an overabundance of these hydrated moieties together with the release of catabolic and processing enzymes (MMPs including collagenases,

Heparan sulfate** (HS)

Chondroitin sulfate* (CS)

Dermatan sulfate* (DS)

Hyaluronan (HA)

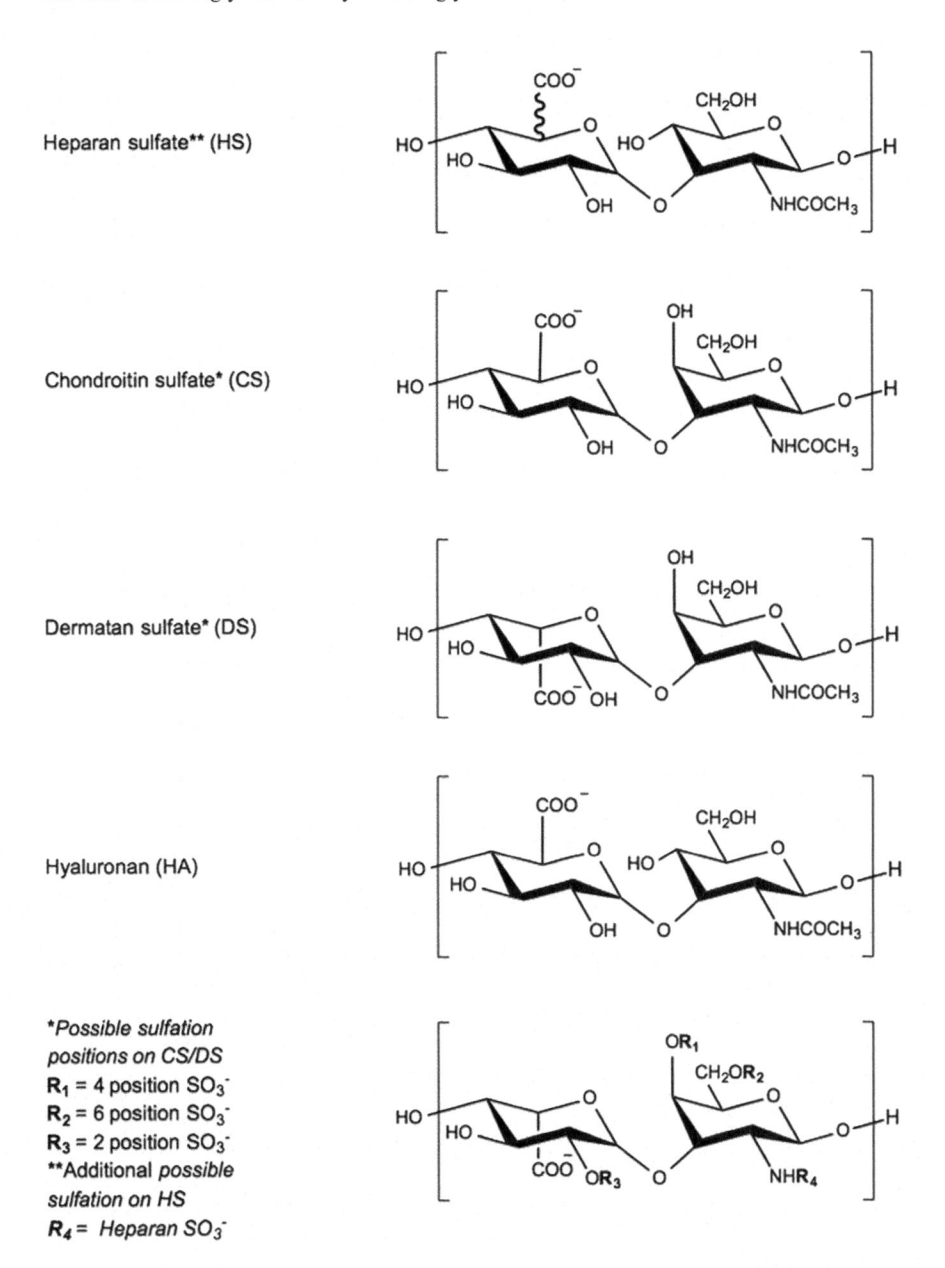

*Possible sulfation
positions on CS/DS
$R_1$ = 4 position $SO_3^-$
$R_2$ = 6 position $SO_3^-$
$R_3$ = 2 position $SO_3^-$
**Additional *possible*
sulfation on HS
$R_4$ = Heparan $SO_3^-$

**Fig. 1** CS GAGs consist of repeats of glucuronate and N-acetyl galactosamine disaccharide, whereas DS GAGs harbor repeats of iduronate and N-acetyl galactosamine disaccharide

elastases, gelatinases, and ADAMTSs) without net increases in ECM synthesis lead to expansion of the spongiosa, and ultimately valve weakening [15]. On the other hand, smaller PGs with largely 4-sulfated CS/DS GAGs such as decorin and biglycan are found in abundance in tensile load bearing regions, e.g., the ventricular inflow side of aortic valve, suggesting the presence of GAGs in not only the spongiosa layer, but in other layers of the valve as well [16].

It is well established that PGs interact with other ECM structural proteins such as elastic fibers and collagen; these interactions may profoundly influence valve biomechanical and molecular processes [17]. Versican, for example, is known to interfere with elastic fiber assembly by preventing tropoelastin from binding to the microfibrillar scaffold through the elastin-binding protein [18, 19]. The CS GAGs of versican may bind to the galactolectin-binding site of elastin-binding protein, reducing the protein's affinity for tropoelastin and disrupting elastic fiber formation. In the diseased aortic valve, elastin insufficiency causes increased versican synthesis together with versican processing defects, resulting in compromised valve tissue stiffness [20]. Elastic fibers are not restricted just to the ventricularis; they form an interweaving network in the PG-rich spongiosa of aortic valves, impacting local tissue biomechanics [21].

Furthermore, biglycan and decorin bind tropoelastin and elastin, suggesting a role for these PGs in elastogenesis, and are consistent with co-localization of these proteins in the ventricularis [22]. These PGs bind tropoelastin (biglycan) or microfibrils (decorin) via their protein cores and promote elastic fiber assembly, with CS/DS and transforming growth factor (TGF) β possibly mediating these regulatory processes [23, 24]. Decorin is highly prevalent in the collagen-rich fibrosa layer where it facilitates collagen fiber formation by binding to and protecting the fibrils from cleavage by collagenases [25, 26]. Additionally, they mediate force transmission between adjacent collagen fibrils, thereby providing structural and mechanical integrity to the tissue [27]. Intriguingly, biglycan is known to interact with collagen during tissue development where it facilitates early collagen fibrillogenesis and fibril assembly. Biglycan expression tapers off with tissue maturation; however, decorin expression persists in order to stabilize the fiber growth [28, 29]. Not just the valve fibrosa, but the ventricular inflow surface of aortic valve, which experiences high shear, laminar flow and tensile forces, also demonstrate decorin and biglycan expression to necessitate collagen fibrillogenesis and resist these forces. Other SLRPs associated with collagen fiber assembly such as lumican and fibromodulin have also recently shown to be critical in regions where valves anchor to the muscle, promoting structural and mechanical integrity at the base of the leaflets [30]. Taken together, site-specific expression and co-localization patterns of PGs reveal key information about their roles in regional valve biomechanics.

## 3 Glycosaminoglycan Stability Affects Tissue Biomechanics in Bioprostheses

Valve disease is a significant cause of cardiovascular morbidity and mortality, and over 100,000 bioprosthetic heart valve (BHV) replacement surgeries are performed each year throughout the world [31]. Glutaraldehyde (GLUT)-based fixation of these tissues prior to implantation remains the gold standard, as it ameliorates immunogenicity of the tissue and permits the clinical sterilization and safe storage of BHVs. In the long term, however, these valves exhibit calcification and mechanical degeneration leading to loss of durability and eventual failure in over 50% of young patients [32–34]. Although GLUT cross-links and protects collagen fibers against enzymatic degradation, GAGs lack the amine functionalities necessary for cross-linking, and as a result, their stability is severely compromised [35].

An important early biomechanical study of porcine BHVs investigated the degradation of GAG content due to tissue fatigue [36]. When subjected to in vitro accelerated fatigue (5–500 million cycles of fatigue testing), these valves demonstrated decreased leaflet bending strength and loss of flexural rigidity. Further, there was a significant reduction in GAG levels (after only ten million cycles of fatigue testing) that preceded structural alterations in type I collagen (at 50 million cycles). Later reports also accounted for progressive GAG loss in BHVs even during normal fixation and regular storage in GLUT, indicating their intrinsic vulnerability [35]. Dysregulation of the spongiosa layer in these valves due to GAG depletion also increased the susceptibility of the leaflet to buckling during flexure, which heightened the potential for mechanical failure [37]. In vivo analysis revealed that, when GLUT-fixed porcine aortic valves were implanted subdermally for 3 weeks in a rat model, GAG-degrading enzymes such as hyaluronidase, chondroitinase (GAGases), and proteases that cleave the protein core were heightened in the explanted fixed valves compared to native and freshly fixed valves [38]. Interestingly, these enzymes were present during the early phases of valve tissue preparation for use as BHVs, as well as after BHV implantation, suggesting that enhanced enzyme activity during initial BHV preparation (which could take up to 24 h) led to the prefixation GAG degradation; this was followed by a phase of enzyme overexpression after implantation, possibly due to a host-induced response.

Simultaneously, other biochemical studies described the loss of 6-sulfated CS and hyaluronan from explanted porcine BHVs, indicating loss of versican aggregates [39]. Since GAGs do not react with GLUT, the hypothesis is that versican aggregates remain un-cross-linked and unbound, and hence leach out of the valve leaflets gradually after BHV implantation. Surprisingly, the 4-sulfated CS/DS such as that found in decorin and biglycan were better presented in these valves, possibly due to the relatively well-exposed core protein of these PGs being located adjacent to collagen fibrils, thereby increasing their susceptibility to fixation. Taken together, while the function of spongiosa is to primarily allow shearing between the fibrosa and ventricularis, when GAG depletion occurs in fixed valves, shearing is resisted by the remaining fibrillar ECM components, consequently increasing tissue rigidity. Increase in leaflet rigidity due to compromised hydration, compressive resistance and viscosity resulting from GAG loss leads to biomechanical deterioration and contributes to calcification in vivo.

Recent studies have demonstrated the effectiveness of neomycin, a hyaluronidase inhibitor, in enhancing GLUT cross-linking to stabilize GAGs [40]. After subjecting the BHVs to the same fatigue cycles as described previously [36], it was observed that the neomycin-glutaraldehyde (NG) cross-linking provided resistance to fatigue-induced GAG loss and insulated the cross-linked ECM from enzymatic degradation via GAGases such as hyaluronidase and chondroitinase. Elastin degradation was still prevalent. The NG cross-linked leaflets were functionally similar to GLUT-fixed tissues prior to GAGase treatment; after GAGase treatment, however, these tissues showed increased areal compliance and reduced hysteresis. Therefore, GLUT-associated alterations in valve mechanical behavior were prevented by the NG treatment, and inclusion of neomycin helped maintain biomechanical properties of the cross-linked tissue. More recently, interesting new combinations of treatments and novel fabrication/fixation methods have been assessed, such as incorporation of neomycin and pentagalloyl glucose (PGG) into glutaraldehyde cross-linking to create a neomycin-PGG-glutaraldehyde (NPG) approach that stabilizes both GAGs and elastin in porcine aortic valves, and the creation of a tri-cross-linked carbodiimide/neomycin/PGG fixative for improved prosthetic stability [41, 42].

## 4 Theoretical and Experimental Determination of Glycosaminoglycan Biomechanics in Native Valve Tissue

While the mechanical behaviors of the fibrosa and ventricularis layers of the aortic valve leaflet are well characterized, little information exists on how their mechanical interactions are mediated by the GAG-rich central spongiosa layer. The spongiosa dampens vibrations from closing, confers flexibility to the leaflet, and lubricates shear between the outer layers, all functions that are derived from the ability of GAGs to retain water [11, 43]. From a mechanistic perspective, GAGs are hypothesized to contribute substantially to the time-dependent behavior of the bulk tissue at lower physiological force levels through mechanical interactions achieved by direct bonding with the partially undulating collagen fibers. This hypothesis has been used to explain why the magnitude of stress relaxation is reduced at higher tissue stress levels, wherein the bulk tissue response is dominated by the fully loaded and straightened collagen fibers [44]. Additionally, despite the heterogeneous valve architecture, the leaflet layers function as a bonded structure during low-strain flexure due to interlayer interactions. In this low-strain environment, the spongiosa may mechanically act as a contiguous extension of the outer two layers [45]. Interestingly, GAG accumulation has been shown to lead to significant stress concentration and intralamellar Donnan swelling pressures during thoracic aortic aneurysm and dissection, and these may disrupt the normal cell–matrix interactions that are fundamental to tissue homeostasis and delaminate the layered microstructure [46].

Prior to biomechanical characterization of GAGs in heart valves, Schmidt et al. explored the influence of PGs on the static and kinetic behavior in articular cartilage [47], a tissue with similar ECM composition and developmental pathways to that of valves [48]. Viscoelastic creep and slow constant-rate uniaxial tension tests were performed on cartilage treated with chondroitinase ABC, or chondroitinase ABC with trypsin and hyaluronidase. Creep results showed that the GAG-degraded tissue demonstrated a rapid increase in strain following load application and reached equilibrium within 100 s while untreated tissues exhibited either rapid or gradual increase in strain but did not reach equilibrium even after 25,000 s. However, uniaxial tension tests showed that intrinsic tensile stiffness, strength, or the collagen network of the cartilage tissue remained unaffected. These findings suggested that through their physical restraints on collagen, the bulk of PGs present in the tissue acted to retard fibrillar reorganization and alignment under tensile loading, thereby effectively preventing sudden extension of the collagen network. This also has implications in understanding the differential, functional roles of PGs in largely immobile vs. mobile tissues, which is not well unknown.

In one of the early studies aimed at determining the relative roles of hydration and GAG content on the viscoelasticity of valve tissues, porcine aortic valve leaflets were subjected to GAG depletion using 0.1 M NaOH, and stress relaxation and failure tests were performed [49]. GAG depletion affected the viscoelastic response of aortic valve tissue. However, the major portion of this effect correlated with water content, which was significantly removed along with GAGs as well. Intriguingly, removal of GAGs did not independently reduce the tissue viscoelasticity proving that, regardless of the presence of absence of GAGs, the main contributor to soft tissue viscoelasticity was its water content.

The roles of GAGs in the tensile and viscoelastic properties of leaflets and how GAGs contribute and relate to the observed valvular tissue behavior at high tensile loading remained unclear until recently. In order to quantify the mechanical contribution of GAGs, the tensile and 45° loading (with respect to circumferential axis) of porcine aortic valves were compared between valves with GAGs present and valves with GAGs removed through enzymatic digestion by chondroitinase and hyaluronidase [50]. After GAG removal, there was a significant reduction in hysteresis at low force ranges (from ~35% in control to ~25% after GAG removal) but hardly any change in the tensile properties, indicating that the viscoelastic behaviors of valve tissues manifested only at low strains. Further, this relationship implicated GAGs in providing important damping mechanisms to reduce leaflet flutter when the leaflet is not under high tensile stress, and associated GAGs with fiber–fiber and fiber–matrix interactions at low force levels. In a similar study, removal of GAGs using a cocktail of chondroitinase ABC, hyaluronidase, and keratinase had an effect on viscoelastic properties (relaxation kinematics and percentage relaxation) of fresh porcine aortic valves, but no effect on tensile properties (stiffness), which may depend more on collagen and elastic structure [51]. More specifically, removal of GAGs decreased the relaxation percentage, slowed the relaxation rates of the valve tissue, and reduced the natural anisotropic (directionally dependent) viscoelasticity of the valve.

In contrast, Tseng et al. showed that GAG concentration in porcine aortic valves played a significant role in leaflet tensile behavior by lubricating tissue motion and reducing stresses, but not in viscoelasticity [52]. Upon treatment with hyaluronidase, the elastic modulus, maximum stress, and hysteresis significantly increased validating previous reports [35, 53], whereas extensibility and stress relaxation behavior did not change. Extensibility is strongly linked to the structure of collagen fiber network, and since hyaluronidase does not degrade collagen, the enzyme treatment had no effect. Further, since GAG degradation has previously been shown to affect valve stress relaxation [49, 51], the absence of any change in tissue stress relaxation here was contradictory; however, this result was attributed to water retention in these leaflets even at high hyaluronidase concentrations. Likewise, the enhanced stiffness values observed in these hyaluronidase-treated leaflets varied from the tensile properties of GAGase-treated leaflets reported in the studies above [50, 51], highlighting the differential involvement of broad spectrum vs. specific enzymes in valve regional GAG digestion, and their consequent effect on tissue biomechanics. On the whole, current information on experimental determination of GAG biomechanical properties is scant, and the dependence of valve leaflet viscoelasticity on GAG concentration is controversial and warrants further investigation.

## 5  Bioreactor Studies to Evaluate Valve Tissue Biomechanics and ECM Remodeling

In order to recapitulate in vivo valve biomechanical properties or GAG response to biochemical factors during development or disease, many successful in vitro experimental models have been developed. Although animal models, such as mice, can be used to examine underlying disease mechanisms in vivo, these studies are invasive, expensive, and time consuming. Cell studies offer a less expensive alternative; however cells grown on monolayers are not truly representative of the in vivo environment, i.e., the layered structure of the valve tissue. Further, external factors such as mechanics—as in how mechanotransduction from global and multimodal stresses (planar tension, flexure, and shear stresses) on valves translate to cellular and subcellular levels—are unclear. Therefore, organ cultures that maintain valve cells within the native physical structure of the tissue are being utilized. Heart valve tissues are particularly appropriate for organ culture research because the cells within adult valves are nourished primarily through diffusive transport of oxygen and nutrients (as opposed to through vascularization). Hence, designing a valvular organ culture system is much simpler than designing systems for more complicated fully vascularized tissues.

Most of the valve organ culture studies investigated tissue biomechanics and/or ECM remodeling after multi-day leaflet culture (Table 1). Merryman et al. used a tension bioreactor to subject porcine aortic valve leaflets to mechanical and cytokine stimuli. They analyzed valve tissues cultured for 1 and 2 weeks in each of four conditions: 15% cyclic stretch (physiologic tension), exogenous TGFβ1, both, and neither [55]. The "neither," "TGFβ1," and "Tension + TGFβ1" groups demonstrated loss of spongiosa with no apparent PGs. Interestingly, there was a loss of PGs in the group receiving no mechanical stimulation, which was unexpected as tension was believed to compress the spongiosa region. However, this finding was validated by another recent study showing downregulation of PGs in statically cultured porcine

**Table 1** Overview of different organ culture bioreactor systems to assess ECM remodeling and PG/GAG phenotype in cultured aortic/mitral valve tissues

| Tissue type | Bioreactors for organ culture | GAG phenotype | References |
|---|---|---|---|
| Porcine aortic valve, animal age: unknown | Cyclic pressure (80–120, 120–160, 150–190 mm), pulse frequency (0.5, 1.17, 2 Hz), culture time: 48 h | • Sulfated GAG synthesis increased with the magnitude of applied cyclic pressure<br>• GAG synthesis increased at 0.5 but not at 1.17 and 2 Hz | [54] |
| Porcine aortic valve, animal age: ~10 months | Tension (15% cyclic stretch), culture time: 2 weeks | • Loss of spongiosa with no PGs (static conditions)<br>• Trilaminar structure maintained but with more intense elastic fibers (15% strain) (Fig. 2) | [55] |
| Porcine mitral valve, animal age: ~6 months | Splashing bioreactor (normal and dynamic shear forces), culture time: 2 weeks | • Less defined spongiosa with loss of PG/GAGs and increase in collagen (static)<br>• PG retained. Heightened versican and biglycan synthesis (dynamic) | [57] |
| Porcine aortic valve, animal age: 6–12 months | Stretch-pressure bioreactor, 10 or 15% stretch, 120/80 mm or 140/100 mm pressure, culture time: 48 h | • Spongiosa thickness increased at elevated pressure (140/100) and stretch (15%) (Fig. 2) | [56] |
| Porcine aortic valve, animal age: 12–24 months | Tensile (10% cyclic stretch), culture time: 72 h | • Sulfated GAGs increased with cyclic stretch<br>• Upregulation of versican, biglycan, lumican, and aggrecan | [58] |
| Sheep mitral valve, animal age: 18–24 months | Tensile (strains of 10%, 20%, and 30%), culture time: 72 h | • Sulfated GAG content increased with increased magnitude of strain<br>• Disruption of spongiosa, GAG deposition throughout leaflet at high strains | [59] |
| Porcine mitral valve, animal age: 3–6 months | Flow loop tension system with physiologic waveforms and pressures, culture time: 1 week | • Mitral valve prolapsed tissues had elevated GAG levels<br>• Functional mitral regurgitation conditioned tissues demonstrated more collagen and decorin | [60] |

aortic valve leaflets [61]. It appeared from histology that either the PGs dissociated on their own, unstimulated VICs enzymatically digested them, or VICs were not able to actively synthesize new PGs in the absence of applied tension. Together, these suggested that proper preimplant mechanical conditioning of an engineered tissue is required prior to implantation in order to retain native architecture.

In mitral valves, using a splashing rotating bioreactor, normal and shear forces were applied for 2 weeks, and the ECM architecture in these valves was compared to statically cultured valves [57]. The spongiosa became less defined in static cultures with a loss of PG/GAGs and an increase in collagen. With dynamic stimulation, versican expression (quite low in static) was higher in the chordae, whereas decorin remained constant between the two groups. Biglycan expression in the chordae was modest and high in static and dynamic cultured groups, respectively. In contrast with the chordae, the leaflets demonstrated normal expression levels of versican and decorin, and reduced expression of biglycan, in both static and dynamic culture conditions. The apparent collagen upregulation and rapid disorganization of the tissue collagen layers in static conditions was attributed to the absence of mechanical stretch-induced perfusion. Moreover, the dissipation of PG/GAGs was speculated to be caused by tissue disorganization allowing these water-binding moieties to leach out over time. Therefore, mechanical stimulation via perfusion and stretching maintained tissue integrity.

In another study, porcine aortic valve leaflets to combinations of normal and pathological stretch and pressure magnitudes using an ex vivo cyclic stretch and pressure bioreactor [56]. Under pathologically elevated pressure (140/100 mmHg) and stretch (15%), VIC phenotype was abnormal and the thickness of spongiosa and fibrosa layers increased, suggestive of maladaptive ECM remodeling. Remarkably, in the absence of stretch, cyclic pressure upregulated sulfated GAG expression in porcine aortic valve leaflets, while constant static pressure did not [54]. GAG synthesis increased with the magnitude of applied cyclic pressure with 170 mmHg exhibiting a greater effect compared to 100 and 140 mmHg pressures. Pulse frequency influenced GAG homeostasis as well. Furthermore, when high magnitude and high frequency were combined, significant increase in GAG synthesis was observed; however, pressure magnitude seemed to have a greater influence on the tissue than frequency.

Similar to porcine valves, cultured sheep (mitral) valve leaflets were cyclically stretched under static, 10% (low physiologic strain), 20% (high physiologic), and 30% (pathologic) cyclic strain at 0.5 Hz for 1 and 3 days [59]. The valves were clamped at the leaflet annulus and free edge at the attachment of primary chordae tendineae into the mounting device, and radial strain was applied. The measured tensile load necessary to maintain strain on mitral valves decreased exponentially over the first 30 h of cyclic strain. The magnitude of decrease in tensile load was greater at higher strain levels, consistent with greater weakening of the ECM with higher strain levels. Histological analysis revealed increased GAG deposition throughout the leaflet and disruption of elastin fiber organization with increasing levels of cyclic strain. Sulfated GAG content in mitral valves increased with increased magnitude of cyclic strain, and was highest at 30% strain after 24 h, and at both 20 and 30% strain levels after 72 h.

Excessive serotonin levels are known to cause increased VIC proliferation and upregulation of valve ECM production [62–65]. The controlled effects of cyclic stretch and adding serotonin/inhibiting serotonin transporter on ECM remodeling was investigated in porcine aortic valve leaflets loaded in a tensile stretch bioreactor [58]. The leaflets were subjected to 10% cyclic stretch for 3 days. Addition of serotonin resulted in reduced amounts of sulfated GAGs. However, GAG levels increased in the presence of cyclic stretch, with significant upregulation of biglycan, versican, lumican, and aggrecan, and serotonin transport blockage together with cyclic stretch resulted in greater GAG biosynthesis.

Mitral valve disease can manifest as mitral valve prolapse (MVP), with mechanical weakening of valve tissue due to an accumulation of PGs and GAGs, or functional mitral regurgitation (FMR), with distortion of the valve annulus and papillary muscle leading to higher tension forces throughout the leaflet. To probe these two conditions in vitro, a flow loop system was developed and contained porcine mitral valves with altered geometry (MVP—5 mm displacement of papillary muscles toward the annulus; FMR—5 mm apical, 5 mm lateral papillary muscle displacement, 65% larger annular area) to mimic the diseased conditions [60]. MVP conditioned tissues were less stiff, weaker, and had elevated levels of type III collagen and GAGs, suggesting myxomatous remodeling. FMR conditioned tissues were stiffer, more brittle, less extensible, and with more collagen synthesis, cross-linking-related enzymes and PGs, including decorin, suggesting fibrotic remodeling. These findings confirmed that valve ECM was differentially remodeled in response to altered mechanical homeostasis resulting from disease hemodynamics.

Overall, organ culture is a valuable tool for simultaneous assessment of tissue biomechanics and ECM remodeling in valves. However, there are some compelling comparisons and contrasting aspects among these studies (Table 1). The duration of these organ culture studies vary from 48 h to 2 weeks. Therefore, the effect of culture time on tissue ECM remodeling needs to be elucidated. In addition, having some standardization across bioreactor and tissue mechanical stimulation studies would improve consistency between various studies. Further areas worth investigating, due to differences across previous studies, include the effect of animal species and age. Particularly relating to age, it may be the case that young animals are better models for understanding disease specificity and progression (e.g., early genetic predisposition) whereas older animals may be better models for studying latent disease severity and manifestation. Importantly, after valve culture, there tends to be collagen accumulation (fibrotic) in some of these tissues but GAG accumulation (myxomatous) in others (Fig. 2); the mechanism during these outcomes warrants further study.

## 6 Correlating Glycosaminoglycan Levels with Tissue Biomechanics (Normal vs. Disease)

The knowledge of composition and distribution of PGs and GAGs within valves is essential for understanding and recapitulating the complex mechanics of valve leaflets [11]. During valve disease, cell-mediated ECM remodeling, often involving PGs and GAGs, can lead to gross tissue biomechanical changes. Biochemical studies on valves have well established that region-specific PG/GAG distribution is related to localized compressive loads experienced by the tissue [16]. For example, in human mitral valves, the regions that experience compression, such as the posterior leaflet and the free edge of anterior leaflet, contain significantly more water,

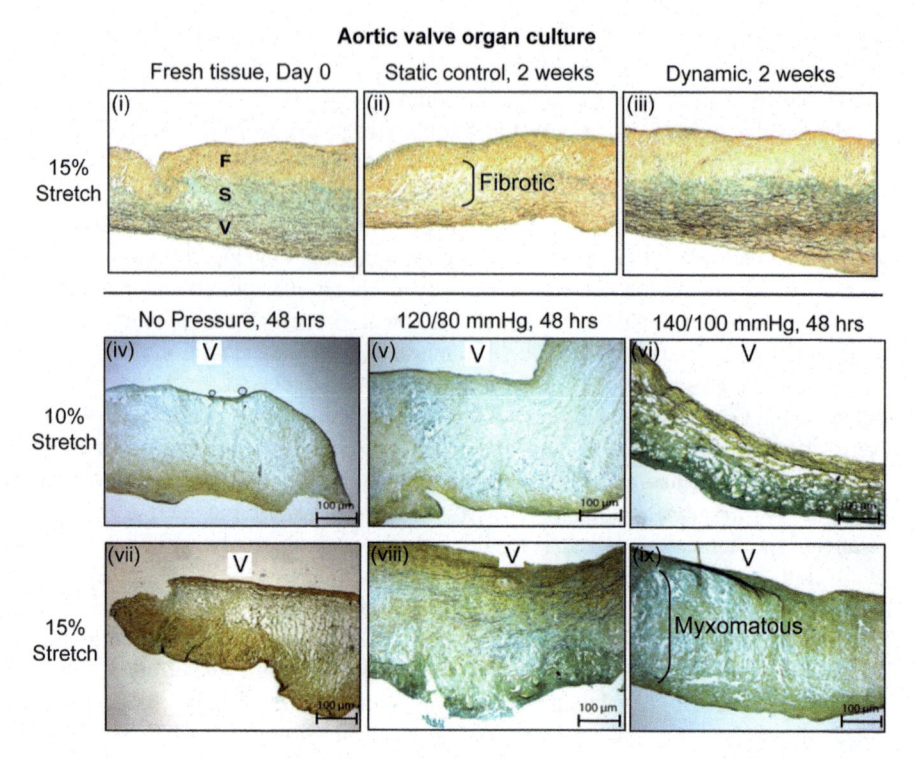

**Fig. 2** Top panels (**i–iii**): fibrotic phenotype with loss of layer stratification is observed in statically cultured aortic valves whereas the trilayered structure is maintained with the application of 15% stretch (from [55]). Bottom panels (**iv–ix**): with heightened stretch (15%) and pressure (140/100 mmHg), expansion of the spongiosa with increased GAGs is observed, whereas spongiosa narrowing occurred at 10% stretch (from [56]). *F* fibrosa, *S* spongiosa, *V* ventricularis

hyaluronan, and high 6-sulfated to 4-sulfated CS ratio with longer chain lengths of 80–90 disaccharides, indicative of versican. The regions that experience tension, such as the chordae and the central portion of anterior leaflet, contain less water, less hyaluronan, and mainly 4-sulfated DS with chain lengths of 50–70 disaccharides, indicative of decorin and biglycan.

While Grande-Allen et al. showed 80% 6-sulfated disaccharide and 20% 4-sulfated disaccharide in normal posterior mitral leaflet, in studies by Dainese et al., the posterior leaflet demonstrated higher percentage of 4-sulfated (38%) than the 6-sulfated (23%) disaccharide, suggesting discrepancies due to different analytical approaches [66]. In particular, the relative percentage of non-sulfated (hyaluronan and non-sulfated CS/DS) disaccharides and the 4-/6-sulfated disaccharides were calculated using HPLC after treatment with chondroitinase ABC, whereas Grande-Allen et al. evaluated these disaccharides by fluorophore-assisted carbohydrate electrophoresis (FACE) analysis after treatment with hyaluronidase and chondroitinase ABC or chondroitinase ACII and further drying and fluorotagging the disaccharides. Further, the age of subjects ranged from 20 to 75 years in studies by Grande-Allen et al., whereas the patient group in Dainese et al. was a narrow age range of approximately 11–13 years. Additionally, age-induced differences in human anterior and posterior mitral leaflet are known to exist; the anterior leaflet progressively becomes stiffer than posterior leaflet, and both the leaflets are stiffer at advanced age than at younger stages, consistent with PG infiltration and other ECM abnormalities [67].

The chordae tendineae prevent prolapse of the closed mitral valve into the left atrium and are thus essential for proper valve function and ventricular geometry [68]. When the chordae become diseased and elongate/rupture, valvular prolapse and regurgitation ensue. In healthy chordae, Liao and Vesely showed that the concentrations of total GAGs, unsulfated CS/DS, and 6- and 4-sulfated CS were highest in the marginal chordae, and significantly decreased in the order of marginal, basal, and strut chordae [69]. Marginal chordae were found to have a greater collagen fibril density and thus more GAG-mediated fibril-to-fibril linkages. Fibrils with fewer PG connections would thus slip more and reorganize more than would those with abundant PG connections. In other words, the greater number of PG linkages could prevent the slippage of fibrils with respect to each other, and might in turn reduce tissue stress relaxation. The thicker strut chordae were found to have more PG connections and they consequently relaxed more and faster than the thinner marginal chordae; the different viscoelastic properties of mitral valve chordae could therefore be explained morphologically.

Like the valve leaflets, the mitral valve chordae show compromised mechanical behavior during disease. While studies have shown that normal chordae contained primarily 4-sulfated DS in chains approximately 50–60 disaccharides long, myxomatous chordae had a greater proportion of 6-sulfated CS and hyaluronan, indicative of versican abundance, a reduced proportion of 4-sulfated DS, and significantly longer CS/DS chains [70]. Therefore, the GAG profile of myxomatous chordae was less like normal chordae and more like normal and myxomatous leaflets; i.e., these chordae were more extensible and less stiff with a biomechanical shift from a normal "chordal phenotype" (involving high-tensile loads) toward a "leaflet phenotype" (involving a combination of compressive and low-tensile loads) [71].

In aortic valves, the leaflet belly is associated with heightened compressive loading and the commissure is associated with tensile loading during cyclic flexure and bending. While CS and DS are the main sulfated GAGs in human aortic valves, in patients with severe aortic insufficiency, aortic valves demonstrate increased sulfated GAGs and increased 4-sulfation/6-sulfation ratio, suggesting altered hydration due to tissue pathology during disease and mechanical weakening [72]. In normal aortic valves, there were differences in 4-sulfation/6-sulfation ratio between the belly and commissure regions. Regions under tension such as the commissure had abundant 4-sulfated DS whereas regions under compressive forces, like the belly, contained hyaluronan and 6-sulfated CS. In addition, hyaluronan levels were lower in pathologic aortic valves compared to controls, especially in the leaflet attachment zones where compression was higher, which most likely lowers the ability of this region to resist compression.

The characterization of GAGs in porcine mitral and aortic valves is largely consistent with human valves. In porcine mitral leaflets, the free edge, which experiences compression, was shown to contain an abundance of water and

GAGs, particularly 6-sulfated CS and hyaluronan. In comparison, the central tensile loading region of the anterior mitral leaflet demonstrated reduced GAG content and chain length, a larger fraction of 4-sulfated iduronate, and a lower fraction of 6-sulfation [11]. With increasing animal age, there was an upregulation of PGs in the mitral valve, possibly due to increased left atrial pressure. The central loading region exhibited a dramatic increase in the ratio of 4- to 6-sulfated DS and aging accentuated the tensile tissue phenotype in mitrals overall. In aortic valves, the aging process downregulated the ratio of CS to DS, and increased the ratio of 4- to 6-sulfated DS. In other words, these valves progressively transformed from a compressive (at younger age) to a tensile phenotype (at advanced age); this may reflect the increased tensile loading on these tissues that would accompany an age-related decrease in compliance. Therefore, differences in loading patterns, as well as in geometry, may explain the differences in composition between porcine aortic and mitral valves. Recently identified porcine models of valve disease with PG abnormalities could be investigated in future, for assessment of GAG fine structure modifications and tissue biomechanical alterations [73, 74].

Given the robust insight regarding pathogenesis that transgenic manipulation provides, targeted mutagenesis mice provide an opportunity to combine engineering and molecular approaches to study PGs in valve disease. Mouse models of aortic and mitral valve diseases with postnatal PG dysregulation demonstrate abnormalities in valve tissue stiffness or systolic/diastolic function as early as 4 months of age, caused by an altered tissue collagen/GAG ratio [20, 75–77]. Further, micromechanical approaches such as atomic force microscopy and micropipette aspiration coupled with histology reveal that even wild-type mice demonstrate differences in regional valve tissue stiffness with age, with focal collagen deposition in the leaflet and PG infiltration at the valve hinge [78, 79]. While myxomatous valve phenotype with PG accumulation and valve weakening is mostly thought to be a hallmark of mitral valve disease (and not aortic valve disease, which predominantly results in fibrocalcification and valve stiffening), studies have shown that even aortic valves undergo myxomatous transformation in humans as well as mice; and PG defects have been observed in both fibrotic and myxomatous valve states [17, 80, 81]. Future studies need to combine characterization of GAG fine structure with measurement of GAG biomechanics in mouse models to determine the correlation between regulation of GAG homeostasis and valve function during different manifestations of valve disease.

## 7 Effect of Cell Stimulation on Proteoglycan and Glycosaminoglycan Response

VICs are mechanosensitive, and highly responsive to the hemodynamically active environment to which the valve tissue is normally constantly exposed, and to the altered hemodynamic state during disease [82]. VICs experience uniaxial, biaxial, or multiaxial strains in vivo; furthermore, normal and disease states exert different strain patterns. Because these cells are connected to the valve ECM via integrins, mechanical stresses/strains in the ECM result in cell mechanotransduction, and may alter the regulation of ECM gene expression and GAG homeostasis [83]. Therefore, while organ culture is important to investigate global tissue remodeling response, cell culture studies represent a more mature and detailed approach to quantify mechanistic alterations following biomechanical stimulation.

Gupta et al. examined the effect of uniaxial and biaxial static tension on PG and GAG synthesis by VICs from porcine mitral valve leaflet and chordae using a porous 3D collagen scaffold [84, 85]. The three predominant valve PGs—decorin, biglycan, and versican—were present in the collagen gels as well as the conditioned medium; however, in constrained conditions, more 4-sulfated CS/DS GAGs, decorin and biglycan but less versican were retained within the collagen gel (due to the close association between these SLRPs and collagen fibrils in the scaffold), whereas abundant hyaluronan was leaching out into the surrounding medium. The presence or absence of tensile strain and the magnitude of contraction also produced distinct effects on the VIC production of different GAG classes. Further, there was no difference in total GAGs between collagen gels seeded with chordal cells and those seeded with leaflet cells. However, constrained collagen gels containing leaflet cells retained more decorin and biglycan than did those containing chordal cells. Overall, mechanical loading of VICs influence GAG production and localization during mitral valve remodeling.

After static tensile testing, the effect of alternating cyclic stretch-relaxation on PG/GAG synthesis by VICs seeded on collagen gels was analyzed at 24 h periods for a week [86]. Cyclic stretch induced upregulation of total GAG content as well as the individual GAG classes secreted into the culture medium. Leaflet cells demonstrated altered proportions of various GAG classes they secreted during the culture duration, but showed a delayed response to stretching compared to chordal cells. In a parallel study, VICs were stretched at different strains of 2, 5, or 10% at 1.16 Hz for 48 h using a custom-built stretching device [87]. Cyclic strains reduced the total GAGs retained within collagen gels in a strain-magnitude-dependent manner for both leaflet and chordal cells. With increasing strain, interestingly, secretion of total GAGs into the medium was reduced for leaflet cells but elevated for chordal cells. Retention of 4-sulfated GAGs increased with increasing strain for both cell types; for the chordal samples, retention of 6-sulfated GAGs was reduced at higher strain. Compared to statically constrained or unconstrained conditions, the application of cyclic strain reduced the secretion of 6-sulfated GAGs by both cell types, and elevated the secretion of 4-sulfated GAGs by leaflet cells only. Retention of biglycan and secretion of decorin was significantly reduced at 10% strain compared to 2% strain.

More recently, porcine aortic VIC cultures stretched to 5% cyclic strain for 48 h using Flexcell, demonstrated heightened cell activation and sulfated GAG expression, notably in the presence of exogenous TGFβ1 and at advanced animal age [83]. Examination of the markers of hyaluronan homeostasis revealed that CD44, a hyaluronan receptor, and the ratio of hyaluronidase (HYAL) to hyaluronan synthase (HAS) enzymes, were increased with the combined strain and TGFβ1 treatment in older VICs, suggestive of hyaluronan degradation. These findings were consistent with human valve disease pathology and established that valve hyaluronan homeostasis was affected by VIC biomechanical stimulation. Overall, these results affirm that mechanical strain conditions regulate GAG and PG production by VICs in vitro.

In the context of tissue engineered heart valves, Engelmeyr et al. evaluated the effect of unidirectional cyclic three point flexure (physiological frequency, amplitude) on ovine vascular smooth muscle cells that were seeded on polyglycolic acid and poly-L-lactic acid (PGA-PLLA) scaffolds, to estimate ECM remodeling [88]. At the 30 h time point, there was significant secretion of sulfated GAG but not collagen or elastin, suggesting that sulfated GAG secretion precedes collagen and elastin synthesis in tissue engineered valve development, potentially similar to that observed during fetal semilunar valve development. After 3 week incubation, sulfated GAG concentration decreased drastically. Interestingly, there was not much difference between the static and flex groups after 3 weeks. In sum, studies evaluating the effect of biomechanical factors such as cyclic mechanical strain or fluid flow (pulsatile or steady state shear) on 2D or 3D valve cell culture are important for understanding cellular response and mechanisms of GAG synthesis or degradation; these have profound implications in ECM remodeling during normal and disease states, and for tissue engineering applications.

## 8 Summary

Historically, GAGs are the least studied ECM component in valve tissue. Furthermore, the biomechanics of the valve spongiosa layer, of which GAGs are a dominant component, has not been as thoroughly explored as have the fibrosa or ventricularis layers. More recently, however, there is growing interest in developing a combination of experimental and theoretical approaches to estimate the complexity of GAGs, their contribution to the valve structure-function relationships, and their molecular interactions with distinct ECM and cellular components, and the mechanical environment. These GAG studies utilize a wide range of approaches employing cultured cells, native or fixed valve tissues, tissue engineered scaffolds, and simple through complex bioreactor systems to apply mechanical stimulation. The experiments provided great insight into GAG function and biomechanics of GAG dysregulation during valve remodeling in native as well as tissue engineered valves. In the future, mechanisms that control GAG synthesis and degradation need to be elucidated in organ cultured valves; this information is important for understanding pathways that control fibrotic (stiff valve) versus myxomatous (floppy valve) phenotypes in normal and disease states. In addition, cocultures of VIC and valve endothelial cells better mimic the in vivo valve cell milieu and may offer an alternative to standard VIC cultures for investigating the effect of mechanotransduction on GAG homeostasis. Further, determination of alterations in GAG fine structure together with GAG biomechanics in animal models of valve disease during the origin, onset, progression, and overt manifestation of disease will be valuable in the development of novel therapeutics and durable tissue prostheses. Finally, examining the association between GAG biomechanics, composition, and mechanisms is common to various tissues, and their results may be generalizable, which could impact other areas of research.

**Acknowledgements** The authors would like to thanks Dr. Jennifer Petsche Connell and Andy Zhang for assistance in the preparation of this chapter.

# References

1. Latif N, Sarathchandra P, Taylor PM, Antoniw J, Yacoub MH. Localization and pattern of expression of extracellular matrix components in human heart valves. J Heart Valve Dis. 2005;14:218–27.
2. Sacks MS, David Merryman W, Schmidt DE. On the biomechanics of heart valve function. J Biomech. 2009;42:1804–24.
3. Gupta V, Barzilla JE, Mendez JS, Stephens EH, Lee EL, Collard CD, Laucirica R, Weigel PH, Grande-Allen KJ. Abundance and location of proteoglycans and hyaluronan within normal and myxomatous mitral valves. Cardiovasc Pathol. 2009;18:191–7.
4. Hinton RB Jr, Lincoln J, Deutsch GH, Osinska H, Manning PB, Benson DW, Yutzey KE. Extracellular matrix remodeling and organization in developing and diseased aortic valves. Circ Res. 2006;98:1431–8.
5. Lockhart M, Wirrig E, Phelps A, Wessels A. Extracellular matrix and heart development. Birth Defects Res A Clin Mol Teratol. 2011;91:535–50.
6. Taylor PM, Batten P, Brand NJ, Thomas PS, Yacoub MH. The cardiac valve interstitial cell. Int J Biochem Cell Biol. 2003;35:113–8.
7. Iozzo RV, Goldoni S, Berendsen A, Young MF. Small leucine-rich proteoglycans. In: Mecham RP, editor. Extracellular matrix: an overview, biology of extracellular matrix. Berlin Heidelberg: Springer; 2011.
8. Wight TN, Toole BP, Hascall VC. Hyaluronan and the aggregating proteoglycans. In: Mecham RP, editor. Extracellular matrix: an overview, biology of extracellular matrix. Berlin Heidelberg: Springer; 2011.
9. Esko JD, Kimata K, Lindahl U. Proteoglycans and sulfated glycosaminoglycans. In: Varki A, Cummings RD, Esko JD, et al., editors. Essentials of glycobiology. 2nd ed. Cold Spring Harbor: Cold Spring Harbor Laboratory Press; 2009.
10. Nkomo VT, Gardin JM, Skelton TN, Gottdiener JS, Scott CG, Enriquez-Sarano M. Burden of valvular heart diseases: a population-based study. Lancet. 2006;368:1005–11.
11. Stephens EH, Chu CK, Grande-Allen KJ. Valve proteoglycan content and glycosaminoglycan fine structure are unique to microstructure, mechanical load and age: relevance to an age-specific tissue-engineered heart valve. Acta Biomater. 2008;4:1148–60.
12. Chang Y, Yanagishita M, Hascall VC, Wight TN. Proteoglycans synthesized by smooth muscle cells derived from monkey (Macaca nemestrina) aorta. J Biol Chem. 1983;258:5679–88.
13. Schonherr E, Jarvelainen HT, Sandell LJ, Wight TN. Effects of platelet-derived growth factor and transforming growth factor-beta 1 on the synthesis of a large versican-like chondroitin sulfate proteoglycan by arterial smooth muscle cells. J Biol Chem. 1991;266:17640–7.
14. Zimmermann DR, Ruoslahti E. Multiple domains of the large fibroblast proteoglycan, versican. EMBO J. 1989;8:2975–81.
15. Rabkin E, Aikawa M, Stone JR, Fukumoto Y, Libby P, Schoen FJ. Activated interstitial myofibroblasts express catabolic enzymes and mediate matrix remodeling in myxomatous heart valves. Circulation. 2001;104:2525–32.

16. Grande-Allen KJ, Calabro A, Gupta V, Wight TN, Hascall VC, Vesely I. Glycosaminoglycans and proteoglycans in normal mitral valve leaflets and chordae: association with regions of tensile and compressive loading. Glycobiology. 2004;14:621–33.
17. Krishnamurthy VK, Godby RC, Liu GR, Smith JM, Hiratzka LF, Narmoneva DA, Hinton RB. Review of molecular and mechanical interactions in the aortic valve and aorta: implications for the shared pathogenesis of aortic valve disease and aortopathy. J Cardiovasc Transl Res. 2014;7:823–46.
18. Hinek A, Rabinovitch M. 67-kD elastin-binding protein is a protective "companion" of extracellular insoluble elastin and intracellular tropoelastin. J Cell Biol. 1994;126:563–74.
19. Wu YJ, La Pierre DP, Wu J, Yee AJ, Yang BB. The interaction of versican with its binding partners. Cell Res. 2005;15:483–94.
20. Krishnamurthy VK, Opoka AM, Kern CB, Guilak F, Narmoneva DA, Hinton RB. Maladaptive matrix remodeling and regional biomechanical dysfunction in a mouse model of aortic valve disease. Matrix Biol. 2012;31:197–205.
21. Tseng H, Grande-Allen KJ. Elastic fibers in the aortic valve spongiosa: a fresh perspective on its structure and role in overall tissue function. Acta Biomater. 2011;7:2101–8.
22. Reinboth B, Hanssen E, Cleary EG, Gibson MA. Molecular interactions of biglycan and decorin with elastic fiber components: biglycan forms a ternary complex with tropoelastin and microfibril-associated glycoprotein 1. J Biol Chem. 2002;277:3950–7.
23. Dupuis LE, Kern CB. Small leucine-rich proteoglycans exhibit unique spatiotemporal expression profiles during cardiac valve development. Dev Dyn. 2014;243:601–11.
24. Hwang JY, Johnson PY, Braun KR, Hinek A, Fischer JW, O'Brien KD, Starcher B, Clowes AW, Merrilees MJ, Wight TN. Retrovirally mediated overexpression of glycosaminoglycan-deficient biglycan in arterial smooth muscle cells induces tropoelastin synthesis and elastic fiber formation in vitro and in neointimae after vascular injury. Am J Pathol. 2008;173:1919–28.
25. Geng Y, McQuillan D, Roughley PJ. SLRP interaction can protect collagen fibrils from cleavage by collagenases. Matrix Biol. 2006;25:484–91.
26. Kalamajski S, Oldberg A. The role of small leucine-rich proteoglycans in collagen fibrillogenesis. Matrix Biol. 2010;29:248–53.
27. Redaelli A, Vesentini S, Soncini M, Vena P, Mantero S, Montevecchi FM. Possible role of decorin glycosaminoglycans in fibril to fibril force transfer in relative mature tendons--a computational study from molecular to microstructural level. J Biomech. 2003;36:1555–69.
28. Zhang G, Chen S, Goldoni S, Calder BW, Simpson HC, Owens RT, McQuillan DJ, Young MF, Iozzo RV, Birk DE. Genetic evidence for the coordinated regulation of collagen fibrillogenesis in the cornea by decorin and biglycan. J Biol Chem. 2009;284:8888–97.
29. Zhang G, Ezura Y, Chervoneva I, Robinson PS, Beason DP, Carine ET, Soslowsky LJ, Iozzo RV, Birk DE. Decorin regulates assembly of collagen fibrils and acquisition of biomechanical properties during tendon development. J Cell Biochem. 2006;98:1436–49.
30. Dupuis LE, Doucette L, Rice AK, Lancaster AE, Berger MG, Chakravarti S, Kern CB. Development of myotendinous-like junctions that anchor cardiac valves requires fibromodulin and lumican. Dev Dyn. 2016;245:1029–42.
31. Manji RA, Menkis AH, Ekser B, Cooper DK. Porcine bioprosthetic heart valves: the next generation. Am Heart J. 2012;164:177–85.
32. Schoen FJ. Evolving concepts of cardiac valve dynamics: the continuum of development, functional structure, pathobiology, and tissue engineering. Circulation. 2008;118:1864–80.
33. Siddiqui RF, Abraham JR, Butany J. Bioprosthetic heart valves: modes of failure. Histopathology. 2009;55:135–44.
34. Zilla P, Brink J, Human P, Bezuidenhout D. Prosthetic heart valves: catering for the few. Biomaterials. 2008;29:385–406.
35. Lovekamp JJ, Simionescu DT, Mercuri JJ, Zubiate B, Sacks MS, Vyavahare NR. Stability and function of glycosaminoglycans in porcine bioprosthetic heart valves. Biomaterials. 2006;27:1507–18.

36. Vyavahare N, Ogle M, Schoen FJ, Zand R, Gloeckner DC, Sacks M, Levy RJ. Mechanisms of bioprosthetic heart valve failure: fatigue causes collagen denaturation and glycosaminoglycan loss. J Biomed Mater Res. 1999;46:44–50.
37. Shah SR, Vyavahare NR. The effect of glycosaminoglycan stabilization on tissue buckling in bioprosthetic heart valves. Biomaterials. 2008;29:1645–53.
38. Simionescu DT, Lovekamp JJ, Vyavahare NR. Glycosaminoglycan-degrading enzymes in porcine aortic heart valves: implications for bioprosthetic heart valve degeneration. J Heart Valve Dis. 2003;12:217–25.
39. Grande-Allen KJ, Mako WJ, Calabro A, Shi Y, Ratliff NB, Vesely I. Loss of chondroitin 6-sulfate and hyaluronan from failed porcine bioprosthetic valves. J Biomed Mater Res A. 2003;65:251–9.
40. Friebe VM, Mikulis B, Kole S, Ruffing CS, Sacks MS, Vyavahare NR. Neomycin enhances extracellular matrix stability of glutaraldehyde crosslinked bioprosthetic heart valves. J Biomed Mater Res B Appl Biomater. 2011;99:217–29.
41. Tam H, Zhang W, Feaver KR, Parchment N, Sacks MS, Vyavahare N. A novel crosslinking method for improved tear resistance and biocompatibility of tissue based biomaterials. Biomaterials. 2015;66:83–91.
42. Tripi DR, Vyavahare NR. Neomycin and pentagalloyl glucose enhanced cross-linking for elastin and glycosaminoglycans preservation in bioprosthetic heart valves. J Biomater Appl. 2014;28:757–66.
43. Stella JA, Sacks MS. On the biaxial mechanical properties of the layers of the aortic valve leaflet. J Biomech Eng. 2007;129:757–66.
44. Thornton GM, Frank CB, Shrive NG. Ligament creep behavior can be predicted from stress relaxation by incorporating fiber recruitment. J Rheol. 2001;45:493–507.
45. Buchanan RM, Sacks MS. Interlayer micromechanics of the aortic heart valve leaflet. Biomech Model Mechanobiol. 2014;13:813–26.
46. Roccabianca S, Ateshian GA, Humphrey JD. Biomechanical roles of medial pooling of glycosaminoglycans in thoracic aortic dissection. Biomech Model Mechanobiol. 2014;13:13–25.
47. Schmidt MB, Mow VC, Chun LE, Eyre DR. Effects of proteoglycan extraction on the tensile behavior of articular cartilage. J Orthop Res. 1990;8:353–63.
48. Lincoln J, Lange AW, Yutzey KE. Hearts and bones: shared regulatory mechanisms in heart valve, cartilage, tendon, and bone development. Dev Biol. 2006;294:292–302.
49. Bhatia A, Vesely I. The effect of glycosaminoglycans and hydration on the viscoelastic properties of aortic valve cusps. Conf Proc IEEE Eng Med Biol Soc. 2005;3:2979–80.
50. Eckert CE, Fan R, Mikulis B, Barron M, Carruthers CA, Friebe VM, Vyavahare NR, Sacks MS. On the biomechanical role of glycosaminoglycans in the aortic heart valve leaflet. Acta Biomater. 2013;9:4653–60.
51. Borghi A, New SE, Chester AH, Taylor PM, Yacoub MH. Time-dependent mechanical properties of aortic valve cusps: effect of glycosaminoglycan depletion. Acta Biomater. 2013;9:4645–52.
52. Tseng H, Kim EJ, Connell PS, Ayoub S, Shah JV, Grande-Allen KJ. The tensile and viscoelastic properties of aortic valve leaflets treated with a hyaluronidase gradient. Cardiovasc. Eng Technol 2013;4(2):151–160. https://doi.org/10.1007/s13239-013-0122-1.
53. Talman EA, Boughner DR. Effect of altered hydration on the internal shear properties of porcine aortic valve cusps. Ann Thorac Surg. 2001;71:S375–8.
54. Xing Y, Warnock JN, He Z, Hilbert SL, Yoganathan AP. Cyclic pressure affects the biological properties of porcine aortic valve leaflets in a magnitude and frequency dependent manner. Ann Biomed Eng. 2004;32:1461–70.
55. Merryman WD, Lukoff HD, Long RA, Engelmayr GC Jr, Hopkins RA, Sacks MS. Synergistic effects of cyclic tension and transforming growth factor-beta1 on the aortic valve myofibroblast. Cardiovasc Pathol. 2007;16:268–76.

56. Thayer P, Balachandran K, Rathan S, Yap CH, Arjunon S, Jo H, Yoganathan AP. The effects of combined cyclic stretch and pressure on the aortic valve interstitial cell phenotype. Ann Biomed Eng. 2011;39:1654–67.
57. Barzilla JE, McKenney AS, Cowan AE, Durst CA, Grande-Allen KJ. Design and validation of a novel splashing bioreactor system for use in mitral valve organ culture. Ann Biomed Eng. 2010;38:3280–94.
58. Balachandran K, Bakay MA, Connolly JM, Zhang X, Yoganathan AP, Levy RJ. Aortic valve cyclic stretch causes increased remodeling activity and enhanced serotonin receptor responsiveness. Ann Thorac Surg. 2011;92:147–53.
59. Lacerda CM, Kisiday J, Johnson B, Orton EC. Local serotonin mediates cyclic strain-induced phenotype transformation, matrix degradation, and glycosaminoglycan synthesis in cultured sheep mitral valves. Am J Physiol Heart Circ Physiol. 2012;302:H1983–90.
60. Connell PS, Azimuddin AF, Kim SE, Ramirez F, Jackson MS, Little SH, Grande-Allen KJ. Regurgitation hemodynamics alone cause mitral valve remodeling characteristic of clinical disease states in vitro. Ann Biomed Eng. 2016;44(4):954–67. https://doi.org/10.1007/s10439-015-1398-0.
61. Sapp MC, Krishnamurthy VK, Puperi DS, Bhatnagar S, Fatora G, Mutyala N, Grande-Allen KJ. Differential cell-matrix responses in hypoxia-stimulated aortic versus mitral valves. J R Soc Interface. 2016;13(125):20160449.
62. Connolly JM, Bakay MA, Fulmer JT, Gorman RC, Gorman JH 3rd, Oyama MA, Levy RJ. Fenfluramine disrupts the mitral valve interstitial cell response to serotonin. Am J Pathol. 2009;175:988–97.
63. Hafizi S, Taylor PM, Chester AH, Allen SP, Yacoub MH. Mitogenic and secretory responses of human valve interstitial cells to vasoactive agents. J Heart Valve Dis. 2000;9:454–8.
64. Jian B, Xu J, Connolly J, Savani RC, Narula N, Liang B, Levy RJ. Serotonin mechanisms in heart valve disease I: serotonin-induced up-regulation of transforming growth factor-beta1 via G-protein signal transduction in aortic valve interstitial cells. Am J Pathol. 2002;161:2111–21.
65. Rajamannan NM, Caplice N, Anthikad F, Sebo TJ, Orszulak TA, Edwards WD, Tajik J, Schwartz RS. Cell proliferation in carcinoid valve disease: a mechanism for serotonin effects. J Heart Valve Dis. 2001;10:827–31.
66. Dainese L, Polvani G, Barili F, Maccari F, Guarino A, Alamanni F, Zanobini M, Biglioli P, Volpi N. Fine characterization of mitral valve glycosaminoglycans and their modification with degenerative disease. Clin Chem Lab Med. 2007;45:361–6.
67. Pham T, Sun W. Material properties of aged human mitral valve leaflets. J Biomed Mater Res A. 2014;102:2692–703.
68. Obadia JF, Casali C, Chassignolle JF, Janier M. Mitral subvalvular apparatus: different functions of primary and secondary chordae. Circulation. 1997;96:3124–8.
69. Liao J, Vesely I. Relationship between collagen fibrils, glycosaminoglycans, and stress relaxation in mitral valve chordae tendineae. Ann Biomed Eng. 2004;32:977–83.
70. Grande-Allen KJ, Griffin BP, Ratliff NB, Cosgrove DM, Vesely I. Glycosaminoglycan profiles of myxomatous mitral leaflets and chordae parallel the severity of mechanical alterations. J Am Coll Cardiol. 2003;42:271–7.
71. Lincoln J, Alfieri CM, Yutzey KE. BMP and FGF regulatory pathways control cell lineage diversification of heart valve precursor cells. Dev Biol. 2006;292:292–302.
72. Dainese L, Guarino A, Micheli B, Biagioli V, Polvani G, Maccari F, Volpi N. Aortic valve leaflet glycosaminoglycans composition and modification in severe chronic valve regurgitation. J Heart Valve Dis. 2013;22:484–90.
73. Porras AM, Shanmuganayagam D, Meudt JJ, Krueger CG, Hacker TA, Rahko PS, Reed JD, Masters KS. Development of aortic valve disease in familial hypercholesterolemic swine: implications for elucidating disease etiology. J Am Heart Assoc. 2015;4:e002254.
74. Sider KL, Zhu C, Kwong AV, Mirzaei Z, de Lange CF, Simmons CA. Evaluation of a porcine model of early aortic valve sclerosis. Cardiovasc Pathol. 2014;23:289–97.

75. Gould RA, Sinha R, Aziz H, Rouf R, Dietz HC 3rd, Judge DP, Butcher J. Multi-scale biomechanical remodeling in aging and genetic mutant murine mitral valve leaflets: insights into Marfan syndrome. PLoS One. 2012;7:e44639.
76. Hinton RB, Adelman-Brown J, Witt S, Krishnamurthy VK, Osinska H, Sakthivel B, James JF, Li DY, Narmoneva DA, Mecham RP, Benson DW. Elastin haploinsufficiency results in progressive aortic valve malformation and latent valve disease in a mouse model. Circ Res. 2010;107:549–57.
77. Wilson CL, Gough PJ, Chang CA, Chan CK, Frey JM, Liu Y, Braun KR, Chin MT, Wight TN, Raines EW. Endothelial deletion of ADAM17 in mice results in defective remodeling of the semilunar valves and cardiac dysfunction in adults. Mech Dev. 2013;130:272–89.
78. Krishnamurthy VK, Guilak F, Narmoneva DA, Hinton RB. Regional structure-function relationships in mouse aortic valve tissue. J Biomech. 2011;44:77–83.
79. Sewell-Loftin MK, Brown CB, Baldwin HS, Merryman WD. A novel technique for quantifying mouse heart valve leaflet stiffness with atomic force microscopy. J Heart Valve Dis. 2012;21:513–20.
80. Roberts WC, Honig HS. The spectrum of cardiovascular disease in the Marfan syndrome: a clinico-morphologic study of 18 necropsy patients and comparison to 151 previously reported necropsy patients. Am Heart J. 1982;104:115–35.
81. Stephens EH, Saltarrelli JG Jr, Balaoing LR, Baggett LS, Nandi I, Anderson KM, Morrisett JD, Reardon MJ, Simpson MA, Weigel PH, Olmsted-Davis EA, Davis AR, Grande-Allen KJ. Hyaluronan turnover and hypoxic brown adipocytic differentiation are co-localized with ossification in calcified human aortic valves. Pathol Res Pract. 2012;208:642–50.
82. Liu CA, Joag VR, Gotlieb AI. The emerging role of valve interstitial cell phenotypes in regulating heart valve pathobiology. Am J Pathol. 2007;171:1407–18.
83. Krishnamurthy VK, Stout AJ, Sapp MC, Matuska B, Lauer ME, Grande-Allen KJ. Dysregulation of hyaluronan homeostasis during aortic valve disease. Matrix Biol. 2017;62:40–57.
84. Gupta V, Werdenberg JA, Blevins TL, Grande-Allen KJ. Synthesis of glycosaminoglycans in differently loaded regions of collagen gels seeded with valvular interstitial cells. Tissue Eng. 2007;13:41–9.
85. Gupta V, Werdenberg JA, Mendez JS, Jane Grande-Allen K. Influence of strain on proteoglycan synthesis by valvular interstitial cells in three-dimensional culture. Acta Biomater. 2008;4:88–96.
86. Gupta V, Werdenberg JA, Lawrence BD, Mendez JS, Stephens EH, Grande-Allen KJ. Reversible secretion of glycosaminoglycans and proteoglycans by cyclically stretched valvular cells in 3D culture. Ann Biomed Eng. 2008;36:1092–103.
87. Gupta V, Tseng H, Lawrence BD, Grande-Allen KJ. Effect of cyclic mechanical strain on glycosaminoglycan and proteoglycan synthesis by heart valve cells. Acta Biomater. 2009;5:531–40.
88. Engelmayr GC Jr, Rabkin E, Sutherland FW, Schoen FJ, Mayer JE Jr, Sacks MS. The independent role of cyclic flexure in the early in vitro development of an engineered heart valve tissue. Biomaterials. 2005;26:175–87.

# On the Unique Functional Elasticity and Collagen Fiber Kinematics of Heart Valve Leaflets

**Jun Liao and Michael S. Sacks**

**Abstract** With the growing prevalence of heart valve diseases, it is important to better understand the biomechanical behavior of normal and pathological heart valve tissues. Recent studies showed that heart valve leaflets exhibited a unique functionally elastic behavior, in which valvular tissues exhibited minimal hysteretic and creep behaviors under biaxial loading, yet allowed stress relaxation similar to other types of collagenous tissues. This unique behavior is in favor of heart valve function, enabling leaflets to bear peak physiological loading without time-dependent deformation. To explore the underlying micromechanical mechanisms, we used small angle X-ray scattering (SAXS) under biaxial stretch to explore the collagen fibril kinematics in stress relaxation and creep. We found that collagen fibril reorientation/realignment did not contribute to stress relaxation and creep. In stress relaxation, collagen fibril strain released largely during the first 20 min and the remaining collagen fibril strain stayed relatively constant in the remaining relaxation time. The overall reduction rate of the collagen fibril strain was much larger than the stress decay rate at the tissue level. When the leaflet tissue experienced negligible time-dependent deformation under constant load (negligible creep), the collagen fibril strain was maintained at a constant level during the time course. This difference in collagen fibril kinematics implies the mechanisms responsible for creep and stress relaxation in the leaflet tissue are functionally independent. We thus speculate some type of fibril-level "locking" mechanism exists in leaflet tissue that allows for stress release under constant strain condition, yet does not allow for continued straining under a constant stress. We speculate that the degenerated ECM components in

---

J. Liao (✉)
Tissue Biomechanics and Bioengineering Laboratory, The Department of Bioengineering, The University of Texas at Arlington, Arlington, TX, USA
e-mail: jun.liao@uta.edu

M. S. Sacks
The Oden Institute and the Department of Biomedical Engineering, The University of Texas at Austin, Austin, TX, USA
e-mail: msacks@ices.utexas.edu

diseased valvular tissues might cause changes in these quasi-elastic behaviors and thus contribute to malfunction of heart valves.

**Keywords** Heart valves · Valve leaflets · Viscoelasticity · Stress relaxation · Creep · Negligible creep · Collagen fibril · *D*-period · Collagen fibril kinematics · Small angle X-ray scattering · Biaxial stretching system · Quasilinear viscoelastic model · Quasi-elastic behavior

# 1 Introduction

## 1.1 Function of Heart Valve Leaflets

Heart valves are the essential apparatuses in heart function and blood circulation. Four valves (i.e., mitral valve, tricuspid valve, aorta valve, and pulmonary valve) accurately coordinate with the sequential contraction and relaxation of heart muscles, thus ensuring the unidirectional flow of blood among four heart chambers [1]. Often referred to as atrioventricular valves, the mitral valve and the tricuspid valve are located between the atria and ventricles and guide the blood from the left atrium to the left ventricle and from the right atrium to the right ventricle, respectively. Of the two semilunar valves, the aortic valve is situated between the left ventricle and the aorta and directs blood flow from the left ventricle to the aorta, while the pulmonary valve, located between the right ventricle and the pulmonary artery, guides blood flow from the right ventricle to the pulmonary artery.

As the heart pumps about 7000 L of blood each day, the heart valves coordinate the contraction and relaxation of the four heart chambers by opening and closing in a precise, sequential manner for ~$3 \times 10^9$ times (nonstop cycles) during a person's lifetime [2, 3]. Looking closely at the mechanical function of the heart valves, one can notice the very demanding working loads the heart valves experience in response to blood pressure changes in the upstream and downstream locations. The three leaflets of the pulmonary and aortic valves, the anterior, septal, and posterior leaflets of the tricuspid valve, and the anterior and posterior leaflets of the mitral valve all experience multiple loading states and large deformations [2–6]. The loading states include multidirectional tension due to backward pressure gradients during the closing phase, as well as bending and shearing due to forward pressure gradients and blood flow during the opening phase [2–4, 7, 8].

## 1.2 Composition and Ultrastructure and Heart Valve Leaflets

To cope with the complicated loading states and large deformation, as well as minimize internal structural damage/fatigue, the valve leaflets have adapted unique, optimal tissue composition and ultrastructure [1–3, 6, 9]. Three extracellular matrices (ECM), i.e., collagen, elastin, and proteoglycans (PGs), are the major structural components of the four types of heart valve leaflets, which all exhibit layered structure. The optimal, natural designs of leaflet tissues are reflected by the amount of collagen, elastin, and PGs, as well as the network configuration of collagen fibers and elastin fibers in each layer [1–3, 10]. From a structural-functional perspective, the morphology, layered structure, composition, and ultrastructure of heart valves are most likely correlated to their anatomic configuration, mechanical loading, and boundary conditions [7, 11, 12]. It is hence understandable that the mitral valve and tricuspid valve leaflets share similarities while the aortic valve and pulmonary valve leaflets share similarities [1].

The mitral valve and tricuspid valve have scallop-shaped leaflets which attach to the heart wall via a fibrous annulus and connect to the ventricular wall via the supporting chordae tendineae system/papillary muscles [12, 13]. The mitral valve has two leaflets, a large anterior leaflet and a small posterior leaflet; the tricuspid valve has a large anterior leaflet, a small posterior leaflet, and a septal leaflet with size in-between [14, 15]. The layered structure in the mitral valve and tricuspid valve are similar. A thin atrialis layer mainly composed of elastin fibers faces the atrial chamber; a thick fibrosa layer consisting of dense collagen fibers faces the

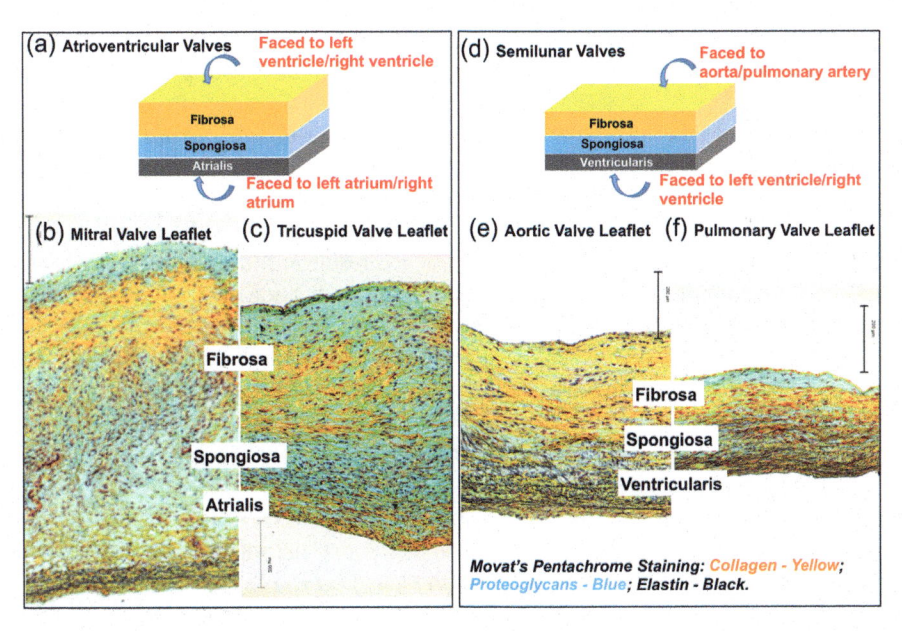

Fig. 1 Layered structure of four types of heart valves. (a) Schematic illustration of layered structure of atrioventricular valves. Movat's pentachrome staining shows the fibrosa, spongiosa, and atrialis layers of mitral valve leaflet (b) and tricuspid valve leaflet (c). (d) Schematic illustration of layered structure of semilunar valves. Movat's pentachrome staining shows the fibrosa, spongiosa, and ventricularis layers of aortic valve leaflet (e) and pulmonary valve leaflet (f). Scale bar = 200 μm. Note that the average thickness of leaflets decreases in the order of mitral valve, tricuspid valve, aortic valve, and pulmonary valve [6]

ventricular chamber; and a spongiosa layer made of mainly PGs serves as a cushion between the atrialis and the fibrosa (Fig. 1) [13, 15–19]. Both the aortic and pulmonary valves have three semilunar-shaped leaflets with similar size, while the overall size and thickness of the pulmonary valve leaflets are slightly smaller and thinner compared to aortic valve leaflets. Both aortic and pulmonary valve leaflets share a similar trilayered structure. The ventricularis faces the ventricular chambers and consists of mainly radially aligned elastin fibers. The fibrosa layer faces the aorta/pulmonary artery and is dominantly composed of Type I collagen fibers which have a circumferential orientation. Between the ventricularis and the fibrosa is the spongiosa layer, which is largely made of PGs and has a distinguishing cushion layer morphology (Fig. 1) [9, 20, 21].

## 1.3 Valvular Diseases and Viscoelastic Properties of Leaflet Tissues

Statistics from the American Heart Association show that more than 20,000 people died from valvular heart disease in 2009, and overall valvular heart disease affects 2.5% of the American population, with the prevalence increasing along with age [22, 23]. Valve stenosis and valve regurgitation are two common valvular diseases. Valve stenosis restricts blood flow due to the narrowing of the valve opening, while valve regurgitation causes leakage of blood through the leaflets when the valve leaflets cannot close correctly [22, 24]. The pathological causes of valvular diseases can be aging, congenital defects, infective endocarditis, rheumatic fever, Marfan syndrome, etc. [25–27]. For each pathological cause, valve tissue composition and ultrastructure are altered accordingly. Those compositional and ultrastructural alterations change the mechanical behavior of valve tissues, which result in the malfunction of heart valves. For example, fibrotic tissue formation and calcification of valve leaflets cause the thickening and stiffening of valve tissues and are often the underlying reason of valve stenosis [28]. As another example, valve regurgitation can result from the myxoid degeneration of valve tissue, in which valve tissue experiences weakening due to the excessive accumulation of PGs/glycosaminoglycans (GAGs) and the disruption of the collagen fiber network [13, 29–31].

With the high prevalence of heart valve diseases as a concern, it is important to better understand the biomechanical behavior of heart valves, identify the structural-functional relations of the valvular tissues, and reveal the intrinsic cause-effect of heart valve malfunction at various hierarchical levels [4, 32–35]. To achieve those understandings, one must obtain the accurate and thorough description of the mechanical behavior of native valvular tissue, reveal the intriguing contributions of various ECMs at different hierarchical levels, and comprehend the effectiveness and efficiency of natural designs. As we mentioned above, those natural designs allow the optimal performance of heart valves under complicated boundary conditions, as well as provide durability for heart valve tissues that are able to withstand ~3 billion opening and closing cycles per lifetime.

With those biomechanical understandings of the native heart valve tissues as a foundation, the inferior performance of the diseased valvular tissues can be traced down to the pathological alterations in ECM composition and ultrastructure, and the factors that cause the fatigue-prone nature of bioprosthetic heart valves can be better identified. Furthermore, the tissue-specific structural-mechanical behavior, as well as its underlying structural mechanism, sets goals/standards for valve surgical repairing, next-generation bioprosthetic valve design, and heart valve tissue engineering. In this chapter, we report our findings on the unique viscoelasticity of heart valve leaflets, the essence of this behavior in valve functionality, and the underlying collagen fibril level kinematics.

## 2 Conventional Concepts on Viscoelasticity of Soft Tissues

Soft tissues, such as skin, tendon/ligament, blood vessel, and fascia, are inherently viscoelastic due to their constituents, which are always cells, scaffolding materials (solid materials), and water [36]. The solid materials consist of collagen, elastin, and other small structural proteins such as laminin and fibronectin. Water, the fluid material, is often mediated by proteoglycans (i.e., ground substance) which consist of highly negatively charged glycosaminoglycan chains capable of attracting and entrapping water molecules. Hence, the biomechanical behavior of soft tissues exhibits viscoelastic properties, of which the elastic behavior can be attributed to solid compositions and the viscous behavior to water. The three major characteristics of viscoelastic materials are (1) strain rate sensitivity, (2) stress relaxation, and (3) creep. Those three characteristic behaviors are observations of experimentalists during their efforts to capture the full mechanical behavior of viscoelastic materials.

Briefly, if soft tissues have a stiffer stress-strain behavior when loaded at a higher deformation rate, the soft tissues are strain rate sensitive. Stress relaxation is always observed in soft tissues; the soft tissues show a time-dependent stress (load) decay when loaded and held at constant deformation. Many studies on soft tissues such as skin, carotid artery, and tendon/ligament have shown that creep, a time-dependent deformation if the soft tissues are under the action of a constant load, is always accompanying the stress relaxation behavior of those soft tissues [36]. This coexisting of stress relaxation and creep gives a symmetrical appearance of those two viscoelastic characteristics (Fig. 2a, b). The coexisting of stress relaxation and creep behavior was not only observed experimentally, but also described by theoretical models such as linear viscoelasticity theory, in which creep can be predicted from the stress relaxation [36].

In-depth studies have been carried out to look into the nonlinear nature of the viscoelasticity of soft tissues. For instance, both Provenzano et al. [37] and Thornton et al. [38] reported the coexisting of stress relaxation and creep in ligament tissues, while relaxation proceeds faster than creep—a phenomenon that cannot be explained by linear viscoelasticity. Lakes et al. [39] and

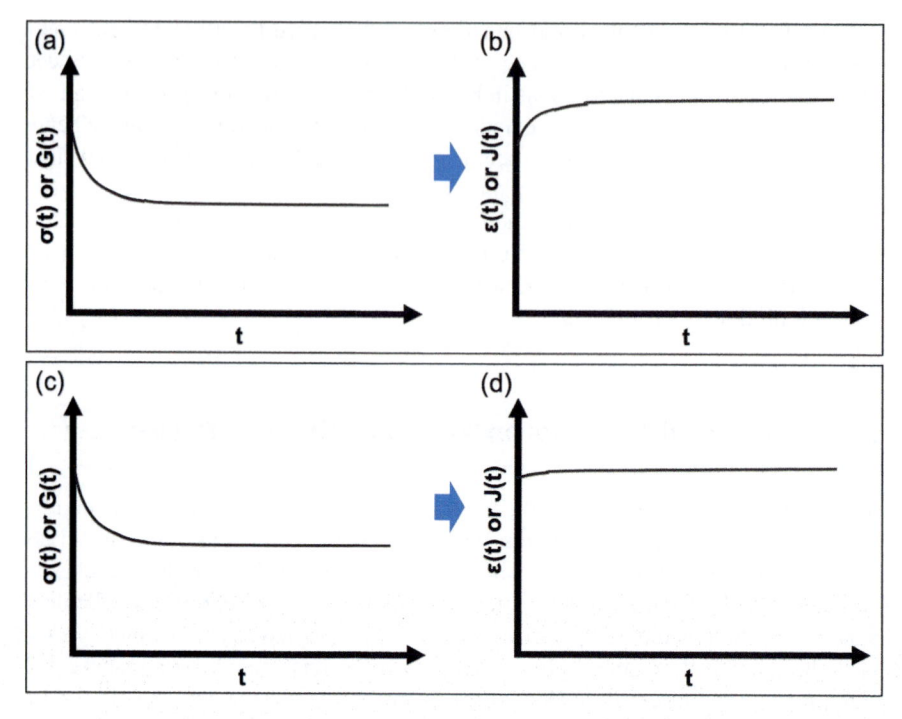

**Fig. 2** Schematic illustration of typical stress relaxation behavior (**a**) and creep behavior (**b**) for soft tissues such as skin, carotid artery, and tendons/ligaments. The coexisting of stress relaxation and creep behavior was not only observed experimentally, but also described by theoretical model such as linear viscoelasticity theory, in which creep can be predicted from the stress relaxation. Schematic illustration of stress relaxation behavior (**c**) and creep behavior (**d**) of heart valve leaflets. Heart valve leaflets show significant stress relaxation similar to other types of soft tissues, but exhibit negligible creep behavior when subjected to a constant load. Note that $X$ axis represents time ($t$); $Y$ axis for stress relaxation is often presented as stress $\sigma(t)$ or normalized stress relaxation $G(t)$, in which all values were normalized to the initial stress at time zero. $Y$ axis for creep is often presented as strain $\varepsilon(t)$ or normalized creep $J(t)$, in which all values were normalized to the initial strain at time zero

Thornton et al. [40] showed that this observation can be explained by taking into consideration of the nonlinearities, the interrelationship of creep and relaxation, as well as the fiber recruitment mechanism. Those interesting observations and developments in nonlinear theories of ligament viscoelasticity reflect the vision of Dr. Y.C. Fung. In his classic biomechanics book (p. 287) [36], Dr. Fung puts his thoughts into the following words: "I have a feeling that creep is fundamentally more nonlinear, and perhaps does not obey the quasi-linear hypothesis. The microstructural process taking place in a material undergoing creep could be quite different from that undergoing relaxation or oscillation."

## 3 The Unique Viscoelastic Properties of Heart Valve Tissues: Normal Stress Relaxation but Negligible Creep

The nonlinear, anisotropic mechanical behavior of heart valve leaflets has been widely studied. For all four types of heart valves, the stress-strain curves share overall characteristics. Using mitral valve leaflets and aortic valve leaflets as examples, the stress-strain curves can be divided into three regions: (1) a long "toe" region with low slope, (2) a nonlinear transitional region, followed by (3) a linear region with high slope [31, 41–44]. In the long "toe" region, the collagen fibers mainly experience uncrimping and hence give rise to a very low tensile modulus. In the transitional region, collagen fibers are gradually stretched and recruited to bear the load. In the high slope linear region, most of the collagen fibers have been straightened and recruited, providing a very stiff stress-strain behavior and serving as a lock-up mechanism to prevent excessive deformation [41].

Biaxial testing shows that, for the leaflets of atrioventricular valves and the leaflets of semilunar valves, the tissues are always stiffer and less extensible in the circumferential direction than in the radial direction, which reflects the fact that the dominant orientation of collagen fibers is along the circumferential direction of valve leaflets. The leaflet extensibilities along the circumferential direction and radial direction, as well as the degree of anisotropy, are various among the different types of valves. Moreover, differences in leaflet extensibilities and anisotropy also exist between the anterior leaflet and the posterior leaflet of the mitral valve, and among the anterior, septal, and posterior leaflets of the tricuspid valve.

Grashow et al. looked into the viscoelastic behavior of the mitral valve leaflets and reported that the stress-strain responses of mitral valve leaflets were independent of a range of strain rates and the hysteresis is relatively small (~12%), indicating a low level of energy dissipation during leaflet opening and closing cycles [42]. Moreover, a unique viscoelastic behavior has been identified in mitral valve leaflets, i.e., the mitral valve leaflets showed significant stress relaxation similar to other types of collagenous tissues, but exhibited negligible creep when subjected to a constant load at a physiological level over a 3-h period (Fig. 3a, b) [45]. Sacks et al. conducted an in vitro dynamic study to understand the closure behavior of mitral valve leaflets [46]. They found that in the closing cycle the anterior leaflet experienced large, anisotropic strains and peak stretch rates reached 500–1000%/s. After the rapid stretching, a plateau phase was observed, implying relatively constant strain state of the leaflet tissue [46]. This minimal creep behavior was further confirmed in an in vivo study by Sacks et al. [47]. In the in vivo study, nine 1-mm hemispherical piezoelectric transducers were implanted onto the anterior leaflet of Dorsett sheep in a 15-mm square array [47]. The 3D spatial positions of the nine transducers were monitored via a sonomicrometry system, and the transducer displacements were then used to calculate the in-surface Eulerian strain tensor [47]. The in vivo data showed that the mitral valve anterior leaflet showed a constant deformation state during the period of valve leaflet closure. From a functional perspective, the capability to

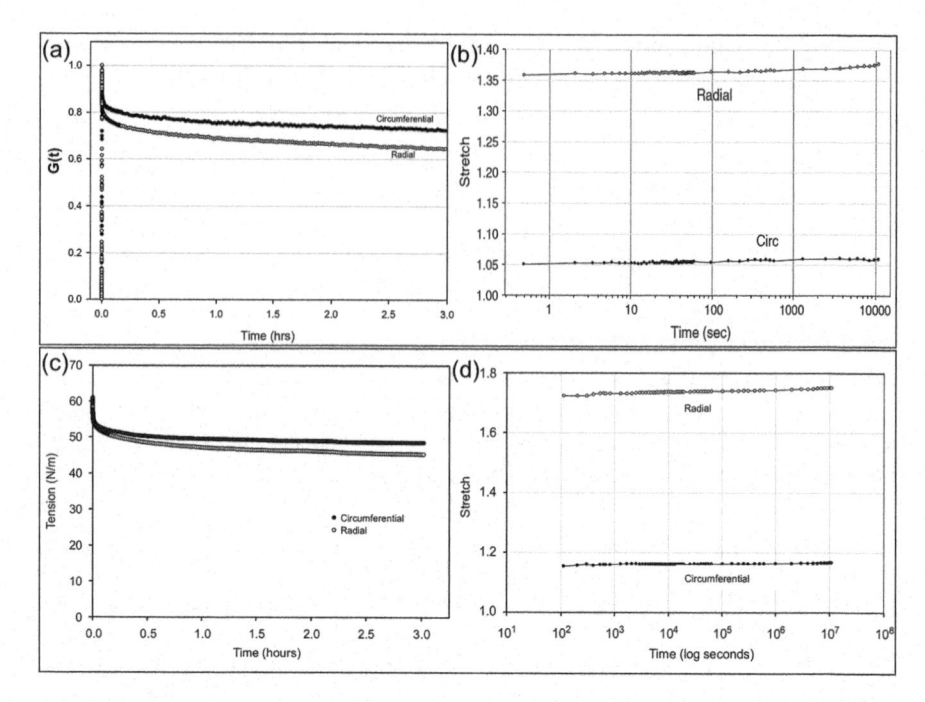

**Fig. 3** (a) Representative normalized stress relaxation curves for a biaxial stress-relaxation experiment on mitral valve leaflet. Leaflet tissue was biaxially loaded to 90 N/m membrane tension and kept at that strain level for stress decay monitoring. Typical stress relaxation behavior similar to other types of collagenous tissues was observed throughout the 3-h time frame [42]. (b) Representative stretch versus time curves for a biaxial creep experiment on mitral valve leaflet. Note the anisotropic leaflet behavior exhibited by the higher radial stretch required to maintain the 90 N/m membrane tension. Only very slight increases in strain were observed past 1000 s [42]. (c) Representative stress relaxation curves for a biaxial stress-relaxation experiment on aortic valve leaflet. Leaflet tissue was biaxially loaded to 60 N/m membrane tension and kept at that strain level for stress decay monitoring. Typical stress relaxation behavior similar to other types of collagenous tissues was observed throughout the 3-h time frame [48]. (d) Representative stretch versus time curves for a biaxial creep experiment on aortic valve leaflet. Note the anisotropic leaflet behavior exhibited by the higher radial stretch required to maintain the 60 N/m membrane tension. Only very slight increases in strain were observed past 1000 s [48]

quickly reach a peak strain and then resist further deformation during valve coaptation is a favorable design for mitral valve leaflets [47].

A similar study has been performed on aortic valve leaflets. Stellar et al. reported that the stress-strain responses are independent of strain rate, along with a low hysteresis at ∼17%. It was found that the aortic valve leaflets exhibited typical significant stress relaxation, but showed negligible creep over the 3 h test duration (Fig. 3c, d) [48]. The observation of negligible creep in aortic valve leaflets echoes with the observation in mitral valve leaflets, indicating that this unique viscoelastic behavior is likely a characteristic behavior of heart valve leaflets.

Note that Glashow et al. and Stellar et al. performed their biomechanical characterizations on mitral valve leaflet tissues and aortic valve leaflet tissues harvested from pigs [45, 48]. The porcine mitral valve apparatus was also used for in vitro dynamic testing [46]. The piezoelectric transducer-assisted in vivo functional characterization, however, was carried out on the mitral valve apparatus of a sheep model [47]. Both the in vitro dynamic testing and in vivo functional characterization demonstrated the minimal creep of leaflets during the stage of peak pressure loading. Recently, a study has been conducted on bovine mitral valve leaflets and aortic valve leaflets to understand the structural and biomechanical alterations during pregnancy [49]. Pierlot et al. showed that the bovine-originated mitral valve leaflets and aortic valve leaflets, both in nonpregnant status and pregnant status, demonstrated a minimal creep; moreover, the amount of creep of mitral valve leaflets decreased even further with pregnancy [49].

The abovementioned studies were performed on mitral valve leaflets/aortic valve leaflets that originated from porcine, ovine, and bovine species. The observations on strain rate sensitivity, hysteresis, stress relaxation, and creep behavior all point to a fact that heart valve leaflets cannot be treated as conventional viscoelastic materials. The heart valve leaflets, indeed, exhibit a unique viscoelastic property that is able to minimize creep and hysteresis while still allowing occurrence of typical stress relaxation. Both ex vivo and in vivo functional assessments indicate that this unique behavior enables the valve leaflets to withstand peak physiological loading without time-dependent deformation [46, 47]. Overall, the heart valve leaflets have adapted a mechanical behavior more as "quasi-elastic" biological materials, which is in favor of valve functionality [46–48].

## 4 Collagen Fibril Kinematics in Valve Leaflet Stretch, Stress Relaxation, and Creep

### 4.1 Hierarchical Structure of Collagen in Heart Valve Leaflets

We have briefly described the ECM components and overall structure of heart valve leaflets in Sect. 1.2. In order to understand the mechanistic causes of the unique viscoelasticity of heart valve leaflets, the knowledge of hierarchical structure of collagen needs to be introduced in detail. The elaboration of the collagen molecule, collagen fibril (with $D$-period structure), collagen fibril interactions via small PG linking bridges, collagen fiber, and tissue-specific collagen network are of importance if we want to reveal and picture the underlying mechanisms of the unique viscoelastic behavior of heart valve leaflets.

As the major load-bearing structural protein in heart valve tissues, collagen (Type I and Type III) has a nature-designed hierarchical structure spanning from nm building units (amino acids) to μm fibers. Three helical amino acid chains are packed together to form one collagen molecule with a very high aspect ratio (~280 nm in length and ~1.4 nm in diameter) [50]. The amino acid chain exhibits a pattern of Glycine-X-Y, which results in a helical twist of the individual amino acid chains. The three amino acid chains of the triple helix can be either the same (homotrimer) or different (heterotrimer), and are well stabilized by hydrogen bonds (inter- and intra-chain interactions) [51]. At the next hierarchical level, the very thin, rod-like collagen molecules experience self-assembly via covalent bonds to form long fibrils, i.e., collagen fibrils with 64–68 nm periodic banding appearance that reflects the pattern of self-assembly (Fig. 4b) [36, 50, 52, 53]. This periodic structure along the collagen fibril is called the $D$-period (or $D$-spacing), a pattern that resulted from self-assembling of collagen molecules via a "quarter-stagger" formation (Fig. 4). The Hodge-Petruska model (also called "quarter-stagger" model) described the "quarter-stagger" formation, i.e., (1) how the ~280 nm long collagen molecule packs together with great accuracy like "crystallinity"; (2) how the gaps between the axially neighboring collagen molecules and the shifting arrangement of laterally neighboring collagen molecules structurally result in the $D$-period with a gap region of ~0.6 $D$ and an overlap region 0.4 $D$ (Fig. 4b) [50]. The $D$-periods on the collagen fibril can be clearly observed under a transmission electron microscope (TEM) as bands with switching dark and light contrasts due to the mass variations of the gap region versus the overlap region. The $D$-period structure can also be observed by high-resolution scanning electron microscopy (SEM) and atomic force microscopy (AFM). Moreover, the variations in mass (electron density) of the $D$-period can serve as a periodic structure that is able to scatter the X-ray beam passing through it, and the length of $D$-period can hence be measured. This X-ray scattering mechanism will be described in more detail in Sect. 4.2.1 as we used it as a means to monitor collagen fibril elongation and orientation in the viscoelastic behavior of valve leaflets.

The cross-sections of normally developed collagen fibrils show an overall round-shaped feature, with a diameter ranging from ~10 nm (small fibril) to >100 nm (large fibril). Another noteworthy characteristic is that the collagen fibrils are discontinuous with extremely large aspect ratios (length of collagen fibril/diameter of collagen fibril), reported as high as ~2000 [46, 47]. Those discontinuous, very long collagen fibrils then form collagen fibers with a diameter of ~1 μm, which are often distinguishable under SEM or light microscopy. TEM studies have clearly shown that the parallel-aligned collagen fibrils are bonded together by small PGs to form the collagen fibers (Fig. 4c) [54, 55]. The small PG is composed of a long glycosaminoglycan (GAG) chain, of which two ends of the chain bind to two horseshoe-shaped protein cores via a covalent bond. Via ionic and polar interactions, the concave surface of the protein core has the capability to bind to specific sites of the $D$-period (on d, e bands), and hence the small PG serves as a crosslink (linking bridge) among the adjacent collagen fibrils [54, 55]. The role of small PGs in mediating interfibrillar interaction is supported by numerous experimental observations [56–69], as well as atomic-level theoretical modeling of the PG-collagen fibril configuration under loading [70–72]. According to shear-lag theory, even though the PG crosslink is relatively weak, a very large number of small PG connections, if

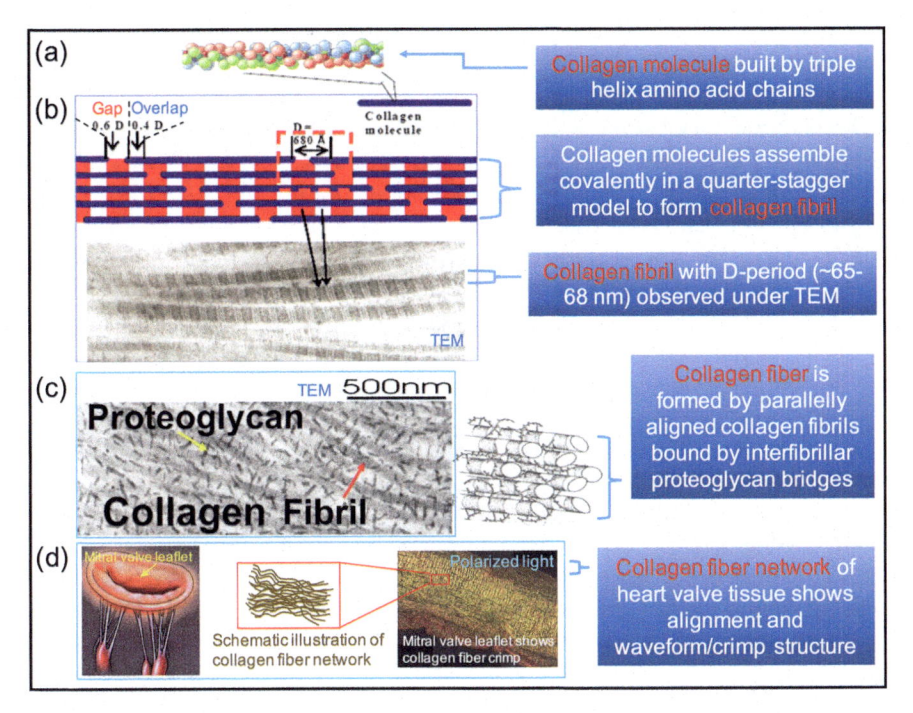

**Fig. 4** Schematic illustration showing the hierarchical structure of collagen. (**a**) Collagen molecule built by triple helix amino acid chains. (**b**) Collagen fibril self-assembled from collagen molecules via covalent bond and a "quarter-stagger" formation (The Hodge-Petruska model). Blue rods represent collagen molecules and their schematic arrangment shows how the gaps between the axially neighboring collagen molecules and the shifting arrangement of laterally neighboring collagen molecules structurally result in the $D$-period with a gap region of ~0.6 $D$ and an overlap region 0.4 $D$. Note that the $D$-period (~65–68 nm) structure can be observed with transmission electron microscopy (TEM). (**c**) Collagen fiber is formed by parallel-aligned collagen fibrils bounded by interfibrillar small proteoglycan bridges. Left panel is the TEM image showing collagen-small PG configurations in leaflet tissue; right panel is the schematic illustration. (**d**) Lastly, a collagen network, which takes a tissue-specific 3D arrangement, is constructed from collagen fibers. In heart valve leaflets, the wavy collagen fibers are aligned multi-directionally, with a dominant orientation along the circumferential direction of the leaflet. Left panel shows a mitral valve apparatus [12]; right panel shows collagen fiber network and collagen fiber alignment and crimp structure in mitral valve leaflet [13]

employed simultaneously, have the capability to facilitate load transfer among the discontinuous collagen fibrils [56, 70–73].

Although the direct load transfer role of small PGs is still debated [74, 75], it is clear that (1) small PGs are the entities connecting adjacent collagen fibrils and mediating interfibrillar interaction; (2) small PG crosslinks are mostly perpendicular to the collagen fibrils under zero load but become skewed when the tissue is under load (PG skewness angle increases with the applied load); (3) the PG skewness under load directly verifies the existence of interfibrillar slippage through the ground substance (small PG crosslinks) when the external load is applied [68]. The concept of interfibrillar slippage is important to help us understand the ultrastructural behavior in valve leaflet viscoelasticity.

Lastly, the collagen fibers form a collagen network, which takes a tissue-specific 3D arrangement. For example, in chordae tendineae of mitral and tricuspid valves, the collagen fibers take wavy pathways and have an overall alignment along the chordal longitudinal direction—the chordal load-bearing direction when valves close. However, in heart valve leaflets the collagen fibers are aligned multi-directionally, with a wavy structure and a dominant orientation along the circumferential direction in the fibrosa layer, a configuration helping to bear the pressure loading during leaflet coaptation [12, 41, 42, 76–78].

## 4.2 SAXS, Biaxial Device for SAXS Beamline, and Experimental Protocols

### 4.2.1 Small Angle X-ray Scattering for Collagenous Tissues

Due to the hierarchical complexity of collagenous tissues, it is important to have characterization methods to reveal the mechanical behavior at various hierarchical levels, i.e., from the tissue level, to collagen fiber level, to collagen fibril level, to collagen molecule level. For example, the very essential collagen fiber uncrimping mechanism, which enables the tissue extensibility of tendons/ligaments (hence joint flexibility), was revealed by polarized light microscopy [79, 80]. The periodic distinguishing bands of polarized light of the collagenous tissues are found to be associated with the crimp (wavy) pattern of collagen fibers. Along with the straightening of the collagen crimp in response to external load, the polarized light distinguishing bands disappear [79, 80]. Furthermore, the mechanical behavior of collagen fibers, including fiber alignment, rotation, and reorientation, has been studied in intact collagenous tissues using small angle light scattering (SALS) [81, 82].

As we mentioned above, the characteristic $D$-period of the collagen fibril is indeed the periodic variations in mass (electron density) along the fibril and is able to scatter the X-ray beam (Sect. 4.1). Hence, researchers have applied X-ray scattering [83–86] techniques, specifically small angle X-ray scattering (SAXS) [59, 87–90], to investigate the elongation and reorientation of collagen fibrils when connective tissues are under stretch. As shown by Fig. 5a, the axial fibril $D$-period generates a series of diffraction rings. The maxima of the diffraction series, called Bragg reflections, were governed by the equation, $2\sin(\theta) = \lambda n/D$, where $2\theta$ is the scattering angle, $\lambda$ is the X-ray wavelength, $n$ is the order of the reflection, and here $D$ is the fibril $D$-period in real space (Fig. 5c).

From the Bragg reflection equation, one can see that the larger the $D$ period is in real space, the closer the diffraction rings are in the reciprocal space (diffraction image) (Fig. 5c). By analyzing the diffraction series, we can accurately calculate the $D$-period value in real space. Details of how to analyze the diffraction profile and

Heart Valve Functional Elasticity 93

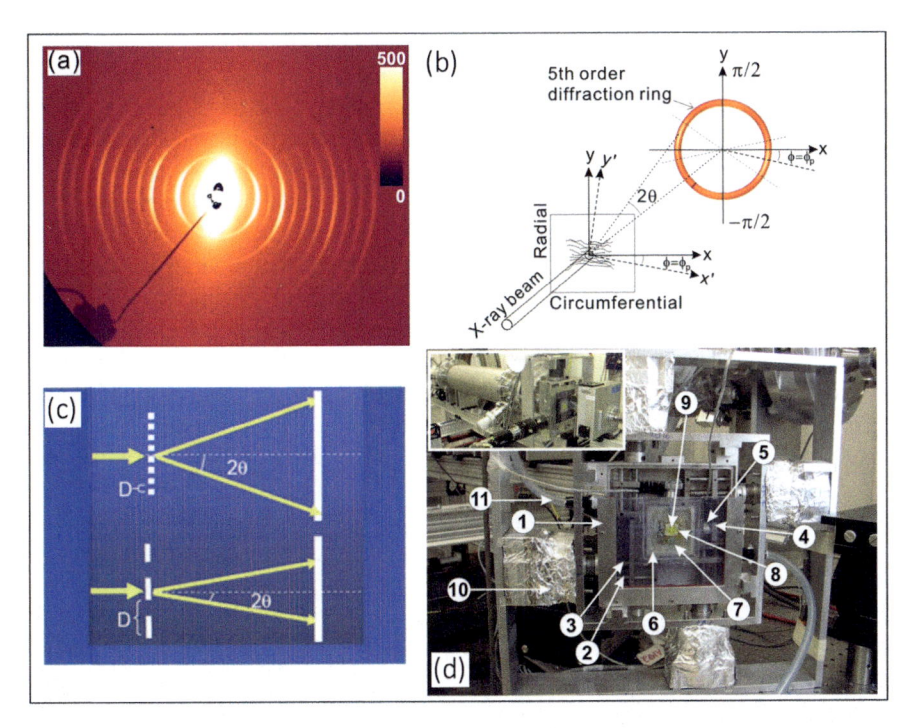

**Fig. 5** (**a**) SAXS image from the mitral valve anterior leaflet. The axial fibril $D$-period generates a series of diffraction rings. The maxima of diffraction series were used to calculate the $D$-period value in real space using Bragg reflection equation. (**b**) The brightest fifth order diffraction ring was used to obtain the angular distribution of collagen fibril orientation. The higher the diffraction intensity, the more collagen fibrils are aligned along that direction. The analysis can estimate the preferred orientation angle ($\phi_p$) and angular distribution profile of collagen fibrils. (**c**) Schematic illustration shows that X-ray beam is scattered by period structure such as D period. The larger the D period is in real space, the closer the diffraction rings are in the reciprocal space (diffraction image). (**d**) Biaxial stretch device custom made for the synchrotron SAXS beamline in X21 endstation at the NSLS. Numerical labels: (*1*) specimen chamber, (*2*) guide shaft, (*3*) lead screw, (*4*) tissue holder, (*5*) submersible load cell, (*6*) concave stage, (*7*) suture, (*8*) tissue in PBS solution, (*9*) small X-ray window, (*10*) motor, and (*11*) PBS solution guiding tube. Inserted image on top-left: the experimenting photo of biaxial stretching device on synchrotron SAXS beamline [44]

calculate the $D$-period value can be found in our previous papers [44, 91]. If defining $D_0$ as the collagen $D$-period value at the reference state, the $D$-period strain $\varepsilon_D$ can be computed using $\varepsilon_D = (D - D_0)/D_0$. We can further analyze the angular distribution of intensity along the diffraction ring to obtain the collagen fibril orientations (Fig. 5b). In our study, we plot the angular intensity distribution of the brightest fifth order diffraction ring (Fig. 5b). The higher the diffraction intensity, the more collagen fibrils are aligned along that direction. To better quantify the collagen fibril orientation, we used an $x$–$y$ Cartesian coordinate system with the origin point coinciding with the center of diffraction rings, and the $x$ axis aligned to the horizontal direction (Fig. 5b). Via a series of analyses of the angular intensity distribution data, we were able to estimate the preferred orientation angle ($\phi_p$) and angular distribution profile of collagen fibrils.

## 4.2.2 Biaxial Stretching System Designed for SAXS Beamline

*SAXS biaxial stretching device.* We have designed a biaxial stretching device specifically for the SAXS beamline. The biaxial stretching device consists of four major components: (1) sample loading actuators with capability to monitor loads biaxially, (2) motion control system, (3) X-ray transparent sample chamber, and (4) camera system to monitor biaxial strains on the sample surface (Fig. 6). Details of the *SAXS biaxial stretching system* can be found in our previous publications [44, 91]. Briefly, a vertically configured biaxial stretching sample chamber was made using an aluminum alloy, with actuator motion provided by two pairs of orthogonally positioned stepper motors. Motion control was provided by four micro-stepping drives, an interconnection module, and a motor controller card. Five hundred grams of submersible load cells, along with two amplifiers and a multifunction DAQ card, were used to record the biaxial loads. Biaxial strains were

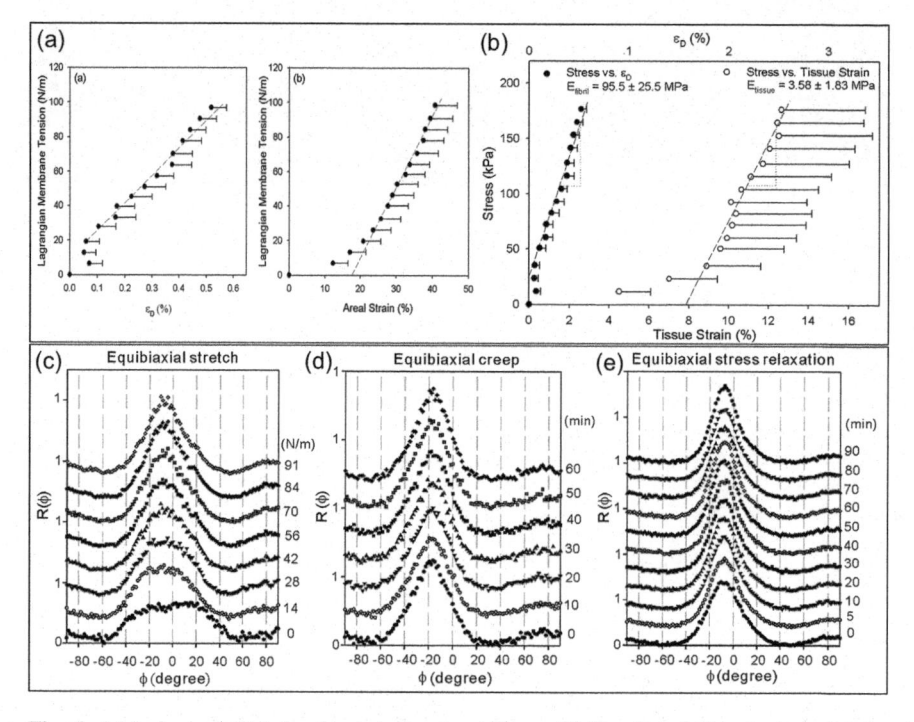

**Fig. 6** (a) Left panel: Relationship between the collagen fibrillar $D$-period strain ($\varepsilon_D$) and the applied equibiaxial tissue tension; right panel: tissue level equibiaxial tension-areal strain curve. (b) The relationship between stress, collagen fibril $D$-period strain ($\varepsilon_D$), and tissue strain along the fibril preferred direction. The tensile modulus of the MVAL collagen fibrils was found to be 95.5 ± 25.5 MPa along the preferred fibril direction, with the tissue tensile modulus 3.58 ± 1.83 MPa. Data presented as mean ± SEM. Representative angular distribution of collagen fibrils in biaxial stretch (c), creep (d), and stress relaxation (e). A peak value of 90 N/m equibiaxial tension was maintained during the creep tests and 90 N/m equibiaxial tension was also used as the initial stress level for stress relaxation [44]

tracked with four graphite markers on the sample. A transparent polyvinyl chloride window was used to allow the monitoring of markers with a CCD camera. A 45° mirror with a centered hole (3 mm diameter) reflected the marker images to the CCD camera positioned perpendicular to the SAXS beam line. Note that the centered 3 mm hole on the mirror allowed the X-ray beam to pass through. Lastly, a 1 mm diameter entrance and 3 mm diameter exit holes were made on the two PVC covers of the sample chamber, and Kapton film (high X-ray transmittance) was mounted over the holes to facilitate X-ray passage. The whole SAXS biaxial stretching system was controlled by a custom written LabVIEW program for delivering stretching protocols, recording biaxial loads, and monitoring biaxial strains.

*SAXS beamline parameters and experimental protocols.* The collagen fibril kinematic study was performed using the synchrotron SAXS beamline at X21 endstation in the National Synchrotron Light Source at Brookhaven National Laboratory. SAXS beamline parameters were as follows: 1.24 Å X-ray wavelength, 0.4 mm beam size, 1.33 m sample-to-detector distance, and a Mar CCD detector with AgBH (Silver Behenate) sample with known 58.376 Å spacing was used as calibration of reciprocal space [44]. As we mentioned above, the $D$-period of the collagen fibrils and the fibril orientations were extracted from the reciprocal spacing between the Bragg peaks and the angular intensity distribution of the diffraction ring, respectively.

For the collagen fibril kinematic study, fresh porcine mitral valve anterior leaflets (MVAL, $N = 5$) were dissected into ~15 mm square specimens with sides parallel to the circumferential and radial direction of the leaflet. After affixed with graphite markers, the square specimen was mounted on the SAXS biaxial stretching device using suture loops with stainless steel hooks. The circumferential direction of the specimen was aligned parallel to the horizontal direction and the radial direction was aligned to the vertical direction (Fig. 5d). Each MVAL specimen was then subjected to biaxial stretch, stress relaxation, and creep testing protocols, in which SAXS images of collagen fibrils were recorded [44]. During the testing, the specimen was immersed in PBS in the biaxial sample chamber at room temperature.

1. *SAXS-biaxial stretching protocol.* After ten preconditioning cycles to a maximum tension of 90 N/m and acquiring the initial SAXS pattern in the reference state, a sequence of quasi-static biaxial loading and SAXS measurements were then applied, with an increment of stepwise tension at 6.5 N/m (10 g) to reach 90 N/m peak tension. The specimen was then unloaded for a recovery period of 5 min.

2. *SAXS-stress relaxation protocol.* The specimen was loaded to 90 N/m equibiaxial tension and held at that fixed strain level. SAXS images were recorded along the time course of stress relaxation at 1 min intervals in the first 10 min, then at 10 min intervals from 10 to 90 min. Biaxial loads were recorded at 1 min intervals in the first 5 min, then at 5 min intervals up to 20 min, and finally at 10 min intervals up to 90 min. Afterwards, the specimen was unloaded for recovery for 5 min.

3. *SAXS-creep protocol*. Lastly, biaxial creep tests were performed with each specimen loaded and maintained at 90 N/m equibiaxial tension. Tissue strain and SAXS patterns were monitored in the following 60 min at an interval of 5 min. For both stress relaxation and creep protocols, the ramp time for sample was ~30 s.

## 4.3 Collagen Fibril Kinematics in MV Leaflets

### 4.3.1 Collagen Fibril Kinematics in Biaxially Stretched MV Leaflets

The MV leaflet behavior at the tissue level showed a typical nonlinear relationship of collagenous tissue, as shown by tension-areal strain responses (Fig. 6a). The nonlinear region was found to end at ~20 N/m tension, followed by a relative linear region. MV leaflet anisotropy is also revealed by tissue level stress-strain curves, with a maximum circumferential stretch at $1.117 \pm 0.046$ and a maximum radial stretch at $1.261 \pm 0.041$.

SAXS measurements showed that the $D$-period of collagen fibrils ($D_0$) was $658.8 \pm 0.4$ Å at the reference state. During the biaxial stretching, the average $\varepsilon_D$ remained almost unchanged for tensions lower than ~20 N/m (Fig. 6a). Beyond 20 N/m, the average $\varepsilon_D$ increased linearly with the applied equibiaxial tension (Fig. 6a). We further investigated the relationship among stress, collagen fibril $D$-period strain ($\varepsilon_D$), and tissue strain along the fibril preferred direction (Fig. 6b). Again, the curves showed that the onset point of $D$-period increase started at the end of the nonlinear region and the $\varepsilon_D$ increased linearly with the increase of stress. Tracking the collagen fibril angular distribution, we showed that fibril orientation/alignment changes mainly happened at tension levels smaller than ~28 N/m, i.e., in the nonlinear or "toe" region (Fig. 6c). At tension levels higher than ~28 N/m, the overall collagen fibril orientation (~circumferential direction) was maintained, while the degree of alignment increased along with the increased tension (Fig. 6c).

### 4.3.2 Collagen Fibril Kinematics in Stress Relaxation and Creep of MV Leaflets

As we just mentioned, the collagen fibrils in MV leaflets experienced major orientation/alignment changes during the biaxial stretch. For both the stress relaxation and creep tests, the MV leaflets were biaxially loaded to the linear region and maintained at constant stretch or tension, respectively, in which the collagen fibrils took a certain orientation and alignment corresponding to that stretch or tension level. We found that, during the time courses of stress relaxation and creep, the angular distributions of collagen fibrils exhibited no changes (Fig. 6d, e), indicating that there was no occurrence of reorientation/realignment of collagen fibrils.

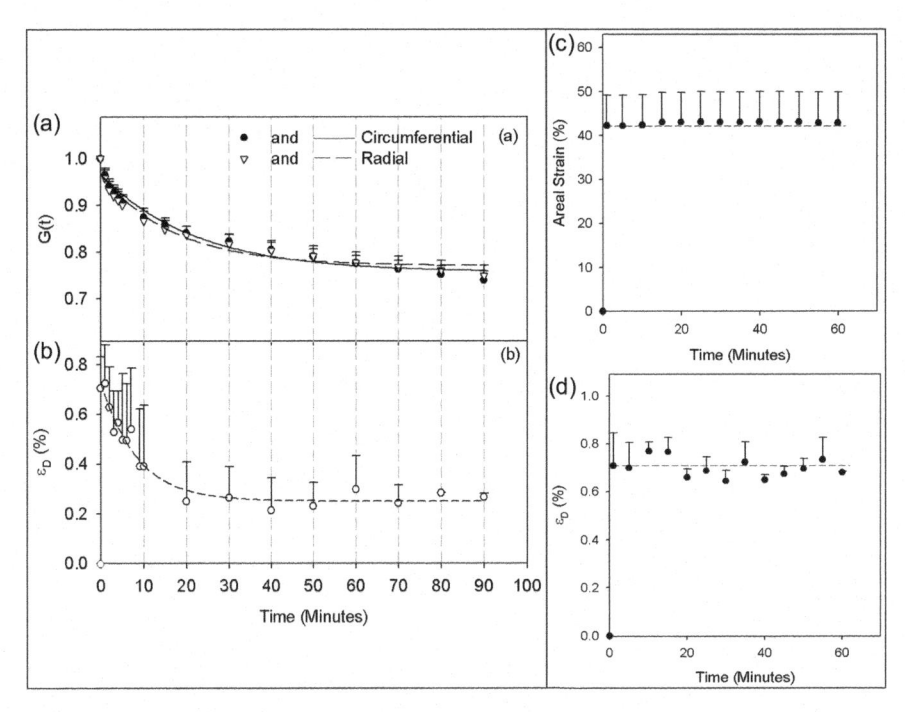

**Fig. 7** (**a**) Biaxial stress relaxation of mitral valve leaflet. Normalized stress relaxation $G(t)$ of both circumferential and radial directions were plotted with respect to time. (**b**) Reduction of $D$-period strain ($\varepsilon_D$) in the stress relaxation process. (**c**) Tissue areal strain versus time during creep tests, showing no detectable change. (**d**) Collagen fibril $D$-period strain ($\varepsilon_D$) versus time during creep tests. Data presented as mean ± SEM [44]

In the stress relaxation of MV leaflets, we found that ~16.1% of stress relaxed within 20 min and ~25.5% of stress relaxed at 90 min (Fig. 7a). Although the orientation and alignment of collagen fibrils experienced no changes, the collagen fibril strain, $\varepsilon_D$, experienced reduction during the stress relaxation, in which ~68.3% of fibril strain was released within the first 20 min and the remaining ~31.7% of fibril strain stayed relatively constant from 20 to 90 min (Fig. 7b). In the first 20 min, the overall reduction rate of $\varepsilon_D$ was found to be larger than the reduction rate of the stress at tissue level. The creep testing of MV leaflets verified the lack of measurable tissue strain during a 90 min timespan (Fig. 7c). The collagen $D$-period measurement showed that $\varepsilon_D$ was maintained at a constant level under biaxial creep (Fig. 7d).

### 4.3.3 Comparison with Other Collagenous Tissues

Folkhard et al. performed a SAXS study on rat tail tendon under uniaxial stress relaxation and creep. For stress relaxation, they reported two major observations: (1) the $D$-period in rat tail tendon decreased during the time course of stress relaxation; (2) a fast $D$-period decrease occurred in the beginning followed by a

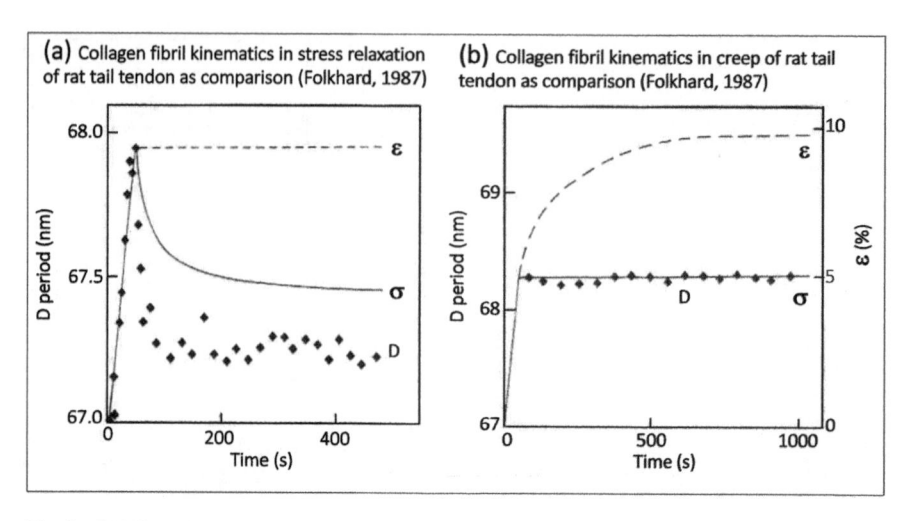

**Fig. 8** SAXS study on stress relaxation and creep of rat tail tendon by Folkhard as a comparison. (**a**) Tissue level stress ($\sigma$) and collagen $D$-period length ($D$) versus time in seconds. (**b**) Tissue level strain ($\varepsilon$) and collagen $D$-period length ($D$) versus time in seconds [87]

relatively slow decrease, and the overall reduction rate of fibril strain was faster than the stress decay (Fig. 8a) [87]. Those observations were consistent with what we reported in MV leaflets. The decrease of $D$-period suggests the stress relaxation in MV leaflets resulted from the release of load from collagen fibrils. Note that the collagen fibrils in MV leaflets, although they had an extremely large aspect ratio, were still discontinuous fibrils bounded together by interfibrillar small proteoglycans. The large reduction rate of collagen fibril strain during initial stress relaxation implied that, while the overall collagen fibril orientation and alignment were maintained during the time course, the interfibrillar slippage took place and the release of load from the collagen fibril were likely accompanied by part of the load redistributed to other structural ECM materials. The redistribution mechanisms could possibly be changes in interfibrillar small proteoglycan connections, interactions with elastin, or changes at larger structural scales, which are currently under investigation [44].

For creep, Folkhard [87] demonstrated that the rat tail tendon exhibited significant creep and that the tissue deformed continuously to maintain the constant load (Fig. 8b). Interestingly, the $D$-period of collagen fibrils in the rat tail tendon was found to remain constant. In a study on bovine Achilles tendon, which had typical tissue level creep, Sasaki et al. showed that, under 4–8 MPa stress, the $D$-period of collagen fibrils increased immediately in response to the application of the target load, but there was no further change during the time course of maintaining constant load [89]. Sasaki et al. also reported that, under 13–18 MPa, the $D$-period showed slight initial creep (20–30 s), then kept a constant value during the time course of maintaining constant load [89]. Based on these observations, the tissue deformation in the creep of rat tail tendon and bovine Achilles tendon could be ascribed to the deformation of the interfibrillar ECM, possibly interfibrillar small proteoglycan connections and others.

One interesting phenomenon noticed in tendons/ligaments was that the creep rates were smaller than those predicted from stress relaxation [38, 89]. Thornton et al. [38] speculated that the progressive recruitment of relaxed fibers during creep contributed to the lower creep rates. This gradual relaxed-fiber recruitment model was later incorporated into the inverse stress relaxation function to capture the lower creep rates [40]. While this gradual relaxed-fiber recruitment model explained the lower creep rates in tendons/ligaments, it does not explain our findings that no creep occurs in heart valve leaflets. It is worthy to mention that the complete lack of creep in heart valve leaflets appears to be a unique tissue behavior in the soft tissue literature. From a functional perspective, this unique behavior enables valvular tissues to withstand significant loading without time-dependent creep effect. We speculate that the negligible creep in valvular tissues is due to as yet an unidentified mechanism related to the composition and micro-architecture of valvular tissues that are different from tendons/ligaments and other soft collagenous tissues.

## 5 Final Remarks

Heart valve leaflets showed a unique viscoelastic behavior, in which valvular tissue allowed significant stress relaxation similar to other types of collagenous tissues, but exhibited negligible creep when subjected to a constant load. This unique tissue behavior well serves the function of the valve leaflets, i.e., bearing peak physiological loading without time-dependent deformation.

Our SAXS study on mitral valve leaflets demonstrated that collagen fibril kinematics are different in stress relaxation and creep. The first collagen fibril behavior we noticed was that the orientation and alignment of the collagen fibril network experienced no change in both stress relaxation and creep, indicating that collagen fibril reorientation/realignment did not contribute to stress relaxation and creep. However, in stress relaxation, collagen fibril strain released largely during the first 20 min and the remaining collagen fibril strain stayed relatively constant in the remaining relaxation time. The overall reduction rate of the collagen fibril strain was much larger than the reduction rate of stress at the tissue level. Our observation indicated that the stresses on collagen fibrils were redistributed to non-collagenous ECM under a constant strain condition, which also explained why, under stress relaxation, the collagen fibril strain decreased more quickly than the stress decay rate at the tissue level. Under a constant load (creep protocol), leaflet tissue experienced negligible time-dependent deformation (negligible creep), and the collagen fibril strain was maintained at a constant level during the time course.

This difference in collagen fibril kinematics implies the mechanisms responsible for creep and stress relaxation in the leaflet tissue are functionally independent. We thus speculate some type of fibril-level "locking" mechanism exists in leaflet tissue that allows for stress release under constant strain condition, yet does not allow for continued straining under a constant stress [44]. This fibril-level "locking" mechanism is possibly related to the composition and micro-architecture of leaflet tissues, which is currently under investigation.

In short, recent findings and our studies showed that the valvular tissues cannot be modeled as a quasilinear viscoelastic biological tissue [36] but rather quasi-elastically, and are intrinsically "designed" to minimize hysteretic and creep-like behaviors [44]. While the exact changes in these behaviors in heart valve diseases are still under investigation, we speculate that degenerated ECM components in diseased valvular tissues might cause changes in these quasi-elastic behaviors and thus contribute to malfunction of heart valves.

**Acknowledgments** The authors would like to thank the support from AHA BGIA-0565346, GRNT17150041, NIH 1R01EB022018-01, and UT STARS.

# References

1. Schoen F, Edwards W. Valvular heart disease: general principles and stenosis. Cardiovasc Pathol. 2001;3:403–42.
2. Sacks MS, David Merryman W, Schmidt DE. On the biomechanics of heart valve function. J Biomech. 2009;42(12):1804–24.
3. Sacks MS, Yoganathan AP. Heart valve function: a biomechanical perspective. Philos Trans R Soc B Biol Sci. 2007;362(1484):1369–91.
4. Mendelson K, Schoen FJ. Heart valve tissue engineering: concepts, approaches, progress, and challenges. Ann Biomed Eng. 2006;34(12):1799–819.
5. Merryman WD, Engelmayr GC Jr, Liao J, Sacks MS. Defining biomechanical endpoints for tissue engineered heart valve leaflets from native leaflet properties. Prog Pediatr Cardiol. 2006;21(2):153–60.
6. Brazile B, Wang B, Wang G, Bertucci R, Prabhu R, Patnaik SS, Butler JR, Claude A, Brinkman-Ferguson E, Williams LN, Liao J. On the bending properties of porcine mitral, tricuspid, aortic, and pulmonary valve leaflets. J Long-Term Eff Med Implants. 2015;25(1–2):41–53.
7. Vesely I, Boughner D. Analysis of the bending behaviour of porcine xenograft leaflets and of natural aortic valve material: bending stiffness, neutral axis and shear measurements. J Biomech. 1989;22(6/7):655–71.
8. Mirnajafi A, Raymer J, Scott MJ, Sacks MS. The effects of collagen fiber orientation on the flexural properties of pericardial heterograft biomaterials. Biomaterials. 2005;26(7):795–804.
9. Christie GW, Barratt-Boyes BG. Mechanical properties of porcine pulmonary valve leaflets: how do they differ from aortic leaflets? Ann Thorac Surg. 1995;60:S195–S9.
10. Doehring TC, Kahelin M, Vesely I. Mesostructures of the aortic valve. J Heart Valve Dis. 2005;14(5):679–86.
11. Vesely I. Reconstruction of loads in the fibrosa and ventricularis of porcine aortic valves. ASAIO J. 1996;42(5):M739–46.
12. Kunzelman KS, Cochran RP, Murphree SS, Ring WS, Verrier ED, Eberhart RC. Differential collagen distribution in the mitral valve and its influence on biomechanical behaviour. J Heart Valve Dis. 1993;2(2):236–44.
13. Grande-Allen KJ, Liao J. The heterogeneous biomechanics and mechanobiology of the mitral valve: implications for tissue engineering. Curr Cardiol Rep. 2011;13(2):113–20.

14. Roberts WC. Morphologic features of the normal and abnormal mitral valve. Am J Cardiol. 1983;51(6):1005–28.
15. Anderson R, Becker A. Anatomy of the heart. Stuttgart, NY: Thieme; 1982.
16. Gross L, Kugel M. Topographic anatomy and histology of the valves in the human heart. Am J Pathol. 1931;7(5):445.
17. Ho S. Anatomy of the mitral valve. Heart. 2002;88(Suppl 4):iv5–iv10.
18. Bezerra A, DiDio L, Prates J. Dimensions of the left atrioventricular valve and its components in normal human hearts. Cardioscience. 1992;3(4):241–4.
19. Misfeld M, Sievers H-H. Heart valve macro-and microstructure. Philos Trans R Soc B Biol Sci. 2007;362(1484):1421–36.
20. Thubrikar M, Klemchuk PP. The aortic valve. Boca Raton, FL: CRC Press; 1990.
21. Joyce EM, Liao J, Schoen FJ, Mayer JE Jr, Sacks MS. Functional collagen fiber architecture of the pulmonary heart valve cusp. Ann Thorac Surg. 2009;87(4):1240–9.
22. Go AS, Mozaffarian D, Roger VL, Benjamin EJ, Berry JD, Borden WB, Bravata DM, Dai S, Ford ES, Fox CS. Heart disease and stroke statistics—2013 update a report from the American Heart Association. Circulation. 2013;127(1):e6–e245.
23. Nkomo VT, Gardin JM, Skelton TN, Gottdiener JS, Scott CG, Enriquez-Sarano M. Burden of valvular heart diseases: a population-based study. Lancet. 2006;368(9540):1005–11.
24. Roger VL, Go AS, Lloyd-Jones DM, Benjamin EJ, Berry JD, Borden WB, Bravata DM, Dai S, Ford ES, Fox CS. Heart disease and stroke statistics—2012 update a report from the American Heart Association. Circulation. 2012;125(1):e2–e220.
25. Nishimura RA. Aortic valve disease. Circulation. 2002;106(7):770–2.
26. Lilly LS. Pathophysiology of heart disease: a collaborative project of medical students and faculty. Philadelphia: Wolters Kluwer Health; 2012.
27. Takkenberg JJ, Rajamannan NM, Rosenhek R, Kumar AS, Carapetis JR, Yacoub MH. The need for a global perspective on heart valve disease epidemiology the SHVD working group on epidemiology of heart valve disease founding statement. J Heart Valve Dis. 2008;17(1):135.
28. Peeters F, Meex SJR, Dweck MR, Aikawa E, Crijns H, Schurgers LJ, Kietselaer B. Calcific aortic valve stenosis: hard disease in the heart: a biomolecular approach towards diagnosis and treatment. Eur Heart J. 2018;39(28):2618–24.
29. Katsi V, Georgiopoulos G, Oikonomou D, Aggeli C, Grassos C, Papadopoulos DP, Thomopoulos C, Marketou M, Dimitriadis K, Toutouzas K, Nihoyannopoulos P, Tsioufis C, Tousoulis D. Aortic Stenosis, Aortic Regurgitation and Arterial Hypertension. Current Vascular Pharmacology 2018;16:1. https://doi.org/10.2174/1570161116666180101165306.
30. Grande-Allen KJ, Griffin BP, Calabro A, Ratliff NB, Cosgrove DM 3rd, Vesely I. Myxomatous mitral valve chordae. II: Selective elevation of glycosaminoglycan content. J Heart Valve Dis. 2001;10(3):325–32; discussion 32–3
31. Stephens EH, Timek TA, Daughters GT, Kuo JJ, Patton AM, Baggett LS, Ingels NB, Miller DC, Grande-Allen KJ. Significant changes in mitral valve leaflet matrix composition and turnover with tachycardia-induced cardiomyopathy. Circulation. 2009;120(11 Suppl):S112–9.
32. Breuer CK, Mettler BA, Anthony T, Sales VL, Schoen FJ, Mayer JE. Application of tissue-engineering principles toward the development of a semilunar heart valve substitute. Tissue Eng. 2004;10(11–12):1725–36.
33. Schoen FJ. Evolving concepts of cardiac valve dynamics the continuum of development, functional structure, pathobiology, and tissue engineering. Circulation. 2008;118(18):1864–80.
34. Rabkin-Aikawa E, Mayer JE Jr, Schoen FJ. Heart valve regeneration. Regenerative Medicine II. Berlin: Springer; 2005. p. 141–79.
35. Katz A. Physiology of the heart. Philadelphia, PA: Wolters Kluwer Health; 2011.
36. Fung YC. Biomechanics: mechanical properties of living tissues. 2nd ed. New York: Springer; 1993. 568 p
37. Provenzano P, Lakes R, Keenan T, Vanderby R Jr. Nonlinear ligament viscoelasticity. Ann Biomed Eng. 2001;29(10):908–14.

38. Thornton GM, Oliynyk A, Frank CB, Shrive NG. Ligament creep cannot be predicted from stress relaxation at low stress: a biomechanical study of the rabbit medial collateral ligament. J Orthop Res. 1997;15(5):652–6.
39. Lakes RS, Vanderby R. Interrelation of creep and relaxation: a modeling approach for ligaments. J Biomech Eng. 1999;121(6):612–5.
40. Thornton GM, Frank CB, Shrive NG. Ligament creep behavior can be predicted from stress relaxation by incorporating fiber recruitment. J Rheol. 2001;45(2):493–507.
41. May-Newman K, Yin FC. Biaxial mechanical behavior of excised porcine mitral valve leaflets. Am J Phys. 1995;269(4 Pt 2):H1319–27.
42. Grashow JS, Yoganathan AP, Sacks MS. Biaxial stress-stretch behavior of the mitral valve anterior leaflet at physiologic strain rates. Ann Biomed Eng. 2006;34(2):315–25.
43. Kunzelman KS, Cochran RP. Stress/strain characteristics of porcine mitral valve tissue: parallel versus perpendicular collagen orientation. J Card Surg. 1992;7(1):71–8.
44. Liao J, Yang L, Grashow J, Sacks MS. The relation between collagen fibril kinematics and mechanical properties in the mitral valve anterior leaflet. J Biomech Eng. 2007;129(1):78–87.
45. Grashow JS, Sacks MS, Liao J, Yoganathan AP. Planar biaxial creep and stress relaxation of the mitral valve anterior leaflet. Ann Biomed Eng. 2006;34(10):1509–18.
46. Sacks MS, He Z, Baijens L, Wanant S, Shah P, Sugimoto H, Yoganathan AP. Surface strains in the anterior leaflet of the functioning mitral valve. Ann Biomed Eng. 2002;30(10):1281–90.
47. Sacks MS, Enomoto Y, Graybill JR, Merryman WD, Zeeshan A, Yoganathan AP, Levy RJ, Gorman RC, Gorman JH III. In-vivo dynamic deformation of the mitral valve anterior leaflet. Ann Thorac Surg. 2006;82(4):1369–77.
48. Stella JA, Liao J, Sacks MS. Time-dependent biaxial mechanical behavior of the aortic heart valve leaflet. J Biomech. 2007;40(14):3169–77.
49. Pierlot CM, Moeller AD, Lee JM, Wells SM. Biaxial creep resistance and structural remodeling of the aortic and mitral valves in pregnancy. Ann Biomed Eng. 2015;43(8):1772–85.
50. Hodge AJ, Petruska JA. Recent studies with the electron microscope on ordered aggregates of the tropocollagen molecule. London: Academic Press; 1963.
51. Nimni ME. The molecular organization of collagen and its role in determining the biophysical properties of the connective tissues. Biorheology. 1980;17:51–82.
52. Silver FH, Freeman JW, Seehra GP. Collagen self-assembly and the development of tendon mechanical properties. J Biomech. 2003;36(10):1529–53.
53. Chapman JA, Hulmes DJS. Electron microscopy of the collagen fibril. In: Ruggeri A, Motto PM, editors. Ultrastructure of the connective tissue matrix. Boston: Martinus Nijhoff; 1984. p. 1–33.
54. Scott JE. Proteoglycan: collagen interactions in connective tissues. Ultrastructural, biochemical, functional and evolutionary aspects. Int J Biol Macromol. 1991;13(3):157–61.
55. Weber IT, Harrison RW, Iozzo RV. Model structure of decorin and implications for collagen fibrillogenesis. J Biol Chem. 1996;271(50):31767–70.
56. McBride DJ, Trelstad RL, Silver FH. Structural and mechanical assessment of developing chick tendon. Int J Biol Macromol. 1988;10:194–200.
57. Kastelic J, Palley I, Baer E. A structural mechanical model for tendon crimping. J Biomech. 1980;13(10):887–93.
58. Silver FH, Kato YP, Ohno M, Wasserman AJ. Analysis of mammalian connective tissue: relationship between hierarchical structures and mechanical properties. J Long-Term Eff Med Implants. 1992;2(2–3):165–98.
59. Fratzl P, Misof K, Zizak I, Rapp G, Amenitsch H, Bernstorff S. Fibrillar structure and mechanical properties of collagen. J Struct Biol. 1998;122(1–2):119–22.
60. Screen HR, Lee DA, Bader DL, Shelton JC. An investigation into the effects of the hierarchical structure of tendon fascicles on micromechanical properties. Proc Inst Mech Eng H. 2004;218 (2):109–19.

61. Elliott DM, Robinson PS, Gimbel JA, Sarver JJ, Abboud JA, Iozzo RV, Soslowsky LJ. Effect of altered matrix proteins on quasilinear viscoelastic properties in transgenic mouse tail tendons. Ann Biomed Eng. 2003;31(5):599–605.
62. Scott JE. Supramolecular organization of extracellular matrix glycosaminoglycans, in vitro and in the tissues. FASEB J. 1992;6(9):2639–45.
63. Grande-Allen KJ, Griffin BP, Ratliff NB, Cosgrove DM, Vesely I. Glycosaminoglycan profiles of myxomatous mitral leaflets and chordae parallel the severity of mechanical alterations. J Am Coll Cardiol. 2003;42(2):271–7.
64. Barber JE, Kasper FK, Ratliff NB, Cosgrove DM, Griffin BP, Vesely I. Mechanical properties of myxomatous mitral valves. J Thorac Cardiovasc Surg. 2001;122(5):955–62.
65. Dahners LE, Lester GE, Caprise P. The pentapeptide NKISK affects collagen fibril interactions in a vertebrate tissue. J Orthop Res. 2000;18(4):532–6.
66. Nishimura M, Yan W, Mukudai Y, Nakamura S, Nakamasu K, Kawata M, Kawamoto T, Noshiro M, Hamada T, Kato Y. Role of chondroitin sulfate-hyaluronan interactions in the viscoelastic properties of extracellular matrices and fluids. Biochim Biophys Acta. 1998;1380(1):1–9.
67. Liao J, Vesely I. A structural basis for the size-related mechanical properties of mitral valve chordae tendineae. J Biomech. 2003;36(8):1125–33.
68. Liao J, Vesely I. Skewness angle of interfibrillar proteoglycans increases with applied load on mitral valve chordae tendineae. J Biomech. 2007;40(2):390–8.
69. Liao J, Vesely I. Relationship between collagen fibrils, glycosaminoglycans, and stress relaxation in mitral valve chordae tendineae. Ann Biomed Eng. 2004;32(7):977–83.
70. Jeronimidis G, JFV V. Composite materials. In: Hukins DWL, editor. Connective. Tissue matrix. Weinheim: Verlag Chemie; 1984. p. 187–210.
71. Redaelli A, Vesentini S, Soncini M, Vena P, Mantero S, Montevecchi FM. Possible role of decorin glycosaminoglycans in fibril to fibril force transfer in relative mature tendons--a computational study from molecular to microstructural level. J Biomech. 2003;36(10):1555–69.
72. Vesentini S, Redaelli A, Montevecchi FM. Estimation of the binding force of the collagen molecule-decorin core protein complex in collagen fibril. J Biomech. 2005;38(3):433–43.
73. Cox HL. The elasticity and strength of paper and other fibrous materials. Br J Appl Phys. 1952;3:72–9.
74. Fessel G, Snedeker JG. Equivalent stiffness after glycosaminoglycan depletion in tendon—an ultra-structural finite element model and corresponding experiments. J Theor Biol. 2011;268(1):77–83.
75. Rigozzi S, Muller R, Snedeker JG. Collagen fibril morphology and mechanical properties of the Achilles tendon in two inbred mouse strains. J Anat. 2010;216(6):724–31.
76. Biological Materials SFH. Structure, mechanical properties, and modeling of soft tissues. New York and London: New York University Press; 1987.
77. Billiar KL, Sacks MS. A method to quantify the fiber kinematics of planar tissues under biaxial stretch. J Biomech. 1997;30(7):753–6.
78. Gilbert TW, Sacks MS, Grashow JS, Woo SLY, Chancellor MB, Badylak SF. Fiber kinematics of small intestinal submucosa under uniaxial and biaxial stretch. J Biomech Eng. 2006;128(6):890–8.
79. Hilbert SL, Sword LC, Batchelder KF, Barrick MK, Ferrans VJ. Simultaneous assessment of bioprosthetic heart valve biomechanical properties and collagen crimp length. J Biomed Mater Res. 1996;31(4):503–9.
80. Hansen KA, Weiss JA, Barton JK. Recruitment of tendon crimp with applied tensile strain. J Biomech Eng. 2002;124(1):72–7.
81. Kronick PL, Buechler PR. Fiber orientation in calfskin by laser light scattering or X-ray diffraction and quantitative relation to mechanical properties. J Am Leather Chem Assoc. 1986;81:221–9.
82. Sacks MS, Smith DB, Hiester ED. A small angle light scattering device for planar connective tissue microstructural analysis. Ann Biomed Eng. 1997;25(4):678–89.

83. Farkasjahnke M, Synecek V. Small-angle X-ray diffraction studies on rat-tail tendon. Acta Physiol Acad Sci. 1965;28(1):1–17.
84. Bigi A, Incerti A, Leonardi L, Miccoli G, Re G, Roveri N. Role of the orientation of the collagen fibers on the mechanical properties of the carotid wall. Boll Soc Ital Biol Sper. 1980;56 (4):380–4.
85. Aspden RM, Bornstein NH, Hukins DW. Collagen organisation in the interspinous ligament and its relationship to tissue function. J Anat. 1987;155:141–51.
86. Sasaki N, Odajima S. Stress-strain curve and Young's modulus of a collagen molecule as determined by the X-ray diffraction technique. J Biomech. 1996;29:655–8.
87. Folkhard W, Geercken W, Knorzer E, Mosler E, Nemetschek-Gansler H, Nemetschek T, Koch MH. Structural dynamic of native tendon collagen. J Mol Biol. 1987;193(2):405–7.
88. Sasaki N, Odajima S. Elongation mechanism of collagen fibrils and force-strain relations of tendon at each level of structural hierarchy. J Biomech. 1996;29(9):1131–6.
89. Sasaki N, Shukunami N, Matsushima N, Izumi Y. Time resolved X-ray diffraction from tendon collagen during creep using synchrotron radiation. J Biomech. 1999;32:285–92.
90. Purslow PP, Wess TJ, Hukins DW. Collagen orientation and molecular spacing during creep and stress-relaxation in soft connective tissues. J Exp Biol. 1998;201 .(Pt 1:135–42.
91. Liao J, Yang L, Grashow J, Sacks MS. Molecular orientation of collagen in intact planar connective tissues under biaxial stretch. Acta Biomater. 2005;1(1):45–54.

# Tricuspid Valve Biomechanics: A Brief Review

William D. Meador, Mrudang Mathur, and Manuel K. Rausch

**Abstract** The mechanics of the tricuspid valve are poorly understood. Today's unsatisfying outcomes of tricuspid valve surgery, at least in part, may be due to this lack of knowledge. Therefore, the tricuspid valve in general, and its mechanics specifically, have recently received an increasing interest. This chapter briefly summarizes what we currently know about tricuspid valve mechanics. To this end, we separately review tricuspid leaflet mechanics, annular mechanics, and the chordae's mechanics. Moreover, we categorize our discussion by the experimental environment in which these tissues were studied: in vivo, in vitro, and in silico. Finally, we make suggestions as to which areas of tricuspid valve mechanics should receive additional attention from the biomechanics community.

**Keywords** Atrioventricular heart valve · Morphology · Microstructure · Constitutive behavior · Dynamics · Stress · Strain · Leaflet · Chordae tendineae · Annulus

## 1 Introduction

Once considered the "forgotten valve," the tricuspid valve has received increased attention over the past decade. This interest is largely driven by the high prevalence of tricuspid regurgitation, or leakage of the tricuspid valve. Also, current success rates of surgery for tricuspid regurgitation are far from optimal with as many as 30%

W. D. Meador
Department of Biomedical Engineering, University of Texas at Austin, Austin, TX, USA

M. Mathur
Department of Mechanical Engineering, University of Texas at Austin, Austin, TX, USA

M. K. Rausch (✉)
Department of Aerospace Engineering and Engineering Mechanics, University of Texas at Austin, Austin, TX, USA
e-mail: manuel.rausch@utexas.edu

of patients developing recurrent regurgitation 5 years after surgery [1, 2]. Being historically understudied, the hope is that a better understanding of tricuspid valve morphology, microstructure, constitutive behavior, and dynamics (referred to here as "biomechanics") will ultimately lead to better clinical management of regurgitation through improved repair techniques and novel devices.

This brief review summarizes the relatively sparse data on tricuspid valve biomechanics. To this end, the chapter discusses these data separately for leaflets, annulus, and chordae tendineae and includes reports from in vivo, in vitro, and in silico studies.

## 2 Tricuspid Leaflets

### 2.1 Morphology and Nomenclature

The tricuspid valve, as opposed to its left heart counterpart, the mitral valve, has three leaflets. These leaflets insert into the myocardium and connective tissue of the right atrioventricular junction, at the tricuspid annulus. Similar to the mitral valve, chordal attachments to the right ventricular endocardium prevent the tricuspid valve leaflets from prolapsing into the right atrium.

The leaflets are denoted as anterior (or superior), posterior (or inferior), and septal (or medial) according to their anatomical positions adjacent to the anterior right ventricular free wall, posterior right ventricular free wall, and the interventricular septum, respectively. In humans, the anterior leaflet is usually the largest leaflet, followed by the septal leaflet, and the posterior leaflet. The latter is frequently organized into two or more scallops, while the other two leaflets present with only one scallop (with few exceptions, [3, 4]). Additionally, there are interspecies differences. For example, in sheep only one scallop is usually observed in the posterior leaflet. Independent of species, the leaflets themselves are divided into a "rough" zone which extends from the free edge to the coaptation line, and a "clear" zone, which is identified by its translucency. A basal zone, between the clear zone and the annulus, is only found in the posterior leaflet in humans.

### 2.2 In Vivo Studies

To the best of our knowledge there are no data available on the in vivo mechanics of the tricuspid leaflets. This paucity of data is likely due to the small thickness of the tricuspid leaflets (~300–700 μm), which prevents reliable identification with most imaging modalities especially in a dynamic environment such as the beating heart. Thus, there is an urgent need for investigation of the tricuspid leaflet mechanics using fiduciary marker techniques in the animal, for example.

## 2.3 In Vitro Studies

Khoiy et al. investigated the mechanics of the septal tricuspid leaflet in situ by suturing sonomicrometry crystals to the leaflet surface and pressurizing an explant porcine heart [5]. They found mean stretches across the leaflet at peak systolic pressure (30 mmHg) of 1.1 (maximum principal), 1.05 (circumferential), and 1.04 (radial). They concluded that the in vitro deformation of the septal tricuspid leaflet is similar to that of the in vivo anterior mitral leaflet [6]. In a similar setup, Pant et al. also studied microstructural changes in the tricuspid leaflets as a function of transvalvular pressure. To this end, they used a similar in vitro apparatus to Khoiy et al.'s previous study but replaced the pressurizing fluid with 0.5% glutaraldehyde. Subsequently they fixed the tricuspid leaflets in either an undeformed (unpressurized) configuration or a deformed (pressurized) configuration [7]. Using small angle light scattering, they then quantified the microstructural anisotropy of all tricuspid leaflets as a function of loading state. They concluded that transvalvular pressure results in strain-induced microstructural organization. In a similar study to Pant et al.'s, Hamed Alavi et al. [8] investigated the organization of the tricuspid valve leaflet matrix under uniaxial and biaxial loading using 2-Photon microscopy. They observed that the collagen fiber orientation in the relaxed state varies with depth and re-orients in response to uniaxial and biaxial loading in a depth-dependent and loading scenario-dependent manner.

In a right-heart in vitro simulator, Spinner et al. used dual camera photogrammetry to track visual marker points on the anterior and posterior leaflets of explant porcine hearts throughout the cardiac cycle [9]. In contrast to the studies by Khoiy et al., Spinner et al. isolated the tricuspid valves and fixed them to an artificial annulus rather than testing the tissue in situ. In this setup, they found that the anterior and posterior leaflets undergo large deformations throughout the cardiac cycle with mean maximum principal stretches of 1.22 and 1.53, respectively. Additionally, they tested the effect of (1) saddle shape (saddle vs. no saddle), (2) papillary muscle displacement (10 mm), and (3) annular dilation (100%) on leaflet stretches but found no significant differences between conditions. Given the relatively small sample number ($n = 8$) and large standard deviations, lack of significance may have been a result of type II error. However, if true these data would imply that the tricuspid annular saddle shape, in contrast to the mitral annular saddle shape, does not minimize leaflets stretches (and therefore stresses) questioning the saddle's teleological origin on the right side of the heart [10]. Noteworthy is the discrepancy between Spinner et al.'s findings and those of Khoiy et al. Spinner et al.'s stretches are significantly larger than those of Khoiy et al., which may be due to the differences in setups (isolated valve versus in situ) and the fact that they studied different leaflets.

Additionally, Pham et al. characterized the constitutive behavior of isolated human tricuspid valve leaflets. They used cadaveric tissue and executed a biaxial protocol to derive the stress-strain behavior of each leaflet. Additionally, they fit a Fung-type material model to their data [11]. They found that the tricuspid valve leaflets exhibit a highly nonlinear response and large degrees of anisotropy. By comparison to the leaflets of the other three heart valves, they additionally determined that the tricuspid leaflets are the most extensible and isotropic of all heart valve leaflets and that tricuspid leaflet extensibility decreases with age. Khoiy et al. also executed a biaxial protocol to investigate the constitutive behavior of isolated porcine tricuspid valve leaflets [12]. In their study, they confirm the large degrees of nonlinearity and anisotropy of the tricuspid valve leaflets reported above and found that the posterior leaflet is the most anisotropic of the three.

## 2.4 In Silico Studies

We have been able to identify only two numerical studies on the mechanics of the tricuspid valve [13, 14]. Kamensky et al. used the tricuspid valve as an application for a novel contact algorithm. Thus, it is not a detailed report on tricuspid valve mechanics. Stevanella et al., on the other hand, performed detailed analyses of tricuspid valve mechanics. In Stevanella et al.'s study, due to a lack of matched data, the authors combined information from sonomicrometry studies on the ovine tricuspid annulus [15], tricuspid leaflet morphological information from cadaveric analyses, and constitutive parameters from mitral valves [16] to develop a model of an isolated tricuspid valve. Using the finite element method to solve for leaflet deformation and stresses provided transvalvular pressure and kinematic boundary conditions assigned to the annulus, they found that the motion and stresses of the tricuspid leaflets are "almost insensitive" to the leaflet constitutive model. Stress peaks were found in the anterior and septal leaflets close to the annulus. Additionally, the authors reported mean circumferential leaflet stretches in the range of 1.11–1.25 for the anterior leaflet, 1.09–1.22 for the posterior leaflets, and 1.09–1.25 for the septal leaflet, depending on the hyperelastic leaflet material parameter choice. Thus, their leaflet stretch values are in a similar range to those reported by Khoiy et al., which were derived from an in vitro model.

## 3 Tricuspid Annulus

### 3.1 Morphology and Nomenclature

Albeit not always referred to as a saddle, the tricuspid valve annulus, similar to the mitral annulus, has a distinct three-dimensional topology reminiscent of a saddle. High points at the antero-septal annulus and the postero-lateral segment of the annulus as well as low points at the postero-septal annulus and the anterior annulus yield this shape [17]. Its two-dimensional projection lacks the symmetry of the mitral annulus in that it shows a clear deviation from an oval. Anatomically, the tricuspid annulus is situated more apically than the mitral valve by a few millimeters [3].

## 3.2 In Vivo Studies

In contrast to the tricuspid leaflets, the tricuspid annulus has been explored in vivo in detail, both in humans and in animals. Magnetic resonance imaging and echocardiography have been used to characterize the shape of the tricuspid annulus and its dynamics in patients with and without tricuspid regurgitation [18–21]. One of the earliest reports on annular dynamics in a beating human heart dates back to Tei et al. who used 2D echocardiography [22]. More recently, Fukuda et al. used 3D echocardiography to report the shape of the annulus throughout the cardiac cycle also in humans [17]. They describe the annulus as having distinct peaks and valleys. However, they refrain from describing the annulus as "saddle-shaped." Furthermore, they found the annulus to undergo significant cinching throughout the cardiac cycle with the orifice area decreasing from diastole to systole by ~29% in healthy patients and ~22% in patients with tricuspid regurgitation. This reduction in area was driven by length changes in the tricuspid annulus of approximately 15% in healthy patients, and approximately 10% in patients with tricuspid regurgitation.

The most common animal model used in the study of tricuspid annular dynamics is sheep. Hiro et al. were the first to implant sonomicrometry crystals on the tricuspid annulus of sheep and to record their locations in the beating heart [15]. They found that the mean tricuspid valve orifice area changes dynamically by 21.3% throughout the cardiac cycle. Fawzy et al. performed a similar analysis also using sonomicrometry crystals in sheep [23]. However, care must be taken, because their reported values for annular area differ by a factor of five from both Hiro et al. and Malinowski et al., who confirmed Hiro's findings [24]. Malinowski et al. also investigated the dynamics of the ovine tricuspid annulus under acute pulmonary hypertension and found that acute pulmonary hypertension lengthens the tricuspid valve annulus by 12% and reduces annular contractility locally. Additionally, they studied the effect of acute left ventricular mechanical unloading on tricuspid annular shape and dynamics but found little effect [25].

Lastly, Rausch et al. reanalyzed data by Malinowski et al. on the normal dynamics of the ovine tricuspid annulus employing mechanical metrics strain and curvature [26]. They found that strain and curvature change significantly throughout the cardiac cycle with focal minima and maxima of both metrics driving the dynamics of the annulus.

## 3.3 In Vitro Studies

Few data are available on the in vitro mechanics of the tricuspid annulus, likely due to the abundance of in vivo data. An exception are the studies by Basu et al. [27, 28]. In their first study, they used an intricate experimental setup to measure "annulus tension" in isolated porcine tricuspid valves. Specifically, they attached 10 wires to the annulus, each radially connecting to a surrounding force transducer.

In this configuration, they placed isolated porcine tricuspid valves in a right heart simulator. Pressurizing the valve to a systolic pressure of 40 mmHg, they measured wire forces and calculated annulus tension (wire force divided by annulus segment length) under three configurations: (1) normal, (2) papillary muscles displaced (15 mm) and annulus dilated (70%), and (3) after "clover" repair [29]. The study concluded that papillary muscle displacement and annular dilation increases annulus tension fourfold, implying that disease-induced annular forces may counteract and "decelerate" pathological annular dilation in patients. Additionally, they found that clover repair does not further impact annulus tension and therefore would not aid in reverse remodeling in patients. In their second study, Basu et al. investigated the mechanical properties and histological composition of the isolated porcine annulus. To this end, they explanted tricuspid valves, isolated the annular tissue, and divided it into anterior, posterior, and septal regions. Subsequently, they performed uniaxial tensile tests and histological tests on each region. They found that the septal annulus in pigs is the stiffest, likely due to high collagen contents, followed by the posterior, and, lastly, the anterior annulus.

Moreover, Adkins et al. studied the suture force necessary to cinch the dilated tricuspid annulus in an in vitro heart preparation, where dilation was achieved via phenol injection [30]. The authors found that phenol injection increases tricuspid annular area by 8.82% in vitro. The mean suture force necessary to reestablish normal valve area in the pressurized heart was measured to be 0.03 N.

## 3.4 In Silico Studies

To the best of our knowledge no numerical studies exist that focus on the tricuspid annulus alone.

## 4 Tricuspid Chordae Tendineae

### 4.1 Morphology and Nomenclature

The chordae tendineae insert into the leaflets' free-edge, rough zone, clear zone, and the basal region, and are classified accordingly. Silver et al. identifies additional "fan-like" chordae which insert into the commissural regions between leaflets. In contrast to the mitral chordae, the tricuspid chordae are generally less well organized and form complex networks connecting the leaflets to insertion sites on the endocardial wall, which are often distinct from either of the three papillary muscle heads [3, 31, 32].

## 4.2 In Vivo Studies

Similar to the tricuspid leaflets, there are currently few data available on the in vivo mechanics of the tricuspid chordae tendineae likely due to the challenges of imaging small anatomic details in a dynamic environment. One exception is the study by Fawzy et al. who implanted sonomicrometry crystals in sheep on the tips of the papillary muscles as well as the free edge of each leaflet. From these crystal pairs, they determine that the mean peak chordal deformation in the beating heart is 14%, 16.9%, and 5.2% for the anterior, posterior, and septal chordae, respectively.

## 4.3 In Vitro Studies

The most complete investigation of the in vitro mechanical properties of chordae tendineae goes back to Lim [31]. They performed uniaxial tensile tests on human tricuspid chordae tendineae from three patients (58+ years). Additionally, they performed "ultrastructural" analyses via scanning and transmission electron microscopy. From the uniaxial tensile test data, they derived that chordae tendineae follow the classic nonlinear stress-strain behavior observed in many other collagenous tissues with a linear pre-toe region, a transition region, and a linear post-transition region [33]. They suggested that tricuspid chordae tendineae are less extensible than those of the mitral valve. They attributed this difference to observed variations in the ultrastructure of the chordae, namely different collagen fibril diameters, different collagen fibril density, and different percentage of cross-sectional area covered by collagen fibrils. However, it must be noted that a subsequent study by the same authors performed the same analyses on a different subset of tricuspid chordae and found elastic properties that varied significantly from above findings [32]. Thus, care must be taken when interpreting either findings.

## 4.4 In Silico Studies

Although we are not aware of any numerical studies solely focusing on tricuspid chordae tendineae, Stevanella et al. reported papillary muscle reaction forces (the sum of all chordal forces for each papillary muscle) and force ranges for classes of chordae in their finite element study of the isolated tricuspid valve [14]. They found that reaction forces are similar between anterior, posterior, and septal papillary muscles and invariant to leaflet material models (<1 N). Additionally, they found that "marginal" chordae experience higher forces than "second order" chordae by a factor of approximately three.

## 5  Most Recent Studies

After submission of this chapter, a number of pertinent papers were published, which we briefly list here for completeness. Jett et al. [34] published a comprehensive in vitro study on the anisotropic mechanical properties and anatomical structure of porcine atrioventricular heart valve tissue. Moreover, Khoiy et al. [35] expanded on their previous work on the mechanical behavior of tricuspid leaflet mechanical properties and informed a hyperelastic constitutive law of the same tissue. On the modeling side, Kong et al. [36] developed detailed finite element models of human tricuspid valves based on computed tomography data and studied in vivo leaflet stress. Rausch et al. [37] used a sonomicrometry-based approach to delineate the in vivo mechanics of the tricuspid annulus under acute pulmonary hypertension, while Malinowski et al. [38] used the same technique to study tricuspid annular dynamics in explant, beating human hearts. Finally, Madukauwa-David et al. [39] investigated in vitro the effect of collagen content in human tricuspid annuli on suture dehiscence.

## 6  Conclusion

Although in the past decade significant effort has been invested into elucidating the biomechanics of the tricuspid valve, our understanding of the forgotten valve is still lacking behind that of the mitral valve. After reviewing the existing literature, we identify three areas of research that we believe require more attention: (1) in vivo studies of the tricuspid leaflet and chordae mechanics, (2) in silico studies of tricuspid leaflets, annulus, and chordae, and (3) studies of the remodeling potential of the tricuspid valve that are currently absent. Future studies will hopefully fill these gaps in our knowledge and bring us closer to a more complete understanding of the tricuspid valve that may translate into better clinical management of tricuspid regurgitation.

## References

1. Mangieri A, Montalto C, Pagnesi M, Jabbour RJ, Rodés-Cabau J, Moat N, Colombo A, Latib A. Mechanism and implications of the tricuspid regurgitation: from the pathophysiology to the current and future therapeutic options. Circ Cardiovasc Interv. 2017;10:1–13. https://doi.org/10.1161/CIRCINTERVENTIONS.117.005043.
2. Tang GHL, David TE, Singh SK, Maganti MD, Armstrong S, Borger MA. Tricuspid valve repair with an annuloplasty ring results in improved long-term outcomes. Circulation. 2006;114:I-577–81. https://doi.org/10.1161/CIRCULATIONAHA.105.001263.
3. Silver MD, Lam JHC, Ranganathan N, Wigle ED. Morphology of the human tricuspid valve. Circulation. 1971;43:333–48. https://doi.org/10.1161/01.CIR.43.3.333.

4. Tretter JT, Sarwark AE, Anderson RH, Spicer DE. Assessment of the anatomical variation to be found in the normal tricuspid valve. Clin Anat. 2016;29:399–407. https://doi.org/10.1002/ca.22591.
5. Amini Khoiy K, Biswas D, Decker TN, Asgarian KT, Loth F, Amini R. Surface strains of porcine tricuspid valve septal leaflets measured in ex vivo beating hearts. J Biomech Eng. 2016;138:111006. https://doi.org/10.1115/1.4034621.
6. Sacks MS, Enomoto Y, Graybill JR, Merryman WD, Zeeshan A, Yoganathan AP, Levy RJ, Gorman RC, Gorman JH. In-vivo dynamic deformation of the mitral valve anterior leaflet. Ann Thorac Surg. 2006;82:1369–77. https://doi.org/10.1016/j.athoracsur.2006.03.117.
7. Pant AD, Thomas VS, Black AL, Verba T, Lesicko JG, Amini R. Pressure-induced microstructural changes in porcine tricuspid valve leaflets. Acta Biomater. 2017;67:248–58. https://doi.org/10.1016/j.actbio.2017.11.040.
8. Hamed Alavi S, Sinha A, Steward E, Milliken JC, Kheradvar A. Load-dependent extracellular matrix organization in atrioventricular heart valves: differences and similarities. Am J Physiol Hear Circ Physiol. 2015;309:276–84. https://doi.org/10.1152/ajpheart.00164.2015.
9. Spinner EM, Buice D, Yap CH, Yoganathan AP. The effects of a three-dimensional, saddle-shaped annulus on anterior and posterior leaflet stretch and regurgitation of the tricuspid valve. Ann Biomed Eng. 2012;40:996–1005. https://doi.org/10.1007/s10439-011-0471-6.
10. Salgo IS, Gorman JH, Gorman RC, Jackson BM, Bowen FW, Plappert T, St John Sutton MG, Edmunds LH. Effect of annular shape on leaflet curvature in reducing mitral leaflet stress. Circulation. 2002;106:711–7. https://doi.org/10.1161/01.CIR.0000025426.39426.83.
11. Pham T, Sulejmani F, Shin E, Wang D, Sun W. Quantification and comparison of the mechanical properties of four human cardiac valves. Acta Biomater. 2017;54:345–55. https://doi.org/10.1016/j.actbio.2017.03.026.
12. Amini Khoiy K, Amini R. On the biaxial mechanical response of porcine tricuspid valve leaflets. J Biomech Eng. 2016;138:104504. https://doi.org/10.1115/1.4034426.
13. Kamensky D, Xu F, Lee C-HH, Yan J, Bazilevs Y, Hsu M-CC. A contact formulation based on a volumetric potential: application to isogeometric simulations of atrioventricular valves. Comput Methods Appl Mech Eng. 2018;330:522–46. https://doi.org/10.1016/j.cma.2017.11.007.
14. Stevanella M, Votta E, Lemma M, Antona C, Redaelli A. Finite element modelling of the tricuspid valve: a preliminary study. Med Eng Phys. 2010;32:1213–23. https://doi.org/10.1016/j.medengphy.2010.08.013.
15. Hiro ME, Jouan J, Pagel MR, Lansac E, Lim KH, Lim H-S, Duran CM. Sonometric study of the normal tricuspid valve annulus in sheep. J Heart Valve Dis. 2004;13:452–60.
16. May-Newman K, Yin FC. A constitutive law for mitral valve tissue. J Biomech Eng. 1998;120:38–47. https://doi.org/10.1115/1.2834305.
17. Fukuda S, Saracino G, Matsumura Y, Daimon M, Tran H, Greenberg NL, Hozumi T, Yoshikawa J, Thomas JD, Shiota T. Three-dimensional geometry of the tricuspid annulus in healthy subjects and in patients with functional tricuspid regurgitation a real-time, 3-dimensional echocardiographic study. Circulation. 2006;114:I492–8. https://doi.org/10.1161/CIRCULATIONAHA.105.000257.
18. Leng S, Jiang M, Zhao XD, Allen JC, Kassab GS, Ouyang RZ, Le TJ, He B, Tan RS, Zhong L. Three-dimensional tricuspid annular motion analysis from cardiac magnetic resonance feature-tracking. Ann Biomed Eng. 2016;44:3522–38. https://doi.org/10.1007/s10439-016-1695-2.
19. Maffessanti F, Gripari P, Pontone G, Andreini D, Bertella E, Mushtaq S, Tamborini G, Fusini L, Pepi M, Caiani EG. Three-dimensional dynamic assessment of tricuspid and mitral annuli using cardiovascular magnetic resonance. Eur Heart J Cardiovasc Imaging. 2013;14:986–95. https://doi.org/10.1093/ehjci/jet004.
20. Owais K, Taylor CE, Jiang L, Khabbaz KR, Montealegre-Gallegos M, Matyal R, Gorman JH, Gorman RC, Mahmood F. Tricuspid annulus: a three-dimensional deconstruction and reconstruction. Ann Thorac Surg. 2014;98:1536–42. https://doi.org/10.1016/j.athoracsur.2014.07.005.

21. Tsakiris AG, Mair DD, Seki S, Titus JL, Wood EH. Motion of the tricuspid valve annulus in anesthetized intact dogs. Circ Res. 1975;36:43–8. https://doi.org/10.1161/01.RES.36.1.43.
22. Tei C, Pilgrim JP, Shah PM, Ormiston JA, Wong M. The tricuspid valve annulus: study of size and motion in normal subjects and in patients with tricuspid regurgitation. Circulation. 1982;66:665–71. https://doi.org/10.1161/01.CIR.66.3.665.
23. Fawzy H, Fukamachi K, Mazer CD, Harrington A, Latter D, Bonneau D, Errett L. Complete mapping of the tricuspid valve apparatus using three-dimensional sonomicrometry. J Thorac Cardiovasc Surg. 2011;141:1037–43. https://doi.org/10.1016/j.jtcvs.2010.05.039.
24. Malinowski M, Wilton P, Khaghani A, Langholz D, Hooker V, Eberhart L, Hooker RL, Timek TA. The effect of pulmonary hypertension on ovine tricuspid annular dynamics. Eur J Cardiothorac Surg. 2016b;49:40–5. https://doi.org/10.1093/ejcts/ezv052.
25. Malinowski M, Wilton P, Khaghani A, Brown M, Langholz D, Hooker V, Eberhart L, Hooker RL, Timek TA. The effect of acute mechanical left ventricular unloading on ovine tricuspid annular size and geometry. Interact Cardiovasc Thorac Surg. 2016a;23:391–6. https://doi.org/10.1093/icvts/ivw138.
26. Rausch MK, Malinowski M, Wilton P, Khaghani A, Timek TA. Engineering analysis of tricuspid annular dynamics in the beating ovine heart. Ann Biomed Eng. 2017;46:443. https://doi.org/10.1007/s10439-017-1961-y.
27. Basu A, He Z. Annulus tension on the tricuspid valve: an in-vitro study. Cardiovasc Eng Technol. 2016;7:270–9. https://doi.org/10.1007/s13239-016-0267-9.
28. Basu A, Lacerda C, He Z. Mechanical properties and composition of the basal leaflet-annulus region of the tricuspid valve. Cardiovasc Eng Technol. 2018;9:217. https://doi.org/10.1007/s13239-018-0343-4.
29. Alfieri O, De Bonis M, Lapenna E, Agricola E, Quarti A, Maisano F. The "clover technique" as a novel approach for correction of post-traumatic tricuspid regurgitation. J Thorac Cardiovasc Surg. 2003;126:75–9. https://doi.org/10.1016/S0022-5223(03)00204-6.
30. Adkins A, Aleman J, Boies L, Sako E, Bhattacharya S. Force required to cinch the tricuspid annulus: an ex-vivo study. J Heart Valve Dis. 2015;24:644.
31. Lim KO. Mechanical properties and ultrastructure of normal human tricuspid valve chordae tendineae. Jpn J Physiol. 1980;30:455–64. https://doi.org/10.2170/jjphysiol.30.455.
32. Lim KO, Boughner DR, Perkins DG. Ultrastructure and mechanical properties of chordae tendineae from a myxomatous tricuspid valve. Jpn Heart J. 1983;24:539–48. https://doi.org/10.1536/ihj.24.539.
33. Weinberg EJ, Kaazempur-Mofrad MR. On the constitutive models for heart valve leaflet mechanics. Cardiovasc Eng. 2005;5:37–43. https://doi.org/10.1007/s10558-005-3072-x.
34. Jett S, Laurence D, Kunkel R, Babu AR, Kramer K, Baumwart R, Towner R, Wu Y, Lee CH. An investigation of the anisotropic mechanical properties and anatomical structure of porcine atrioventricular heart valves. J Mech Behav Biomed Mater. 2018;87:155–71. https://doi.org/10.1016/j.jmbbm.2018.07.024.
35. Khoiy KA, Pant AD, Amini R, Asme M. Quantification of material constants for a phenomenological constitutive model of porcine tricuspid valve leaflets for simulation applications. J Biomech Eng. 2018. https://doi.org/10.1115/1.4040126.
36. Kong F, Pham T, Martin C, McKay R, Primiano C, Hashim S, Kodali S, Sun W. Finite element analysis of tricuspid valve deformation from multi-slice computed tomography images. Ann Biomed Eng. 2018;46:1112–27. https://doi.org/10.1007/s10439-018-2024-8.
37. Rausch MK, Malinowski M, Meador WD, Wilton P, Khaghani A, Timek TA. The effect of acute pulmonary hypertension on tricuspid annular height, strain, and curvature in sheep. Cardiovasc Eng Technol. 2018;9:365–76. https://doi.org/10.1007/s13239-018-0367-9.
38. Malinowski M, Jazwiec T, Goehler M, Quay M, Bush J, Jovinge S, Rausch M, Timek T. Sonomicrometry derived three-dimensional geometry of the human tricuspid annulus. J Thorac Cardiovasc Surg. 2018. https://doi.org/10.1016/j.jtcvs.2018.08.110.
39. Madukauwa-David ID, Pierce EL, Sulejmani F, Pataky J, Sun W, Yoganathan AP. Suture dehiscence and collagen content in the human mitral and tricuspid annuli. Biomech Model Mechanobiol. 2018. https://doi.org/10.1007/s10237-018-1082-z.

# Measurement Technologies for Heart Valve Function

**Morten O. Jensen, Andrew W. Siefert, Ikechukwu Okafor, and Ajit P. Yoganathan**

**Abstract** Experimental measurement technologies have been critical to advancing scientific knowledge and to the development of prosthetic heart valve devices. A myriad of innovative measurement technologies has been successfully utilized within in vivo, ex vivo, and in vitro models. Within these models, these technologies have been used to evaluate the function of native heart valves, models of heart valve disease, and prosthetic devices. These evaluations have focused on quantifying heart valve geometry, dynamics, tissue deformation, transvalvular flow, and valve and device mechanics. Knowledge gained from these studies has advanced reconstructive surgical techniques, implanted device function, and next generation devices. Understanding the application and relative advantages of these measurement technologies is important not only for scientific research but also in quantifying device function per international standards and regulatory guidance.

**Keywords** Heart Valve · Measurement · Pressure · Flow · Biomechanics · In Vitro · In Vivo

---

M. O. Jensen
Department of Biomedical Engineering, University of Arkansas, Fayetteville, AR, USA
e-mail: mojensen@uark.edu

A. W. Siefert
Cardiac Implants LLC, Tarrytown, NY, USA

I. Okafor
School of Chemical and Biomolecular Engineering, Georgia Institute of Technology, Atlanta, GA, USA

A. P. Yoganathan (✉)
School of Chemical and Biomolecular Engineering, Georgia Institute of Technology, Atlanta, GA, USA

Wallace H. Coulter Department of Biomedical Engineering, Georgia Institute of Technology, Atlanta, GA, USA
e-mail: ajit.yoganathan@bme.gatech.edu

© Springer Nature Switzerland AG 2018
M. S. Sacks, J. Liao (eds.), *Advances in Heart Valve Biomechanics*,
https://doi.org/10.1007/978-3-030-01993-8_6

# 1 Introduction

This chapter is dedicated to identifying and describing heart valve measurement technologies. The first section will describe the experimental models used for which measurement technologies are used. Next, the model tools available for assessment of heart valve geometry and dynamics will be described. The flows and pressures surrounding them largely govern the function of heart valves, hence flow and pressure characteristics are the topics for the following section. Finally, the mechanics of the atrioventricular valves and their repair devices will be summarized, serving as an example of how force measurements and other mechanical assessments of valvular function can help complete the picture of how the heart valve functions.

# 2 Experimental Models

## 2.1 Aortic and Pulmonary Pulse Duplicators

Pulse duplicators are a foundation for evaluating the hemodynamic performance of prosthetic heart valves, explanted semi-lunar valves, and bicuspid valve models [1–12]. The measurement technologies used in these models aid in evaluating transvalvular pressure differences, effective orifice area, geometric orifice area, regurgitation, and leaflet opening and closing characteristics. When paired with relevant measurement technologies, detailed flow features through the valve [5, 13, 14], around the valve [15], in the hinges of mechanical heart valves [9], leaflet deformation [16], valve loading forces [7], and device deployment [14] can be evaluated. In general, these systems consist of a pump, polymeric diaphragm, ventricular chamber, valve mounting plate, test valve, valve sinus, outflow chamber, compliance chambers, resistance valves, reservoirs, and fluid heating element (Fig. 1).

Opposed to right or left heart models, pulse duplicators may be commercially sourced through a variety of laboratories and manufacturers [17–20]. If custom-built systems are intended to be used for device development, these systems are recommend to meet or exceed the operational ranges and measurement accuracies recommended by the informative sections of ISO 5840-3 (Annex N).

## 2.2 Right and Left Heart Bench Models

Benchtop simulators of the right and left heart generate significant data for investigations in tricuspid and mitral valve function. The measurement technologies used in these models aim to characterize leaflet opening and closure, transvalvular hemodynamics, valve tissue mechanics, and post device valve function under simulated

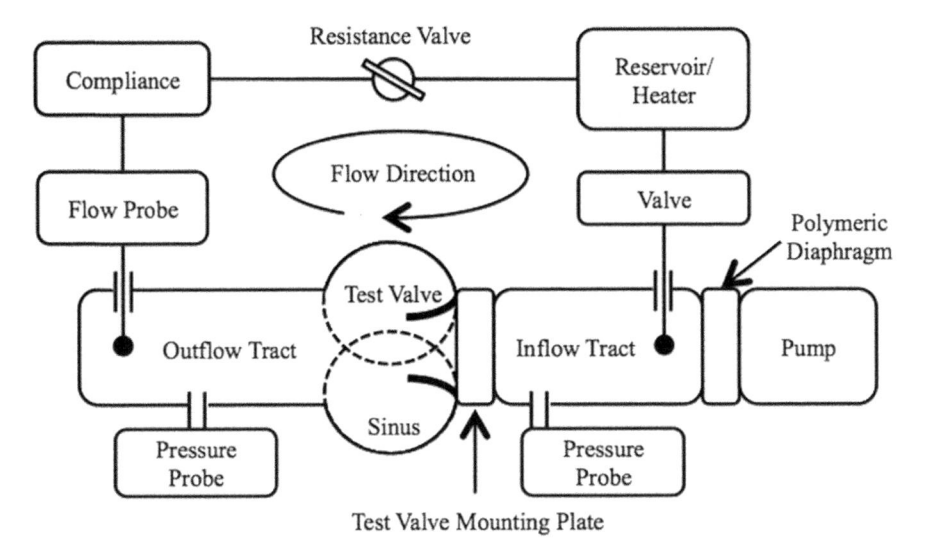

Fig. 1 Schematic diagram of a pulse duplication system with components identified

healthy and disease valve geometries. These metrics are commonly evaluated at population averaged right and left heart hemodynamics that include a cardiac output of 5 L/min, beat rate of 70 beats/min, and peak transvalvular pressure difference of 40 (tricuspid) mmHg or 120 mmHg (mitral) for the tricuspid and mitral valve, respectively. Characterizing the aforementioned metrics and their variation require critical decisions in left and right heart model attributes that balance mechanistic insight with physiologic relevance. The most distinct of these attributes has been used to characterize these models into the three groups that include rigid, flexible, and passive heart models (Fig. 2).

Rigid heart models are fluid-mechanical systems capable of testing explanted tricuspid or mitral valves under static and/or pulsatile hemodynamics. In these systems, an explanted valve is mounted to static and/or adjustable annular and papillary muscle fixtures within an enlarged and non-physiologic rigid-walled ventricular chamber [21–29]. Passive heart models consist of explanted hearts whose left and/or right side is placed in series with a static pressure head or pulsatile flow-loop [29–33]. Passive heart models are advantageous for preserving the native heart geometry and rapid evaluations of static valve function.

The least used of these systems are flexible heart models [34]. These models are similar to rigid heart models except the explanted valve's papillary muscles are partially mounted to an externally actuated polymeric ventricle (see Fig. 2). When immersed in fluid, the polymeric ventricle can be hydrodynamically stretched and compressed, creating pulsatile flow within the flow loop. While the polymeric ventricle can provide interventricular flow patterns that more closely approximate clinical observations [8, 35], this model is burdened with selecting and mounting valves that geometrically fit the polymeric ventricle. Moreover, the polymeric

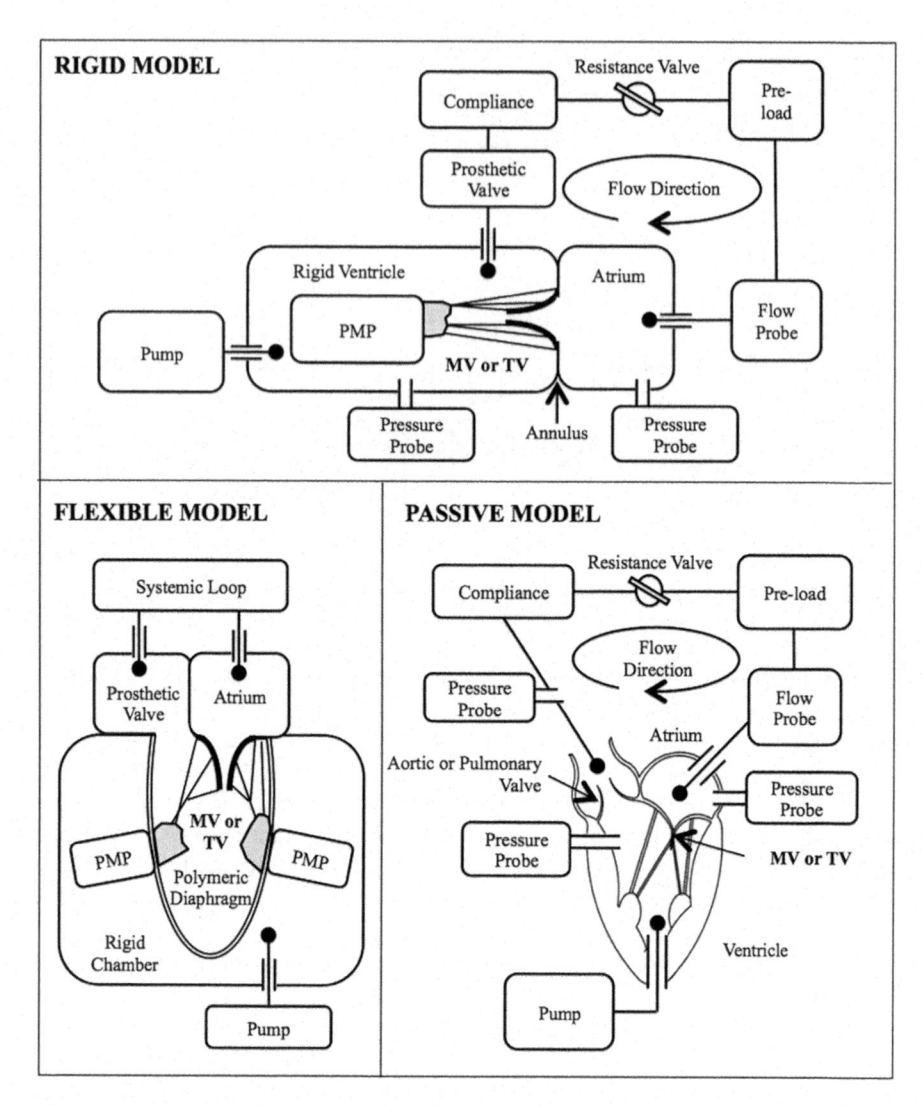

**Fig. 2** Schematic diagrams of common ventricular bench models used in mitral valve (MV) and tricuspid valve (TV) evaluations where PMP denotes papillary muscle positioners (Note: the systemic loop components of the Flexible Model are akin to the Rigid and Passive Models)

ventricle can impede papillary muscle positioning providing further difficulty in establishing user-defined experimental conditions. For these reasons, the flexible heart model has been less popular than rigid and passive heart models for evaluations of explanted mitral and tricuspid valve function.

## 2.3 Steady Backpressure Model

Steady backpressure models are used to evaluate prosthetic heart valve back flow leakage testing. This test aids in characterizing valve leaflet regurgitation and paravalular leakage under anticipated pressures within the pediatric or adult heart. Steady backpressure models may consist of a steady flow pump, upstream chamber (aortic or ventricular chamber), valve mounting plate, test valve, downstream chamber (ventricular or atrial chamber), resistance valves, and fluid heating elements. These components or others should be capable of achieving constant steady backpressures on the test valve at levels appropriate for the valve's intended operating environment. Similar to pulse duplicators, these systems are recommend to meet the operational ranges and measurement accuracies recommended by the informative sections of ISO 5840-3 (Annex N).

## 2.4 Restored Contractility Model

Select investigators have aimed to create a beating heart model by restoring myocardial contractility to explanted porcine hearts using controlled perfusion and controlled loading [36, 37]. This model was adapted and translated from the Langendorff heart model which has been used in pharmacological evaluations [38]. Using porcine hearts, this model has been demonstrated to produce physiologic flow and pressure for time periods sufficient for experimental measurements. During this time period, differing heart valve repair and replacement techniques may be assessed. These may include simulated device deployment or assessing post repair valve function. The ability to conduct these studies in a beating heart is significant. However the complexity associated with model development, operation, and sourcing fresh heart tissue perhaps dampens its application and widespread adoption.

## 2.5 Large Animal Models

Significant preclinical insight to heart valve and device function has been gained from large animal models. Large animal models have primarily included canine, porcine, and ovine species. Both healthy and disease models have been described. The significance of these models and their methods may be found in select publications among others [39–46]. As related to the evaluation of heart valve function, large animal models have aided investigators in evaluating cyclic heart valve geometry, leaflet opening and closure characteristics, transvalvular hemodynamics, detailed flow features through and around the valve, implanted device performance and deformation, and tissue mechanics (annulus, leaflets, chordae, papillary muscles).

While large animal models have provided significant insight to understanding valve and device function, these models are not without their limitations. Compared to human heart valve diseases, only select diseases may be modeled in large animals. Ovine and porcine models exhibit select presentations of ischemic heart disease while canine models exhibit a presentation of myxomatous disease. Towards the development of transcatheter devices, a large animal model mimicking aortic valve calcification has yet to be identified in literature. Porcine models present a more anterior heart position in the chest which can challenge imaging and testing of devices designed to be delivered trans-septally to the human mitral valve [47]. Ovine animals on the other hand are known to exhibit less chordae and thinner leaflets than human patients, which can challenge select experimental designs. While other limitations have been described [39–46], large animal models are critical to scientific knowledge and will continue to be a significant platform for preclinical heart valve device development and evaluation.

## 2.6 Patients

Human patients provide a final platform for evaluating and measuring native and prosthetic heart valve function. Heart valve patients present with a broad spectrum of age and disease whose valvular function can be qualitatively or quantitatively assessed. The measurement technologies used in patients are predominately noninvasive but include select invasive techniques. Noninvasive image-based technologies have allowed investigators to evaluate cyclic valve geometry, leaflet opening and closing characteristics, transvalvular flow, detailed flow features through and around the valve, and implanted device function and deformation. Comparably more invasive catheter-based technologies have been used to evaluate local static pressures, pressure-volume loops, and guidewire position. Given the importance of assessing patient heart valve function, significant resources have been dedicated to improving clinical measurement modalities.

# 3 Assessing Valve Geometry, Dynamics, and Tissue Deformation

## 3.1 Echocardiography

2D and 3D real-time echocardiography is used from bench to bedside to qualitatively and quantitatively assess heart valve dynamics and geometry. In this method, a probe transmits high-frequency acoustic waves to a region of interest that either reflect or pass through the encompassed bodies or boundaries. The waves reflected back to the probe are used to calculate the distance from the probe to the boundary. The distances and intensities of the reflected waves are displayed on a monitor forming 2D or 3D representative temporal images of the plane or volume under examination.

Images acquired from 2D and 3D echocardiography normally allow for all of the valve's structures to be visualized with exception to most of the atrioventricular valve's chordae tendineae and commissural leaflet cusps. Due to limitations in spatial resolution, some larger chordae (strut, intermediary, some basal) may be identified but should not be expected to be regularly visualized. These qualitative images can aid in categorizing valve dysfunctions and assessing valve dynamics and competence. From 2D and 3D echocardiography, quantitative measurements of valvular dimensions, valve orifice area, annular area, annular dynamics, planarity, coaptation length, coaptation depth, tenting area, flail height, flail width, and billowing height, among other related measures, may be quantified.

Echocardiography is a powerful tool when used with bench models. An echocardiography probe can be pressed directly against the simulator's walls near the functioning valve. In this position, the scan settings can be adjusted to obtain images with acceptable spatial and temporal resolution. These images may be used to assess valve geometry, motion, and closure characteristics (fluid measurements are described later). Potential limitations for the use of echocardiography with bench models are the presence of fluid microbubbles and acoustic reflections. Fluid microbubbles can make it appear as if the valve is being imaged through a snowstorm while acoustic reflections can create bright image streaks hindering valve imaging and quantitative measurements.

Similar to bench models, echocardiography probes may be pressed directly against the epicardium of large animals to provide excellent near-field heart valve images. Specific to heart valves, echocardiography has been exceptional for aiding investigators in quantifying valve motion, leaflet coaptation, orifice areas, and planarity and characterizing valve dysfunctions. In another application, echocardiography has been used to quantify prosthetic heart valve stent deflections to estimate device loading magnitudes and their directions [48].

Real-time echocardiography is the standard of care for patient cardiac imaging. Transthoracic or transesophageal imaging windows may be used to directly visualize and quantify heart valve function. The pros and cons of each imaging approach have been extensively discussed, including the advantages and disadvantages of 2D versus 3D real-time echocardiography [49]. Based on current knowledge, the spatial and temporal resolution of echocardiography continues to advance and will foreseeably improve in the future. See Table 1 for a summary of quantitative assessment of valve geometry, dynamics, or both across varying models.

## 3.2 *Magnetic Resonance Imaging*

Cardiac magnetic resonance (CMR) imaging is a nonionizing modality used in large animal research and clinical patients. In CMR, a combination of transmitted radiofrequency pulses and magnetic gradients in the presence of a strong external

**Table 1** Valve geometry (G), dynamics (D), or both (B) can be qualitatively and quantitatively assessed across varying models (marked with the letter "X") using a number of experimental and clinical equipment

|  | Valve anatomy ||||  Available model |||
|---|---|---|---|---|---|---|---|
| Modality | Annulus | Leaflets | Chordae tendineae | Papillary muscles | In vitro | Large animal | Patient |
| 2D echocardiography | B | B |  | B | X | X | X |
| 3D echocardiography | B | B | * | B | X | X | X |
| Magnetic resonance imaging | B | B |  | G |  | X | X |
| Computed tomography | B | B | * | B |  | X | X |
| Micro computed tomography | G | G | G | G | X |  |  |
| Sonomicrometry | B | B |  | B | X | X |  |
| High speed photography | D | D |  | D | X |  |  |
| Stereophotogrammetry | G | B |  | B | X |  |  |
| Trigonometry | G | G |  | G | X |  |  |
| Strobe light |  | D |  |  | X |  |  |

* Some chordae may be imaged with the particular modality

magnetic field are processed to obtain image slices of the heart's valves. CMR may be used to qualitatively evaluate the anatomy and geometry of the patient's valve and surrounding structures [43, 50, 51]. The fluid measuring capabilities of CMR are discussed in subsequent sections.

CMR has not been routinely employed within benchtop heart valve models for assessing heart valve geometry and dynamics. CMR however has been more widely used in large animals but has yet to eclipse the utilities of echocardiography for evaluating structural valve function. Beyond large animals, CMR is widely used in the clinic for assessing a wide array of cardiac and valvular dysfunctions. CMR has been demonstrated to be sufficient in quantifying annular area, planarity, papillary muscle position, geometric orifice area, regurgitant orifice area, leaflet mobility, and leaflet coaptation characteristics [43, 50–52]. Regardless of model, CMR images can be less susceptible to poor patient imaging windows and can provide great insight to valve function.

High Seven Tesla MRI has recently been used in vitro to obtain a superior isotropic voxel resolution of at least 80 $\mu m^3$ in fluid to image heart valves. This enables a true zero-stress and neutrally buoyant configuration for the tissue that is essential for computational modeling [53].

## 3.3 Cardiac Computed Tomography and Micro Computed Tomography

Computed tomography (CT) is an ionizing imaging modality predominately used in large animals and clinical patients to qualitatively and quantitatively assess heart valve dynamics and geometry. In this modality, a series of ECG gated X-ray image slices may be used in isolation or computationally stitched to evaluate 2D or 3D temporal images of the heart's valves. These images exhibit the highest resolution of available imaging modalities. These high-resolution images may be used to qualitatively evaluate valve motion, valve dimensions, orifice area, annular area, planarity, coaptation length, coaptation depth, tenting area, flail height, flail width, billowing height, and regurgitant orifice area.

Beyond measures of valve function, cardiac CT has been used to evaluate implanted device deformation. Temporal image slices may be used to quantify the spatial deflection of implanted devices. These deflections may be used as boundary conditions within computational models to estimate the stress and strain distributions within the device's structure. With caution, these data may be further used to estimate a range of forces and force directions acting on the device. Force data should only be interpreted as a range of values and directions, as with some devices, a resultant stress-strain distribution can be approximated by differing combinations of directional forces.

While cardiac CT has largely been utilized for large animals and clinical patients, micro-CT has been recently used in a benchtop rigid heart model [28]. In this method, excised mitral valve's annular and subvalvular geometry are scanned in statically pressurized and unpressurized conditions. Using mirco-CT in this manner allows the valve to be scanned at an approximate resolution of 20 μm, far exceeding the resolution available in cardiac CT. Due to this high fidelity, scans resulting from this method have been used to create geometric computational models and to assess the accuracy of surgical planning models [54, 55]. More recently, these images have been used to compute valve tissue deformation [56]. Radiopaque makers can be affixed to the mitral valve's tissues and the valve is scanned in a non-pressurized and pressurized state. The resulting scans and marker positions are used to quantify valve tissue deformation and strain. The resulting data provides a whole field strain measurement that as a result exhibits significant value for evaluating finite element model predictions and multi-physics computational model development.

## 3.4 Sonomicrometry

Sonomicrometry is a quantitative spatial measurement technique used within bench and large animal models to quantify valve geometry, dynamics, and tissue deformation [57–61]. In this method, piezoelectric crystals are placed in user-defined positions on the valve annulus, leaflets, papillary muscles, or surrounding structures.

Once the crystals are securely implanted, acoustic waves are transmitted and received between the localized crystals. Based on fluid-material properties and time span between acoustic transmission and reception, the distance between crystals can be calculated. Crystal positions may be used to establish physiologic or device specific coordinate systems. From these coordinates, crystal position data can be used to compute linear distances or alternatively used in least-square-fit algorithms to compute 2D and 3D profiles. These methods have been used to quantify the valve's annular and subvalvular geometry and dynamics, leaflet strain, device deformation [59–62], and tissue velocities [63]. Sampling rates from a commonly used commercial supplier reportedly range from 16 to 1600 Hz with a relative spatial resolution of 0.024 mm [64].

## 3.5 Biplane Videofluoroscopy

Similar to sonomicrometry, biplane fluoroscopy with radiopaque markers has been utilized in large animal models to quantify mitral valve annular and subvalvular geometry, leaflet strain, annuloplasty ring deflection and strain, and annular dynamics and tissue strain [65–71]. In this method, radiopaque tantalum markers are fixed to the large animal's valve tissues under cardiopulmonary bypass and cardioplegic arrest. In a beating heart, the makers are imaged under biplane videofluoroscopy at 60 Hz. These images are subsequently processed using computer algorithms to compute the 4D coordinates of the markers. From these marker coordinates, interpolated shapes and direct distances can be computed to quantify the aforementioned metrics of valve function.

## 3.6 Stereophotogrammetry

Stereophotogrammetry is a method used across varying disciplines to quantify the 3D positions of fiduciary markers on a moving and or deforming surface. In this method, two synchronized cameras are used to capture images of stationary or moving markers. Through calibration, the 3D position of the markers is calculated using direct linear transformation. The spatial positions can then be used to determine maker distances and geometry. By comparing marker positions with other time points, tissue deformation and strain may be calculated. For heart valves, stereophotogrammetry has been used to quantify the leaflet deformation and strain of the tricuspid, mitral, and aortic valves [16, 72–74]. This method has been additionally used to quantify strain in select chordae and their insertion to the mitral valve anterior leaflet [75, 76].

This method has been limited to right and left heart models and pulse duplicators as these models most aptly provide visual access to the test valve. In these systems, the imaging window must be sufficiently large for two cameras to both visualize the fiduciary markers and the calibration target. Oversizing the simulator's chambers therefore significantly aids in visual access. As some simulators contain non-rectangular chambers, care must be taken to account for potential errors introduced by image distortion within the image-strain processing software. Camera, lens, and lighting combinations should not be overlooked with this method. If the imaging target is anticipated to move sufficiently in and out of the calibration plane (example is imaging the mitral leaflets from an enface position), sufficient zoom, a higher f-stop (lower aperture), and brilliant lighting will be required to extend or maintain image focus throughout the cardiac cycle. In terms of application, applied markers should be as small as possible to reduce errors associated with defining the marker centroid in the acquired images. With these challenges in mind, this method can yield excellent temporal resolution with high-speed cameras with spatial resolutions less than 100 μm [16, 56, 72–76].

## 3.7 High-Speed Photography and Videography

High-speed photography and videography may be used within benchtop simulators to assess leaflet opening and closing behavior, geometric orifice area, leaflet bending, pin-wheeling, cavitation and other phenomenon associated with the effect of flow and pressure on the leaflets [77, 78]. The cameras used in these methods are typically placed external to the simulator's chambers at a position directly upstream or downstream from the test valve's mounted position. In a passive heart model, the camera may be directly mounted to a borescope to image the valve from within heart's chambers. When quantitative measurements are sought, a scale or object with known dimensions is inserted into the simulator for image calibration. These images can be manually processed in simple tools such as Image J (National Institute of Health, Bethesda, MD) or may be automatically processed via more advanced methods using MATLAB (Mathworks, Natick, MA), LabVIEW IMAQ/Vision (National Instruments, Austin, TX), or similar programming languages.

## 3.8 Trigonometry

Without other available modalities, investigators with heart bench models have used the relative position of the prosthetic annulus and papillary muscle positioners to determine the explanted tricuspid or mitral valve's geometry. This method has been used by investigators in modeling annular dilatation and papillary muscle displacement in functional or ischemic regurgitation [25, 79, 80]. Further, this method has been used in simulating repair procedures which augment the valve's geometry. These studies have included evaluating the effects of annuloplasty, edge-to-edge repair, chordal repair, tissue fixation with glutaraldehyde, and papillary muscle repositioning on mitral valve function [81–85].

## 4 Assessing Flow Characteristics

The purpose of native and prosthetic heart valves is to control forward flow between and from the heart's chambers. This requires a valve to allow for forward flow with an acceptably small transvalvular pressure gradient while minimizing leakage or retrograde flow during valve closure. Beyond these basic requirements, heart valves are desired to minimize blood shearing in order to reduce the potential for thromboembolic events. Quantifying these parameters requires determining both bulk and detailed flow characteristics through and around a heart valve. These parameters can be quantified using some of the most simple and most complex measurement technologies available to heart valve investigators.

### 4.1 Bulk Flow Characteristics

#### 4.1.1 Flow Probes

The majority of bulk flow measurement technologies used in heart valve research are based on electromagnetic or Doppler ultrasound principles. Flow probes based on these techniques are available in widely varying sizes that can provide high sensitivity and adequate frequency response without impeding fluid (blood or analog solution) flow. The operating principles of these technologies are now described.

Electromagnetic flow probes operate on the principle of electromagnetic induction and measure the disturbance of a magnetic field by the flow of ions through that field (using Faraday's Law of Electromagnetic Induction). The probe associated with the flow meter contains an electromagnet, which produces a magnetic field across the vessel. Electrolytes are minerals in the blood stream that carry an electric charge. If a blood-analog fluid is used in an in vitro flow-loop, salt may be added to ionize the solution. Motion of the ionized fluid through the magnetic field generates an induced voltage. For a given vessel diameter, the induced voltage is proportional to the volumetric bulk flow rate. These probes need to be calibrated and verified using the chosen fluid.

Transit time ultrasonic flow meters measure the change in time that it takes acoustic ultrasound waves to travel through the flow field. The physical principle of measuring this change is accomplished two piezoelectric elements alternating emission and reception of ultrasonic waves. The emitted waves pass through the fluid, are redirected by a reflector placed on the opposite side of the vessel or tube, and are subsequently received by the other crystal and vice versa. This method enables the system to calculate the change in time of ultrasound waves travel in both directions, which is calibrated to derive the flow rate. These probes can measure flow rates above 50 L/min with an absolute accuracy of approximately 10% of the reading.

### 4.1.2 Static Pressure Transducers

Fluid static pressure transducers used in cardiovascular applications measure pressure by way of measuring the deflection of a diaphragm that is in contact with the subject fluid. The deflection of the diaphragm as a response to a pressure can be measured by strain gauges mounted to the diaphragm. As the diaphragm deforms, the strain gauge is stretched or compressed, and by means of calibration, the strain gauges' electrical output voltage can be converted to measured pressure. These transducers can measure pressures of more than 300 mmHg with an accuracy of $\pm 1$ mmHg.

Static pressure transducers are characterized into using direct and fluid-filled-catheter sensing techniques. The latter is performed with a fluid-filled pressure catheter and is the most commonly used technique clinically. A fluid-filled tube connects the sensor position to the actual measurement location. This catheter type needs priming, as air bubbles present in the fluid line can be detrimental to signal quality. Examples of fluid-filled pressure catheters are the Edwards Lifesciences TruWave® (Edwards Lifesciences, Irvine, CA) or the Utah Medical Products Disposable Pressure Transducers (Utah Medical Products, Inc., Midvale, UT).

The frequency response of the fluid-filled catheter-based pressure sensors are much lower than the direct sensing ones, which are not commonly used clinically but more so in cardiovascular research applications. In direct sensing techniques, the diaphragm is in direct touch with the fluid at the measurement location. The frequency response of these direct sensing devices is very high (in the kHz range), which make them great to use for research applications. For in vivo heart valve pressure sensing applications, the placement of this pressure sensor is critical, as the sensing membrane of the catheter should not be touching any moving parts of the heart tissue. If that happens, an erroneous signal known as catheter whip is confounding the actual fluid pressure signal. An example of a direct-sensing pressure catheter is the Millar Mikro-Tip® Research Pressure System (Millar, Houston, TX).

Differential pressure transducers where the diaphragm is part of an inductor can be used to measure the pressure across heart valves with one transducer. As with the strain-gauge-based devices, these devices measure the deflection of a diaphragm. However, it is an inductance change that measures the diaphragm deflection. The diaphragm is indirectly subjected to the fluid pressure through a fluid-filled tube that is more stiff and shorter than those used clinically, making the frequency response of this system better than the clinically used fluid-filled pressure catheters. An example of differential pressure transducers where the diaphragm is part of an inductor that can be used for measuring pressure across heart valve is the Validyne DP15-24 (Validyne, Northridge, CA).

## 4.2 Doppler Echocardiography

Doppler echocardiography for fluid velocity measurements uses the principle of measuring the change in frequency of a sound wave that is emitted and received as an echo. The measured change in frequency is used to calculate the instantaneous velocity of a particle in the fluid field [86]. In Doppler echocardiography, the emitter is stationary and the object of interest is moving. The location of the emitter is the same as the receiver, and this is where the Doppler frequency shift is measured. This modality has been successfully used in bench models, large animals, and clinical patients. A few limitations however exist. The Nyquist Theorem determines the maximum velocity that a Doppler system is able to measure without aliasing. The further away from the ultrasound transducer the flow velocity is measured, the lower the maximum velocity can be measured with the same carrier frequency [87]. Also, Doppler systems provide the component of the velocity towards or away from the transducer. The orientation of the Doppler ultrasound transducer away from a perfect parallel to the flow-field of interest is called the insonation angle. To adjust the displayed velocity, a factor of cosine to the insonation angle is multiplied to the apparent measured velocity. However, an error is inherent to this approach: at 40°, a 5% error in angle estimation results in an error in velocity measurement of approximately 10% [88].

### 4.2.1 Continuous Wave Doppler

Continuous wave (CW) Doppler is an operating mode of most echocardiography machines that measures the maximum velocity along a line within 2D flow field or plane [89, 90]. In the CW mode, the piezo-electric elements within the echocardiograph probe transmit ultrasonic waves continuously along a user-defined line. Simultaneously, the transducer receives all echoes from along that same line, with the highest velocity of particles passing through the line determined. Beyond measuring peak fluid velocities, CW may be used to determine the pressure gradient across a stenotic heart valve, which is found [unit: mmHg] with the modified Bernoulli's formula as four times the maximum velocity [unit: meters/second] squared. The velocity time integral may also be used to find the effective orifice area of a heart valve, which uses the principle of mass conservation of an incompressible fluid [91].

### 4.2.2 Pulsed Wave Doppler

The pulsed wave (PW) Doppler mode is available in select 2D echocardiography systems aiming to measure flow velocity at a singular point within a 2D flow field [92]. The location of the point is obtained by knowing the time of flight of the ultrasound beam to reach that point. The fluid velocity at that point is found by the same principle as CW, but for that point only. Knowing the velocity in different locations along an assumed streamline can be used with the modified Bernoulli equation to quantify a pressure difference between locations in the flow field.

### 4.2.3 Color Doppler

Color Doppler echocardiography has revolutionized qualitative blood flow imaging the way clinicians assess the flow properties of the cardiovascular system and heart valves, in patients and the unborn fetus. The flow is imaged with color code for direction and velocity. This principle is the same as a pulsed wave, but instead of focusing at a single location, the line is swept across the 2D or 3D image-field and the location along the line is shifted, such that the flow velocity at all locations in the selected field of interest is calculated and displayed.

### 4.2.4 Fast Vector Echocardiography

In this modality, a vector flow ultrasound technique uses radar principles to get full flow-field as true 2D vectors information, which is comparable to particle imaging velocimetry. The best way to understand the technique is by looking at a linear array of regular color Doppler transducers and combine the information from each of these into an algorithm that provides the 2D flow vectors (Fig. 3). This technique is still in its early stages of development, but exhibits significant potential. The current most significant challenge for this modality is penetration depth for in vivo cardiac use, although the first reports of intraoperative imaging of heart valves with this measurement technology has been published [93]. The frame rates for live imaging are in the 15–25 Hz range.

**Fig. 3** Fast vector flow ultrasound in the heart. Reproduced from [93] with permission from Elsevier Limited, Kidlington, Oxford, UK

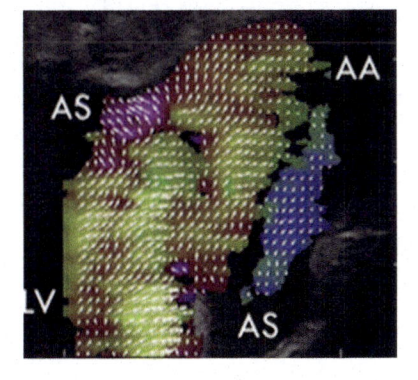

### 4.2.5 Proximal Isovelocity Surface Area (PISA) and 3D Methods (3D-FOM) for Regurgitant Flow Measurements

Clinical assessment of heart valve regurgitation is most commonly performed with a technique called two-dimensional PISA [94, 95]. The effective regurgitant orifice area and regurgitant volume are quantified. One assumption of this technique is that the convergence zone of the regurgitant flow upstream of the valve is a true hemispheric shape. Additionally, regurgitant orifices that are split into two or three separate orifices further challenge the application. In recent years, three-dimensional echocardiography has spawned methods for further improvement of regurgitant flow measurements [96–98]. These new techniques are promising in terms of less user input, faster processing, and improved reliability compared with PISA, and as the 3D echocardiography equipment improves any of these may become the method of choice.

## 4.3 Detailed Flow Field Characterization

### 4.3.1 Optical Flow Visualization

Flow visualization is an optical technique that utilizes neutrally buoyant tracer particles (such as pliolite) seeded inside a fluid to qualitatively assess flow structures present within a fluid field. This technique involves illuminating the seeded tracer particles by means of a light sheet and capturing the particle motion using a high-speed camera. In select applications, flow visualization can also involve the use of dyes to visualize flow structures within the fluid field. Although most flows found in the cardiovascular system are three dimensional, flow visualization is limited to 2D planes that are illuminated by the light sheet. This technique is mostly used to analyze the flow structures generated by valvular prosthetic devices, as well as flow structures generated by other prostheses such as ventricular assist devices [99, 100]. Flow visualization is used as a first pass check for flow structures and regimes which could be undesirable, such as recirculation zones, valve leakage, and regions of flow stasis (Fig. 4).

Flow visualization is only performed on bench models with sufficient imaging access. The advantage of flow visualization is the ease at which it can be performed without the pre- and post-processing required for quantitative techniques. Like other optical techniques, flow visualization is susceptible to errors resulting from optical refraction. These errors include, but are not limited to, over- or underestimation of particle displacement due to light crossing a curved surface or moving between mediums.

**Fig. 4** Use of flow visualization to observe flow structures generated at peak systole by a bioprosthetic heart valve in an idealized aortic geometry

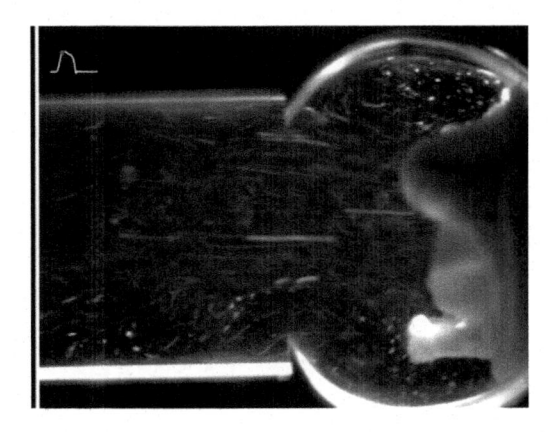

### 4.3.2 Particle Image Velocimetry (PIV)

PIV is an optical technique used to qualitatively and quantitatively analyze a particle-seeded fluid domain. It utilizes a pulsed laser to illuminate a fluid region of interest and frame-straddling cameras to capture two successive instantaneous images separated by a known time interval (a function of the velocity of the flow being measured). Using cross-correlation statistical analysis, the velocity of the particles within the domain captured by the cameras is calculated.

There are many variations of the PIV technique including, but not limited to, 2-dimensional-2-component (velocity in the $x$ and $y$-direction), stereoscopic (2-dimensional-3-component), tomographic, high-speed, micro, and plenoptic PIV. Each of these variations has their respective advantages and disadvantages [101]; however, the most commonly used PIV technique is 2-dimensional-2-component PIV, simply because of its quick setup time and cost relative to the other PIV techniques.

In optically accessible bench models, PIV is predominantly used to evaluate the hydrodynamic performance of prosthetic heart valves [100]. This method has additionally been used to understand bicuspid aortic flow fields [5, 102], hinge flow in mechanical heart valves [9, 103], valve induced ventricular filling patterns [35, 104], thromboembolic potential [105], and diseased aortic valves, such as aortic stenosis and regurgitation [106, 107]. From the extracted velocity fields, parameters such as velocity magnitudes, velocity profiles, flow stasis, shear stresses, relative pressure fields, elevated turbulence, and much more can be extracted. These metrics have increased the understanding of valvular regurgitation and/or stenosis, blood damage, and thromboembolic potential—all of which provide insight on the performance of the prosthetic heart valve being tested (Fig. 5).

Two major requirements of PIV are (a) visual access to the region of interest and (b) the fluid domain being studied has to be seeded with neutrally buoyant particles of some form. These requirements make it currently impossible for PIV to be performed in vivo. Within bench models, optical techniques are susceptible to errors arising from changes in refractive indices between mediums (for example, air to

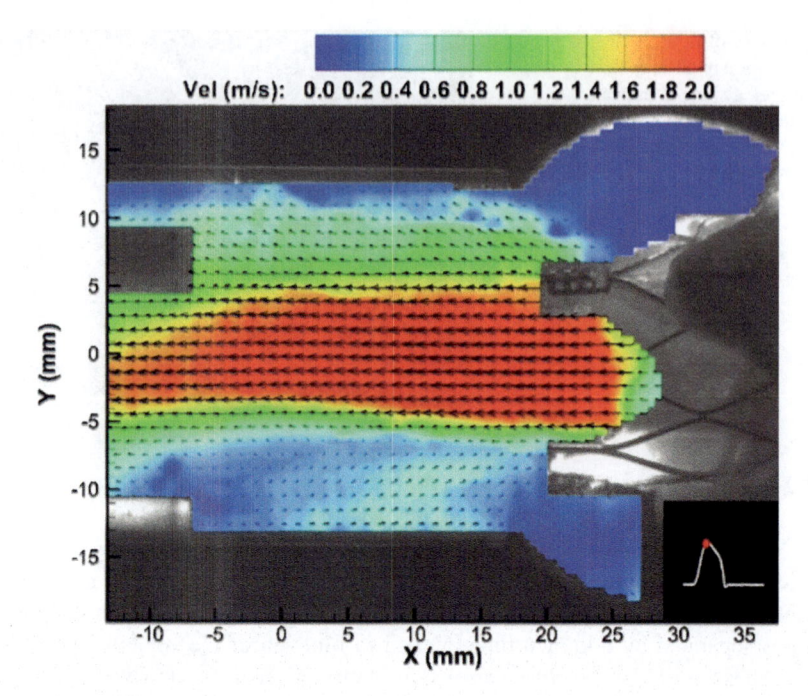

**Fig. 5** PIV used to visualize and quantify flow structures generated at peak systole by a transcatheter aortic heart valve in an idealized aortic geometry

water); therefore, in some cases (especially for small fields of view or high curvatures between mediums), index matching is necessary. For further minimization of errors, PIV also requires the seeded particles to freely follow the fluid medium without interfering with the flow. This means that the densities of the particles must be very close to that of the fluid. Smaller particles can be used to extract finer flow features; however, a camera with a higher resolution must be used in order to capture the smaller particles.

Recently, a new PIV technique is in development known as echocardiographic PIV [108]. It combines contrast echocardiography with PIV processing techniques to extract a velocity field. This PIV technique attempts to overcome the optical access limitation of all other PIV techniques by using ultrasonic waves. Certain challenges exist in order for this technology to become clinically useful on a large scale. For example, the spatial resolution of the echocardiography images still needs to be improved. In addition, having to use contrast agents in certain echo-PIV systems is a significant limitation [109, 110].

### 4.3.3 Laser Doppler Velocimetry

Laser Doppler Velocimetry (LDV) utilizes coherent, monochromatic, collimated, and intersecting laser beams to measure the velocity (either 2 or 3-components) of fluids at a single point. Common lasers that are used include Argon ion, Helium-Neon, and laser diodes. The intersecting laser beams, generally obtained from a single source beam to ensure coherence, interfere and generate a set of fringes. It is common practice that the laser beams are set to intersect at their respective focal points, which is the point where the fluid velocity will be measured. As neutrally buoyant, seeded particles cross the intersection of the laser beams, they reflect light, which fluctuates in intensity, and is collected by optical sensors (usually photodiodes). The frequency of the intensity fluctuations in the collected light is correlated to the Doppler shift between the incident and reflected light which can then be used to determine the velocity of the moving particle. Using this technique, all three components of velocity can be determined at a high frequency.

LDV has historically been used in bench models to noninvasively measure velocities around prosthetic heart valve devices [111, 112]; however, this technology is slowly being replaced by PIV. Because LDV is a point measurement, its usage is currently focused on regions of interest that remain difficult to access via PIV, such as near wall measurements [113, 114], and regions difficult to view via cameras [115]. These are both the advantage and disadvantage of LDV in comparison to PIV—accessibility to difficult regions of interest, and single point measurements, respectively. Like PIV, LDV is also an optical technique, meaning that not only is it susceptible to errors arising from refractive index changes, the fluid under observation must be either transparent or semitransparent.

### 4.3.4 Phase Contrast Magnetic Resonance Imaging

Phase Contrast Magnetic Resonance Imaging (PCMRI) is a flow visualization and quantification technique used to analyze blood flow in large animals and clinical patients. PCMRI relies on magnetic fields and radio waves to alter the alignment of the spinning hydrogen atoms in the body. To measure blood flow, phase contrast pulse sequences use bipolar gradients to encode the velocity of the spins. Blood that is in motion experience a difference in magnitude between the two gradients because of its change in spatial location. This experienced phase shift is then correlated to the velocity of the blood.

PCMRI is commonly used to extract the velocity of blood in either one or two dimensions within a 2D image slice. PCMRI has been used to assess blood flow across heart valves [116–118], within the heart chambers [119], and within the greater vessels of the cardiovascular system [120, 121]. The velocity fields from this technique can be used to evaluate the in vivo performance of prosthetic devices by identifying unwanted phenomena such as regurgitation, stenosis, regions of low flow, and high residence times. Beyond in vivo applications, select investigators have used PCMRI on the bench to evaluate and isolate these phenomena [8, 122] (Fig. 6).

Advances in encoding techniques has made it possible for PCMRI to be used to acquire 3D, time resolved, velocity information. This allows for further analyses to be carried out on the extracted information, such as the 3D visualization and

**Fig. 6** Evaluation of flow through the native mitral valve using PCMRI

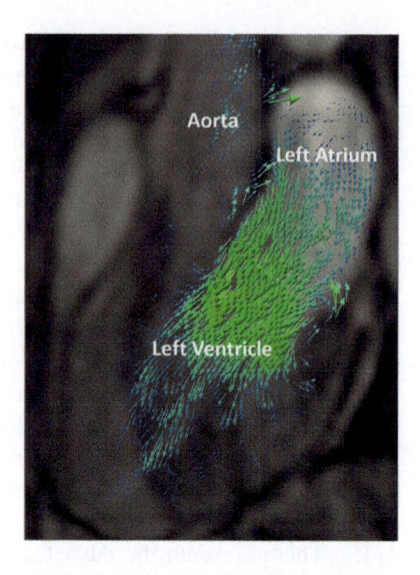

characterization of the temporal evolution of blood flow across valves or within the heart chambers [123], and the calculation of relative pressure fields [124]. As more functionality is added to PCMRI acquisition, like moving from 2D to 3D, trade-offs between spatial/temporal resolutions, scan time, field of view, and signal-to-noise ratio must be considered for a given evaluation.

When compared to bench-based modalities such as PIV and LDV, PCMRI exhibits inferior spatial and temporal resolution. Due to the nature of the measurement technique, PCMRI is susceptible to errors arising from eddy currents and gradient field distortions [125, 126]. Despite the limitations of PCMRI on the bench, this modality is currently the best technique for evaluating blood velocities within a 2D or 3D region near or through native and prosthetic heart valves.

## 5 Technologies for Quantifying Atrioventricular Valve and Repair Device Mechanics

Normal function of the atrioventricular (AV) valves consists of a complex interplay between the annulus, leaflets, subvalvular apparatus, and right or left ventricle [127, 128]. Clinical visualization techniques such as echocardiography and magnetic resonance imaging (described previously) have provided insight into the 3D dynamic behavior of the AV valves. However, the biomechanics of these complex valves must be seen through the light of the components making up the force balance. These dynamics are the result of a three-dimensional force balance that consists of a complex and time-varying combination of interacting tissue mechanics.

Limitations observed in implanted devices have aided the development of new measurement technologies to elucidate biomechanical phenomenon and supplement

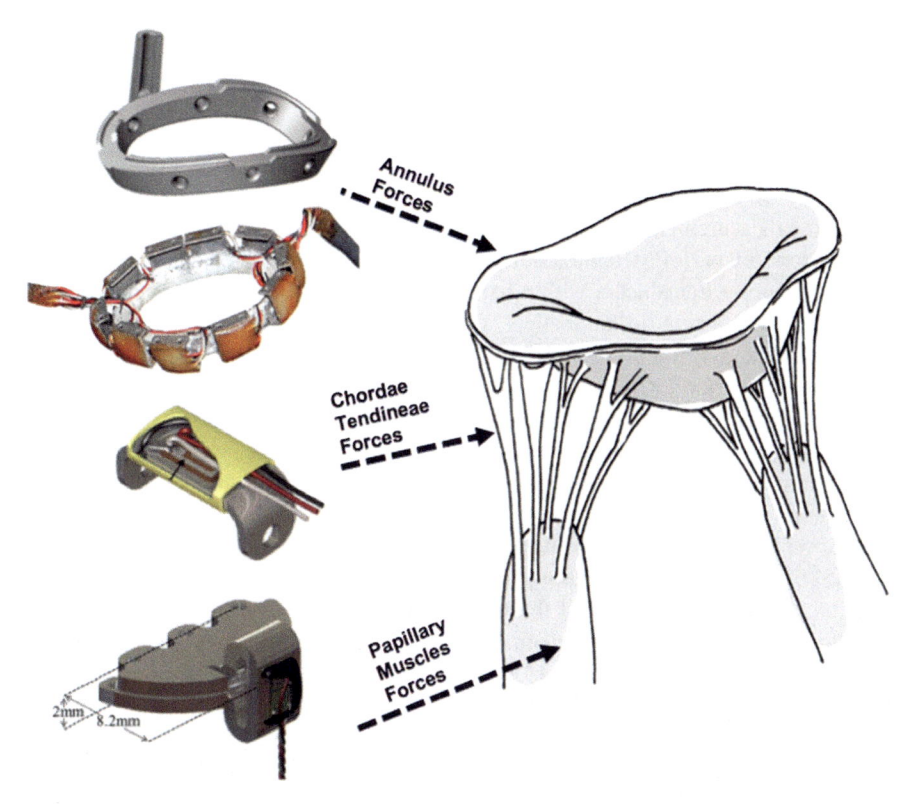

**Fig. 7** Select examples of transducers used to quantify AV valve annulus, chordal, and papillary muscle based forces

device development. Many of these failures may be attributed to unbalanced forces, which have been measured to improve the development of surgical techniques and implantable devices. Results from using heart valve measurement technologies have taught the importance of considering the AV valves as part of the ventricular myocardium at the subvalvular apparatus as well as the atria-ventricular-aortic interaction at the annulus [128, 129]. This has motivated a more in-depth analysis of important components of the force balance in the valve apparatus, see Fig. 7.

## 5.1 *Assessing Subvalvular Mechanics of the Atrioventricular Valves*

Starting from the apex of the heart, the PMs are the first interface between the right or left ventricle and the atrioventricular valves. Force measurements of the chordae tendineae and Papillary Muscles (PMs) have been performed to assess healthy valve function, severity of disease, and biomechanical impact of repair on the AV valve force balance. This has been performed in models of both degenerative and functional situations as seen from the subvalvular apparatus.

### 5.1.1 Papillary Muscle Force Transducers

PM force measurements were first performed in vitro by use of strain-gauge-based transducers. In vitro PM contraction is absent in this model, and therefore all PM force measuring transducers within bench models provided a surrogate measure of pressure forces acting on the leaflets. These strain-gauge-based transducers consist of instrumented PM holder rods that relate strain to force reported as a three-dimensional vector [80] or as a one-dimensional magnitude of the force vector along the principal axis [130]. The PM from the explanted valve was mounted on the tip of these devices, and this tip allowed the PMs to rotate freely, aligning with the principal force axis of the chordae tendineae.

A large animal in vivo papillary muscle force transducer was developed as a platform plate that connected directly to the fibrous membrane of the PM head and the rest of the muscle (Fig. 7 above left) to provide a one-dimensional magnitude of the force along the direction of the transducer. This in vivo transducer provides a reaction force between the left ventricle and the transvalvular pressure force acting on the mitral leaflets [60, 131, 132]. As with all force measurements in the AV apparatus, and in particular with an in vivo transducer of this caliper, the challenge is to minimize the impact that the sensor has on the overall valve functionality. For this particular transducer, that was performed by minimizing the height of the part of the transducer placed under the fibrous membrane.

For both the in vitro and in vivo transducers, whenever possible the force was acquired as a function of time and synchronized with a number of other measurement technologies (pressure, flow, images of the valve, etc.). The in vivo PM force has shown to be a combination of the fluid pressure on the AV leaflets *and* the valvular–ventricular interaction force: These measurements provided insight to timing of these forces with the function of the valves. Careful combination of the results from in vivo and in vitro measurement technologies adjusting to valve area and trans-leaflet pressure has indirectly indicated the valvular–ventricular interaction force through the PMs onto the subvalvular apparatus [60, 128]. In addition, the synchronization technology with other physiological measurements has provided the important difference in the timing of this force between the valve closing in a passive setup (in vitro) and an active setup with PM contraction (in vivo) [128].

### 5.1.2 Chordal Force Transducers

For Chordal Tendineae (CT) force measurements, several approaches have been used to quantify chordal tension. The most commonly used is a "C"-shaped frame, which has been instrumented with strain gauges to provide the force in individual CT [133]. The transducer is sutured onto the chord on each end of the "C," and the chordae tissue in-between the ends are cut. This transducer design and implantation procedure minimizes the amount of tissue that is replaced by the transducer and ensures that all force is directed through the transducer and no force is partially transferred through the chordae. The size of the transducer limits the number of transducers that can be placed. The maximum number of instrumented chordae that have been used is six for the mitral and tricuspid valves in vitro [134, 135] and four in vivo [136, 137]. An alternative version of this transducer mounts the transducer frame through the chord basically splitting the chord into two halves. As the force in the chord is increased, the frame is compressed, and strain-gauge-based measurement equipment provides the calibrated force [138].

Over the years, the C-based transducer has answered many questions within bench and large animal models. Examples are the difference in forces for different chordal types [139], the difference in force balance of the chordae between a stiff, flat annulus and a flexible annulus model [85], the force in chordae following leaflet alteration [140, 141], the force in ePTFE artificial neo-chordae replacement [82, 142], and the force in specific CT as a response to functional disease and repair with different types of annuloplasty [136, 143].

## 5.2 Assessing Annular-Based Mechanics

The loads that act on devices implanted to the mitral valve annulus result from a combination of fluid pressure, tension within the valve leaflets, myocardial muscle fiber contraction, and expansion of the aortic root. These forces in combination can apply compressive, tensile, bending, and torsional loads on implanted mitral annular devices, and have been demonstrated to exhibit non-uniformity through the annular perimeter. These forces are inherently three dimensional. Annular-based force transducers aim to understand the forces acting within healthy native and diseased valves and implanted devices, and the implications of these forces to device design and repair strategies.

### 5.2.1 Annuloplasty Ring Mechanics

Annuloplasty is a reconstructive surgical technique that constrains the shape and size of the annulus to a prosthetic ring. Force measurements in annuloplasty rings can be ambiguous, and the reporting of forces need to refer back to a calibration setup to lower (but not entirely remove) the dependency of the reported forces on the transducer frame design and material. In addition, cross talk elimination and analysis of the data with dependent variable considerations is necessary. The force in the mitral annulus has been described in vitro as a pure response to the pressure acting on the AV valve leaflets [144]. However, since the active myocardial interaction with the annulus and any device attached to this part of the valve is crucial, in vivo measurements of annulus-based forces with devices that attach to the active annulus have provided a more realistic picture of how these forces may interact with a device [128].

Out-of-Plane Annular Force Transducers

The first direct force measurement in the mitral annulus in an animal model was performed with a transducer modeled after the traditional Carpentier-Edwards Classic annuloplasty ring (Fig. 7 above left). Strain gauges were placed at four locations of the transducer frame, and the measurements were reported as forces with relation to a dedicated calibration setup. This approach was used for rings with different shapes, identifying that the out-of-plane saddle shape experienced the lowest force, which have been adopted in future annuloplasty ring designs [145–147].

Indirect measurement of the mechanics in annuloplasty rings has also been reported by utilizing fluoroscopy and tantalum markers to follow the annulus movements before and after ring implantation (technique described elsewhere in this chapter). The knowledge of the particular ring design and material properties enables the "reverse engineering" method of deriving the stress distribution in the particular device [148].

In-Plane and Multiple-Plane Annular Force Transducers

The majority of studies seeking to measure mitral and tricuspid annular forces have focused on quantifying cyclic compression and tension. Hasenkam and colleagues were the first to describe the forces generated by the myocardium on a mechanical heart valve [149]. This was accomplished by adhering strain gauge rosettes to a 29 mm Edwards-Duromedics mitral valve. For calibration, strain gauge output voltages were calibrated to known applied forces. These calibrations were used to convert strain gauge output voltages measured in large animals to estimated forces. Later, Shandas et al. used 3D ultrasound to measure the deformation of St. Jude Medical Biocor® stented prosthetic MVs in two healthy porcine subjects [48]. Measured ring deformations were used as boundary conditions for finite element analysis to estimate the maximum directional forces acting on the tissue valve. This technique is similar to the use of CT described above.

These studies were further supplemented and supported by research conducted with an x-shaped, strain gauge-based transducer aiming to isolate forces in the septal-lateral and transverse directions of the mitral annulus [150, 151]. Compared to previous studies, this transducer demonstrated differences in directional annular forces and their variation with left ventricular pressure and ischemic mitral regurgitation. To avoid cross talk between the annular and apical planes, additional strain gauges have been mounted on an additional transducer setup on the segments attached to the annulus. The signals from the strain gauge elements are acquired synchronously and careful calibration with the application of cross-talk elimination algorithms developed during calibration of the device enabling directional annular force measurements [152].

In-plane forces have also been reported as relating to a dedicated calibration setup of a full D-Shaped annuloplasty ring with the strain gauges mounted on the inner surface towards the center of the annulus [153]. Simultaneous in- and out-of-plane forces were measured with a similar type of ring with gauges placed both on the top and inner side of the transducer [154].

### 5.2.2 Forces in Annuloplasty Ring Sutures

The anchoring of annuloplasty rings to the surrounding tissue is of crucial importance to short-term as well as long-term success of the intervention. Dehiscence of annuloplasty rings is becoming a recognized problem [155]. Hence, investigations of the force that the annuloplasty ring suture experiences as a response to ring size and left ventricular pressure has been measured with a novel system that interfaces the entire annuloplasty ring [156] (Fig. 7 above, left). Used in an ovine model, this system is used to investigate the mechanics of annuloplasty ring dehiscence and impact on suture force from annular downsizing, and it has been shown that although the dehiscence most typically is experienced in the muscular portion of the annulus, the highest suture forces are experienced in the fibrous part of the annulus.

### 5.2.3 Forces in Valve Leaflet Approximation Procedures

The system described above to measure forces in the chordae tendineae has been adapted to measure the force in the leaflet approximation edge-to-edge repair technique referred to as the "Alfieri-stitch" in bench [157] and large animal studies [158]. The studies have demonstrated that diastolic force in the Alfieri stitch increases with increases in the septal-lateral dimension of the mitral annulus.

### 5.2.4 Coronary Cinching Forces

The force necessary to cinch the mitral annulus through the coronary sinus has been measured in an ex vivo model with a hydrostatic pressure applied to the left ventricle [30]. The measurement technology was built around a one-dimensional load cell that measures a tensile force from the cinching system, which intend to compresses the mitral annulus in an attempt to restore a competent valve.

## 5.3 Heart Valve Force Transducers with External Anchoring

Measuring forces in heart valve structures can be challenging when the strain sensing element is directly embedded among the valvular structures as described in several of the technologies above. An alternative approach is exteriorizing the anchoring of the strain sensing element externally which also minimizes the ambiguity of calibration since the fixation of the transducer is well defined.

One advantage of applying these exteriorization techniques for measuring annular forces is that locating a rigid fixation outside of the heart minimizes the ambiguity of calibration. As with any force transducer in the AV valves where minimizing this ambiguity is part of the interpretation, careful consideration of the free-body diagram of the device interacting with the mitral annulus in the apical direction is important [159].

### 5.3.1 Papillary Muscle Relocation Forces

Recent reparative techniques for functional AV valve insufficiency include papillary muscle relocation. A device to measure forces generated on traction sutures utilized for this purpose was designed based on a modified caliper with strain gauges attached. The system was designed to facilitate investigation of the effects of shortening GoreTex traction suture that was extended from near the fibrous portions of the AV valve through the papillary muscles. The suture was exteriorized out through the left ventricle in a porcine setup and attached to the dedicated device for simultaneous papillary muscle relocation and traction suture force measurement. Peak force was seen at the onset of the systolic isovolumic contraction [159, 160].

### 5.3.2 Apically Tethered Annular Force Transducer

The force required to anchor a transcatheter mitral valve replacement (TMVR) to the left ventricular apex within a porcine model was recently demonstrated [161]. This force was measured by the use of a load cell mounted external to the beating heart. This device was used to exteriorize the anchor and measure the tethering force between the TAVR and the left ventricular apex.

### 5.3.3 Diametric Annular Cinching

Annular cinching has been described as a method of restoring AV valve competence. The type and amount of cinching necessary to obtain the equivalent functionality of an annuloplasty ring has been documented by several animal experiments using fluoroscopy and tantalum markers [162]. An original approach to measure the cinching force reuse the above-described transducer that was originally developed for chordal force measurements [61]. The transducer was mounted on non-flexible suture that was anchored to the fibrous part of the AV annulus and exteriorized through the annulus on the muscular side with a system that could stepwise decrease the size of the annulus. The system provided an experimental basis for determining how segmental annular cinching forces are related to valve function. This information is important for designing diametric annular cinching systems.

## 6 Additional Modalities for Assessing Prosthetic Heart Valves

Beyond the measurement technologies previously discussed in this chapter, additional modalities exist to evaluate the function of heart valve devices and their components under simulated use conditions: Servo-hydraulic or electromechanical load frames that for example may be used to assess device component material properties, Stress/Life (S/N) characterization, Strain/Life ($\varepsilon$/N) characterization, fatigue crack growth (da/dN) characterization, crush resistance, and component testing. More information regarding these analyses may be found in the following representative publication, standard, and guidance [10–12]. Support structure creep testing may additionally be performed using custom loading frame techniques. In the case of stent-based devices, specialized iris-based systems may be used to record and characterize stent radial stiffness and strength, chronic outward force during expansion, and the radial reactive force during compression. Simple experimental setups may additionally be used to evaluate post expansion device recoil and dimensional verification. Device corrosion may additionally be assessed. These modalities among all others supplement and aid in heart valve device development.

## 7 General Discussion and Final Remarks

Heart valve measurement technologies will continue to evolve as the functionality and complexity of repair strategies and devices increases. The technologies described in this chapter have all been successfully utilized within bench, large animal, and clinical patients as part of evaluating native and diseased heart valve function as well as devices for repair or replacement. Each of the technologies focus on the evaluation of heart valve performance via measurements of blood flow, pressure, and tissue force characteristics.

Bench models can provide optimal access for manipulation of the heart valve being tested as well as precise control of the individual components of the cardiovascular system: Pressure, flow, geometries, etc. The advantage of large animal models is that the entire dynamics of the living beating heart is included. The decision of which to choose that involves several factors such as time and funding available to perform the experiments. For device development, which phase in the product development cycle is also a determining factor.

Comparing in vitro and in vivo experimentation with heart valve measurement technologies will be increasingly common in the future. As it has been demonstrated previously, these comparisons may provide mechanistic insight of heart valve function that was otherwise not possible by any of the two techniques individually.

## References

1. Bluestein D, Rambod E, Gharib M. Vortex shedding as a mechanism for free emboli formation in mechanical heart valves. J Biomech Eng. 2000;122(2):125–34.
2. Falahatpisheh A, Kheradvar A. High-speed particle image velocimetry to assess cardiac fluid dynamics in vitro: from performance to validation. Eur J Mech B Fluids. 2012;35:2–8.
3. Herbertson LH, Deutsch S, Manning KB. Modifying a tilting disk mechanical heart valve design to improve closing dynamics. J Biomech Eng. 2008;130(5):054503.
4. Moore B, Dasi LP. Spatio-temporal complexity of the aortic sinus vortex. Exp Fluids. 2014;55(7):1770.
5. Saikrishnan N, et al. In vitro characterization of bicuspid aortic valve hemodynamics using particle image velocimetry. Ann Biomed Eng. 2012;40(8):1760–75.
6. Kheradvar A, Gharib M. On mitral valve dynamics and its connection to early diastolic flow. Ann Biomed Eng. 2009;37(1):1–13.
7. Reul H, Potthast K. Durability/wear testing of heart valve substitutes. J Heart Valve Dis. 1998;7(2):151–7.
8. Okafor IU, et al. Cardiovascular magnetic resonance compatible physical model of the left ventricle for multi-modality characterization of wall motion and hemodynamics. J Cardiovasc Magn Reson. 2015;17:51.
9. Jun BH, et al. Effect of hinge gap width of a St. Jude medical bileaflet mechanical heart valve on blood damage potential—an in vitro micro particle image velocimetry study. J Biomech Eng. 2014;136(9):091008.
10. 5840:2005, A.A.I., Cardiovascular implants—Cardiac valve prostheses. 2005.
11. Administration, F.a.D., Draft Guidance for Industry and FDA Staff: Heart Valves—Investigational Device Exemption (IDE) and Premarket Approval (PMA) Applications, Submitted for Comment, January 20, 2010. 2010.
12. Kelley TA, Marquez S, Popelar CF. In vitro testing of heart valve substitutes. In: Heart valves. New York: Springer; 2013. p. 283–320.
13. Browne P, et al. Experimental investigation of the steady flow downstream of the St. Jude bileaflet heart valve: a comparison between laser Doppler velocimetry and particle image velocimetry techniques. Ann Biomed Eng. 2000;28(1):39–47.
14. Gunning PS, et al. An in vitro evaluation of the impact of eccentric deployment on transcatheter aortic valve hemodynamics. Ann Biomed Eng. 2014;42(6):1195–206.
15. King M, et al. A three-dimensional, time-dependent analysis of flow through a bileaflet mechanical heart valve: comparison of experimental and numerical results. J Biomech. 1996;29(5):609–18.
16. Gunning PS, Vaughan TJ, McNamara LM. Simulation of self expanding transcatheter aortic valve in a realistic aortic root: implications of deployment geometry on leaflet deformation. Ann Biomed Eng. 2014;42(9):1989–2001.
17. Laboratories, D. MP3 Pulse Duplicator. 2015 [cited 2015 10/2/2015]. http://dynateklabs.com/mp3-heart-valve-tester/.

18. Labs, V. ViVitro Labs Pulse Duplicator. 2015 [cited 2015 10/2/2015]. http://vivitrolabs.com/product/pulse-duplicator/.
19. LLC, B.D.C.L. Heart Valve HDT-500 Pulse Duplicator System. 2015 [cited 2015 10/2/2015]. http://www.bdclabs.com/testing-equipment/pulse-duplicator-system/.
20. Laboratories, M.I.T. Heart Valve Pulse Duplicator. 2015 [cited 2015 10/2/2015]. http://www.medicalimplanttestinglab.com/products.html.
21. Boronyak SM, Merryman WD. Development of a simultaneous cryo-anchoring and radiofrequency ablation catheter for percutaneous treatment of mitral valve prolapse. Ann Biomed Eng. 2012;40(9):1971–81.
22. Gao B, et al. Effects of papillary muscle position on anterior leaflet stretches under mitral valve edge-to-edge repair. J Heart Valve Dis. 2009;18(2):135–41.
23. Gheewala N, Grande-Allen KJ. Design and mechanical evaluation of a physiological mitral valve organ culture system. Cardiovasc Eng Technol. 2010;1(2):123–31.
24. Ostli B, et al. In vitro system for measuring chordal force changes following mitral valve patch repair. Cardiovasc Eng Technol. 2012;3(3):263–8.
25. Siefert AW, et al. In vitro mitral valve simulator mimics systolic valvular function of chronic ischemic mitral regurgitation ovine model. Ann Thorac Surg. 2013;95(3):825–30.
26. Spinner EM, et al. In vitro characterization of the mechanisms responsible for functional tricuspid regurgitation. Circulation. 2011;124(8):920–9.
27. Vismara R, et al. A pulsatile simulator for the in vitro analysis of the mitral valve with tri-axial papillary muscle displacement. Int J Artif Organs. 2011;34(4):383–91.
28. Rabbah JP, Saikrishnan N, Yoganathan AP. A novel left heart simulator for the multi-modality characterization of native mitral valve geometry and fluid mechanics. Ann Biomed Eng. 2013;41(2):305–15.
29. Siefert AW, Siskey RL. Bench models for assessing the mechanics of mitral valve repair and percutaneous surgery. Cardiovasc Eng Technol. 2014;6(2):193–207.
30. Bhattacharya S, et al. Tension to passively cinch the mitral annulus through coronary sinus access: an ex vivo study in ovine model. J Biomech. 2014;47(6):1382–8.
31. Leopaldi AM, et al. In vitro hemodynamics and valve imaging in passive beating hearts. J Biomech. 2012;45(7):1133–9.
32. Richards AL, et al. A dynamic heart system to facilitate the development of mitral valve repair techniques. Ann Biomed Eng. 2009;37(4):651–60.
33. Yamauchi H, et al. Right ventricular papillary muscle approximation as a novel technique of valve repair for functional tricuspid regurgitation in an ex vivo porcine model. J Thorac Cardiovasc Surg. 2012;144(1):235–42.
34. He S, et al. Mitral leaflet geometry perturbations with papillary muscle displacement and annular dilatation: an in-vitro study of ischemic mitral regurgitation. J Heart Valve Dis. 2003;12(3):300–7.
35. Pierrakos O, Vlachos PP, Telionis DP. Time-resolved DPIV analysis of vortex dynamics in a left ventricular model through bileaflet mechanical and porcine heart valve prostheses. J Biomech Eng. 2004;126(6):714–26.
36. Modersohn D, et al. Isolated hemoperfused heart model of slaughterhouse pigs. Int J Artif Organs. 2001;24(4):215–21.
37. de Hart J, et al. An ex vivo platform to simulate cardiac physiology: a new dimension for therapy development and assessment. Int J Artif Organs. 2011;34(6):495–505.
38. Broadley K. The Langendorff heart preparation—reappraisal of its role as a research and teaching model for coronary vasoactive drugs. J Pharmacol Methods. 1979;2(2):143–56.
39. Dixon JA, Spinale FG. Large animal models of heart failure: a critical link in the translation of basic science to clinical practice. Circ Heart Fail. 2009;2(3):262–71.
40. Edmunds Jr LH, Gorman III JH, Gorman RC. Sheep models of postinfarction left ventricular remodeling. In: Cardiac remodeling and failure. Springer; 2003. p. 231–43.
41. Fomovsky GM, et al. Anisotropic reinforcement of acute anteroapical infarcts improves pump function. Circ Heart Fail. 2012;5(4):515–22.

42. Jensen H, et al. Three-dimensional assessment of papillary muscle displacement in a porcine model of ischemic mitral regurgitation. J Thorac Cardiovasc Surg. 2010;140(6):1312–8.
43. Kalra K, et al. Temporal changes in interpapillary muscle dynamics as an active indicator of mitral valve and left ventricular interaction in ischemic mitral regurgitation. J Am Coll Cardiol. 2014;64(18):1867–79.
44. Komeda M, et al. Geometric determinants of ischemic mitral regurgitation. Circulation. 1997;96(9 Suppl):Ii-128–33.
45. Llaneras MR, et al. Large animal model of ischemic mitral regurgitation. Ann Thorac Surg. 1994;57(2):432–9.
46. Pedersen HD, Häggström J. Mitral valve prolapse in the dog: a model of mitral valve prolapse in man. Cardiovasc Res. 2000;47(2):234–43.
47. Maisano F, et al. A translational "humanised" porcine model for transcatheter mitral valve interventions: the neo inferior vena cava approach. EuroIntervention. 2015;11(1):92–5.
48. Shandas R, et al. A general method for estimating deformation and forces imposed in vivo on bioprosthetic heart valves with flexible annuli: in vitro and animal validation studies. J Heart Valve Dis. 2001;10(4):495–504.
49. Hung J, et al. 3D echocardiography: a review of the current status and future directions. J Am Soc Echocardiogr. 2007;20(3):213–33.
50. Capoulade R, Pibarot P. Assessment of aortic valve disease: role of imaging modalities. Curr Treat Options Cardiovasc Med. 2015;17(11):49.
51. Sucha D, et al. Multimodality imaging assessment of prosthetic heart valves. Circ Cardiovasc Imaging. 2015;8(9):e003703.
52. Spinner EM, et al. Altered right ventricular papillary muscle position and orientation in patients with a dilated left ventricle. J Thorac Cardiovasc Surg. 2011;141(3):744–9.
53. Stephens SE, et al. High resolution imaging of the mitral valve in the natural state with 7 Tesla MRI. PLoS One. 2017 Aug 30;12(8):e0184042.
54. Neumann D, et al. Multi-modal pipeline for comprehensive validation of mitral valve geometry and functional computational models. In: Statistical atlases and computational models of the heart. Imaging and modelling challenges. Cham: Springer; 2014. p. 188–95.
55. Toma M, et al. Fluid-structure interaction analysis of papillary muscle forces using a comprehensive mitral valve model with 3D chordal structure. Ann Biomed Eng. 2016;44(4):942–53.
56. Pierce EL, et al. Novel method to track soft tissue deformation by micro-computed tomography: application to the mitral valve. Ann Biomed Eng. 2016;44(7):2273–81.
57. Fawzy H, et al. Complete mapping of the tricuspid valve apparatus using three-dimensional sonomicrometry. J Thorac Cardiovasc Surg. 2011;141(4):1037–43.
58. Gorman JH 3rd, et al. Dynamic three-dimensional imaging of the mitral valve and left ventricle by rapid sonomicrometry array localization. J Thorac Cardiovasc Surg. 1996;112(3):712–26.
59. Kalejs M, et al. Comparison of radial deformability of stent posts of different aortic bioprostheses. Interact Cardiovasc Thorac Surg. 2013;16(2):129–33.
60. Askov JB, et al. Significance of force transfer in mitral valve-left ventricular interaction: in vivo assessment. J Thorac Cardiovasc Surg. 2013;145(6):1635–41.. 1641 e1
61. Jensen MO, et al. Mitral valve annular downsizing forces: implications for annuloplasty device development. J Thorac Cardiovasc Surg. 2014;148(1):83–9.
62. Redmond J, et al. In-vivo motion of mitral valve annuloplasty devices. J Heart Valve Dis. 2008;17(1):110.
63. Uemura K, et al. Peak systolic mitral annulus velocity reflects the status of ventricular-arterial coupling-theoretical and experimental analyses. J Am Soc Echocardiogr. 2011;24(5):582–91.
64. Sonometrics. Basic principles of sonomicrometry. 2015 [cited 2015 10/20/2015]. http://www.sonometrics.com/index-p.html.
65. Bothe W, et al. Rigid, complete annuloplasty rings increase anterior mitral leaflet strains in the normal beating ovine heart. Circulation. 2011;124(11 Suppl):S81–96.
66. Bothe W, et al. How do annuloplasty rings affect mitral annular strains in the normal beating ovine heart? Circulation. 2012;126(11 Suppl 1):S231–8.

67. Rausch MK, et al. In vivo dynamic strains of the ovine anterior mitral valve leaflet. J Biomech. 2011;44(6):1149–57.
68. Rausch MK, et al. Characterization of mitral valve annular dynamics in the beating heart. Ann Biomed Eng. 2011;39(6):1690–702.
69. Dagum P, et al. Coordinate-free analysis of mitral valve dynamics in normal and ischemic hearts. Circulation. 2000;102(19 Suppl 3):III62–9.
70. Timek TA, et al. Annular height-to-commissural width ratio of annulolasty rings in vivo. Circulation. 2005;112(9 Suppl):I423–8.
71. Timek TA, et al. Aorto-mitral annular dynamics. Ann Thorac Surg. 2003;76(6):1944–50.
72. Spinner EM, et al. The effects of a three-dimensional, saddle-shaped annulus on anterior and posterior leaflet stretch and regurgitation of the tricuspid valve. Ann Biomed Eng. 2012;40 (5):996–1005.
73. Jimenez JH, et al. A saddle-shaped annulus reduces systolic strain on the central region of the mitral valve anterior leaflet. J Thorac Cardiovasc Surg. 2007;134(6):1562–8.
74. Weiler M, et al. Regional analysis of dynamic deformation characteristics of native aortic valve leaflets. J Biomech. 2011;44(8):1459–65.
75. Padala M, et al. Mechanics of the mitral valve strut chordae insertion region. J Biomech Eng. 2010;132(8):081004.
76. Ritchie J, et al. The material properties of the native porcine mitral valve chordae tendineae: an in vitro investigation. J Biomech. 2006;39(6):1129–35.
77. Stearns G, et al. Transcatheter aortic valve implantation can potentially impact short-term and long-term functionality: an in vitro study. Int J Cardiol. 2014;172(3):e421–2.
78. Dolensky JR, et al. In vitro assessment of available coaptation area as a novel metric for the quantification of tricuspid valve coaptation. J Biomech. 2013;46(4):832–6.
79. Siefert AW, et al. Isolated effect of geometry on mitral valve function for in silico model development. Comput Methods Biomech Biomed Engin. 2015;18(6):618–27.
80. Jensen MO, Fontaine AA, Yoganathan AP. Improved in vitro quantification of the force exerted by the papillary muscle on the left ventricular wall: three-dimensional force vector measurement system. Ann Biomed Eng. 2001;29(5):406–13.
81. Siefert AW, et al. Quantitative evaluation of annuloplasty on mitral valve chordae tendineae forces to supplement surgical planning model development. Cardiovasc Eng Technol. 2014;5 (1):35–43.
82. Padala M, et al. Comparison of artificial neochordae and native chordal transfer in the repair of a flail posterior mitral leaflet: an experimental study. Ann Thorac Surg. 2013;95(2):629–33.
83. Rabbah JP, et al. Effects of targeted papillary muscle relocation on mitral leaflet tenting and coaptation. Ann Thorac Surg. 2013;95(2):621–8.
84. Jensen MO, et al. Harvested porcine mitral xenograft fixation: impact on fluid dynamic performance. J Heart Valve Dis. 2001;10(1):111–24.
85. Jimenez JH, et al. Effects of a saddle shaped annulus on mitral valve function and chordal force distribution: an in vitro study. Ann Biomed Eng. 2003;31(10):1171–81.
86. Nichols WW, O'Rourke MF, Vlachopoulos C. McDonald's blood flow in arteries, sixth edition: theoretical, experimental and clinical principles. New York: CRC Press; 2011. p. 768.
87. Jensen JA. Range/velocity limitations for time-domain blood velocity estimation. Ultrasound Med Biol. 1993;19(9):741–9.
88. Holland CK, et al. Lower extremity volumetric arterial blood flow in normal subjects. Ultrasound Med Biol. 1998;24(8):1079–86.
89. Webster JG. Medical instrumentation: application and design. New York: Wiley; 2008.
90. Shiota T. Clinical application of 3-dimensional echocardiography in the USA. Circ J. 2015;79 (11):2287–98.
91. Saikrishnan N, et al. Accurate assessment of aortic stenosis: a review of diagnostic modalities and hemodynamics. Circulation. 2014;129(2):244–53.

92. Tsujino H, et al. Combination of pulsed-wave Doppler and real-time three-dimensional color Doppler echocardiography for quantifying the stroke volume in the left ventricular outflow tract. Ultrasound Med Biol. 2004;30(11):1441–6.
93. Hansen KL, et al. First report on intraoperative vector flow imaging of the heart among patients with healthy and diseased aortic valves. Ultrasonics. 2015;56:243–50.
94. Cape EG, et al. A new method for noninvasive quantification of valvular regurgitation based on conservation of momentum. In vitro validation. Circulation. 1989;79(6):1343–53.
95. Zoghbi WA, et al. Recommendations for evaluation of the severity of native valvular regurgitation with two-dimensional and Doppler echocardiography. J Am Soc Echocardiogr. 2003;16(7):777–802.
96. Yap CH, et al. Novel method of measuring valvular regurgitation using three-dimensional nonlinear curve fitting of Doppler signals within the flow convergence zone. IEEE Trans Ultrason Ferroelectr Freq Control. 2013;60(7):1295–311.
97. Yosefy C, et al. Direct measurement of vena contracta area by real-time 3-dimensional echocardiography for assessing severity of mitral regurgitation. Am J Cardiol. 2009;104(7):978–83.
98. Pierce EL, et al. Three-dimensional field optimization method: gold-standard validation of a novel color Doppler method for quantifying mitral regurgitation. J Am Soc Echocardiogr. 2016;29(10):917–25.
99. Hochareon P, et al. Diaphragm motion affects flow patterns in an artificial heart. Artif Organs. 2003;27(12):1102–9.
100. Yoganathan AP, et al. Bileaflet, tilting disk and porcine aortic-valve substitutes—in vitro hydrodynamic characteristics. J Am Coll Cardiol. 1984;3(2):313–20.
101. Raffel M, et al. Particle image velocimetry: a practical guide. Berlin: Springer; 2013.
102. Seaman C, Akingba AG, Sucosky P. Steady flow hemodynamic and energy loss measurements in Normal and simulated calcified tricuspid and bicuspid aortic valves. J Biomech Eng Trans ASME. 2014;136(4):11.
103. Jun BH, Saikrishnan N, Yoganathan AP. Micro particle image velocimetry measurements of steady diastolic leakage flow in the hinge of a St. Jude Medical® Regent™ mechanical heart valve. Ann Biomed Eng. 2014;42(3):526–40.
104. Westerdale JC, et al. Effects of bileaflet mechanical mitral valve rotational orientation on left ventricular flow conditions. Open Cardiovasc Med J. 2015;9:62–8.
105. Bellofiore A, Quinlan NJ. High-resolution measurement of the unsteady velocity field to evaluate blood damage induced by a mechanical heart valve. Ann Biomed Eng. 2011;39(9):2417–29.
106. Kadem L, et al. Flow-dependent changes in Doppler-derived aortic valve effective orifice area are real and not due to artifact. J Am Coll Cardiol. 2006;47(1):131–7.
107. Manning KB, et al. Regurgitant flow field characteristics of the St. Jude bileaflet mechanical heart valve under physiologic pulsatile flow using particle image velocimetry. Artif Organs. 2003;27(9):840–6.
108. Kheradvar A, et al. Echocardiographic particle image velocimetry: a novel technique for quantification of left ventricular blood vorticity pattern. J Am Soc Echocardiogr. 2010;23(1):86–94.
109. Kim HB, Hertzberg JR, Shandas R. Development and validation of echo PIV. Exp Fluids. 2004;36(3):455–62.
110. Westerdale J, et al. Flow velocity vector fields by ultrasound particle imaging velocimetry: in vitro comparison with optical flow velocimetry. J Ultrasound Med. 2011;30(2):187–95.
111. Maymir JC, et al. Mean velocity and Reynolds stress measurements in the regurgitant jets of tilting disk heart valves in an artificial heart environment. Ann Biomed Eng. 1998;26(1):146–56.
112. Simon HA, et al. Comparison of the hinge flow fields of two bileaflet mechanical heart valves under aortic and mitral conditions. Ann Biomed Eng. 2004;32(12):1607–17.

113. Herbertson LH, Deutsch S, Manning KB. Near valve flows and potential blood damage during closure of a bileaflet mechanical heart valve. J Biomech Eng Trans ASME. 2011;133(9):7.
114. Yap CH, et al. Experimental measurement of dynamic fluid shear stress on the aortic surface of the aortic valve leaflet. Biomech Model Mechanobiol. 2012;11(1–2):171–82.
115. Ellis JT, Travis BR, Yoganathan AP. An in vitro study of the hinge and near-field forward flow dynamics of the St. Jude Medical® Regent™ bileaflet mechanical heart valve. Ann Biomed Eng. 2000;28(5):524–32.
116. Pennekamp W, et al. Determination of flow profiles of different mechanical aortic valve prostheses using phase-contrast MRI. J Cardiovasc Surg. 2011;52(2):277–84.
117. Kvitting JPE, et al. In vitro assessment of flow patterns and turbulence intensity in prosthetic heart valves using generalized phase-contrast MRI. J Magn Reson Imaging. 2010;31(5):1075–80.
118. Kim SJ et al. Dynamic 3-dimensional phase-contrast technique in MRI: application to complex flow analysis around the artificial heart valve. In: Conference on physiology and function from multidimensional images—medical imaging 1998. 1998. San Diego, CA: SpIE-International Society of Optical Engineering; 1998
119. Kumar R, et al. Assessment of left ventricular diastolic function using 4-dimensional phase-contrast cardiac magnetic resonance. J Comput Assist Tomogr. 2011;35(1):108–12.
120. Kvitting JPE, et al. Flow patterns in the aortic root and the aorta studied with time-resolved, 3-dimensional, phase-contrast magnetic resonance imaging: implications for aortic valve-sparing surgery. J Thorac Cardiovasc Surg. 2004;127(6):1602–7.
121. Yokosawa S, et al. Quantitative measurements on the human ascending aortic flow using 2D cine phase-contrast magnetic resonance imaging. JSME Int J Ser C Mech Syst Mach Elem Manufact. 2005;48(4):459–67.
122. Jackson M, et al. Development of a multi-modality compatible flow loop system for the functional assessment of mitral valve prostheses. Cardiovasc Eng Technol. 2014;5(1):13–24.
123. Carlsson M, et al. Quantification of left and right ventricular kinetic energy using four-dimensional intracardiac magnetic resonance imaging flow measurements. Am J Phys Heart Circ Phys. 2012;302(4):H893–900.
124. Delles M et al. Non-invasive computation of aortic pressure maps: a phantom-based study of two approaches. In: Conference on medical imaging—biomedical applications in molecular, structural, and functional imaging. San Diego, CA: SPIE-International Society of Optical Engineering; 2014.
125. Markl M, et al. Generalized reconstruction of phase contrast MRI: analysis and correction of the effect of gradient field distortions. Magn Reson Med. 2003;50(4):791–801.
126. Bernstein MA, et al. Concomitant gradient terms in phase contrast MR: analysis and correction. Magn Reson Med. 1998;39(2):300–8.
127. Levine RA, et al. Mitral valve disease-morphology and mechanisms. Nat Rev Cardiol. 2015;12(12):689–710.
128. Ingels NB, Karlsson M. Mitral valve mechanics. Dropbox. 2014. p. 1.
129. Smerup M, et al. Strut chordal-sparing mitral valve replacement preserves long-term left ventricular shape and function in pigs. J Thorac Cardiovasc Surg. 2005;130(6):1675–82.
130. Ropcke DM, et al. Papillary muscle force distribution after total tricuspid reconstruction using porcine extracellular matrix: in-vitro valve characterization. J Heart Valve Dis. 2014;23(6):788–94.
131. Ropcke DM, et al. Functional and biomechanical performance of stentless extracellular matrix tricuspid tube graft: an acute experimental porcine evaluation. Ann Thorac Surg. 2016;101(1):125–32.
132. Askov JB, et al. Papillary muscle force transducer for measurement in vivo. Cardiovasc Eng Technol. 2011;2(3):196–202.
133. Nielsen SL, et al. Miniature C-shaped transducers for chordae tendineae force measurements. Ann Biomed Eng. 2004;32(8):1050–7.

134. Troxler LG, Spinner EM, Yoganathan AP. Measurement of strut chordal forces of the tricuspid valve using miniature C ring transducers. J Biomech. 2012;45(6):1084–91.
135. Jimenez JH, et al. Mitral valve function and chordal force distribution using a flexible annulus model: an in vitro study. Ann Biomed Eng. 2005;33(5):557–66.
136. Nielsen SL, et al. Imbalanced chordal force distribution causes acute ischemic mitral regurgitation: mechanistic insights from chordae tendineae force measurements in pigs. J Thorac Cardiovasc Surg. 2005;129(3):525–31.
137. Lomholt M, et al. Differential tension between secondary and primary mitral chordae in an acute in-vivo porcine model. J Heart Valve Dis. 2002;11(3):337–45.
138. He Z, Jowers C. A novel method to measure mitral valve chordal tension. J Biomech Eng. 2009;131(1):014501.
139. Nielsen SL, et al. Chordal force distribution determines systolic mitral leaflet configuration and severity of functional mitral regurgitation. J Am Coll Cardiol. 1999;33(3):843–53.
140. Granier M, et al. Consequences of mitral valve prolapse on chordal tension: ex vivo and in vivo studies in large animal models. J Thorac Cardiovasc Surg. 2011;142(6):1585–7.
141. Levine RA, et al. Mechanistic insights into functional mitral regurgitation. Curr Cardiol Rep. 2002;4(2):125–9.
142. Jensen H, et al. Transapical neochord implantation: is tension of artificial chordae tendineae dependent on the insertion site? J Thorac Cardiovasc Surg. 2014;148(1):138–43.
143. Nielsen SL, et al. Mitral ring annuloplasty relieves tension of the secondary but not primary chordae tendineae in the anterior mitral leaflet. J Thorac Cardiovasc Surg. 2011;141(3):732–7.
144. He Z, Bhattacharya S. Mitral valve annulus tension and the mechanism of annular dilation: an in-vitro study. J Heart Valve Dis. 2010;19(6):701–7.
145. Medtronic, Simulus® Flexible Annuloplasty System. Marketing Literature. 2015. p. 2.
146. Edwards Lifesciences. Carpentier-Edwards Physio II Annuloplasty Ring. Marketing Literature. 2015. p. 1.
147. Jensen MO, et al. Saddle-shaped mitral valve annuloplasty rings experience lower forces compared with flat rings. Circulation. 2008;118(14 Suppl):S250–5.
148. Rausch MK, et al. Mitral valve annuloplasty: a quantitative clinical and mechanical comparison of different annuloplasty devices. Ann Biomed Eng. 2012;40(3):750–61.
149. Hasenkam JM, et al. What force can the myocardium generate on a prosthetic mitral valve ring? An animal experimental study. J Heart Valve Dis. 1994;3(3):324–9.
150. Siefert AW, et al. Dynamic assessment of mitral annular force profile in an ovine model. Ann Thorac Surg. 2012;94(1):59–65.
151. Siefert AW, et al. In-vivo transducer to measure dynamic mitral annular forces. J Biomech. 2012;45(8):1514–6.
152. Skov SN, et al. New mitral annular force transducer optimized to distinguish annular segments and multi-plane forces. J Biomech. 2016;49(5):742–8.
153. Kragsnaes ES, et al. In-plane tricuspid valve force measurements: development of a strain gauge instrumented annuloplasty ring. Cardiovasc Eng Technol. 2013;4(2):131–8.
154. Skov SN, et al. Simultaneous in- and out-of-plane mitral valve annular force measurements. Cardiovasc Eng Technol. 2015;6(2):185–92.
155. Chitwood WR Jr, et al. Robotic mitral valve repairs in 300 patients: a single-center experience. J Thorac Cardiovasc Surg. 2008;136(2):436–41.
156. Siefert AW, et al. Suture forces in undersized mitral annuloplasty: novel device and measurements. Ann Thorac Surg. 2014;98(1):305–9.
157. Jimenez JH, et al. Effects of annular size, transmitral pressure, and mitral flow rate on the edge-to-edge repair: an in vitro study. Ann Thorac Surg. 2006;82(4):1362–8.
158. Timek TA, et al. Mitral annular size predicts Alfieri stitch tension in mitral edge-to-edge repair. J Heart Valve Dis. 2004;13(2):165–73.
159. Jensen MO, et al. External approach to in vivo force measurement on mitral valve traction suture. J Biomech. 2012;45(5):908–12.

160. Jensen H, et al. Impact of papillary muscle relocation as adjunct procedure to mitral ring annuloplasty in functional ischemic mitral regurgitation. Circulation. 2009;120(11 Suppl): S92–8.
161. Pokorny S, et al. In vivo quantification of the apical fixation forces of different mitral valved stent designs in the beating heart. Ann Biomed Eng. 2015;43(5):1201–9.
162. Timek TA, et al. Septal-lateral annular cinching ('SLAC') reduces mitral annular size without perturbing normal annular dynamics. J Heart Valve Dis. 2002;11(1):2–9.

# Part II
# Heart Valve Disease and Treatment

# Calcific Aortic Valve Disease: Pathobiology, Basic Mechanisms, and Clinical Strategies

Payal Vyas, Joshua D. Hutcheson, and Elena Aikawa

**Abstract** Calcific aortic valve disease (CAVD) is a leading cause of cardiovascular morbidity and mortality, and its prevalence is expected to increase in the aging population of the developed world. Currently, no noninvasive therapeutic strategies exist to prevent or treat CAVD. Though the advent of new valve replacement technologies have improved clinical outcomes, these techniques remain suboptimal for the two populations most at risk for valvular complications—pediatric and elderly patients. Recent advances in basic research have shown that CAVD arises through active cellular mechanisms, offering hope that drugs can be developed to target relevant pathways and provide new clinical options for CAVD patients. Translating these benchtop discoveries to clinical realities, however, will require both a holistic understanding of how targetable cellular level processes affect valve tissue function and the ability to identify early CAVD development in patients. This chapter addresses this translation by reviewing the current state of CAVD research and the ongoing efforts to meet the clinical need.

**Keywords** Aortic valve · Calcification · Aortic stenosis · Calcific aortic valve disease · Aortic valve remodeling · Disease mechanisms · Fibrocalcific remodeling

## 1 Introduction

Calcific aortic valve disease (CAVD) is the most common valvular heart disease in developed countries. CAVD increases in prevalence with age with its early stage—aortic sclerosis present in approximately 25% of individuals over 65 years of age and late stage—aortic stenosis in 2–9% of adult population [1, 2]. With increasing life

---

P. Vyas · J. D. Hutcheson · E. Aikawa (✉)
Department of Cardiovascular Medicine, Brigham and Women's Hospital, Harvard Medical School, Boston, MA, USA
e-mail: eaikawa@bwh.harvard.edu

span in developed and rapidly developing nations, the projected disease burden is expected to rise from 2.5 million in 2000 to 4.5 million in 2030.

Aortic sclerosis, defined as increased leaflet thickness without restriction of leaflet motion, progresses to aortic stenosis that is characterized by excessive leaflet thickening and stiffness due to scarring and formation of calcified nodules on the surface of the leaflet proximal to the aorta. Aortic stenosis, therefore, results in significantly reduced systolic opening causing left ventricular outflow obstruction and increasing left ventricular pressure load and insufficient coaptation of leaflets to prevent backflow from the aorta to the left ventricle.

Symptoms of aortic stenosis usually develop gradually after a long asymptomatic latent period usually of 10–20 years. However, onset of symptoms such as exertional dyspnoea, angina, or dizziness and syncope portends poor outcomes [3]. It is the characteristic harsh systolic murmur that defines initial diagnosis of calcific aortic stenosis and guides further diagnostic workup. Two-dimensional (2D) Doppler echocardiography is the imaging modality of choice to diagnose and estimate the severity of aortic stenosis by calculating valve area, pressure gradients and localize the level of obstruction [4, 5]. Patients with severe aortic stenosis have a survival rate as low as 50% at 2 years and 20% at 5 years without treatment [6, 7].

Surgical aortic valve replacement (AVR) with a mechanical or bioprosthetic valve or transcatheter aortic valve replacement (TAVR) is the only viable treatment option for CAVD. AVR is performed approximately 85,000 times annually in the United States and 275,000 cases performed worldwide [4, 5]. Limitations of surgical replacement include hemorrhagic complications ascribed to anticoagulation therapy especially in case of mechanical valves and failure of bioprosthetic valves due to calcification, which often results in the need of a second operation [8, 9]. There is approximately 30–60% incidence of structural degeneration of the valve at 15 years of implantation [10, 11]. Surgical AVR is a high-risk option for the elderly due to increased age and comorbidities [12]. TAVR is a less invasive and therefore an attractive option for this patient population [13, 14] as evident by a 39% reduction in mortality at 1 year when compared to AVR [13]. However, this option is limited by issues such as lack of integration and growth in the host body and potential rapid tissue degradation, major concerns in pediatric patients, and is associated with higher risk of stroke in the TAVR group compared with standard therapy [15, 16]. No pharmacological treatment options exist to reverse or slow the progression of CAVD.

Conventionally CAVD was viewed as a passive result of aging in which years of mechanical stress and hemodynamic forces during normal cardiac function lead to tissue scarring and calcium deposition on the surface of aortic valve leaflets. Limited understanding of the fundamental mechanisms that underlie CAVD may be the reason for lack of targetable points of pharmacological intervention, thus hindering progress towards rational design for therapeutic options. However recent research efforts have replaced this traditional view of CAVD as a "degenerative disease" with that of it being an "active" process regulated by cellular mechanisms, which remodel the valve leaflet in response to biomechanical and molecular stimuli [17]. The valve leaflet undergoes an active disease process marked by inflammatory cell infiltration, degradation and aberrant deposition of extracellular matrix (ECM), neoangiogenesis and osteocalcific changes [18]. CAVD, therefore, progresses from initial changes in the cell biology of the valve leaflets through fibrosis to calcification.

These recent exciting developments in understanding biological mechanisms of CAVD progression have raised hopes of developing tools to identify patients at risk, as well as effective diagnostic and therapeutic interventions for this disease. This chapter discusses our current understanding of the pathophysiology, risk factors, cellular mechanisms, diagnosis, and clinical management of CAVD, and describes areas of future research vital for diagnosing, treating, and potentially preventing this disease.

## 2 Normal Aortic Valve Physiology and Function

The human heart consists of four valves: the two atrioventricular valves called mitral and tricuspid valves and two semilunar valves called the aortic valve and pulmonary valve. The basic function of the heart valves is to maintain unidirectional blood flow through the cardiac chambers and into the pulmonary and systemic circulation by their coordinated opening and closing during the cardiac cycle. The aortic valve is on the left side of the heart, located between the left ventricle and the aorta. The aortic valve opens during systole to allow forward flow of blood from the left ventricle, which is the main pumping chamber of the heart and closes during diastole to prevent blood from backward flow into the ventricle. Heart valves open and close approximately 40 million times a year, and 3 billion times over an average lifetime of 75 years. Their typical motion, flexibility, and mechanical and structural properties enable the aortic valve to function as a unidirectional outflow valve. Changes in these characteristics lead to disruption of valve function and form the basis of CAVD. The following sections elaborate on normal aortic valve anatomy, physiology, and how they relate to the unique function of the aortic valve.

### 2.1 Anatomy of Normal Aortic Valve

Normal aortic valve is anatomically defined in the context of *aortic root*, a bulbous-shaped fibrous structure to which the aortic leaflets are attached. The components of aortic root comprise the annulus, commissures, interleaflet triangles, sinus of Valsalva, sinotubular junction, and leaflets [19]. The anatomical architecture and structural properties of all of these components contribute towards their function to maintain the intermittent, unidirectional channeling of large volumes of blood, while maintaining laminar flow and least possible tissue stress and damage [20, 21]. The aortic root delineates the anatomical boundary between the left ventricle and ascending aorta.

The valve leaflets rise to commissure and descend to its basal attachment to the aortic wall. The *annulus* is demarcated by the virtual aortic ring defined by the nadirs

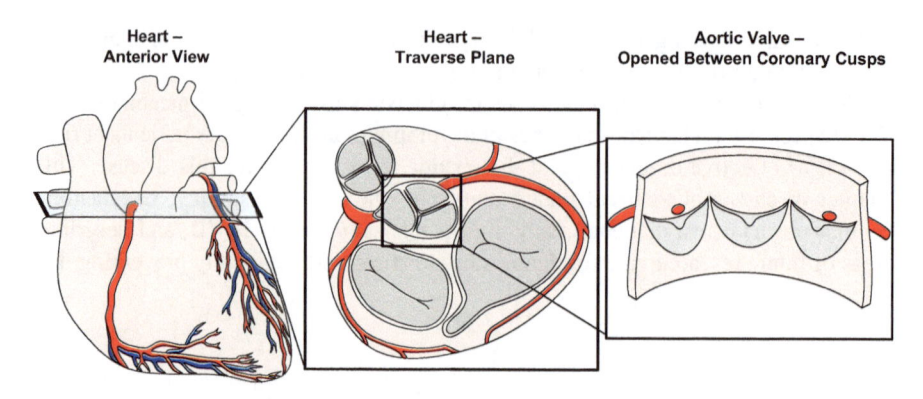

**Fig. 1** Anatomy of the aortic valve. The aortic valve is situated between the left ventricle and aorta and maintains unidirectional blood flow from the ventricles to the systemic circulation. Openings into the cardiac wall in the sinuses of Valsalva of two of the aortic valve leaflets feed oxygenated blood into the main coronary arteries; therefore, these two leaflets are referred to as coronary cusps

of leaflet attachments (Fig. 1). The annulus forms a crown-like structure, which crosses the ventriculo-arterial junction [22, 23]. The three symmetrical bulges of the aortic valve associated with each leaflet are called *sinuses of Valsalva* or aortic sinuses. Coronary arteries originate from two of the three sinuses, thus defining them as left, right, or noncoronary sinuses. *The sinotubular junction* (point of transition of the aortic root to the ascending aorta) is the distal defined ridge of the sinus, while its proximal end is defined by attachments of the valve leaflets.

Normal aortic valves are made up of three cusps or leaflets, which are <1 mm thick extremely pliable structures and appear smooth and opalescent. The semilunar-shaped three valve leaflets form the central structural compartment of the aortic root. Based on their location relative to the coronary artery ostia in the sinuses of Valsalva downstream of the valve, the valve cusps are individually called the left coronary, right coronary, and noncoronary leaflets. The right and left leaflets are usually equal in size, while the posterior cusp is slightly larger in two-thirds of individuals [24]. Each leaflet has two free edges, which they share with the adjacent leaflets. A small fibrous bulge at the center of each free edge is called the *nodule of Arantius*. The rim of the valve leaflet on either side of the nodule is called the *lunula*. The lunulae of adjacent leaflets overlap each other, thereby enabling complete aortic valve sealing during diastole. The remaining non-coapting portion of the leaflet is known as the *belly*, which appears to be almost transparent. The area where the lunulae of two leaflets are attached to the aortic wall at the apex of the annulus is called the *commissure*. There are three commissures. The commissure between the right and left coronary leaflets is positioned anteriorly, the one between the right and noncoronary leaflets is on the right anterior, and the one between the left and noncoronary leaflets is on the right posterior aspect of the aortic root. The commissures are composed of collagenous fibers orientated in a radial direction and provide optimal support to the valve leaflets. The three triangular areas located below the commissures are known as the *interleaflet triangles*. They are extension of left ventricular outflow tract and extend distally to the level of sinotubular junction in the area of the commissures. They separate and mark the three sinuses. Native human aortic valves are virtually avascular, and receive nutrients through hemodynamic convection and diffusion. The organized three-dimensional anatomy of the aortic tri-leaflet cusps, which work in concert with the aortic root, allows for stress sharing between leaflets, the sinuses of Valsalva, and the aortic wall [20, 25].

## 2.2 Function of Normal Aortic Valve

During a normal cardiac cycle, the aortic valve opens when the ventricle contracts during systole and closes when the ventricle relaxes during diastole. The valve closing depends on the changes in valve pressure and the parameters of the vortexes that develop within the sinuses of Valsalva [26]. In the deceleration phase of systole, the opposing axial pressure differences causes the low inertia flow in the developing boundary layer along the aortic wall to decelerate then to reverse direction resulting in vortices in the sinuses behind the aortic valve leaflets. This action forces the belly of the leaflets away from the aortic wall and towards the closed position. The leaflets are stretched via backpressure of 80 mm Hg and are mutually apposed with an overlap at the lunulae, which ensures proper sealing and therefore efficient filling of the left ventricle. The valve annulus expands and prevents collapse by taught pulling of the leaflets. Increase in annular radius at end diastole phase pulls the leaflets from their commissures, resulting in a small stellate orifice even in the absence of transvalvular flow [27]. The valve orifice changes quickly from stellate to triangular, and finally to circular, as the leaflets are forced apart and blood is ejected from the ventricle. The aortic root adjusts from a conical to a cylindrical shape during ejection, providing optimal hemodynamics at the larger flow volume [28].

## 2.3 Aortic Valve Microstructure

The dynamic biomechanical function of the aortic valves is dependent on its organized yet complex and highly differentiated tissue architecture and microstructure, which differs in various stages of valve development and becomes mature in adulthood (Fig. 2). Each leaflet of the aortic valve is composed of three distinct, histologically identifiable micro-layers known as the fibrosa, spongiosa, and ventricularis [29]. The aortic valve leaflet is more mechanically compliant in the radial direction than in the circumferential direction [30]. This ability arises from the distinct composition of the individual layers. The fibrosa, proximal to the aorta, is primarily composed of circumferentially oriented type I and III fibrillar collagen. Collagen is a major load-bearing component throughout many tissues and thus it provides the necessary strength in the valve leaflets to withstand the pressure developed during diastole and prevent leaflet prolapse into the left ventricle. The

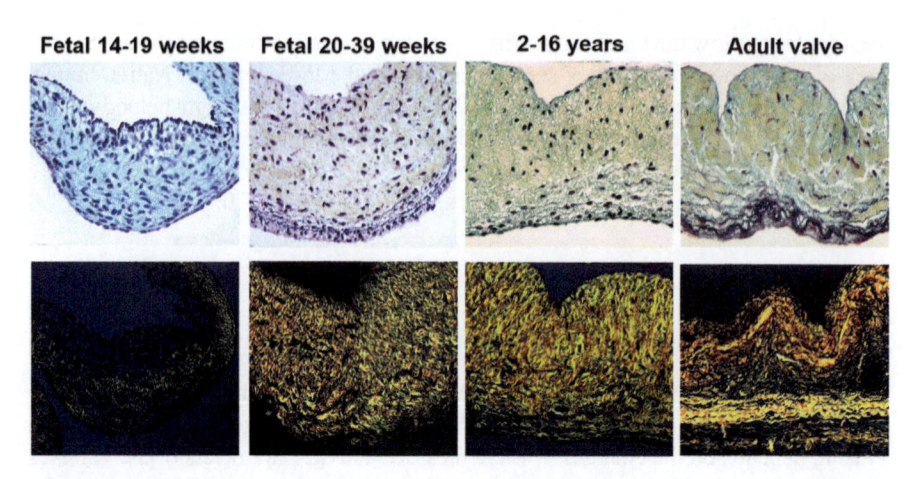

**Fig. 2** Aortic valve microarchitecture in histological sections. During aortic valve thickening in fetal development, collagen accumulates and forms mature fibers within the leaflets as shown by picrosirius red staining in the bottom panel. As the valve matures from childhood to adult, the collagen fibers become aligned in the circumferential direction, and a trilaminar structure is observed by Movat's stain (top panel) with collagen fibers (yellow) in the fibrosa on the aortic side of the valve, glycosaminoglycans of the spongiosa in the middle of the valve (bluish green), and elastin fibers in the ventricularis of the valve leaflets (black). Adapted from [29]

ventricularis layer, which lies on the ventricle side of the leaflet, is composed of elastin fibers that are oriented radially. The ventricularis layer therefore confers elasticity to the leaflets, so that they may stretch radially away from the valve annulus to meet adjacent leaflets and seal the valve [31] and aid in recoil [32]. The spongiosa lies between the fibrosa and ventricularis and is composed primarily of glycosaminoglycans and proteoglycans with scattered collagen fibers. It serves as a lubricating layer between the outer layers of the aortic valve [33]. During diastole, the leaflet behaves relatively inelastically with the diastolic pressure loads supported by fibrosa and elastically in systole, supported by ventricularis. Both layers exhibit distinct micromechanical properties with the fibrosa being stiffer than the ventricularis as shown by tensile tests of normal and enzymatically digested valves as well as biaxial testing of micro-dissected fibrosa and ventricularis layers [34–36]. Differential extacellular matrix (ECM) modulus was noted within the individual layers with distinctly stiff and soft regions in the fibrosa and ventricularis, respectively [37]. The dynamic opening and closing function of the aortic valve relies on the structural integrity and biomechanical properties of the valve microstructure attuned to the relative pressure changes over the course of the heart cycle [38, 39]. The breakdown of the structural and compositional integrity of the valve ECM is thus one of the hallmarks of CAVD.

## 2.4 Aortic Valve Cell Biology

Normal human aortic valves consist of two principal resident cell types: valvular endothelial cells (VECs) and valvular interstitial cells (VICs). These cell types are responsible for maintaining valve homeostasis and integrity. VECs sheathe the leaflet surface and thus form a barrier between the internal tissue components and blood similar to vascular endothelial cells. VICs reside within the valve tissue and regulate valve microstructure through secretion, degradation, and reorganization of ECM components. The populations of smooth muscle cells [40] and nerve cells have also been described [41, 42].

### 2.4.1 Valvular Interstitial Cells (VICs)

All mesenchymal cells within the valve tissue have traditionally been classified as VICs. Five identifiable phenotypes of the VICs have been described: quiescent VICs (qVICs), activated VICs (aVICs), progenitor VICs (pVICs), embryonic mesenchymal cells, and osteoblastic VICs (oVICs) [43–46]. At least three different cell types have been identified in native aortic valves (fibroblasts, myofibroblasts, smooth muscle cells) [47]. VICs maintain a quiescent fibroblast-like cell state in adult valves, but exhibit an activated phenotype gauged by expression of α-smooth muscle actin (α-SMA) in valve interstitium during valve development, in response to injury and in disease [29, 43] (Fig. 3). Under these conditions, VICs can switch to an activated myofibroblast phenotype that measured by positive expression of α-SMA and negative staining for SM1 and SM2, markers of differentiated smooth muscle cells [45]. Only a small subpopulation of α-SMA, SM1 and SM2 expressing cells below the endothelium is reported to be smooth muscle cells. The myofibroblast-like VICs function to remodel the ECM by synthesis of ECM-degrading enzymes such as matrix metalloproteinases (e.g., MMP-1, MMP-2, MMP-9, and MMP-13) and cathepsins (e.g., cathepsins S, K, and B), and tissue inhibitors (e.g., tissue inhibitors

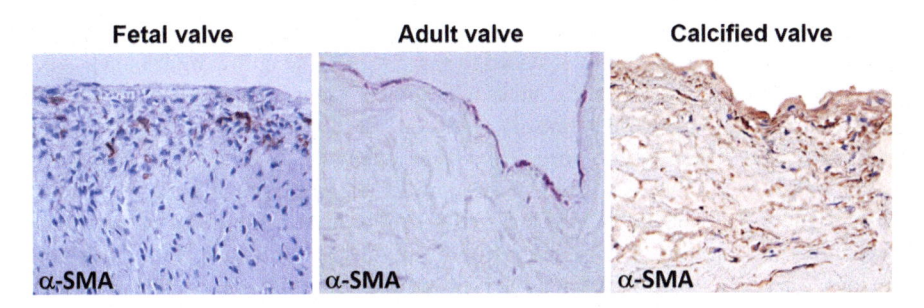

**Fig. 3** Myofibroblast activation of VICs. VICs exhibit an activated myofibroblast phenotype during development as shown by the presence of α-SMA by immunohistochemistry. During adulthood, VICs transition to a quiescent phenotype, but the activated myofibroblast phenotype reappears with the onset of CAVD. Adapted from [29]

of metalloproteinases) [32, 44]. The factors that cause this phenotypic switch are poorly defined, but likely include transforming growth factor-β1 and cytoskeletal tension generated by cell traction [48] or possibly exogenous forces such as stimulation by mechanical loading in order to mediate tissue remodeling for restoration of normal stress profile [29]. VICs isolated from higher pressure left-sided valves (aortic and mitral) were found to be significantly stiffer and had more collagen synthesis than cells isolated from right-sided valves (tricuspid and pulmonary) [49]. Collagen synthesis by VICs was also shown to be dependent upon degree and duration of stretch [50]. Thus the aortic valve responds to repeated mechanical stresses of cardiac cycle by constant remodeling of ECM by VICs [51]. However, when adaptation to changing environmental condition is not possible due to changing mechanical stress and injury, valve remodeling progresses to a pathological state [44, 52].

### 2.4.2 Valvular Endothelial Cells (VECs)

Valvular endothelial cells are fundamentally responsible for maintaining a non-thrombogenic blood contact surface and transmitting nutrient and biochemical signals to VICs [53]. VECs are similar to arterial endothelial cells in the expression of von Willebrand factor and production of nitric oxide and prostacyclin activity [54]. Cell junctions between VECs are also similar to those of arterial endothelial cells. Interestingly, however, VECs have been shown to exhibit transcriptional differences compared to vascular endothelial cells [55]. Also while vascular endothelial cells align parallel to blood flow, VECs are oriented circumferentially across leaflets, perpendicular to the direction of blood flow [56]. VECs show an increased expression of genes involved in chondrogenic differentiation, while arterial endothelial cells more strongly express osteogenic genes [27], suggesting both cell types are prone to calcification, albeit with different mechanisms. Electron micrographs have shown that unlike the vascular endothelium, VECs present cortical filipodia that can overlap with neighboring cells for an unknown role [56, 57]. These differential properties suggest VECs are a phenotypically distinct cell type compared to other endothelial cells.

Aortic valves are constantly exposed to shear stress. The inflow ventricularis side is exposed to unidirectional shear stress with cycle averaged magnitude of 20 dynes/cm$^2$. The outflow or fibrosa surface is exposed to oscillatory shear stress [58]. Consequently, the VECs on the fibrosa and ventricularis surfaces are phenotypically different [59, 60]. Since CAVD has been shown to initiate in the fibrosa layer of the valve [61, 62], it has been hypothesized that the differential valve endothelium phenotype of the fibrosa may contribute to the disease initiation [59]. Emerging evidence suggests that VECs may serve as endothelial progenitor cells that can populate the valve leaflet with osteogenic-like VICs [63–65]. This hypothesis is supported by evidence indicating that mesenchymal differentiation of VECs, through endothelial to mesenchymal transition (EndMT), can be directed towards multi-lineage phenotypes [66]. VEC dysfunction has been implicated in calcification. Being in close proximity with each other, intercellular interactions between VECs and VICs are highly likely. Coculture studies with VECs and VICs have shown that VECs help maintain the quiescent phenotype of VICs. Upon exposure to shear stress, VIC proliferation is stabilized with an increase in ECM synthesis and decrease in glycosaminoglycans [67], and removal of VECs from aortic valve explants promotes calcification [68]. Future research in this direction is required to understand the mechanisms of intercellular interactions between the aortic valve cell types.

## 3 Calcific Aortic Valve Disease Pathobiology

Many of the pathophysiological changes in CAVD and atherosclerotic vascular calcification overlap; however, it should be noted that these diseases are now believed to have distinct etiologies [69]. Not every patient who presents prominent vascular calcification also exhibits signs of CAVD, suggesting that at least an extra set of variables are involved in CAVD pathology. Further, as reviewed in Sect. 2, valve cells are phenotypically different than their vascular counterparts, and therefore different factors seem to be responsible for producing the fibrocalcific endpoint that is apparent in both diseases. Both VICs and VECs have shown a unique susceptibility to changes in biomechanical forces that may distinguish them from their vascular counterparts. Similar to atherosclerosis, early CAVD lesions contain inflammatory cells, with advanced lesions containing calcium deposits [70–72]. The extent of similarity between the two disease etiologies however is being debated [61, 73] and perhaps would be better resolved as we acquire clear understanding of the fundamental mechanisms of CAVD. CAVD does exhibit several distinguishing pathobiological features, in that the underlying cellular mechanisms involved are distinct than those identified for atherosclerosis. In CAVD, the VICs are a quiescent population of cells, which acquire an activated myofibroblastic cell phenotype as the disease progresses [45, 74, 75], as opposed to pathological de-differentiation of contractile smooth muscle cells towards myofibroblastic cells in atherosclerosis [76, 77].

Not only do VICs differ substantially from vascular smooth muscle cells, but VIC heterogeneity between the cardiac valves has also been noted. One of the most remarkable differences between the four cardiac valves is their divergence in disease susceptibility and presentation [78]. The aortic valve on the left side of the heart is much more likely to undergo pathologic remodeling than its mirror image on the right side of the heart, the pulmonary valve [79]. It has been hypothesized and is widely accepted that the reason for this difference lies in the deviation between the biomechanical environments of systemic versus pulmonic circulation. Compared to the short distances required for pulmonary blood flow, pumping newly oxygenated blood such that it reaches the extremities of the body requires large pressures that generate significantly higher forces on the cardiovascular tissues associated with systemic distribution. As a result of these higher forces, the aortic valve is stiffer than

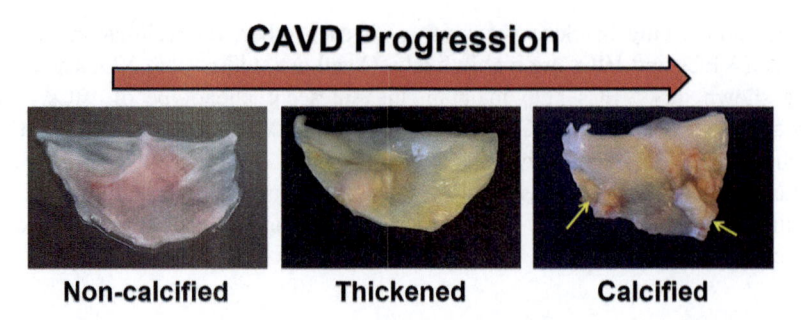

**Fig. 4** CAVD progression and gross anatomical changes. Healthy aortic valve leaflets are thin, membranous structures that are pliable enough to open during systole but exhibit the appropriate strength to resist prolapse and regurgitant blood flow during diastole. CAVD leads to fibrotic thickening of the leaflets, reducing their biomechanical function, and ultimately to the deposition of calcific mineral within the leaflets (arrows)

the pulmonary valve [78]. An increase in the forces placed on the aortic valve due to risk factors such as hypertension and aortic wall stiffening results in pathological remodeling of the valve leaflets [80]. CAVD is characterized by thickened, fibrotic leaflets resulting in overall stiffened leaflet with compromised biomechanical functionality (Fig. 4). The fibrotic thickening of the leaflet is accompanied by inflammation, neoangiogenesis and formation of calcium phosphate nodule and ECM calcification. The accumulation of calcific deposits further aggravates the loss of leaflet compliance as the valve becomes increasingly stenotic. Lipoprotein deposition, lesion formation, and calcifications occur preferentially in the fibrosa layer [81, 82]. The reasons for this asymmetry in the pathophysiology are not fully understood but points towards a role of the distinct microenvironment within the valve to be important in disease initiation and progression [83]. The factors contributing to a unique microenvironment include infiltration of inflammatory cells, lipoprotein deposition, hemodynamic forces [84], mechanical stress [85] or extracellular matrix composition [86], and heterogeneity between cells from the aortic and the ventricular side of the valve [59].

Disruption of the trilaminar structure as a result of disorganized synthesis, misaligned deposition, and degradation of ECM proteins is one of the hallmarks of CAVD [87] in addition to an increase in apoptotic cells [88], myofibroblastic content [45], and calcium deposition [81]. The disease progresses from mildly sclerotic thickening to stenosis leading to improper opening or insufficient coaptation of the valve leaflets to prevent backflow from the ventricle. Ectopic bone and cartilage are also found in diseased aortic valves. An upregulation of bone-related markers such as osteopontin, bone sialoprotein, osteocalcin, alkaline phosphatase (ALP), and osteoblast-specific transcription factor core binding factor $\alpha$-1/runt-related transcription factor 2 (Cbfa-1/Runx2) has been reported in human calcified valves indicating active osteogenic processes [62, 89].

The end-stage valve disease is represented by compromised valve function and compliance due to stiffening of valve leaflets. CAVD can occur in younger and generally healthier population as is the case with biscuspid aortic valves, which can calcify heavily after birth [90].

The early mechanisms that initiate CAVD largely remain unknown. However, recent understanding of the disease based on the in vitro and ex vivo studies coupled with identified risk factors have shed some light on the cellular and molecular mechanisms associated with disease initiation and progression.

## 4 Mechanisms of Calcific Aortic Valve Disease

Despite the high prevalence of CAVD and the increasing morbidity and mortality, very little is known regarding the cellular basis of calcific aortic stenosis. Progress in studying the cell biology of this disease has been limited by the paucity of experimental models available. However recent experimental evidence has led to CAVD being now recognized as an active cellular driven process and not a passive "degenerative" disease. The following sections discuss our current understanding of the mechanisms of CAVD.

### 4.1 Cellular Mechanisms

As mentioned, the heightened forces exerted on the aortic valve result in thickened, stiffer leaflets with greater collagen content than that of the pulmonary valve, and pathological remodeling follows an increase in biomechanical pressures. Transduction of these biomechanical property changes has even been observed down to the cellular levels where aortic VICs were shown to exhibit increased cytoskeletal stiffness compared to pulmonary VICs [78]. Stiffness data correlated well with the level of collagen synthesis by the VICs, indicating that the aortic VICs are responsible for producing more robust ECM proteins, which yields the higher overall stiffness seen in the aortic valve. Further, VECs also exhibit a unique response to biomechanical forces and even show phenotypic differences on either side of the aortic valve leaflets, presumably due to the differences in hemodynamic forces [59]. Therefore, both VICs and VECs may be primed for active ECM remodeling that leads to CAVD when uncontrolled. The cellular contributors to CAVD are discussed in this section and followed by a section overviewing the underlying molecular mechanisms.

#### 4.1.1 Endothelial Dysfunction

Endothelial damage and dysfunction due to increased mechanical stress and reduced shear stress is believed to be the initiating event of CAVD [91]. In response to shear stress, VECs exhibit upregulation of inflammatory receptors such as ILRs and

TNFRs, for binding of circulating cytokines and increased expression of adhesion molecules such as ICAM-1 and VCAM-1 to recruit inflammatory cells [92, 93]. Migration of these cells into the valve interstitium results in activation of local inflammation. A strong correlation has been reported between macrophage accumulation and calcification in aortic valves of apolipoprotein deficient mice [94]. The fibrosa surface is more susceptible to this inflammatory state, as this surface experiences disturbed oscillatory blood flow that associates with VEC dysfunction [59]. The disturbed flow on the aortic side can lead to release of pro-inflammatory cytokines (BMP2/4) and pro-oxidant phenotype (NADH, ROS) in VECs [95, 96]. ROS, including oxidized lipids, can cause VEC injury resulting in loss of cell alignment [97]. VECs have also been indicated as key regulators of CAVD via dysregulation of protective nitric oxide signaling [98, 99]. VEC dysfunction may result in increased permeability to inflammatory molecules while focal denudation often co-localized above valve lesions occur at later stages [100].

### 4.1.2 Endothelial and Mesenchymal Transition (EndMT)

During development EndMT occurs in endocardial cushions where a subset of VECs detach from the endothelium, express the contractile protein α-SMA, and migrate into the valve interstitium of the embryonic valves to become VICs [101–103]. VECs may play an active role in tissue remodeling during CAVD by invading the valve interstitium through EndMT, similar to processes that occur during valve development [104]. Mechanical strain, Wnt/β-catenin signaling pathway and factors such as transforming growth factor-β [105] and transcription factor Msx2 have been implicated as mechanisms of EndMT of VECs, all of which are present in calcified aortic valves [106]. During this transition, VECs lose many common endothelial markers such as endothelial cadherin and gain contractile markers such as those observed in mesenchymal cells or myofibroblast activation of VICs (e.g., α-SMA, type I collagen and vimentin) [64, 107, 108]. Once EndMT occurs, the VECs may contribute to fibrosis. EndMT plays a role in the adaptive pathologic remodeling of mitral valve leaflets in an ovine model of functional mitral regurgitation [109, 110], and these cells are capable of osteogenic and chondrogenic differentiation [63]. In aortic valves, EndMT may precede VEC osteogenesis. EndMT was observed to coexpress with osteogenic markers in a mouse model of aortic valve calcification and in calcified human aortic valves, thus demonstrating that VECs contain the capacity to differentiate into endothelial-derived VICs through an EndMT process [65]. Further studies are required to demonstrate these mechanisms in aortic valve calcification.

### 4.1.3 VIC Plasticity and Fibrocalcific Remodeling

VICs reside within the aortic valve leaflets and directly remodel and maintain the microarchitecture of the leaflets by synthesis and degradation of ECM components.

In the healthy adult valve, VICs remain in a quiescent state with characteristics similar to fibroblasts; however, these specialized cells exhibit a high degree of phenotypic plasticity with VIC differentiation observed during periods of valve development and disease [69]. Notably, activation of VICs to a myofibroblast phenotype—characterized by increased expression of smooth muscle cell protein markers—is observed within developing valve leaflets and within fibrotic and calcific valve leaflets [29]. This activated phenotype is believed to be responsible for ECM synthesis and remodeling, and overactive VICs may be responsible for fibrotic collagen accumulation during the initial stages of CAVD. Markers of osteoblastic differentiation of VICs have also been observed in CAVD and may be responsible for the bone-like calcific nodules observed within CAVD leaflets [111]. The relationship between these activated and osteoblastic phenotypes has been confounding, and it is currently unknown whether the phenotypes arise from a single VIC progenitor population or whether different populations of VICs give rise to the myofibroblastic and osteoblastic cells. Future works that clarify the relationship between these phenotypes may provide insight into the appropriate VIC populations that should be targeted for the treatment of CAVD.

### 4.1.4 Stem and Progenitor Cells

Bone-marrow-derived mesenchymal stem cells have been found in both human and mouse valves [112, 113] and have been suggested to participate in CAVD [114]. Clonal VIC cultures from porcine aortic valve have indicated the presence of mesenchymal progenitor cells based on their ability to undergo differentiation into osteoblast-like cells [115]. Bone-marrow-derived cells expressing markers of myofibroblasts and osteoblasts were found near calcified nodules in aged mice and in humans [88, 116]. Coculturing of bone-marrow-derived cells with explanted calcified valves led to their differentiation towards an osteoblastic lineage suggesting a role of the local microenvironment in differentiation of progenitor cells [117]. Circulating osteogenic progenitor cells have also been identified in diseased aortic valves and are believed to participate in repair of tissue injury [118, 119]. However, whether these progenitor cells that reside in the aortic valve trigger valve calcification has yet to be confirmed. Mesenchymal-like stem cells expressing known stem cell markers, which were isolated from both noncalcified and calcified human aortic valves, were shown to be more susceptible to calcification in vitro. These cells were further identified to be present in higher proportion in calcified human valves relative to noncalcified controls, suggesting a cellular origin of calcification [120]. Although these studies present exciting evidence of the presence of progenitor cells in aortic valve tissue and their possible role in calcification, research in this direction is still in its infancy. The biological role of the progenitor cells in tissue injury, fibrosis, and calcification and the mechanism of their recruitment and their activation and differentiation into other cell types are yet to be uncovered.

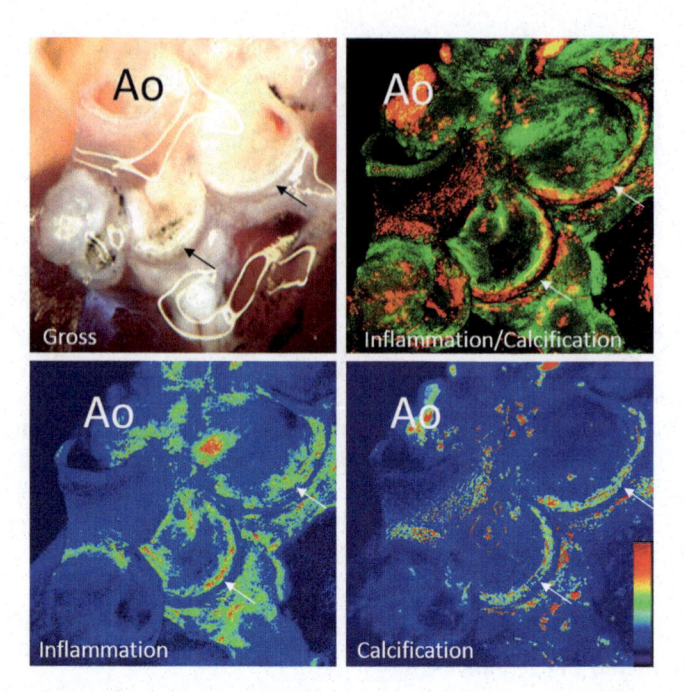

**Fig. 5** Inflammation associates with calcification in CAVD. ApoE$^{-/-}$ mice fed an atherogenic diet exhibit calcification that is spatially associated with inflammation as identified by molecular imaging probes targeted to each process. Similar observations have been made in aortic valves from CAVD patients. Adapted from [94]

### 4.1.5 Inflammation

Increased expression of adhesion molecules (e.g., ICAM-1 and VCAM-1) associated with VEC response to changes in shear stresses and cytokines also contributes to the progression of CAVD through recruitment of circulating leukocytes to the aortic valve leaflets [96]. Migration of these cells into the leaflet interstitium results in activation of local inflammation, which may play a major role in CAVD. Molecular imaging techniques identified a strong correlation between macrophage accumulation and calcification in aortic valves of apolipoprotein E-deficient (ApoE$^{-/-}$) mice [111, 121, 122] (Fig. 5). Additionally, the accumulation of leukocytes and related inflammatory cytokines positively correlates with remodeling in CAVD leaflets and aortic stenosis in human patients [70]. It currently remains unclear whether the infiltrating leukocytes directly contribute to the calcification process or induce calcific responses in VICs and VECs through release of pro-inflammatory cues. One of the major reasons for the uncertainty of the role of circulating cells in CAVD is that these factors are difficult to study in an in vitro setting. However, the strong association between CAVD and inflammation merits further attention through the development of new in vitro models that allow for cellular coculture and more closely recapitulate in vivo conditions.

## 4.2 Molecular Mechanisms

Though a lack of clarity exists surrounding the cellular contributions to CAVD, recent mechanistic studies have identified important pathways that modulate the phenotypes of valvular cells and produce responses similar to those observed in CAVD tissues. Many of the relevant molecular mechanisms associated with pathological changes in VEC and VIC behavior are highlighted in this section. Future studies may build upon these works to connect these pathways to the progression of CAVD within the leaflets and develop means to therapeutically interfere with the pathways.

### 4.2.1 Transforming Growth Factor β (TGF-β) Induced Activation of VIC

Signaling pathways involving TGF-β superfamily of cytokines have a role in myofibroblastic activation of VICs [123]. VICs exhibit increased α-SMA expression and collagen-I production upon exposure to active TGF-β in vitro. Additionally application of TGF-β to aortic valve leaflets in combination with mechanical stress resulted in increased responses related to CAVD [48, 124, 125]. Exposing VICs to mechanical strain after exposure to TGF-β exacerbates formation of dystrophic calcific nodules [126]. TGF-β induces expression of myofibroblastic markers and calcified nodule formation on polyacrylamide based substrate with stiffness similar to the stiffer fibrosa region of the leaflet [127]. Under these culture conditions, Wnt/β-catenin signaling pathway was found to be responsible for increase in myofibroblastic activation. TGF-β initiates signaling through activation of Smad2 and Smad3 proteins. TGF-β can also act through activation of mitogen activated protein (MAP) kinase signaling pathways, specifically by activating the MAP kinases p38 [128, 129] and Erk1/2 [130]. Activation of VICs is associated with increased secretion and degradation of ECM, expression of matrix metalloproteinases (MMPs) and tissue inhibitors of MMPs (TIMPs), processes important for wound repair [131]. ECM remodeling leads to further release of bound TGF-β [125]. Interestingly, combined application of TGF-β and cyclic strain to aortic valve leaflets under osteogenic conditions ex vivo leads to increased expression of BMP-2 and BMP-4, cytokines associated with osteogenesis [132]. These observations suggest that both myofibroblastic and osteogenic pathways may contribute to CAVD.

### 4.2.2 Bone Morphogenic Protein Signaling

Increased BMP2 and BMP4 expression as well as proteins associated with osteogenesis such as osteopontin, bone sialoprotein, and alkaline phosphatase have been observed within stenotic CAVD leaflets [62, 133–135]. These reports suggest that aortic valves undergo osteogenic calcification [89], which may be mediated by BMP2 signaling. Canonical BMP signaling occurs through activation of Smad and Wnt/β-catenin signaling and upregulation of transcription factor Msx2. These pathways in turn converge to activate Runx2, the master regulator of osteoblastogenesis followed by upregulation of osteopontin, osteocalcin, bone sialoprotein II [136]. In calcified valve leaflets, increased Smad 1/5/8 and Runx2 have been detected [137, 138]. Additionally, VICs were shown to exhibit a pro-osteogenic phenotype upon stimulation with BMP2, and this effect was abolished by treatment with a BMP antagonist, Noggin, indicating the role of BMP signaling in CAVD [139]. Msx2, a downstream target of BMP signaling, has also been identified in areas of calcification within valves from experimental models of CAVD [140]. Taken together the evidence suggests BMP signaling pathways are activated during CAVD and are capable of driving VICs to an osteoblast-like phenotype.

### 4.2.3 Wnt/Lrp5 Signaling Pathway

Activation of Wnt/β-catenin signaling pathway by BMP results in increase in expression of alkaline phosphatase, a driver of calcific mineral deposition [141]. Expression of Wnt3a and co-receptor lipoprotein receptor-related protein 5 and nuclear β-catenin has been confirmed in calcified valve tissues and in cell culture models [127, 142–144]. In adult valves, Wnt signaling can promote VIC calcification response, as loss of Wnt signaling through the Lrp5 receptor in ApoE knockout mice results in decreased aortic valve calcification [145].

### 4.2.4 The OPG/RANKL/RANK Signaling Pathway

The binding of receptor activator of nuclear factor-κB ligand (RANKL) to its receptor, RANK stimulates osteoclastogenesis and bone resorption through activation of NFκB [146]. Osteoprotegerin (OPG) acts as a RANK decoy receptor and inhibits this process. RANKL is a member of the TNFα superfamily, and is present on osteoblasts, stromal cells, T-cells, and endothelial cells [147]. Interestingly RANK, RANKL, and NFκB are significantly expressed in calcified valves [148]. In vitro treatment of VICs with RANKL or TNFα induces alkaline phosphatase activity and enhanced expression of Runx2 and formation of calcific nodules [149, 150]. Mice deficient in OPG were found to develop vascular calcification [151], and in hypercholesterolemic mice, OPG inhibits aortic valve calcification and preserves valve function by inhibition of osteogenesis [152]. These findings suggest an inverse correlation between processes of valvular calcification and bone formation.

### 4.2.5 Notch Signaling

Notch signaling is a central pathway in aortic valve development. Activation of Notch signaling suppresses Runx2 activity, through its downstream targets Hey1 and Hey2, thus inhibiting osteogenesis [153]. Notch1 heterozygous mice develop aortic valve calcification with increased BMP/SMAD1/5/8 signaling and Runx2 expression in the aortic valves [154, 155]. Interestingly, mutations in the Notch gene are also associated with developmental aortic valve abnormalities and calcification in humans [156]. These results were confirmed by in vitro culture models in which Notch inhibition promotes calcification of VICs by repressing chondrogenic genes, including Sox9, and inducing expression of osteogenic genes [154, 155]. Combined, evidence indicates Notch signaling as a negative regulator of osteogenic differentiation and that its inhibition or dysregulation can result in CAVD [157].

### 4.2.6 Renin-Angiotensin System

Stenotic aortic valves have significantly increased expression levels of both tissue angiotensin converting enzyme (ACE) and angiotensin II type 1 (AT1) receptor [158, 159]. ACE is produced by monocytes/macrophages and colocalizes with LDL. Blockage of receptor AT1 by an antagonist prevents infiltration by inflammatory cells, endothelial disruption, and VIC activation in rabbit model of valve disease [160]. Patients treated with angiotensin II-receptor blocker showed significantly slower progression of CAVD and lower mortality rates [161]. However, observational studies with ACE inhibitors have yielded conflicting results [160, 162]. Angiotensin signaling pathways may present an appropriate therapeutic target for CAVD treatment, but future works are needed to better understand the potential and appropriate points of intervention.

## 5 Diagnostic and Therapeutic Approaches

### 5.1 Aortic Valve Replacement

The only currently viable clinical option for CAVD patients is replacement of the native valve with mechanical valves or bioprosthetic valves [69]. These options are poor for the most commonly affected patient populations—pediatrics with congenital disease and the elderly. Bioprosthetic valves are produced by decellularizing and fixing valve apparatuses from bovine or porcine hearts. Implantation of these valves requires open chest surgery, which is risky for many pediatric and elderly patients. Cellular debris remaining in these leaflets has been shown to lead to accelerated calcification in bioprosthetic valves in contrast to the active cellular mechanisms that mediate CAVD in native leaflets [163]. The calcific nodules in the bioprosthetic valves nucleate within cellular remnants following fixation, whereas CAVD nucleation appears to occur in the extracellular and pericellular space. Due to problems with calcification and collagen degradation these bioprosthetic valves have a life span of only 15–20 years and do not fully incorporate with host tissue, and therefore do not allow for growth in the developing hearts of pediatric patients [11]. Traditional mechanical valves suffered from poor hemodynamic recapitulation that often led to remodeling in other cardiovascular tissues; however, newer generations of these synthetic valves better mimic native biomechanical function [164]. Additionally, many of these new valves can be implanted in a much more noninvasive manner. TAVR approach may provide a good option for CAVD patients who cannot tolerate invasive surgery [165]. Similar to bioprosthetic valves, these synthetic materials are also incapable of integrating with the growing heart of pediatric patients. In addition, a problem with TAVR approaches is the potential for vascular wall injury during implantation. Patients with CAVD are often most at risk for diseases of the vascular wall such as atherosclerosis, and injury caused by catheterization may increase the likelihood of thrombotic events that lead to myocardial infarction and stroke. However, early studies have indicated similar or reduced complications in TAVR patients compared to traditional AVR [166]. As TAVR becomes more prevalent, these cases will be monitored to ensure patient safety.

## 5.2 Potential Drug Targets and Clinical Trials

Due to the absence of noninvasive therapeutic interventions, clinicians often elect to delay surgical or catheter-based valve replacement until gross changes severely hinder leaflet biomechanics [167]. In the duration, patients must cope with a reduced quality of life due to diminished cardiac function [168]. The asymptomatic progression of CAVD presents a double-edged sword for the discovery and development of therapies. The slow progression of the disease presents a long window for therapeutic intervention, but the disease duration makes it difficult and costly to follow patients over the course of required randomized clinical trials [122]. Further, retrospective studies often focus on easily identifiable acute endpoints such as myocardial infarction, stroke, or death. Slowly progressing deterioration in valve function may be overlooked. Aortic valve leaflets, however, exhibit rapid calcification in chronic renal disease patients undergoing dialysis [169]. It is currently unclear to what extent calcification in these patients shares a common mechanism with that observed in patients with normal serum phosphate [170]. These patients may present an appropriate clinical trial population to test therapeutics designed to inhibit mineral deposition. The chronic renal disease patient population has a well-defined calcification initiation point at the onset of dialysis [169], and the accelerated mineralization circumvents the normal problems associated with the duration of aortic valve calcification. If basic mechanisms of mineral deposition are conserved, therapeutics developed for these patients could be extended to the general population. This patient population may allow testing of therapeutic targets that modulate the molecular mechanisms reviewed in Sect. 4. In addition to these potential targets, recent genome-wide association studies revealed a significant association with a polymorphism in the gene encoding the lipoprotein Lp(a) and CAVD [171]. This association may explain the lack of efficacy of traditional lipid lowering drugs such as statins in treating CAVD as these drugs do not affect Lp(a) levels. The mechanisms through which Lp(a) may mediate CAVD remain unknown, but Lp(a) lowering agents such as niacin and PCSK9 inhibitors may present new therapies to treat CAVD [172]. Further, Lp(a) may represent a surrogate marker for CAVD that could be used to identify patients requiring treatment and monitoring Lp(a) levels could be used as a short-term benchmark to assess the effectiveness of a treatment strategy.

## 5.3 Clinical Imaging and Identification

The development of a non-replacement therapeutic for CAVD dovetails with the need for strategies to identify early changes within the leaflets [122, 173]. Once calcific deposits form within the valve, reversibility is unclear; however, early ECM remodeling responses that presage gross calcific mineral deposition may fall within an appropriate therapeutic window (Fig. 6). Traditional detection of valvular insufficiency using echocardiography is unable to identify early leaflet changes. Recently, PET/CT imaging modalities have emerged as a means to detect inflammation and mineralization within cardiovascular tissues of human patients [174], and optical molecular imaging strategies have been successfully employed to visualize the progression of CAVD in mice [111, 122, 173]. Utilizing these imaging strategies may also give better insight into the temporal aspects of CAVD, and may lead to a better understanding of the required therapeutic strategy at each stage of CAVD.

## 6 Future Perspectives

With the aging baby boomer population and the continued industrialization of the developing world, CAVD will remain a major contributor to cardiovascular morbidity and mortality. Therefore, the development of noninvasive therapeutic strategies to treat CAVD patients is of utmost importance. Meeting this goal will require financial and intellectual investments in holistic approaches that connect cellular molecular mechanisms to clinically identifiable benchmarks in appropriate patient populations. The complex cellular and biomechanical function of the aortic valve necessitates an interdisciplinary team of basic scientists, engineers, and clinicians to

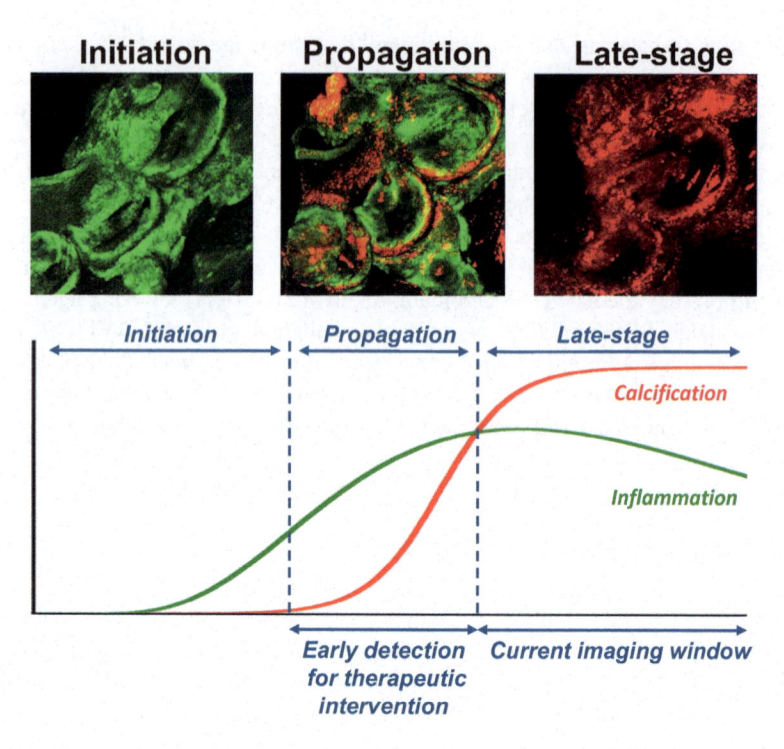

**Fig. 6** CAVD therapeutic window. Inflammation is associated with the initiation of CAVD. The accrual of inflammatory cells within the valve leaflets leads to the activation of calcific processes, which can exacerbate inflammatory processes. These initiation and propagation stages may represent an appropriate therapeutic window for CAVD before the remodeling progresses to an irreversible end-stage calcification. Adapted from [122]

address CAVD underpinnings and develop appropriate interventional strategies. This approach begins with the understanding that the aortic valve is a unique tissue with a distinct population of cells that must be treated differently than other cardiovascular tissues. Treatments that effectively target the pathological differentiation of aortic valve cells and the resultant ECM remodeling will address an important unmet clinical burden.

# References

1. Stewart BF, et al. Clinical factors associated with calcific aortic valve disease. Cardiovascular Health study. J Am Coll Cardiol. 1997;29(3):630–4.
2. Otto CM, et al. Association of aortic-valve sclerosis with cardiovascular mortality and morbidity in the elderly. N Engl J Med. 1999;341(3):142–7.
3. Ross J Jr, Braunwald E. Aortic stenosis. Circulation. 1968;38(1 Suppl):61–7.
4. Nishimura RA, et al. 2014 AHA/ACC Guideline for the Management of Patients with Valvular Heart Disease: executive summary: a report of the American College of

Cardiology/American Heart Association Task Force on Practice Guidelines. Circulation. 2014;129(23):2440–92.
5. Lindman BR, Bonow RO, Otto CM. Current management of calcific aortic stenosis. Circ Res. 2013;113(2):223–37.
6. Otto CM. Timing of aortic valve surgery. Heart. 2000;84(2):211–8.
7. Lloyd-Jones D, et al. Heart disease and stroke statistics--2009 update: a report from the American Heart Association Statistics Committee and Stroke Statistics Subcommittee. Circulation. 2009;119(3):e21–181.
8. Rahimtoola SH. Choice of prosthetic heart valve in adults an update. J Am Coll Cardiol. 2010;55(22):2413–26.
9. Pibarot P, Dumesnil JG. Prosthetic heart valves: selection of the optimal prosthesis and long-term management. Circulation. 2009;119(7):1034–48.
10. Sacks MS, Schoen FJ. Collagen fiber disruption occurs independent of calcification in clinically explanted bioprosthetic heart valves. J Biomed Mater Res. 2002;62(3):359–71.
11. Schoen FJ, Levy RJ. Calcification of tissue heart valve substitutes: progress toward understanding and prevention. Ann Thorac Surg. 2005;79(3):1072–80.
12. Brennan JM, et al. Early anticoagulation of bioprosthetic aortic valves in older patients: results from the Society of Thoracic Surgeons Adult Cardiac Surgery National Database. J Am Coll Cardiol. 2012;60(11):971–7.
13. Leon MB, et al. Transcatheter aortic-valve implantation for aortic stenosis in patients who cannot undergo surgery. N Engl J Med. 2010;363(17):1597–607.
14. Clavel MA, et al. Comparison between transcatheter and surgical prosthetic valve implantation in patients with severe aortic stenosis and reduced left ventricular ejection fraction. Circulation. 2010;122(19):1928–36.
15. Smith CR, et al. Transcatheter versus surgical aortic-valve replacement in high-risk patients. N Engl J Med. 2011;364(23):2187–98.
16. Daneault B, et al. Stroke associated with surgical and transcatheter treatment of aortic stenosis: a comprehensive review. J Am Coll Cardiol. 2011;58(21):2143–50.
17. Rajamannan NM, et al. Calcific aortic valve disease: not simply a degenerative process: a review and agenda for research from the National Heart and Lung and Blood Institute Aortic Stenosis Working Group. Executive summary: calcific aortic valve disease-2011 update. Circulation. 2011;124(16):1783–91.
18. Leopold JA. Cellular mechanisms of aortic valve calcification. Circ Cardiovasc Interv. 2012;5(4):605–14.
19. Misfeld M, Sievers HH. Heart valve macro- and microstructure. Philos Trans R Soc Lond Ser B Biol Sci. 2007;362(1484):1421–36.
20. Thubrikar M, et al. The design of the normal aortic valve. Am J Phys. 1981;241(6):H795–801.
21. Yacoub MH, et al. The aortic outflow and root: a tale of dynamism and crosstalk. Ann Thorac Surg. 1999;68(3 Suppl):S37–43.
22. Anderson RH, et al. The myth of the aortic annulus: the anatomy of the subaortic outflow tract. Ann Thorac Surg. 1991;52(3):640–6.
23. Anderson RH. Clinical anatomy of the aortic root. Heart. 2000;84(6):670–3.
24. Roberts WC. The structure of the aortic valve in clinically isolated aortic stenosis: an autopsy study of 162 patients over 15 years of age. Circulation. 1970;42(1):91–7.
25. Katayama S, et al. The sinus of Valsalva relieves abnormal stress on aortic valve leaflets by facilitating smooth closure. J Thorac Cardiovasc Surg. 2008;136(6):1528–35, 1535 e1.
26. Balachandran K, Sucosky P, Yoganathan AP. Hemodynamics and mechanobiology of aortic valve inflammation and calcification. Int J Inflam. 2011;2011:263870.
27. Butcher JT, Mahler GJ, Hockaday LA. Aortic valve disease and treatment: the need for naturally engineered solutions. Adv Drug Deliv Rev. 2011;63(4–5):242–68.
28. Thubrikar M, Bosher LP, Nolan SP. The mechanism of opening of the aortic valve. J Thorac Cardiovasc Surg. 1979;77(6):863–70.
29. Aikawa E, et al. Human semilunar cardiac valve remodeling by activated cells from fetus to adult: implications for postnatal adaptation, pathology, and tissue engineering. Circulation. 2006;113(10):1344–52.

30. Sacks MS, David Merryman W, Schmidt DE. On the biomechanics of heart valve function. J Biomech. 2009;42(12):1804–24.
31. Schoen FJ. Aortic valve structure-function correlations: role of elastic fibers no longer a stretch of the imagination. J Heart Valve Dis. 1997;6(1):1–6.
32. Schoen FJ. Evolving concepts of cardiac valve dynamics: the continuum of development, functional structure, pathobiology, and tissue engineering. Circulation. 2008;118(18):1864–80.
33. Stephens EH, Chu CK, Grande-Allen KJ. Valve proteoglycan content and glycosaminoglycan fine structure are unique to microstructure, mechanical load and age: relevance to an age-specific tissue-engineered heart valve. Acta Biomater. 2008;4(5):1148–60.
34. Stella JA, Sacks MS. On the biaxial mechanical properties of the layers of the aortic valve leaflet. J Biomech Eng. 2007;129(5):757–66.
35. Vesely I, Noseworthy R. Micromechanics of the fibrosa and the ventricularis in aortic valve leaflets. J Biomech. 1992;25(1):101–13.
36. Merryman WD, et al. The effects of cellular contraction on aortic valve leaflet flexural stiffness. J Biomech. 2006;39(1):88–96.
37. Zhao R, Sider KL, Simmons CA. Measurement of layer-specific mechanical properties in multilayered biomaterials by micropipette aspiration. Acta Biomater. 2011;7(3):1220–7.
38. Sacks MS, Smith DB, Hiester ED. The aortic valve microstructure: effects of transvalvular pressure. J Biomed Mater Res. 1998;41(1):131–41.
39. Stella JA, Liao J, Sacks MS. Time-dependent biaxial mechanical behavior of the aortic heart valve leaflet. J Biomech. 2007;40(14):3169–77.
40. Cimini M, Rogers KA, Boughner DR. Smoothelin-positive cells in human and porcine semilunar valves. Histochem Cell Biol. 2003;120(4):307–17.
41. Chester AH, et al. Localisation and function of nerves in the aortic root. J Mol Cell Cardiol. 2008;44(6):1045–52.
42. Marron K, et al. Innervation of human atrioventricular and arterial valves. Circulation. 1996;94(3):368–75.
43. Liu AC, Joag VR, Gotlieb AI. The emerging role of valve interstitial cell phenotypes in regulating heart valve pathobiology. Am J Pathol. 2007;171(5):1407–18.
44. Rabkin E, et al. Activated interstitial myofibroblasts express catabolic enzymes and mediate matrix remodeling in myxomatous heart valves. Circulation. 2001;104(21):2525–32.
45. Rabkin-Aikawa E, et al. Dynamic and reversible changes of interstitial cell phenotype during remodeling of cardiac valves. J Heart Valve Dis. 2004;13(5):841–7.
46. Yperman J, et al. Molecular and functional characterization of ovine cardiac valve-derived interstitial cells in primary isolates and cultures. Tissue Eng. 2004;10(9–10):1368–75.
47. Taylor PM, et al. The cardiac valve interstitial cell. Int J Biochem Cell Biol. 2003;35(2):113–8.
48. Walker GA, et al. Valvular myofibroblast activation by transforming growth factor-beta: implications for pathological extracellular matrix remodeling in heart valve disease. Circ Res. 2004;95(3):253–60.
49. Merryman WD, et al. Correlation between heart valve interstitial cell stiffness and transvalvular pressure: implications for collagen biosynthesis. Am J Physiol Heart Circ Physiol. 2006;290(1):H224–31.
50. Ku CH, et al. Collagen synthesis by mesenchymal stem cells and aortic valve interstitial cells in response to mechanical stretch. Cardiovasc Res. 2006;71(3):548–56.
51. Schneider PJ, Deck JD. Tissue and cell renewal in the natural aortic valve of rats: an autoradiographic study. Cardiovasc Res. 1981;15(4):181–9.
52. Schoen FJ. Mechanisms of function and disease of natural and replacement heart valves. Annu Rev Pathol. 2012;7:161–83.
53. Frater RW, et al. Endothelial covering of biological artificial heart valves. Ann Thorac Surg. 1992;53(3):371–2.
54. Butcher JT, Simmons CA, Warnock JN. Mechanobiology of the aortic heart valve. J Heart Valve Dis. 2008;17(1):62–73.

55. Butcher JT, et al. Transcriptional profiles of valvular and vascular endothelial cells reveal phenotypic differences: influence of shear stress. Arterioscler Thromb Vasc Biol. 2006;26(1):69–77.
56. Deck JD. Endothelial cell orientation on aortic valve leaflets. Cardiovasc Res. 1986;20(10):760–7.
57. Butcher JT, et al. Unique morphology and focal adhesion development of valvular endothelial cells in static and fluid flow environments. Arterioscler Thromb Vasc Biol. 2004;24(8):1429–34.
58. Kilner PJ, et al. Asymmetric redirection of flow through the heart. Nature. 2000;404(6779):759–61.
59. Simmons CA, et al. Spatial heterogeneity of endothelial phenotypes correlates with side-specific vulnerability to calcification in normal porcine aortic valves. Circ Res. 2005;96(7):792–9.
60. Guerraty MA, et al. Hypercholesterolemia induces side-specific phenotypic changes and peroxisome proliferator-activated receptor-gamma pathway activation in swine aortic valve endothelium. Arterioscler Thromb Vasc Biol. 2010;30(2):225–31.
61. Mohler ER 3rd. Mechanisms of aortic valve calcification. Am J Cardiol. 2004;94(11):1396–402. A6
62. Mohler ER 3rd, et al. Bone formation and inflammation in cardiac valves. Circulation. 2001;103(11):1522–8.
63. Wylie-Sears J, et al. Mitral valve endothelial cells with osteogenic differentiation potential. Arterioscler Thromb Vasc Biol. 2011;31(3):598–607.
64. Bischoff J, Aikawa E. Progenitor cells confer plasticity to cardiac valve endothelium. J Cardiovasc Transl Res. 2011;4(6):710–9.
65. Hjortnaes J, et al. Valvular interstitial cells suppress calcification of valvular endothelial cells. Atherosclerosis. 2015;242(1):251–60.
66. Chakraborty S, et al. Shared gene expression profiles in developing heart valves and osteoblast progenitor cells. Physiol Genomics. 2008;35(1):75–85.
67. Butcher JT, Nerem RM. Valvular endothelial cells regulate the phenotype of interstitial cells in co-culture: effects of steady shear stress. Tissue Eng. 2006;12(4):905–15.
68. Mohler ER 3rd, et al. Identification and characterization of calcifying valve cells from human and canine aortic valves. J Heart Valve Dis. 1999;8(3):254–60.
69. Hutcheson JD, Aikawa E, Merryman WD. Potential drug targets for calcific aortic valve disease. Nat Rev Cardiol. 2014;11(4):218–31.
70. Hjortnaes J, et al. Arterial and aortic valve calcification inversely correlates with osteoporotic bone remodelling: a role for inflammation. Eur Heart J. 2010;31(16):1975–84.
71. Rajamannan NM, Edwards WD, Spelsberg TC. Hypercholesterolemic aortic-valve disease. N Engl J Med. 2003;349(7):717–8.
72. Helske S, et al. Aortic valve stenosis: an active atheroinflammatory process. Curr Opin Lipidol. 2007;18(5):483–91.
73. Mohler ER 3rd. Are atherosclerotic processes involved in aortic-valve calcification? Lancet. 2000;356(9229):524–5.
74. Messier RH Jr, et al. Dual structural and functional phenotypes of the porcine aortic valve interstitial population: characteristics of the leaflet myofibroblast. J Surg Res. 1994;57(1):1–21.
75. Mulholland DL, Gotlieb AI. Cell biology of valvular interstitial cells. Can J Cardiol. 1996;12(3):231–6.
76. Dilley RJ, McGeachie JK, Prendergast FJ. A review of the proliferative behaviour, morphology and phenotypes of vascular smooth muscle. Atherosclerosis. 1987;63(2–3):99–107.
77. Orr AW, et al. Complex regulation and function of the inflammatory smooth muscle cell phenotype in atherosclerosis. J Vasc Res. 2010;47(2):168–80.
78. Merryman WD, et al. Differences in tissue-remodeling potential of aortic and pulmonary heart valve interstitial cells. Tissue Eng. 2007;13(9):2281–9.

79. Freeman RV, Crittenden G, Otto C. Acquired aortic stenosis. Expert Rev Cardiovasc Ther. 2004;2(1):107–16.
80. Merryman WD, Schoen FJ. Mechanisms of calcification in aortic valve disease: role of mechanokinetics and mechanodynamics. Curr Cardiol Rep. 2013;15(5):355.
81. Otto CM, et al. Characterization of the early lesion of 'degenerative' valvular aortic stenosis. Histological and immunohistochemical studies. Circulation. 1994;90(2):844–53.
82. O'Brien KD, et al. Apolipoproteins B, (a), and E accumulate in the morphologically early lesion of 'degenerative' valvular aortic stenosis. Arterioscler Thromb Vasc Biol. 1996;16(4):523–32.
83. Yip CY, Simmons CA. The aortic valve microenvironment and its role in calcific aortic valve disease. Cardiovasc Pathol. 2011;20(3):177–82.
84. Mehrabian M, Demer LL, Lusis AJ. Differential accumulation of intimal monocyte-macrophages relative to lipoproteins and lipofuscin corresponds to hemodynamic forces on cardiac valves in mice. Arterioscler Thromb. 1991;11(4):947–57.
85. Thubrikar MJ, Aouad J, Nolan SP. Patterns of calcific deposits in operatively excised stenotic or purely regurgitant aortic valves and their relation to mechanical stress. Am J Cardiol. 1986;58(3):304–8.
86. Hinton RB Jr, et al. Extracellular matrix remodeling and organization in developing and diseased aortic valves. Circ Res. 2006;98(11):1431–8.
87. Chen JH, Simmons CA. Cell-matrix interactions in the pathobiology of calcific aortic valve disease: critical roles for matricellular, matricrine, and matrix mechanics cues. Circ Res. 2011;108(12):1510–24.
88. Tanaka K, et al. Age-associated aortic stenosis in apolipoprotein E-deficient mice. J Am Coll Cardiol. 2005;46(1):134–41.
89. Rajamannan NM, et al. Human aortic valve calcification is associated with an osteoblast phenotype. Circulation. 2003;107(17):2181–4.
90. Sabet HY, et al. Congenitally bicuspid aortic valves: a surgical pathology study of 542 cases (1991 through 1996) and a literature review of 2,715 additional cases. Mayo Clin Proc. 1999;74(1):14–26.
91. Poggianti E, et al. Aortic valve sclerosis is associated with systemic endothelial dysfunction. J Am Coll Cardiol. 2003;41(1):136–41.
92. Muller AM, et al. Expression of endothelial cell adhesion molecules on heart valves: up-regulation in degeneration as well as acute endocarditis. J Pathol. 2000;191(1):54–60.
93. Shavelle DM, et al. Soluble intercellular adhesion molecule-1 (sICAM-1) and aortic valve calcification in the multi-ethnic study of atherosclerosis (MESA). J Heart Valve Dis. 2008;17(4):388–95.
94. Aikawa E, et al. Arterial and aortic valve calcification abolished by elastolytic cathepsin S deficiency in chronic renal disease. Circulation. 2009;119(13):1785–94.
95. Sorescu GP, et al. Bone morphogenic protein 4 produced in endothelial cells by oscillatory shear stress induces monocyte adhesion by stimulating reactive oxygen species production from a nox1-based NADPH oxidase. Circ Res. 2004;95(8):773–9.
96. Sucosky P, et al. Altered shear stress stimulates upregulation of endothelial VCAM-1 and ICAM-1 in a BMP-4- and TGF-beta1-dependent pathway. Arterioscler Thromb Vasc Biol. 2009;29(2):254–60.
97. Mirzaie M, et al. Evidence of woven bone formation in heart valve disease. Ann Thorac Cardiovasc Surg. 2003;9(3):163–9.
98. Bosse K, et al. Endothelial nitric oxide signaling regulates Notch1 in aortic valve disease. J Mol Cell Cardiol. 2013;60:27–35.
99. Richards J, et al. Side-specific endothelial-dependent regulation of aortic valve calcification: interplay of hemodynamics and nitric oxide signaling. Am J Pathol. 2013;182(5):1922–31.
100. Riddle JM, Magilligan DJ Jr, Stein PD. Surface topography of stenotic aortic valves by scanning electron microscopy. Circulation. 1980;61(3):496–502.
101. Armstrong EJ, Bischoff J. Heart valve development: endothelial cell signaling and differentiation. Circ Res. 2004;95(5):459–70.

102. Combs MD, Yutzey KE. Heart valve development: regulatory networks in development and disease. Circ Res. 2009;105(5):408–21.
103. Person AD, Klewer SE, Runyan RB. Cell biology of cardiac cushion development. Int Rev Cytol. 2005;243:287–335.
104. Frid MG, Kale VA, Stenmark KR. Mature vascular endothelium can give rise to smooth muscle cells via endothelial-mesenchymal transdifferentiation: in vitro analysis. Circ Res. 2002;90(11):1189–96.
105. Paranya G, et al. Aortic valve endothelial cells undergo transforming growth factor-beta-mediated and non-transforming growth factor-beta-mediated transdifferentiation in vitro. Am J Pathol. 2001;159(4):1335–43.
106. Balachandran K, et al. Cyclic strain induces dual-mode endothelial-mesenchymal transformation of the cardiac valve. Proc Natl Acad Sci U S A. 2011;108(50):19943–8.
107. Piera-Velazquez S, Li Z, Jimenez SA. Role of endothelial-mesenchymal transition (EndoMT) in the pathogenesis of fibrotic disorders. Am J Pathol. 2011;179(3):1074–80.
108. Kovacic JC, et al. Epithelial-to-mesenchymal and endothelial-to-mesenchymal transition: from cardiovascular development to disease. Circulation. 2012;125(14):1795–808.
109. Dal-Bianco JP, et al. Active adaptation of the tethered mitral valve: insights into a compensatory mechanism for functional mitral regurgitation. Circulation. 2009;120(4):334–42.
110. Chaput M, et al. Mitral leaflet adaptation to ventricular remodeling: prospective changes in a model of ischemic mitral regurgitation. Circulation. 2009;120(11 Suppl):S99–103.
111. Aikawa E, et al. Multimodality molecular imaging identifies proteolytic and osteogenic activities in early aortic valve disease. Circulation. 2007;115(3):377–86.
112. Deb A, et al. Bone marrow-derived myofibroblasts are present in adult human heart valves. J Heart Valve Dis. 2005;14(5):674–8.
113. Visconti RP, et al. An in vivo analysis of hematopoietic stem cell potential: hematopoietic origin of cardiac valve interstitial cells. Circ Res. 2006;98(5):690–6.
114. Hajdu Z, et al. Recruitment of bone marrow-derived valve interstitial cells is a normal homeostatic process. J Mol Cell Cardiol. 2011;51(6):955–65.
115. Chen JH, et al. Identification and characterization of aortic valve mesenchymal progenitor cells with robust osteogenic calcification potential. Am J Pathol. 2009;174(3):1109–19.
116. Helske S, et al. Possible role for mast cell-derived cathepsin G in the adverse remodelling of stenotic aortic valves. Eur Heart J. 2006;27(12):1495–504.
117. Leskela HV, et al. Calcification and cellularity in human aortic heart valve tissue determine the differentiation of bone-marrow-derived cells. J Mol Cell Cardiol. 2006;41(4):642–9.
118. Gossl M, et al. Role of circulating osteogenic progenitor cells in calcific aortic stenosis. J Am Coll Cardiol. 2012;60(19):1945–53.
119. Egan KP, et al. Role for circulating osteogenic precursor cells in aortic valvular disease. Arterioscler Thromb Vasc Biol. 2011;31(12):2965–71.
120. Nomura A, et al. CD34-negative mesenchymal stem-like cells may act as the cellular origin of human aortic valve calcification. Biochem Biophys Res Commun. 2013;440(4):780–5.
121. Aikawa E, et al. Osteogenesis associates with inflammation in early-stage atherosclerosis evaluated by molecular imaging in vivo. Circulation. 2007;116(24):2841–50.
122. Aikawa E, Otto CM. Look more closely at the valve: imaging calcific aortic valve disease. Circulation. 2012;125(1):9–11.
123. Towler DA. Molecular and cellular aspects of calcific aortic valve disease. Circ Res. 2013;113(2):198–208.
124. Grinnell F, Ho CH. Transforming growth factor beta stimulates fibroblast-collagen matrix contraction by different mechanisms in mechanically loaded and unloaded matrices. Exp Cell Res. 2002;273(2):248–55.
125. Merryman WD, et al. Synergistic effects of cyclic tension and transforming growth factor-beta1 on the aortic valve myofibroblast. Cardiovasc Pathol. 2007;16(5):268–76.
126. Fisher CI, Chen J, Merryman WD. Calcific nodule morphogenesis by heart valve interstitial cells is strain dependent. Biomech Model Mechanobiol. 2013;12(1):5–17.

127. Chen JH, et al. Beta-catenin mediates mechanically regulated, transforming growth factor-beta1-induced myofibroblast differentiation of aortic valve interstitial cells. Arterioscler Thromb Vasc Biol. 2011;31(3):590–7.
128. Zhang YE. Non-Smad pathways in TGF-beta signaling. Cell Res. 2009;19(1):128–39.
129. Hutcheson JD, et al. 5-HT(2B) antagonism arrests non-canonical TGF-beta1-induced valvular myofibroblast differentiation. J Mol Cell Cardiol. 2012;53(5):707–14.
130. Hutcheson JD, et al. Cadherin-11 regulates cell-cell tension necessary for calcific nodule formation by valvular myofibroblasts. Arterioscler Thromb Vasc Biol. 2013;33(1):114–20.
131. Durbin AD, Gotlieb AI. Advances towards understanding heart valve response to injury. Cardiovasc Pathol. 2002;11(2):69–77.
132. Balachandran K, et al. Elevated cyclic stretch induces aortic valve calcification in a bone morphogenic protein-dependent manner. Am J Pathol. 2010;177(1):49–57.
133. O'Brien KD, et al. Osteopontin is expressed in human aortic valvular lesions. Circulation. 1995;92(8):2163–8.
134. Mathieu P, et al. Calcification of human valve interstitial cells is dependent on alkaline phosphatase activity. J Heart Valve Dis. 2005;14(3):353–7.
135. Kaden JJ, et al. Expression of bone sialoprotein and bone morphogenetic protein-2 in calcific aortic stenosis. J Heart Valve Dis. 2004;13(4):560–6.
136. Johnson RC, Leopold JA, Loscalzo J. Vascular calcification: pathobiological mechanisms and clinical implications. Circ Res. 2006;99(10):1044–59.
137. Miller JD, et al. Evidence for active regulation of pro-osteogenic signaling in advanced aortic valve disease. Arterioscler Thromb Vasc Biol. 2010;30(12):2482–6.
138. Alexopoulos A, et al. Bone regulatory factors NFATc1 and Osterix in human calcific aortic valves. Int J Cardiol. 2010;139(2):142–9.
139. Yang X, et al. Pro-osteogenic phenotype of human aortic valve interstitial cells is associated with higher levels of toll-like receptors 2 and 4 and enhanced expression of bone morphogenetic protein 2. J Am Coll Cardiol. 2009;53(6):491–500.
140. Miller JD, et al. Dysregulation of antioxidant mechanisms contributes to increased oxidative stress in calcific aortic valvular stenosis in humans. J Am Coll Cardiol. 2008;52(10):843–50.
141. Bostrom KI, Rajamannan NM, Towler DA. The regulation of valvular and vascular sclerosis by osteogenic morphogens. Circ Res. 2011;109(5):564–77.
142. Caira FC, et al. Human degenerative valve disease is associated with up-regulation of low-density lipoprotein receptor-related protein 5 receptor-mediated bone formation. J Am Coll Cardiol. 2006;47(8):1707–12.
143. Alfieri CM, et al. Wnt signaling in heart valve development and osteogenic gene induction. Dev Biol. 2010;338(2):127–35.
144. Xu S, Gotlieb AI. Wnt3a/beta-catenin increases proliferation in heart valve interstitial cells. Cardiovasc Pathol. 2013;22(2):156–66.
145. Rajamannan NM. The role of Lrp5/6 in cardiac valve disease: experimental hypercholesterolemia in the ApoE−/−/Lrp5−/− mice. J Cell Biochem. 2011;112(10):2987–91.
146. Shiotani A, et al. Regulation of osteoclast differentiation and function by receptor activator of NFkB ligand and osteoprotegerin. Anat Rec. 2002;268(2):137–46.
147. Hofbauer LC, Schoppet M. Clinical implications of the osteoprotegerin/RANKL/RANK system for bone and vascular diseases. JAMA. 2004;292(4):490–5.
148. Steinmetz M, et al. Differential profile of the OPG/RANKL/RANK-system in degenerative aortic native and bioprosthetic valves. J Heart Valve Dis. 2008;17(2):187–93.
149. Kaden JJ, et al. Influence of receptor activator of nuclear factor kappa B on human aortic valve myofibroblasts. Exp Mol Pathol. 2005;78(1):36–40.
150. Kaden JJ, et al. Tumor necrosis factor alpha promotes an osteoblast-like phenotype in human aortic valve myofibroblasts: a potential regulatory mechanism of valvular calcification. Int J Mol Med. 2005;16(5):869–72.
151. Bucay N, et al. Osteoprotegerin-deficient mice develop early onset osteoporosis and arterial calcification. Genes Dev. 1998;12(9):1260–8.

152. Weiss RM, et al. Osteoprotegerin inhibits aortic valve calcification and preserves valve function in hypercholesterolemic mice. PLoS One. 2013;8(6):e65201.
153. Hilton MJ, et al. Notch signaling maintains bone marrow mesenchymal progenitors by suppressing osteoblast differentiation. Nat Med. 2008;14(3):306–14.
154. Nigam V, Srivastava D. Notch1 represses osteogenic pathways in aortic valve cells. J Mol Cell Cardiol. 2009;47(6):828–34.
155. Nus M, et al. Diet-induced aortic valve disease in mice haploinsufficient for the Notch pathway effector RBPJK/CSL. Arterioscler Thromb Vasc Biol. 2011;31(7):1580–8.
156. Garg V, et al. Mutations in NOTCH1 cause aortic valve disease. Nature. 2005;437 (7056):270–4.
157. Acharya A, et al. Inhibitory role of Notch1 in calcific aortic valve disease. PLoS One. 2011;6 (11):e27743.
158. Helske S, et al. Induction of local angiotensin II-producing systems in stenotic aortic valves. J Am Coll Cardiol. 2004;44(9):1859–66.
159. Helske S, et al. Increased expression of profibrotic neutral endopeptidase and bradykinin type 1 receptors in stenotic aortic valves. Eur Heart J. 2007;28(15):1894–903.
160. Arishiro K, et al. Angiotensin receptor-1 blocker inhibits atherosclerotic changes and endothelial disruption of the aortic valve in hypercholesterolemic rabbits. J Am Coll Cardiol. 2007;49(13):1482–9.
161. Capoulade R, et al. Impact of hypertension and renin-angiotensin system inhibitors in aortic stenosis. Eur J Clin Investig. 2013;43(12):1262–72.
162. Rosenhek R, et al. Statins but not angiotensin-converting enzyme inhibitors delay progression of aortic stenosis. Circulation. 2004;110(10):1291–5.
163. Schoen FJ, Levy RJ. SnapShot: calcification of bioprosthetic heart valves. Biomaterials. 2009;30(26):4445–6.
164. Sacks MS. Biomechanics of engineered heart valve tissues. Conf Proc IEEE Eng Med Biol Soc. 2006;1:853–4.
165. Zweng I, et al. Transcatheter versus surgical aortic valve replacement in high-risk patients: a propensity-score matched analysis. Heart Lung Circ. 2016;25:661.
166. Phan K, et al. Transcatheter valve-in-valve implantation versus reoperative conventional aortic valve replacement: a systematic review. J Thorac Dis. 2016;8(1):E83–93.
167. Bach DS. Prevalence and characteristics of unoperated patients with severe aortic stenosis. J Heart Valve Dis. 2011;20(3):284–91.
168. Czarny MJ, Resar JR. Diagnosis and management of valvular aortic stenosis. Clin Med Insights Cardiol. 2014;8(Suppl 1):15–24.
169. Guerraty MA, et al. Relation of aortic valve calcium to chronic kidney disease (from the Chronic Renal Insufficiency Cohort Study). Am J Cardiol. 2015;115(9):1281–6.
170. Rattazzi M, et al. Aortic valve calcification in chronic kidney disease. Nephrol Dial Transplant. 2013;28(12):2968–76.
171. Thanassoulis G, et al. Genetic associations with valvular calcification and aortic stenosis. N Engl J Med. 2013;368(6):503–12.
172. Rogers MA, Aikawa E. A not-so-little role for lipoprotein(a) in the development of calcific aortic valve disease. Circulation. 2015;132(8):621–3.
173. New SE, Aikawa E. Molecular imaging insights into early inflammatory stages of arterial and aortic valve calcification. Circ Res. 2011;108(11):1381–91.
174. Dweck MR, et al. Assessment of valvular calcification and inflammation by positron emission tomography in patients with aortic stenosis. Circulation. 2012;125(1):76–86.

# Remodelling Potential of the Mitral Heart Valve Leaflet

Bruno V. Rego, Sarah M. Wells, Chung-Hao Lee, and Michael S. Sacks

**Abstract** Little is known about how normal valvular tissues grow and remodel in response to altered loading. In the present work, we used the pregnancy state to represent a non-pathological cardiac volume overload that distends the mitral valve (MV), utilizing both extant and new experimental data and a modified form of our MV structural constitutive model. We determined that there was an initial period of permanent set-like deformation where no remodelling occurs, followed by a remodelling phase which resulted in near-complete restoration of homeostatic tissue-level behaviour. In addition, we observed that changes in the underlying MV interstitial cell (MVIC) geometry closely paralleled the tissue-level remodelling events, undergoing an initial passive perturbation followed by a gradual recovery to the pre-pregnant state. Collectively, these results suggest that valvular remodelling is actively mediated by average MVIC deformations (i.e. not cycle to cycle, but over a period of weeks).

**Keywords** Remodelling · Heart valve · Pregnancy · Volume overload · Structural constitutive model · Valve interstitial cell

---

B. V. Rego
James T. Willerson Center for Cardiovascular Modeling and Simulation, Institute for Computational Engineering and Sciences, Department of Biomedical Engineering, The University of Texas at Austin, Austin, TX, USA

S. M. Wells
School of Biomedical Engineering, Dalhousie University, Halifax, NS, Canada

C.-H. Lee
James T. Willerson Center for Cardiovascular Modeling and Simulation, Institute for Computational Engineering and Sciences, Department of Biomedical Engineering, The University of Texas at Austin, Austin, TX, USA

School of Aerospace and Mechanical Engineering, The University of Oklahoma, Norman, OK, USA

M. S. Sacks (✉)
The Oden Institute and the Department of Biomedical Engineering, The University of Texas at Austin, Austin, TX, USA
e-mail: msacks@ices.utexas.edu

# 1 Introduction

A primary hallmark of biological tissues that distinguishes them from abiotic structures is their ability to actively adapt to mechanical, chemical, electrical, and thermal cues from their surroundings. In tissues that respond largely to mechanical stimuli, altered loading can trigger a strategic blend of extracellular matrix (ECM) synthesis, degradation, and remodelling, producing adaptive changes in mechanical properties that lead to restored function. While valvular tissues are known to remodel in response to various pathological and non-pathological conditions [1–6], few studies have elucidated the mechanisms through which heart valve tissues grow and remodel. Most investigations of valvular remodelling also focus on disease conditions and are therefore potentially influenced by factors specific to the pathology of interest. As a result, very little is known about how heart valves grow and remodel purely in response to altered loading. Such information is important not only for our general understanding of heart valve pathophysiology, but also to serve as the basis for improved methods of valvular repair and replacement.

Of the four heart valves, the mitral valve (MV) is under the greatest mechanical demands. Physiologically, the MV functions to prevent the backflow of blood from the left ventricle (LV) to the left atrium when the LV contracts during systole. MV leaflet tissue is composed of four morphologically distinct layers: (1) the atrialis, which faces the left atrium; (2) the spongiosa, which is made up primarily of hydrated glycosaminoglycans and proteoglycans; (3) the fibrosa, consisting mainly of collagen; and (4) the ventricularis, which contains some collagen but also a substantial amount of elastin. Elastin in MV leaflets has been shown to consist of two highly aligned subpopulations of fibres, oriented circumferentially in the ventricularis and radially in the atrialis [7]. Similarly, collagen in MV leaflets is also highly aligned, with the majority of fibres pointing within $\pm 15°$ of the circumferential direction. Collagen fibres are also highly undulated (crimped) and thus bear no load until after they have been stretched enough to be completely straightened [7, 8]. While native valve properties have been extensively studied, we have comparatively little knowledge regarding how these functional aspects of heart valve leaflet tissues adapt and remodel to altered stresses induced by changes in valvular and ventricular geometry from various surgical procedures and cardiac pathologies.

Pregnancy induces a gradual cardiac volume overload, thus providing a natural paradigm for studying growth and remodelling in the non-pathological heart valve in response to altered loading. This is because pregnancy induces a wide array of cardiac changes, as the presence of the foetal-placental unit places additional demands on the maternal heart. Among these effects, some of the most notable in humans include 50% increases in cardiac output and LV mass, a 40% increase in blood volume, a 30% increase in stroke volume, a 20% increase in end-diastolic LV volume, and a 12% increase in MV orifice area [9–11]. At the organ level, the MV is connected to the LV through the chordae tendineae and papillary muscles (PMs) as well as the annulus, making its mechanical behaviour tightly coupled to that of the

**Table 1** Previous findings for MV growth and remodelling during pregnancy

|  | NP average | P average |
|---|---|---|
| *Leaflet stretch (under 60 N/m)* | | |
| Circumferential stretch | 1.37 ± 0.03 | 1.26 ± 0.03 |
| Radial stretch | 1.67 ± 0.06 | 1.67 ± 0.06 |
| *Leaflet surface dimensions* | | |
| Circumferential [cm] | 5.1 ± 0.1 | 5.9 ± 0.2 |
| Radial [cm] | 2.5 ± 0.1 | 3.0 ± 0.1 |
| Thickness [μm] | 1170 ± 75 | 1194 ± 59 |
| *Layer thicknesses [μm]* | | |
| Atrialis/spongiosa | 431 ± 85 | 346 ± 76 |
| Fibrosa | 540 ± 130 | 833 ± 76 |
| Ventricularis | 221 ± 40 | 135 ± 22 |
| *Constituent fractions (% dry wt)* | | |
| Total collagen | 56.7 ± 3.0 | 66.0 ± 3.3 |
| Extractable collagen | 0.6 ± 0.1 | 0.6 ± 0.1 |
| Elastin | 6.2 ± 0.6 | 2.8 ± 0.4 |
| sGAG | 3.3 ± 0.5 | 1.6 ± 0.2 |
| Collagen crimp period [μm] | 22.9 ± 1.2 | 65 ± 15 |
| MVIC population density (per 0.1 mm$^2$) | 10.2 ± 0.7 | 7.6 ± 0.7 |

All values previously reported in [5, 12–14]

LV. Changes in the geometry or contractile motion of the LV, due to factors such as systemic hypertension, myocardial infarction, and pregnancy, result in displacement of the PMs and changes to the shape of the MV annulus. This leads to a net change in MV external loading patterns. In response, the MV undergoes complex remodelling to adapt to its new ventricular environment. Previous animal studies have uncovered significant mechanical, structural, and compositional changes in all four heart valves during pregnancy (Table 1) [5, 12–14]. The mechanical behaviour of the MV anterior leaflet has been shown to undergo a biphasic change during pregnancy, with the apparent stiffness of the leaflet increasing significantly in early pregnancy (EP) but returning to normal by late pregnancy (LP) [12]. In contrast, a more recent investigation into structural changes of MV collagen during pregnancy uncovered no biphasic changes, but instead found only monotonic changes in mass fraction, preferred fibre direction, degree of fibre alignment, and crimp period [13]. The study also reported significant changes in the thickness of each MV leaflet tissue layer, with the fibrosa thickening by about 50%, the spongiosa/atrialis thinning about 30%, and the ventricularis thinning about 45%. We do not yet have a clear understanding of the underlying basis of these phenomena.

The objective of the present study was thus to develop a better understanding of the valvular growth and remodelling process, using the MV pregnancy response as a model system for normal heart valve tissues. Specifically, we sought to draw connections between changes in tissue mechanical properties, composition, structure, and MV interstitial cell (MVIC) geometry. Utilizing these results, we began an

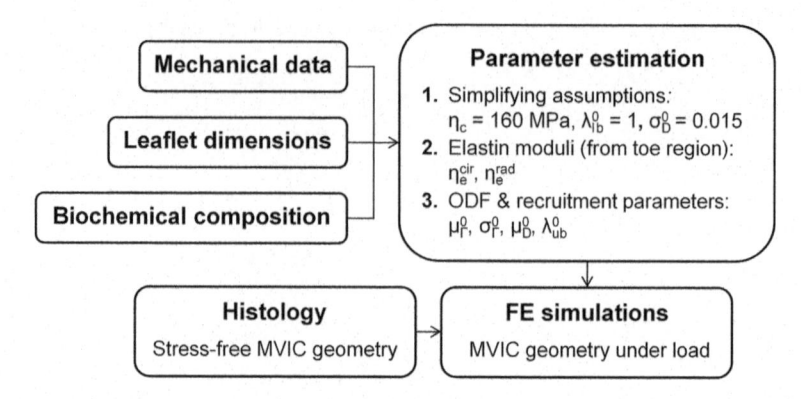

**Fig. 1** Schematic of the workflow in the present study

effort to elucidate the mechanisms through which heart valves are able to adapt under persistent altered loading conditions in vivo. In particular, we wanted to determine if there is a specific homeostatic endpoint that the MV tissue system attempts to re-establish, similar to vascular tissues [15]. To link the observed structural and morphological information, we utilized a recently developed comprehensive structural constitutive model specialized for the MV [16]. Moreover, we linked tissue-level and cellular changes by quantifying how MVIC geometry changes throughout pregnancy (Fig. 1). This additional focus on MVIC geometry was based in part on the role of MVICs (particularly their induced deformations over the cardiac cycle) in MV tissue maintenance [17]. Because both the LV and MV are otherwise healthy throughout pregnancy, our results provided insight into how valvular tissues adapt without the confounding pathological factors that must be considered in other diseased systems.

## 2 Methods

### 2.1 Mechanical Database and Processing

In the present study, we utilized extant biaxial mechanical data from a previous study, for which experimental testing methods have been described previously [12]. Briefly, fresh MV anterior leaflets from 5 never-pregnant heifers (NP) and 10 pregnant cows (P) at a variety of gestational time points were excised, laid flat, and their in-plane dimensions (circumferential length $\ell_{cir}$ and radial length $\ell_{rad}$) were measured using a ruler of 1-mm resolution. Leaflet thicknesses were also determined from light microscopy images of radial Verhoeff-stained histological sections. A rectangular specimen from the belly region of each anterior leaflet was mechanically preconditioned and loaded under a membrane tension-controlled equibiaxial protocol to a peak tension of 60 N/m in both the circumferential and radial directions. The

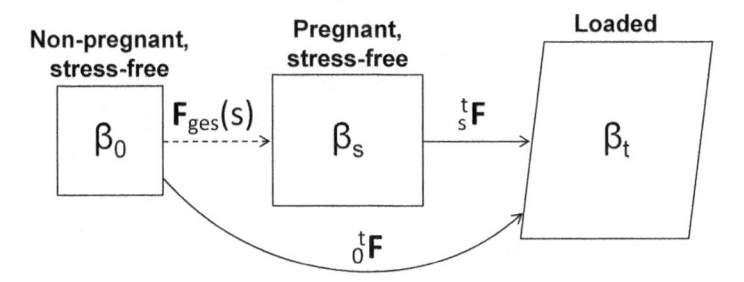

**Fig. 2** Set of possible states in which a leaflet sample may exist. $\beta_0$ is the NP reference configuration used for all structural parameters, $\beta_s(s)$ is the natural stress-free configuration of a P leaflet with gestation time $s$, and $\beta_t$ is the current state of the tissue under load. Corresponding deformation gradient tensors between the reference and loaded states are also shown, as well as $\mathbf{F}_{ges}(s)$, which describes the gestation-induced enlargement due to permanent set and/or growth

specimen geometry and applied forces for each dataset were used to calculate right Cauchy-Green deformation tensors $\mathbf{C} = \mathbf{F}^T\mathbf{F}$, where $\mathbf{F}$ is the deformation gradient, and second Piola-Kirchhoff stress tensors $\mathbf{S}$ at each instant of loading.

## 2.2 Accounting for Changes in Leaflet Dimensions

Permanent set, growth, and true tissue remodelling can all produce substantial changes in the apparent material properties and structure of tissue. To distinguish between these effects, we defined all moduli and structural parameters in our constitutive model with respect to a NP reference state (Sect. 2.3), which we estimated by establishing correspondence between the leaflet's NP geometry and its geometry throughout pregnancy. Specifically, we accounted for both permanent set and growth by defining $\mathbf{F}_{ges}(s)$ as a function of the gestation time $s$, which maps a specimen from its NP stress-free configuration $\beta_0$ to its P stress-free configuration $\beta_s$ (Fig. 2 and Table 2). Since it was not possible to know the actual pre-pregnant dimensions of each P leaflet, $\mathbf{F}_{ges+1}(s)$ was estimated by linear regression fitting of data from Pierlot et al. [14] for circumferential and radial lengths versus gestation time. From the trend lines for $\ell_{cir}(s)$ and $\ell_{rad}(s)$, respectively, $\mathbf{F}_{ges}(s)$ was defined as

$$\mathbf{F}_{ges}(s) = \begin{bmatrix} \dfrac{\ell_{cir}(s)}{\ell_{cir}(0)} & 0 \\ 0 & \dfrac{\ell_{rad}(s)}{\ell_{rad}(0)} \end{bmatrix} \qquad (1)$$

We utilized the approximation that $\ell_{cir}(0)$ and $\ell_{rad}(0)$ are equal to the pre-pregnant values of the circumferential and radial lengths, respectively, and that growth and remodelling do not induce appreciable shearing. For any P leaflet with a known $\beta_s$ and $s$, $\beta_0$ can be estimated using $\mathbf{F}_{ges}^{-1}(s)$. Because growth and remodelling occur on a

**Table 2** Nomenclature

|  | Abbreviation | Definition |
|---|---|---|
| Anatomy | LV | Left ventricle |
|  | PM | Papillary muscle |
|  | MV | Mitral valve |
|  | ECM | Extracellular matrix |
|  | MVIC | Mitral valve interstitial cell |
| Tissue states | NP | Non-pregnant/non-pregnancy |
|  | P | Pregnant |
|  | EP | Early pregnant/pregnancy |
|  | LP | Late pregnant/pregnancy |
|  | $s$ | Gestation time |
|  | $\beta_0$ | Non-pregnant stress-free tissue configuration |
|  | $\beta_s$ | Pregnant stress-free tissue configuration |
|  | $\beta_t$ | Tissue configuration under load |
| Mechanics | $\mathbf{F}_{ges}$ | Change in configuration during gestation |
|  | $\Lambda_{ges}$ | Change in length during gestation |
|  | $\mathbf{F}$ | Deformation gradient |
|  | $\mathbf{C}$ | Right Cauchy-Green deformation tensor |
|  | $\lambda$ | Unidirectional stretch |
|  | $\mathbf{S}$ | Second Piola-Kirchhoff stress |
| Constitutive modelling | MSSCM | Meso-scale structural constitutive model |
|  | $\lambda_{slk}$ | Collagen fibre slack stretch |
|  | $D$ | Slack stretch distribution, recruitment function |
|  | $\lambda_{lb}$ | Lower bound of slack stretch distribution |
|  | $\lambda_{ub}$ | Upper bound of slack stretch distribution |
|  | $\mu_D$ | Mean recruitment stretch |
|  | $\sigma_D$ | Standard deviation of slack stretch distribution |
|  | $\mathbf{n}$ | Collagen ensemble orientation vector |
|  | $\theta$ | Angle from circumferential axis |
|  | ODF, $\Gamma$ | Orientation distribution function |
|  | $\mu_\Gamma$ | Preferred fibre angle (ODF mean) |
|  | $\sigma_\Gamma$ | Fibre splay (ODF standard deviation) |
|  | $\phi_c$ | Collagen fraction |
|  | $\phi_e$ | Elastin fraction |
|  | $\eta_c$ | Collagen modulus |
|  | $\eta_e^{cir}$, $\eta_e^{rad}$ | Circumferential and radial elastin moduli |

time scale much larger than that of the cardiac cycle (i.e. weeks vs. seconds), any changes with respect to $s$ were quantified without considering how tissue geometry and structure change during in vivo loading and unloading. This also implies that for each loading cycle, the tissue can be modelled as an incompressible pseudo-hyperelastic biological material.

## 2.3 Constitutive Model

In order to express the mechanical behaviour of each valve leaflet specimen in terms of physically meaningful parameters, we utilized a recently developed meso-scale structural constitutive model (MSSCM) for MV leaflet tissues [16, 37]. In brief, the total stress in the leaflet in this solid mixture model was assumed to be the weighted linear sum of the stresses borne by the underlying collagen fibre network, elastin fibre network, and the remainder of the tissue matrix, with each stress being scaled by the mass fraction of its corresponding phase. The matrix phase has been shown to have very low tensile stiffness and thus to contribute very little to the mechanical behaviour of the total tissue [16, 37]. Therefore, in the present study, we neglected the matrix phase and instead attributed all the stress in the tissue to the collagen and elastin fibre networks.

A collagen fibre within the MV leaflet is naturally undulated and bears negligible load for stretch values smaller than some critical "slack" stretch $\lambda_{slk}$, above which that fibre is fully straightened [16, 37]. In our MSSCM, the collagen phase was modelled at the level of a fibre ensemble, which exhibits the properties of a unidirectional population of collagen fibres having a bounded distribution of slack stretches with lower bound $\lambda_{lb}$ and upper bound $\lambda_{ub}$. We modelled the distribution of slack stretches $D$, with mean $\mu_D$ and standard deviation $\sigma_D$, using a beta distribution Beta($\alpha$, $\beta$) that is linearly mapped onto [$\lambda_{lb}$, $\lambda_{ub}$] and scaled by $1/(\lambda_{ub} - \lambda_{lb})$ such that $\int_{\lambda_{lb}}^{\lambda_{ub}} D(\lambda_{slk}) \, d\lambda_{slk} = 1$. Functionally, $D$ serves to describe the gradual recruitment (i.e. straightening) of fibres in an ensemble, and is thus referred to as the "recruitment function." It has been found that in normal MV leaflets, $D$ is largely independent of angle [16]. Following anisotropic permanent set and/or growth during pregnancy, however, this assumption is no longer valid if the recruitment function parameters are defined with respect to the natural stress-free reference configuration ($\beta_s$) of the tissue. To account for this, we defined all structural parameters in the MSSCM with respect to $\beta_0$, which we estimated for each P specimen using $\mathbf{F}_{ges}^{-1}(s)$. We then recast $D$ with respect to $\beta_s$ under the assumption of affine mapping, following established push-forward procedures [18]. For clarity, we represent the originally defined recruitment function as $D^0$ and the recast version as $D^s$ (note that superscripts will be used throughout to designate the reference state of a particular quantity).

Homogenizing to the tissue level, we defined the fibre ensemble orientation as $\mathbf{n}^0 = [\cos\theta^0, \sin\theta^0]^T$, where $\theta^0 \in [-\pi/2, \pi/2]$ is the angle from the circumferential axis in $\beta_0$. Additionally, we defined a beta-distributed orientation distribution function (ODF) $\Gamma^0$ supported on $[-\pi/2, \pi/2]$, with mean $\mu_\Gamma^0$ and standard deviation $\sigma_\Gamma^0$, which describes the directional density of collagen fibres at any $\theta^0$. Following established procedures [19], we recast the ODF into $\beta_s$ using

**Table 3** Reference states for mechanical data and model parameters

| Category | Symbols | Reference state $\beta_0$ | $\beta_s(s)$ |
|---|---|---|---|
| Deformation | **F**, **C** | X (NP leaflets) | X (P leaflets) |
| Stress | **S** | X (NP leaflets) | X (P leaflets) |
| Collagen recruitment[a] | $\mu_D^0$, $\sigma_D^0$, $\lambda_{lb}^0$, $\lambda_{ub}^0$ | X | – |
| Collagen orientation[a] | $\mu_\Gamma^0$, $\sigma_\Gamma^0$ | X | – |
| Collagen modulus[b] | $\eta_c$ | – | – |
| Elastin moduli | $\eta_e^{cir}$, $\eta_e^{rad}$ | X | – |

[a]Note that recruitment and orientation parameter results for P leaflets presented with a superscript $s$ (e.g. $\mu_D^s$) have been mapped into $\beta_s$ for ease of interpretation (Sect. 2.5)
[b]The collagen modulus is defined with respect to each fibre's fully straightened stress-free state, and is thus independent of tissue reference configuration [16]

$$\theta^s = \tan^{-1}\left[\frac{F_{ges,21}\cos\theta^0 + F_{ges,22}\sin\theta^0}{F_{ges,11}\cos\theta^0 + F_{ges,12}\sin\theta^0}\right] \text{ and } \Gamma^s(\theta^s) = \Gamma^0(\theta^0)\frac{\Lambda_{ges}(\theta^0)}{\det(\mathbf{F}_{ges})} \quad (2)$$

where $\Lambda_{ges}(\theta^0) = \sqrt{\mathbf{n}^{0,T}\mathbf{F}_{ges}^T\mathbf{F}_{ges}\mathbf{n}^0}$ is the pregnancy-induced length change along $\mathbf{n}^0$. The total stress in the collagen network is the sum of the stresses along all ensemble directions, weighted by the ODF.

The orientation of elastin fibres has similarly been described using two continuous distributions [16], oriented in the circumferential and radial directions in the ventricularis and atrialis, respectively [7]. We found that to capture the effective mechanical contribution of elastin, it was sufficient to model the elastin phase as the superposition of two perfectly aligned orthogonal populations of fibres. Consistent with previous findings [20], elastin fibres were assumed to have a linear **S**-**C** relationship in the present study. To control for changes in the apparent stiffness of the elastin phase due to permanent set, we defined the effective elastin stiffness modulus for each direction with respect to $\beta_0$, although **S** and **C** are always expressed with respect to the natural stress-free configuration of each specimen (Table 3). The sum of the stresses borne by collagen and elastin, weighted by their respective mass fractions $\phi_c$ and $\phi_e$, yields the stress tensor for the entire leaflet specimen as

$$\begin{aligned}\mathbf{S}(\mathbf{C}) = &\phi_c(s)\eta_c\int_{-\pi/2}^{\pi/2}\frac{\Gamma^s(\theta^s;\mu_\Gamma^0,\sigma_\Gamma^0)}{\lambda}\left[\int_1^\lambda D^s(\lambda_{slk}^s;\mu_D^0,\sigma_D^0,\lambda_{lb}^0,\lambda_{ub}^0)\left(\frac{\lambda}{\lambda_{slk}^s}-1\right)d\lambda_{slk}^s\right]\\ &\times\mathbf{n}^s\otimes\mathbf{n}^s\,d\theta^s + \phi_e(s)\left[\eta_e^{cir}\left(\Lambda_{ges}^{cir}\right)^4(\lambda_{cir}^2-1)\mathbf{n}_{cir}^s\otimes\mathbf{n}_{cir}^s\right.\\ &\left.+\eta_e^{rad}\left(\Lambda_{ges}^{rad}\right)^4(\lambda_{rad}^2-1)\mathbf{n}_{rad}^s\otimes\mathbf{n}_{rad}^s\right]\end{aligned} \quad (3)$$

where $\{\eta_c, \eta_e^{cir}, \eta_e^{rad}\}$ are the effective stiffness moduli of collagen, circumferential elastin, and radial elastin, respectively, $\lambda = \lambda(\theta^s) = \sqrt{\mathbf{n}^s\cdot\mathbf{Cn}^s}$ is the stretch for an

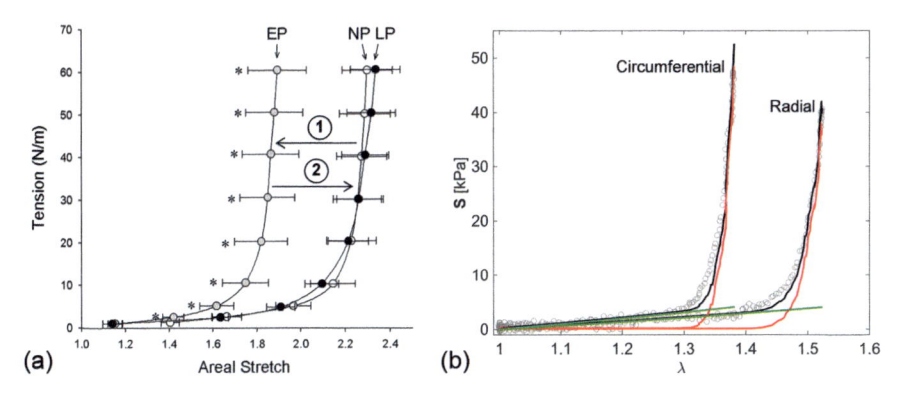

**Fig. 3** (a) Effective equibiaxial mechanical response of the MV leaflet at each pregnancy stage (adapted from [12]). Tension is plotted against the ratio between the current and stress-free specimen areas. The leaflet apparently stiffens in EP, but recovers its original behaviour by LP. (b) Representative fit of the MSSCM to biaxial mechanical data for a MV leaflet specimen, illustrating the individual contributions of collagen (red) and elastin (green). The left curve corresponds to the circumferential response, while the right curve corresponds to the radial response. Model fits achieved $R^2 = 0.940$ on average

ensemble pointing in $\mathbf{n}^s$, $\Lambda_{\text{ges}}^{\text{cir}} = \Lambda_{\text{ges}}(0)$, $\Lambda_{\text{ges}}^{\text{rad}} = \Lambda_{\text{ges}}(\pi/2)$, $\lambda_{\text{cir}} = \lambda(0)$, $\lambda_{\text{rad}} = \lambda(\pi/2)$, and $\{\mathbf{n}_{\text{cir}}^s, \mathbf{n}_{\text{rad}}^s\}$ are the circumferential and radial direction vectors, respectively, cast in $\beta_s$ (Table 2). Note that the model does not require any assumptions to be made about the nature of $\mathbf{F}_{\text{ges}}$ (e.g. negligible shear). The formulation of the MSSCM allows for the convenient decomposition of the contributions by collagen and elastin phases, which can offer insights into the functional roles of each fibre type by highlighting the stress/stretch regimes over which each constituent is dominant (Fig. 3b).

## 2.4 Parameter Estimation

The final MSSCM formulation has nine parameters: the elastin moduli in the circumferential and radial directions $\left(\eta_e^{\text{cir}}, \eta_e^{\text{rad}}\right)$, the collagen modulus $(\eta_c)$, the recruitment function parameters $\left(\mu_D^0, \sigma_D^0, \lambda_{\text{lb}}^0, \lambda_{\text{ub}}^0\right)$, and the collagen ODF parameters $\left(\mu_\Gamma^0, \sigma_\Gamma^0\right)$. These parameters were estimated in the following order either by assumption or via non-linear least squares optimization using a differential evolution algorithm with a seed range of $[\xi/1.1, \ 1.1\xi]$ for any parameter initial guess $\xi$ [21] (Fig. 1). Note that mass fractions of collagen and elastin were not included in the set of estimated parameters, and were instead prescribed from previous biochemical assay results [5, 14]. Mass fraction data were reanalysed to identify potential time-dependent trends, and appropriate values for $\phi_c$ and $\phi_e$ were assigned to each specimen accordingly.

### 2.4.1 Elastin and Collagen Moduli

As done previously [16], elastin moduli values were estimated first, under the approximation that elastin fibres are solely responsible for the mechanical response of the leaflet under low stress (Fig. 3b). In this step, the MSSCM was reduced to just its elastin terms and was then fit to only the "toe region" points of the response curves for all specimens. We hypothesized that all remodelling could be attributed to structural changes in the collagen phase after accounting for changes in mass fractions and leaflet dimensions, without any need for the inherent material properties of the fibre networks (i.e. the moduli) to change.

Next, we noted that for collagen ensembles in the MV, full recruitment does not occur until stress in the ensemble direction exceeds 1.5 MPa [16]. Because the mechanical data used in the present study only spanned a physiological stress range (less than 100 kPa), we were unable to determine the collagen modulus $\eta_c$ directly. Instead, we chose to approximate $\eta_c$ as the average value found by Zhang et al. [16] for the porcine MV anterior leaflet, under the assumption that type I collagen stiffness is well conserved across species. The collagen modulus was thus fixed at $\eta_c = 160$ MPa for all specimens (Fig. 1). Based on our previous biochemical and thermochemical results, which suggested that collagen synthesized during pregnancy undergoes rapid cross-linking and maturation [5, 13], we also assumed that the same collagen modulus could be applied to all groups.

### 2.4.2 ODF and Recruitment Parameters

Due to covariance both among the recruitment function parameters as well as between the recruitment function and ODF parameters, even a search-based optimization algorithm can often fail unless data from numerous loading protocols (with varying applied stress ratios) are collected [22]. This issue can be largely remedied, however, by fixing a subset of the parameters to known values. We chose to approximate the standard deviation of the recruitment function, as it has been shown to be consistently close to $\sigma_D^0 = 0.015$ with little variation between specimens [16]. Further, based on previous results showing that slack stretch distributions in heart valve leaflets are very negatively skewed, we assumed that $\mu_D^0 - 1 \gg \sigma_D^0$ and thus were able to eliminate an additional parameter by setting the recruitment lower bound $\lambda_{lb}^0 = 1$, without significantly changing the shape of the recruitment function (Fig. 1). Note that fixing $\sigma_D^0$ and $\lambda_{lb}^0$ does not lock the effective width of the recruitment function, as $\lambda_{ub}^0$ also greatly affects the shape of $D^0$. Specifically, a linear mapping of $\mu_D^0$ and $\sigma_D^0$ onto [0, 1] shows that the beta distribution Beta($\alpha$, $\beta$), from which $D^0$ is defined, has the mean

$$\mu_B = \frac{\mu_D^0 - \lambda_{lb}^0}{\lambda_{ub}^0 - \lambda_{lb}^0} \tag{4}$$

and the standard deviation

$$\sigma_B = \frac{\sigma_D^0}{\lambda_{ub}^0 - \lambda_{lb}^0} \tag{5}$$

that both depend on $\lambda_{ub}^0$. Moreover, the shape parameters $\alpha$ and $\beta$ are given by

$$\alpha = \frac{-\mu_B \left(\sigma_B^2 + \mu_B^2 - \mu_B\right)}{\sigma_B^2} \text{ and}$$

$$\beta = \frac{(\mu_B - 1)\left(\sigma_B^2 + \mu_B^2 - \mu_B\right)}{\sigma_B^2} \tag{6}$$

and thus, by extension, also depend on $\lambda_{ub}^0$.

Upon fixing $\sigma_D^0$ and $\lambda_{lb}^0$, we had only four remaining parameters ($\mu_\Gamma^0$, $\sigma_\Gamma^0$, $\mu_D^0$, $\lambda_{ub}^0$), which we simultaneously optimized in the final stage of parameter estimation (Fig. 1). The algorithm was executed several times, using different initial guesses for each parameter that were arbitrarily above or below any feasible value. By confirming that the same optimal solution was obtained every time for each specimen, we demonstrated that the optimization was insensitive to seeding when limited to these four parameters.

## 2.5 Interpretation of Parameter Values

All structural parameter values obtained from fitting the MSSCM to P data were defined with respect to an estimated pre-pregnant reference configuration $\beta_0$ (Sect. 2.2; Table 3). In order to draw insights into the relevant changes in collagen structure and to allow for a more physiologically meaningful interpretation of the results, all structural parameters for P specimens were recast into their natural (i.e. post-excision) configuration $\beta_s$, which corresponds more closely to the functional state of the valve. We obtained the "preferred fibre angle" $\mu_\Gamma^s$ and "fibre splay" $\sigma_\Gamma^s$ directly from the corresponding ODF $\Gamma^s$. Due to the angular dependence of $D^s$, we chose to consider $\mu_D^s$ and $\lambda_{ub}^s$ only for an ensemble oriented in the preferred fibre direction, when $\theta^s = \mu_\Gamma^s$. To arrive at the transformed recruitment parameters, we normalized $\mu_D^0$ and $\lambda_{ub}^0$ by the gestational length change along the preferred fibre direction:

$$\mu_D^s = \frac{\mu_D^0}{\Lambda_{\text{ges}}(\mu_\Gamma^0)} \text{ and } \lambda_{\text{ub}}^s = \frac{\lambda_{\text{ub}}^0}{\Lambda_{\text{ges}}(\mu_\Gamma^0)} \tag{7}$$

All structural parameter results for P specimens presented below are therefore expressed with respect to the $\beta_s$ configuration of each leaflet, and can be interpreted as being closely related to the physiological state of the leaflet at their corresponding gestation time $s$.

Because the LV remodels substantially during pregnancy [9–11], changes in the fibre architecture of the MV would occur even in the absence of growth and remodelling in the leaflet itself, since annular dilatation and PM displacement change the geometry of the MV. It is thus crucial to determine the extent to which the structural changes seen in the leaflet are consistent with changes in the leaflet's bulk surface dimensions. To determine how large of an effect growth and remodelling had on the MV fibre architecture, we calculated how each structural parameter *would have* evolved throughout pregnancy in the absence of growth and remodelling. This was accomplished by applying Eq. (2) to the average NP ODF and Eq. (7) to the average NP recruitment parameters for all $s$, which is equivalent to applying a stretch that corresponds to the observed dimensional changes.

## 2.6 MVIC Geometry

To investigate whether structural changes in the ECM coincided with observable changes at the cellular level, we quantified MVIC geometry throughout pregnancy, using the nuclear aspect ratio (NAR) as a metric of overall cell shape (Fig. 1). First, we obtained histological sections showing the circumferential-transmural plane from 3 NP and 6 P leaflets of varying gestation time. Each section was stained with haematoxylin and eosin (H&E), which provided a sharp contrast between nuclei (coloured dark blue) and the surrounding collagen fibres (coloured pale pink). The middle of the fibrosa layer of each section was then imaged using bright-field microscopy at ×40 magnification. To determine the average NAR for each section, a custom-written MATLAB (MathWorks, Inc.) script was used to automatically detect and fit ellipses to 10 nuclei in each image. The NAR of each MVIC was then calculated as the ratio between the major and minor radii of the corresponding ellipse.

While we used our histological data to quantify the stress-free geometry of MVICs throughout pregnancy, we relied on finite element (FE) simulations to estimate how MVIC NAR changed over the cardiac cycle for different pregnancy stages. In order to estimate changes in MVIC deformation throughout pregnancy, we performed FE simulations using Abaqus (Dassault Systèmes Simulia Corp.) for a representative volume element (RVE) of valve leaflet tissue under biaxial tension, following established procedures [17]. Briefly, 10 MVICs were embedded in a $100 \times 100$ μm$^2$ ECM region as Saint Venant-Kirchhoff ellipsoidal inclusions of

~5-kPa stiffness, with an initial NAR equal to the average found for NP specimens (Fig. 10c, inset). The number of cells used matched the MVIC population density previously found by Pierlot et al. [5] (Table 1). Mechanical properties of non-MVIC elements were prescribed using the MSSCM parameter estimates (Sect. 2.4). To simulate the effects of pregnancy-induced dimensional changes at EP, we applied the corresponding pre-stretch to the RVE model and performed the same FE simulation using EP mechanical properties. No separate simulation for LP was performed since the behaviour of LP leaflets is known to be almost identical to NP [12].

## 3 Results

### 3.1 Leaflet Dimensions

Linear trendlines were sufficient to describe the continual changes in leaflet surface dimensions over the course of pregnancy. The circumferential and radial lengths of the excised MV anterior leaflets were found to correlate significantly with gestation time (Fig. 4), with $R^2$ values of 0.510 and 0.226, respectively, and $p < 0.01$ for both cases. Normalizing each dimension by its NP value revealed that relative expansion occurs primarily in the circumferential direction, with the circumferential rate of change doubling its radial counterpart. By 250 days (close to term), circumferential and radial length ratios were found to be 1.4 and 1.2, respectively (Fig. 5a). The observed leaflet dimensions are consistent with several organ-level dimensions previously measured in humans [10], which also increased linearly throughout pregnancy (Fig. 5b). No correlation was found for leaflet thickness, which remained relatively constant throughout pregnancy at $1.183 \pm 0.046$ mm.

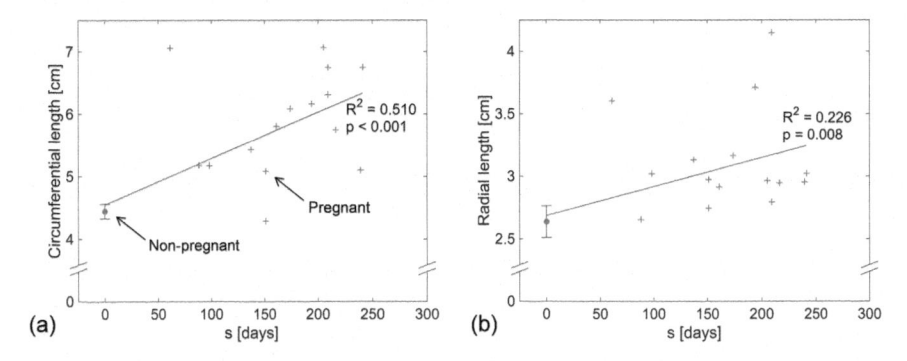

**Fig. 4** (a) Circumferential and (b) radial length plotted versus gestation time, with NP data displayed as mean±standard error at $s = 0$. Linear trend lines were fit to all the data, showing a continuous time-dependent correlation for each dimension (data from [14], with sample sizes of $n_{\text{NP}} = 10$ and $n_{\text{P}} = 15$)

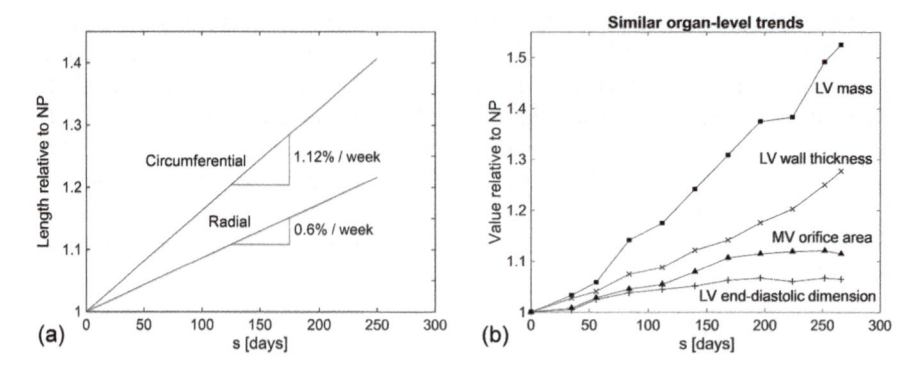

**Fig. 5** (a) Circumferential and radial length ratios (relative to their NP values) over the course of pregnancy. Enlargement occurs primarily in the circumferential direction, at roughly double the rate of that in the radial direction. (b) Similar monotonically increasing dimensional changes that occur at the organ level (data from [10])

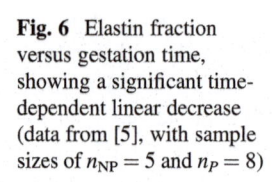

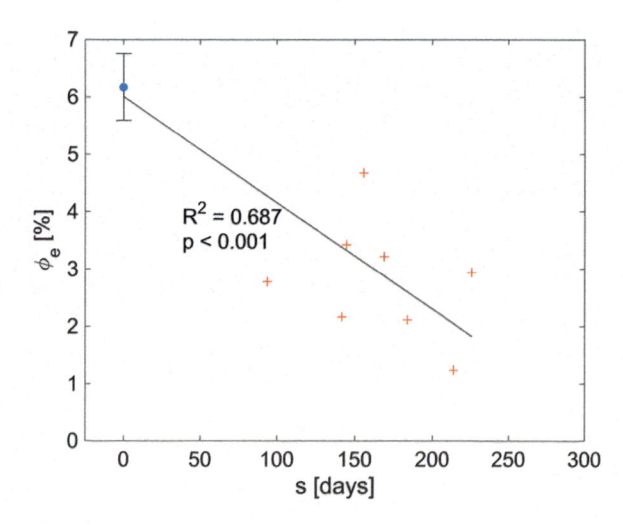

**Fig. 6** Elastin fraction versus gestation time, showing a significant time-dependent linear decrease (data from [5], with sample sizes of $n_{NP} = 5$ and $n_P = 8$)

## 3.2 Collagen and Elastin Fractions

An analysis of mass fraction data from Pierlot et al. [5, 14] showed no correlation between $\phi_c$ and $s$ ($R^2 = 0.098$, $p = 0.073$), but revealed a significant time-dependent linear trend in $\phi_e(s)$ (Fig. 6), with $R^2 = 0.687$ and $p < 0.001$. During model fitting, $\phi_c$ was thus assigned groupwise as 0.567 for all NP specimens and 0.660 for all P specimens (Table 1), while $\phi_e$ was prescribed using the regression curve value corresponding to each specimen's gestation time.

## 3.3 Elastin Moduli

The average effective moduli for circumferential and radial elastin were found to be approximately $\eta_e^{\text{cir}} = 76.6 \pm 5.7$ kPa and $\eta_e^{\text{rad}} = 33.1 \pm 3.4$ kPa, respectively, with no time-dependent trends found ($R^2 = 0.027$ and $p = 0.278$ for $\eta_e^{\text{cir}}$; $R^2 = 0.155$ and $p = 0.073$ for $\eta_e^{\text{rad}}$). When grouped according to pregnancy status, $t$-tests revealed no significant differences between the mean values of the same modulus in NP and P ($p = 0.165$ and $p = 0.169$ for $\eta_e^{\text{cir}}$ and $\eta_e^{\text{rad}}$, respectively). Additional t-tests comparing the moduli in the two directions indicated that the mean $\eta_e^{\text{cir}}$ is significantly greater than that of $\eta_e^{\text{rad}}$ ($p < 0.001$ for both NP and P specimens). Note that this result does not imply that the stiffness of the underlying elastin differs between the two axial directions. The discrepancy in the effective elastin moduli should be interpreted instead as an approximation of the relative amounts of circumferential and radial elastin in the leaflet, since this cannot be captured through direct mass measurements. Specifically, $\eta_e^{\text{cir}}$ correlates with the volume of circumferentially oriented elastin in the ventricularis layer, while $\eta_e^{\text{rad}}$ correlates with the volume of radially oriented elastin in the atrialis layer.

## 3.4 Collagen Fibre Architecture

No significant time-dependent trend was found for the preferred fibre angle $\mu_\Gamma^s$, nor did we find any differences between NP and P groups ($p = 0.295$). The preferred fibre angle in NP leaflets was $0.07 \pm 1.74°$, while in P leaflets it was $0.82 \pm 4.13°$. This suggests that the true average fibre angle remains close to $\mu_\Gamma^s \approx 0°$ throughout pregnancy, consistent with the negligible growth-induced shearing. In contrast, the fibre splay $\sigma_\Gamma^s$ was found to correlate significantly with gestation time, with $R^2 = 0.348$ and $p = 0.036$ (Fig. 7a). We observed that the fibre distribution narrows slightly in EP, while in middle and late pregnancy the splay continuously increased to values greater than $20°$, resulting in a broader ODF (Fig. 7b).

The mean stretch value at which collagen fibres were recruited (i.e. transitioned from a crimped state to a fully straightened state) was found to correlate with gestation time, with $R^2 = 0.572$ and $p = 0.006$ (Fig. 8a). The recruitment upper bound was consistently between one and two standard deviations away from the mean, indicating that the effective width of the recruitment function remains relatively constant throughout pregnancy (Fig. 8b and Table 4). The overall behaviour of the recruitment function during gestation paralleled that of the leaflet's effective mechanical response (Figs. 3a and 8b).

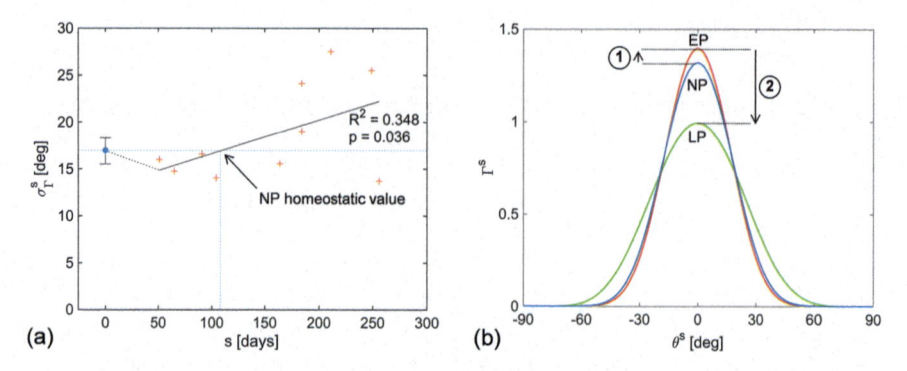

**Fig. 7** (a) Collagen fibre splay plotted versus gestation time, showing a statistically significant time-dependent trend. (b) ODFs representative of NP, EP, and LP. While collagen alignment increases initially, fibres are more uniformly oriented in LP

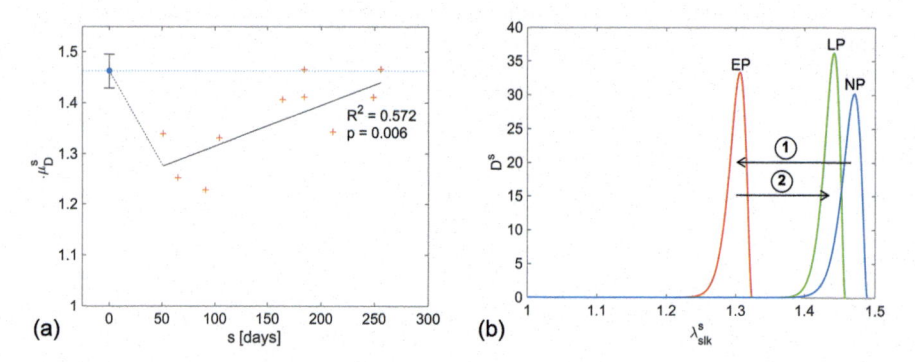

**Fig. 8** (a) Mean recruitment stretch for both NP and P specimens, showing a sharp drop in EP followed by a gradual recovery to the pre-pregnant value. (b) Collagen recruitment functions representative of NP, EP, and LP. While the distribution of slack stretches is drastically shifted in EP, its normal shape and location are essentially recovered by 250 days (close to term). In addition, the width of the recruitment function stays relatively constant throughout. Parameters used to construct the plotted distributions are listed in Table 4

## 3.5 Structural Effects of Growth and Remodelling

To determine the extent to which growth and remodelling activities change the fibre structure of the leaflet compared to strictly passive processes (e.g. permanent set), we compared the actual fibre splay and recruitment stretch results to a prediction of those same parameters based solely on the observed dimensional changes of the leaflet (Fig. 9). We found that in both cases, results for EP specimens were largely consistent with a growth- and remodelling-free permanent set, but results for later stages of pregnancy differed substantially from the growth- and remodelling-free prediction. Specifically, while primarily circumferential enlargement would predict a

**Table 4** Representative recruitment parameters

| Parameter | NP | EP ($s = 80$ days) | LP ($s = 250$ days) |
|---|---|---|---|
| $\mu_D^s$ | $1.462 \pm 0.033$ | $1.299 \pm 0.027$ | $1.435 \pm 0.030$ |
| $\lambda_{ub}^s$ | $1.489 \pm 0.034$ | $1.324 \pm 0.030$ | $1.458 \pm 0.043$ |

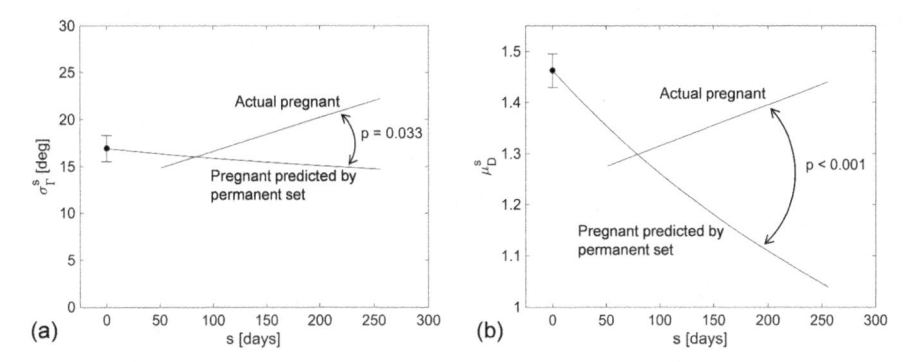

**Fig. 9** Trend lines for (**a**) fibre splay and (**b**) mean recruitment stretch results, plotted together with how each parameter would have evolved in the absence of growth and remodelling. While the bulk dimensional changes in the leaflet alone would cause a continual decrease in both parameters, the actual values diverge from that trend after, demonstrating the magnitude of growth and remodelling effects and the significant role of growth and remodelling in the recovery of normal physiological function

continual decrease in fibre splay, the actual P data show that fibre splay increases during middle and late pregnancy (Fig. 9a). Similarly, continual enlargement in the absence of growth and remodelling would cause a monotonic decrease in recruitment stretch, although the data show a clear recovery towards the NP value by 250 days (Fig. 9b). Linear regressions to the difference between the P data and their predicted values were statistically significant for both fibre splay and recruitment stretch ($p = 0.033$ and $p < 0.001$, respectively). Interestingly, the gestation time at which the permanent set predictions and the P data regressions intersected was close to $s \approx 80$ days for both parameters, suggesting that growth and remodelling effects only become prominent around this time.

## 3.6 Quantitative MVIC Geometry

Differences in the shape of MVICs were observable in the H&E-stained histological images (Fig. 10a). While NP and LP tissues looked morphologically similar with regard to both fibre and cellular geometry, EP leaflets exhibited MVICs that were noticeably pre-stretched and possibly compressed in the transmural direction by the surrounding collagen. In addition, fibres in EP qualitatively appeared less crimped.

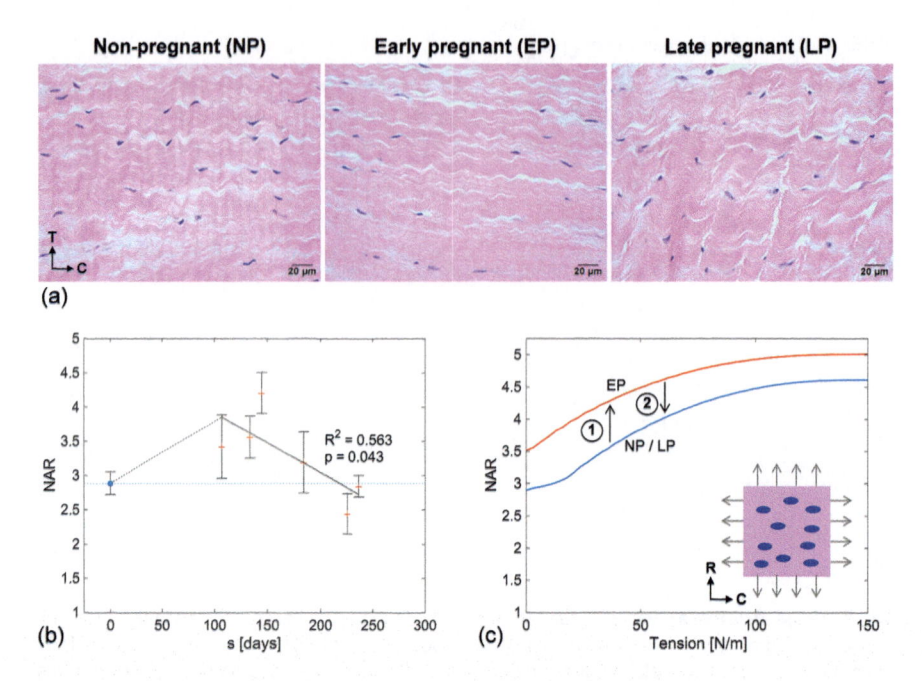

**Fig. 10** (a) Representative histological images of the circumferential (C)-transmural (T) plane for leaflets in NP, EP, and LP. While fibre architecture and MVIC geometry in NP and LP are relatively similar, collagen in EP is less crimped and MVICs are noticeably deformed, possibly triggering growth and remodelling. (b) Measured NAR for both NP and P specimens, showing a sharp rise in EP followed by a gradual recovery to the pre-pregnant value. Average NAR for NP leaflets is shown as the mean and standard error of 30 total VICs (10 from each specimen). Each P sample is shown as the mean and standard error of 10 VICs. (c) FE simulation results for a population of 10 MVICs in a NP/LP microenvironment, compared to an EP microenvironment. MVICs in EP are predicted to experience higher NARs at all tension levels, but particularly at zero tension (diastole, in vivo). (**c**, inset) A visualization of the FE mesh used, showing MVICs (blue) as ellipsoidal inclusions in the ECM (pink)

NAR quantification revealed a biphasic trend that mirrors that of the mean recruitment stretch, suggesting that these two metrics are tightly coupled (Fig. 10b). Our results indicated that NAR increases by about 30% in the first 100 days, but then recovers gradually to within its NP range by 200–250 days ($R^2 = 0.563$, $p = 0.043$).

Similar to our direct imaging results, our FE simulations of a MV leaflet RVE likewise showed that EP MVICs experience higher NARs than NP or LP MVICs under any loading, though the difference is most pronounced at zero tension (Fig. 10c). Specifically, we predicted that EP MVICs experience an increase in NAR greater than 20% when the tissue is unloaded, compared to an increase of less than 10% when the tissue is fully loaded at 150 N/m. This suggested that alterations in leaflet geometry and loading in vivo primarily cause a deviation from homeostatic MVIC geometry during diastole. Under higher loading, when the ECM is much stiffer due to earlier collagen recruitment, MVICs do not bear a significant portion of the total stress and thus experience a milder deviation from their normal systolic geometry.

## 4 Discussion

### 4.1 General Overview

This study, and its antecedents [5, 12–14], represents the first detailed attempt at gaining insight into the complex growth and remodelling processes that occur in non-pathological valvular tissues. In the present study, we have shown evidence for the substantial growth and remodelling that can result in structural and biochemical changes and dramatically alter the morphology of the valve's leaflet tissue. Specifically, collagen fibre alignment, collagen fibre crimp, layer thicknesses, and constituent fractions all change significantly as a result of pregnancy. Our results indicated that growth and remodelling processes in the MV lead to significant structural changes in the collagen phase of the leaflet tissue, both at the tissue and unidirectional fibre ensemble levels. Specifically, we have shown that as pregnancy advances, growth and remodelling alter the orientation distribution of the collagen network as well as the amount of crimp held by the collagen fibres (Figs. 7 and 8). Perhaps the most novel aspect of this study was the observation that these tissue architectural adjustments appear to restore the normal physiological pre-pregnant geometry of MVICs.

A key feature of our methodology was that by quantifying the continually changing dimensions of the leaflet with respect to gestation time, we were able to compensate for the effects of permanent set versus true growth. Moreover, we expressed structural parameters with respect to a single pre-pregnant reference configuration. This approach was vital to establishing correspondence between the mechanical data of NP valves and that of P valves from a wide array of gestational time points. In doing this, we were able to conclude that although the apparent stiffness of the leaflet changes dramatically during pregnancy [12], the underlying collagen and elastin fibre-level properties are not likely to change appreciably during the remodelling process. Instead, our results demonstrated that changes in the apparent leaflet stiffness can be attributed to a continual change in the natural in vivo configurations of the tissue, as a consequence of permanent set, collagen turnover, mass accretion, and elastin degradation. These mechanisms gradually cause the homeostatic mechanical behaviour and interstitial cellular geometry of the MV to be recovered, even in the presence of increasing overload.

At the fibre ensemble level, we hypothesize that collagen crimp is gradually restored at least in part due to the serial addition of fibrillar material into existing fibres (Fig. 11). Our past results suggesting that collagen crimp period increases monotonically in pregnancy [5, 13] imply that the overall collagen crimp would need to be restored via an increase in amplitude, a question which should be addressed in a

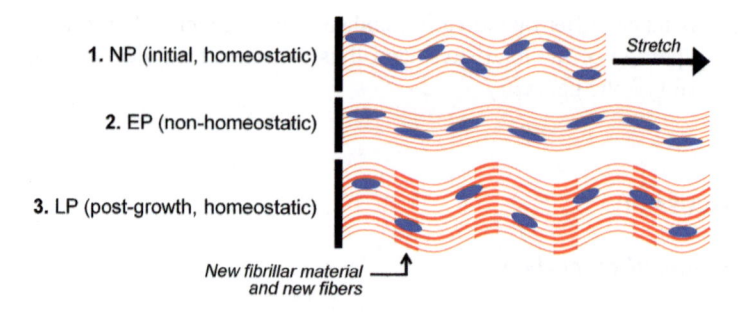

**Fig. 11** Schematic of a possible mechanism for valvular growth via the serial addition of collagenous fibrillar material as well as the deposition of new collagen fibres. As existing fibres elongate, their crimp amplitude and slack are gradually restored, relieving MVIC compression and restoring normal homeostatic function

future study. To bring about the significant widening we observed in the ODF (Fig. 7), we further speculate that new fibres must also be deposited along ensemble directions oriented at larger angles from the circumferential axis. While we do not have exact information on the changes in MV shape during pregnancy, it is likely that in vivo changes in loading make the deposition of more radially oriented fibres adaptive for restoring normal valve function during gestation.

Our model parameters agreed well with the direct structural measurements from Pierlot et al. [13], with both studies finding a decrease in collagen fibre alignment (increase in splay) and an overall decrease in collagen crimp during pregnancy. In addition, our results regarding collagen fibre recruitment showed biphasic changes in mean recruitment stretch and thus the location of the recruitment function overall, which parallels the changes in mechanical behaviour reported by Wells et al. [12] (Fig. 3a). Viewing these results in their entirety, we are confident that through the framework of our constitutive model, we have successfully unified the mechanical and structural results of these previous two studies.

Interestingly, although the MV leaflet was able to regain much of its normal mechanical behaviour by the end of pregnancy, growth and remodelling do not appear to restore the MV ECM back to its pre-pregnant state. While our results show that the leaflet virtually recovers its original properties at the ensemble and cellular scales, the same cannot be said for the entire tissue, since the collagen fibre population is significantly less aligned in LP compared to NP (Fig. 7) and the collagen fraction is elevated in LP [13]. These results suggested that the tissue is limited in its ability to reverse the structural changes brought on by altered loading conditions. Alternatively, it is possible that there are multiple tissue states that provide the same essential interstitial cell microenvironment and tensile behaviour under normal stress levels, and that heart valves may not distinguish between these when remodelling. Moreover, while there is some evidence indicating that the effects of pregnancy on the LV are reversed after birth [11], the extent to which valvular tissues in general return to their pre-pregnant state post-partum is unclear, and should be the subject of a future study.

## 4.2 Delayed Restoration of Homeostasis

Our results suggested that much of the growth and remodelling in the MV leaflet do not begin immediately after the onset of pregnancy, although the exact reasons for this delay are unknown. This contrasts with pregnancy-induced LV remodelling, which has been shown to be significant just 1 month after conception [10]. In our results of fibre splay and recruitment stretch, it is apparent that the trend lines for the actual results begin deviating from that which is predicted by the bulk dimensional changes at a gestation time of about 80 days (Fig. 9). In other words, it seems that the valve is content to obey a trend of mostly passive leaflet enlargement until these parameters reach a critically low level, at which point growth and remodelling are triggered and sustained through middle and late pregnancy.

Taking the case of recruitment stretch specifically, the onset of remodelling at the ensemble level can be associated with a significant difference in the in vivo stretch range that the fibres experience as pregnancy advances. Continual leaflet enlargement leads to increasingly large pre-stretch in all directions, which causes collagen fibres to experience significantly higher stretches compared to pre-pregnancy, especially in diastole as the corresponding ensemble mechanical response curve is robbed of its toe region. The way by which this process triggers growth and remodelling must be closely tied to the deformation of MVICs. It has been shown that collagen synthesis as well as the ratio between matrix metalloproteinases (MMPs) and their tissue inhibitors (TIMPs) is dependent on interstitial cell deformation [23, 24, 33]. These compounds altogether determine the presence and degradation of collagen in the ECM. Because each MVIC is tightly embedded in the collagen network, its deformation will correlate strongly with the stretch of the fibres surrounding it. Indeed, our cell geometry and FE simulation results clearly showed that the cellular and fibre ensemble scales are tightly coupled, with loss of crimp being associated with a substantial increase in MVIC NAR (Figs. 8 and 10). As collagen fibre crimp is lost in EP, MVICs become compressed by surrounding fibre bundles and thus experience increasingly non-homeostatic deformations, causing them to initiate growth and remodelling pathways (Fig. 11). One possible mechanism that would explain the gradual increase in recruitment stretch after EP is the serial addition of fibrillar material within existing collagen fibre bundles, which would increase crimp amplitude and therefore fibre slack. As the insertion of new fibrils causes the fibres to become longer and more undulated, recruitment stretch would steadily increase and normal MVIC geometry would be restored. In direct correlation with local collagen fibre density, it has been shown that MVIC deformation varies across the different tissue layers of the MV leaflet [17, 33]. It is therefore not surprising that MV growth and remodelling is also layer-specific, as demonstrated by the non-uniform changes in layer thicknesses seen in pregnancy [13].

Mechanisms of load-induced stabilization and destabilization of collagen fibres may play an important role in growth and remodelling as well. Several types of collagen are thought to be protected from enzymatic breakdown under low and moderate forces, but are made vulnerable to degradation by higher forces, as large deformations expose sites of protease binding [25–27]. In a collagen network under high pre-stretch (e.g. in EP), it follows that those fibres with the smallest slack stretch (which bear more load and experience higher deformation) are most vulnerable to binding by MMPs and will thus be preferentially degraded. Alongside the deposition of new fibrillar material, this phenomenon would cause the distribution of slack stretches to shift location while maintaining a relatively constant width, which is consistent with our findings (Fig. 8b). In order for net growth to occur, the synthesis of new collagen fibres, mediated by MVIC deformation, must increase sufficiently to outpace degradation. It is therefore likely that both synthesis and degradation increase significantly in pregnancy, similarly to what is seen following tendon overloading in exercise [28–31]. Because fibres are always deposited with respect to a valve's current state, any new fibres will be less pre-stretched and more crimped than the existing population of fibres, thus helping to shift the recruitment function to the right and closer to its pre-pregnant location (Fig. 8b). This chain of growth and remodelling processes, including serial fibril addition, has the potential to explain all the recruitment results reported in the present study and may be a plausible hypothesis for the mechanism by which the MV recovers its normal physiological function during middle and late pregnancy. Future studies should seek to investigate this question using innovative methods that combine microstructural and biochemical measurements related to collagen synthesis and degradation.

## 4.3 Generalization and Special Considerations

Valvular remodelling due to overload (in pregnancy or otherwise) is fundamentally a result of alterations in the boundary conditions to which the valve leaflets are subjected over the cardiac cycle. In the case of the MV, these alterations come primarily in the form of annular dilatation and expansion of the LV cavity, which causes displacement of the PMs [9–11]. As the LV continues to grow and remodel, these changes in boundary conditions become more pronounced and place an increasingly large pre-stretch on the MV leaflets, which eventually triggers growth and remodelling in the valve. Our study suggests that in pregnancy, valvular growth and remodelling becomes significant after about 3 months. Moreover, we have found that by 250 days (close to term), homeostatic MVIC geometry and the pre-pregnant mechanical behaviour of the valve are fully recovered.

It is important to note that pregnancy is a special non-pathological case of volume overload and is unique in several ways. Unlike pathological volume overload brought on by anaemia or valvular regurgitation, pregnancy is a condition in which the ventricles and heart valves remain functional but require remodelling nonetheless due to increased cardiovascular demand. Moreover, pregnancy-induced remodelling is a response to a gradual and continuous onset of volume overload, in contrast with the sporadic stimulus of exercise or certain disease conditions (e.g. infarction). These features make pregnancy a particularly convenient model for studying how heart valves adapt to conditions of increased stretch, since many different magnitudes of overload can be investigated without any influence from disease-specific third variables. In addition to uncovering some of the higher-level mechanisms that drive valvular growth and remodelling, our pregnancy studies provide a first approximation of the rate at which growth and remodelling occurs in response to an increased-stretch condition. This time-course information can potentially be used to inform future studies of valvular growth and remodelling in response to other overload-inducing conditions, such as myocardial infarction [32, 34, 35].

## 4.4 Limitations

The quantitative accuracy of our results may have been affected by multiple factors. First, the mass fractions of collagen and elastin were prescribed in an average sense, since the mechanical and mass fraction data were taken from different studies and therefore came from different groups of animals [5, 12–14]. Because the mass fraction data had relatively high variance, this assumption may have introduced errors in the parameter estimates, since the ODF and recruitment parameters would need to compensate for mass fractions that were too high or too low. Future studies would benefit from controlling individual variations in constituent fractions by performing mass quantification assays on each specimen from which mechanical data are collected. Second, the ODF and recruitment function for each specimen were each modelled using a single beta distribution. While this likely provides reasonable approximations of the overall distribution shapes, we acknowledge that a combination of multiple distributions could more precisely describe the orientation and recruitment functions, since newly synthesized fibres are deposited in a different material state and thus belong to a distinct statistical population. Despite these sources of error, we believe the trends of growth and remodelling elucidated in the present study are accurate. A further limitation of the present study that could impact the generalizability of our results is the potential role of pregnancy-specific hormonal conditions in valvular remodelling, the magnitude of which is currently unknown.

## 4.5 Conclusions

There is a need to develop a mechanistic model of growth and remodelling in heart valve leaflet tissues, as well as a more faithful description of both organ- and tissue-level MV behavior [34, 36]. In the present study, a cellular and tissue-level modelling approach was developed to deduce changes in MV leaflets subjected to pregnancy-induced volume overload, which alters valvular loading and boundary conditions. Collectively, our novel results suggest that valvular growth and remodelling can be understood as a process through which cells seek to restore a normal deformation pattern and thus tissue-level homeostasis. These findings point towards models that incorporate data from both mechanical and biochemical studies to draw a more detailed and biologically informed connection between MVIC deformation and growth and remodelling phenomena [35]. The goal of these future studies should be to quantitatively predict the way in which the valve will return to homeostasis after an imposed alteration in leaflet loading in pathological and surgical repair conditions.

**Acknowledgments** This material is based upon work supported by the National Institutes of Health grant no. R01-HL119297 to MSS, the National Science Foundation grant no. DGE-1610403 to BVR, an American Heart Association Scientist Development Grant Award (16SDG27760143) to CHL, and a Natural Sciences and Engineering Research Council of Canada Discovery Grant to SMW.

# References

1. Butcher JT, Simmons CA, Warnock JN. Review—Mechanobiology of the aortic heart valve. J Heart Valve Dis. 2008;17(1):62–73.
2. Gorman JH, Jackson BM, Enomoto Y, Gorman RC. The effect of regional ischemia on mitral valve annular saddle shape. Ann Thorac Surg. 2004;77(2):544–8.
3. Hinton RB, Lincoln J, Deutsch GH, Osinska H, Manning PB, Benson DW, Yutzey KE. Extracellular matrix remodeling and organization in developing and diseased aortic valves. Circ Res. 2006;98(11):1431–8.
4. Hinton RB, Yutzey KE. Heart valve structure and function in development and disease. Annu Rev Physiol. 2011;73:29.
5. Pierlot CM, Moeller AD, Lee JM, Wells SM. Pregnancy-induced remodeling of heart valves. Am J Phys Heart Circ Phys. 2015;309(9):H1565–78.
6. Walker GA, Masters KS, Shah DN, Anseth KS, Leinwand LA. Valvular myofibroblast activation by transforming growth factor-β implications for pathological extracellular matrix remodeling in heart valve disease. Circ Res. 2004;95(3):253–60.
7. Lee CH, Zhang W, Liao J, Carruthers CA, Sacks JI, Sacks MS. On the presence of affine fibril and fiber kinematics in the mitral valve anterior leaflet. Biophys J. 2015;108(8):2074–87.
8. Liao J, Yang L, Grashow J, Sacks MS. The relation between collagen fibril kinematics and mechanical properties in the mitral valve anterior leaflet. J Biomech Eng. 2007;129(1):78–87.
9. Mone SM, Sanders SP, Colan SD. Control mechanisms for physiological hypertrophy of pregnancy. Circulation. 1996;94(4):667–72.
10. Robson SC, Hunter ST, Boys RJ, Dunlop WI. Serial study of factors influencing changes in cardiac output during human pregnancy. Am J Phys Heart Circ Phys. 1989;256(4):H1060–5.
11. Katz RI, Karliner JS, Resnik RO. Effects of a natural volume overload state (pregnancy) on left ventricular performance in normal human subjects. Circulation. 1978;58(3):434–41.
12. Wells SM, Pierlot CM, Moeller AD. Physiological remodeling of the mitral valve during pregnancy. Am J Phys Heart Circ Phys. 2012;303(7):H878–92.
13. Pierlot CM, Lee JM, Amini R, Sacks MS, Wells SM. Pregnancy-induced remodeling of collagen architecture and content in the mitral valve. Ann Biomed Eng. 2014;42(10):2058–71.
14. Pierlot CM, Moeller AD, Lee JM, Wells SM. Biaxial creep resistance and structural remodeling of the aortic and mitral valves in pregnancy. Ann Biomed Eng. 2015;43(8):1772–85.
15. Hunley SC, Kwon S, Baek S. Influence of surrounding tissues on biomechanics of aortic wall: a feasibility study of mechanical homeostasis. In: ASME 2010 summer bioengineering conference 2010 Jun 16. American Society of Mechanical Engineers; p. 713–4.

16. Zhang W, Ayoub S, Liao J, Sacks MS. A meso-scale layer-specific structural constitutive model of the mitral heart valve leaflets. Acta Biomater. 2016;32:238–55.
17. Lee CH, Carruthers CA, Ayoub S, Gorman RC, Gorman JH, Sacks MS. Quantification and simulation of layer-specific mitral valve interstitial cells deformation under physiological loading. J Theor Biol. 2015;373:26–39.
18. Sacks MS, Zhang W, Wognum S. A novel fibre-ensemble level constitutive model for exogenous cross-linked collagenous tissues. Interface Focus. 2016;6(1):20150090.
19. Fan R, Sacks MS. Simulation of planar soft tissues using a structural constitutive model: finite element implementation and validation. J Biomech. 2014;47(9):2043–54.
20. Fata B, Zhang W, Amini R, Sacks MS. Insights into regional adaptations in the growing pulmonary artery using a meso-scale structural model: effects of ascending aorta impingement. J Biomech Eng. 2014;136(2):021009.
21. Storn R, Price K. Differential evolution—a simple and efficient heuristic for global optimization over continuous spaces. J Glob Optim. 1997;11(4):341–59.
22. Brossollet LJ, Vito RP. A new approach to mechanical testing and modeling of biological tissues, with application to blood vessels. J Biomech Eng. 1996;118(4):433–9.
23. Ku CH, Johnson PH, Batten P, Sarathchandra P, Chambers RC, Taylor PM, Yacoub MH, Chester AH. Collagen synthesis by mesenchymal stem cells and aortic valve interstitial cells in response to mechanical stretch. Cardiovasc Res. 2006;71(3):548–56.
24. Balachandran K, Sucosky P, Jo H, Yoganathan AP. Elevated cyclic stretch alters matrix remodeling in aortic valve cusps: implications for degenerative aortic valve disease. Am J Phys Heart Circ Phys. 2009;296(3):H756–64.
25. Chang SW, Buehler MJ. Molecular biomechanics of collagen molecules. Mater Today. 2014;17(2):70–6.
26. Adhikari AS, Chai J, Dunn AR. Mechanical load induces a 100-fold increase in the rate of collagen proteolysis by MMP-1. J Am Chem Soc. 2011;133(6):1686–9.
27. Huang S, Huang HY. Biaxial stress relaxation of semilunar heart valve leaflets during simulated collagen catabolism: effects of collagenase concentration and equibiaxial strain state. Proc Inst Mech Eng H J Eng Med. 2015;229(10):721–31.
28. Magnusson SP, Langberg H, Kjaer M. The pathogenesis of tendinopathy: balancing the response to loading. Nat Rev Rheumatol. 2010;6(5):262–8.
29. Heinemeier KM, Olesen JL, Haddad F, Langberg H, Kjær M, Baldwin KM, Schjerling P. Expression of collagen and related growth factors in rat tendon and skeletal muscle in response to specific contraction types. J Physiol. 2007;582(3):1303–16.
30. Miller BF, Olesen JL, Hansen M, Døssing S, Crameri RM, Welling RJ, Langberg H, Flyvbjerg A, Kjær M, Babraj JA, Smith K. Coordinated collagen and muscle protein synthesis in human patella tendon and quadriceps muscle after exercise. J Physiol. 2005;567(3):1021–33.
31. Koskinen SO, Heinemeier KM, Olesen JL, Langberg H, Kjaer M. Physical exercise can influence local levels of matrix metalloproteinases and their inhibitors in tendon-related connective tissue. J Appl Physiol. 2004;96(3):861–4.
32. Pfeffer MA, Braunwald E. Ventricular remodeling after myocardial infarction. Experimental observations and clinical implications. Circulation. 1990;81(4):1161–72.
33. Khang A, Buchanan RM, Ayoub S, Rego BV, Lee CH, Ferrari G, Anseth KS, Sacks MS. Mechanobiology of the heart valve interstitial cell: Simulation, experiment, and discovery. In: Verbruggen SW, editor. Mechanobiology in Health and Disease. London: Elsevier; 2018. p. 249–83.
34. Sacks MS, Khalighi A, Rego B, Ayoub S, Drach A. On the need for multi-scale geometric modelling of the mitral heart valve. Healthcare Technology Letters. 2017;4(5):150.
35. Rego BV, Ayoub S, Khalighi AH, Drach A, Gorman JH, Gorman RC, Sacks MS. Alterations in mechanical properties and in vivo geometry of the mitral valve following myocardial infarction. Summer biomechanics, bioengineering and biotransport conference, Tucson, AZ, USA; 2017.

36. Drach A, Khalighi AH, Sacks MS. A comprehensive pipeline for multi-resolution modeling of the mitral valve: validation, computational efficiency, and predictive capability. Int J Numer Methods Biomed Eng. 2017;34:e2921. https://doi.org/10.1002/cnm.2921.
37. Rego BV, Sacks MS. A functionally graded material model for the transmural stress distribution of the aortic valve leaflet. J Biomech. 2017;54:88–95.

# Molecular and Cellular Developments in Heart Valve Development and Disease

Lindsey J. Anstine, Anthony S. Baker, and Joy Lincoln

**Abstract** The prevalence of heart valve disease is estimated at 2.5% and this number is likely to increase with the steady rise in life expectancy and associated degenerative pathology. Affected individuals suffer significant morbidity and mortality, and surgical intervention remains the only effective treatment. Disease of the valves can be congenital associated with structural malformations present at birth as a result of disturbances during embryonic valve development. Alternatively, phenotypes can be acquired later in life following sustained exposure to environmental risk factors including high cholesterol, tobacco-use, hypertension and aging, although there is increasing evidence to suggest that embryonic origins may also play a role. The mechanisms underlying heart valve disease in the human population are largely unknown and this has hindered the development of alternative, more cost-effective therapies. Formation of the valve in the developing embryo is complex and requires the temporal convergence of many signaling pathways in specific cell types. Once the valve structure is established after birth, it must be maintained by additional regulatory networks, many of which are also involved in valvulogenesis. Here, we will highlight the hierarchical pathways known to be essential for establishing and maintaining valve structure–function relationships and discuss how aberrations in their regulation can promote the onset and progression of valve disease. Furthermore, we will focus on how basic science discoveries and tools can provide critical insights into the development of mechanistic-based therapies beyond surgical repair and replacement.

---

L. J. Anstine
Center for Cardiovascular Research and The Heart Center at Nationwide Children's Hospital Research Institute, Columbus, OH, USA

A. S. Baker
Health Sciences Library Medical Visuals, The Ohio State University, Columbus, OH, USA

J. Lincoln (✉)
Center for Cardiovascular Research and The Heart Center at Nationwide Children's Hospital Research Institute, Columbus, OH, USA

Department of Pediatrics, The Ohio State University, Columbus, OH, USA
e-mail: joy.lincoln@nationwidechildrens.org

© Springer Nature Switzerland AG 2018
M. S. Sacks, J. Liao (eds.), *Advances in Heart Valve Biomechanics*,
https://doi.org/10.1007/978-3-030-01993-8_9

**Keywords** Valve endothelial cell · Valve interstitial cell · Extracellular matrix · Signaling pathways · Endocardial cushion · Bicuspid aortic valve · Mitral valve prolapse

# 1 Introduction

Heart valve disease affects approximately 2.5% of the population in the USA and continues to be a public health burden. With the steady rise in life expectancy, population statistics worldwide project a significant increase in prevalence, and therefore morbidity and mortality [1]. At present, there are no effective treatments other than interventional surgery, and procedures to repair or replace diseased valves are the second most common indication for cardiac surgery in the USA. Unfortunately these procedures are often temporary due to inappropriate replacements that fail to grow, remodel, repair, or withstand lifelong mechanical demands in vivo, and therefore alternative approaches are required.

The function of mature heart valves is to regulate unidirectional blood flow through the heart during the cardiac cycle, and in order to do this, valve structure–function relationships must be established during embryonic and postnatal development and maintained in adult life. Congenital malformations present at birth or acquired disease later in life is detrimental to valve structure which in turn attenuates biomechanical properties of the valves preventing them to fully open or close causing obstructive or regurgitant blood flow, respectively. Here, we will review the cellular and molecular mechanisms underlying heart valve disease and discuss the biomechanical influence on disease progression with the goal of identifying potential targets in the development of alternative therapies beyond surgical intervention.

# 2 Structure-Function Relationships of Mature Heart Valves

Heart valves are dynamic structures designed to open and close over 100,000 times a day to maintain unidirectional blood flow through the heart during the cardiac cycle. In healthy individuals this is largely achieved by structure–function relations between the cellular and extracellular components of the mature valve and the hemodynamic environment. There are two sets of cardiac valves: the atrioventricular (AV) valves including the mitral and tricuspid that separate the atria from the ventricles on the left and right sides, respectively, and the aortic and pulmonic semilunar valves that separate the ventricles from the great arteries. Although the function of the valve sets is similar, they are anatomically different. The mitral and tricuspid valves are composed of two and three leaflets, respectively, supported by external fibrous chordae tendineae that attach the underside of the leaflets to papillary muscles within the ventricles and prevent prolapse into the atria during ventricular contraction. The semilunar valve leaflets are often referred to as cusps and lack external chordae but internal support has been described within the aortic roots in the form of the fibrous annulus [2]. It is the coordinated movement of these valvular structures that maintain unidirectional blood flow during the cardiac cycle: In diastole, the papillary muscles are relaxed and high pressure in the atrium causes opening of the mitral (left) and tricuspid (right) valve leaflets to promote blood flow into the respective ventricle. Once ventricular pressure increases during diastole, the chordae "pull" the AV valve leaflets closed and maintain coaptation to prevent eversion of the valve into the atria. As the ventricle contracts, blood exits through the now open semilunar valves and the ventricle relaxes to begin the cycle again. Therefore, throughout the cardiac cycle the heart valve structures are exposed to constant changes in hemodynamic force as a result of pressure differences between systole and diastole. To withstand this complex mechanical environment, the valve leaflets/cusps develop and maintain an intricate and highly organized connective tissue system that provides all the necessary biomechanical properties for efficient function.

## 2.1  Extracellular Matrix

The mature heart valves are connective tissues composed of a stratified extracellular matrix (ECM) interspersed with an interstitial cell population and surrounded by a single layer of endothelial cells (Fig. 1) [3]. The ECM is highly organized into layers and arranged according to blood flow. The fibrosa is located on the ventricular side of the AV valve leaflets and atrial side of the semilunar valves, away from blood flow. This layer is predominantly composed of bundles of collagen fibers aligned along the circumferential direction of the leaflets [4–7]. This arrangement provides tensile strength to the valve leaflet during opening, while transmitting forces to promote coaptation of the leaflets in the closed position [8–10]. Adjacent to the fibrosa is the spongiosa, with a lower abundance of collagens and high prevalence of proteoglycans. This composition provides a more compressible matrix, allowing the valve to geometrically "flex" and absorb high force [7, 11]. Finally, the layer adjacent to blood flow is termed the atrialis (AV) or ventricularis (semilunar) and largely consists of radially orientated elastin fibers that facilitate tissue movement by extending as the valve leaflet opens and recoiling during closure [12–14]. Further support is provided by the fibrous annulus structure at the connection between the leaflets/cusps and the myocardium [13, 15], and the external chordae tendineae in the AV position.

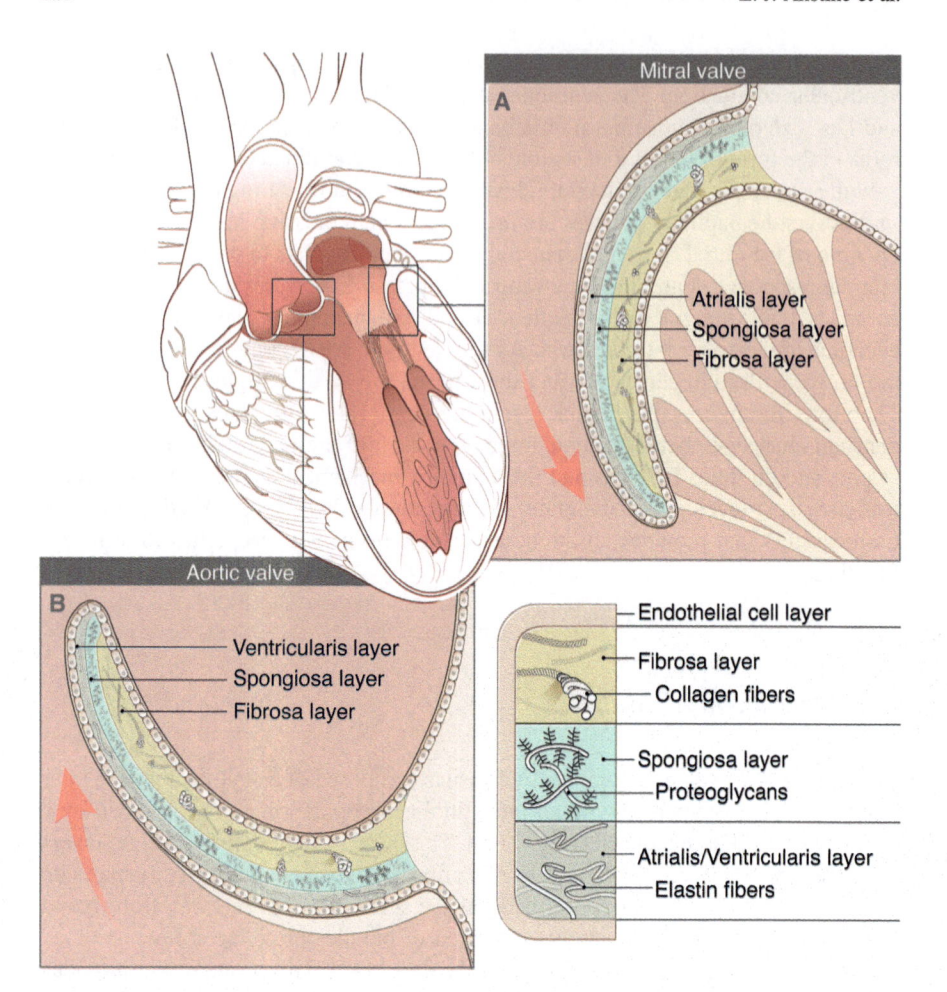

**Fig. 1** Overview of heart valve structure. The atrioventricular (mitral (**A**), tricuspid) and semilunar (aortic (**B**), pulmonic) valves are highly organized structures composed of cellular and extracellular components. Each valve leaflet or cusp is surrounded by a continuous, single layer of valve endothelial cells and interspersed by valve interstitial cells. The extracellular matrix is highly organized and largely composed of elastin fibers in the atrialis/ventricularis (dark gray), proteoglycans in the spongiosa (blue), and collagens (yellow) in the fibrosa, and these are arranged according to blood flow (red arrows)

## 2.2 Valve Cell Populations

The ECM architecture within the mature valve leaflets/cusps is established during embryonic development and postnatal growth, and maintained in the adult by valve interstitial cells (VICs). This cell population is the most abundant within the valve and phenotypically heterogeneous in health and disease [16, 17]. In addition to the

VIC population, the valve leaflet or cusp is encapsulated by a single endothelial cell layer that serves as a physical barrier to protect the ECM and VICs against the constant changes in hemodynamic forces and circulatory risk factors (Fig. 1). VEC function is likely regulated by shear stress and flow, and through mechanosensitive mechanisms VECs transmit signaling pathways to underlying VICs to maintain homeostasis [18, 19].

## 3 Heart Valve Development

### 3.1 Embryonic Precursor Cells

The VIC population within the mature valve leaflets largely originates from endothelial-to-mesenchymal (EMT) in a subpopulation of the endothelial cells lining the atrioventricular canal (AVC) and outflow tract (OFT) regions of the embryonic heart tube. These regions are first identified by localized deposition of proteoglycans and hyaluronan termed cardiac jelly that form swellings known as endocardial cushions (Fig. 2). The endothelial cells overlying these cushions lose cell–cell contact, delaminate, and transform, and resulting mesenchyme cells migrate, proliferate, and invade the cardiac jelly [20, 21]. This essential process is initiated in the chick around embryonic day (E) 3, in the mouse at E9.5, and between E31-E35 in human development [22]. In the AVC, four endocardial cushions form: the superior, inferior, and left and right lateral cushions, while in the OFT, two cushions form in proximal and distal locations [23]. These cushion structures serve as physical barriers to prevent backflow of blood through the primitive heart tube [24], in addition to housing a population of mesenchyme cells that serve as precursors to the mature valve structures [25, 26].

### 3.2 Molecular Regulators of Endocardial Cushion Formation

The process of EMT is complex and each step is tightly regulated by a complex network of growth factors, transcription factors and intermediate signaling molecules that cross talk between multiple cell types.

#### 3.2.1 Transforming Growth Factor (TGF) β Signaling

The Tgfβ superfamily consists of the Tgfβ and bone morphogenetic proteins (BMPs). The canonical signaling pathways function through Tgfβ or BMP ligand binding to a type II receptor, that recruits and phosphorylates the type I receptor

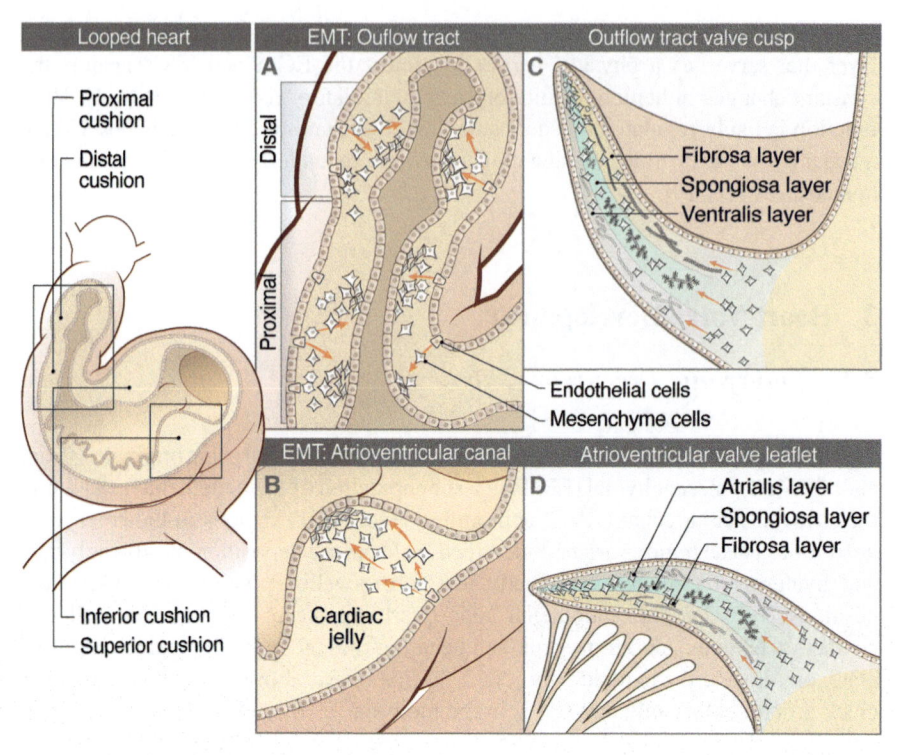

**Fig. 2** Overview of key stages during heart valve development. Heart valve development begins in the embryonic looped heart (left) with formation of endocardial cushions following endothelial-to-mesenchymal transformation (EMT) in the atrioventricular canal (**A**) and outflow tract regions (**B**). (**C, D**) As development progresses, the cushions elongate to form valve primorida, and embryonic valve interstitial cells (square cells) mediate extensive extracellular matrix remodeling until post-natal stages when the trilaminar structure is complete

leading to complex formation with Smad proteins that translocate to the nucleus and function as transcription factors on target genes. Tgfβs, Bmps, and their downstream signaling mediators are the most well characterized regulators of EMT and their signaling is required in multiple cell lineages for the initiation and progression of EMT. Valvular phenotypes observed in mouse models with targeted manipulation of Tgfβ superfamily members are highlighted in Table 1.

Downstream effectors of TGFβ-mediated activation include Smad2/3 and their expression is sufficient to induce the zinc-finger transcription factors Snai1 (Snail) and Snai2 (Slug) [27, 28]. Both these family members are highly expressed in endothelial cells undergoing EMT, as well as in newly transformed mesenchyme cells within the developing cushion [29–31]. Studies in chicken and mice have shown that Snai1 and Snai2 are essential for endocardial cushion EMT [27, 29]. Mice deficient for *Snai2* display hypocellular cushions at E9.5, although this phenotype is compensated by E10.5 due to increased *Snai1* [30]. However,

**Table 1** Bmp signaling in endocardial cushion development

| Target gene | Model | Endocardial cushion phenotype | References |
|---|---|---|---|
| *ALK2* | *Tie2-cre* deletion | Small endocardial cushions at E10.5, remain impaired at E14.5 | [174] |
| *ALK3* | *Tie2-cre* deletion | Absence of AV endocardial cushions at E9.5 | [175] |
| *ALK5* | *Tie2-cre*, *Gata5-cre* deletion | Hypoplastic endocardial cushions | [176] |
| *Bmp2* | *Nkx2.5-Cre* deletion | Impaired endocardial cushion formation at E9.5 | [175] |
| *Bmp4* | *Nkx2.5-Cre* deletion | Reduced endocardial cushion growth at E11.5. Embryonic lethal at E13.5-E18.5 | [177] |
| *Bmp5, 7* | Global double knockout | Absence of endocardial cushions. Embryonic lethal at E9.5-E10.5 | [178] |
| *Bmp6, 7* | Global double knockout | Underdeveloped outflow tract endocardial cushions at E11.5. Embryonic lethal at E9.5-E15.5 | [179] |
| *Bmp7, reduced Bmp4* | Global knockout | Hypoplastic OFT endocardial cushions | [177] |
| *Latent Tgfβ binding protein 1* | Global knockout | Impaired endocardial cushion formation | [180] |
| *Noggin* | Global knockout | Enlarged endocardial cushions | [181] |
|  | Global transgenic | Hypoplastic endocardial cushions | [182] |
| *Smad4* | *Tie2-cre* deletion | Disturbed endocardial cushion formation (decreased mesenchyme cell population) | [183] |
| *Smad6* | Global knockout | Hyperplastic valves | [184] |
| *Tgfβ1* | Global knockout | Disorganized valves. Embryonic lethal at E9.5-E11.5 | [185, 186] |
| *Tgfβ2* | Global knockout | Incomplete valve formation | [187, 188] |
| *TgfβRII* | *cTnT-cre, cGata6-cre, Mlc2v-cre* deletion | Common AV valve | [189, 190] |

*AV* atrioventricular, *E* embryonic day, *OFT* outflow tract

*Snai2* cannot compensate for *Snai1* deficiency and therefore *Snai1*$^{-/-}$ mice die early during development due to defective gastrulation and mesoderm formation prior to endocardial cushion formation [32]. Using a conditional approach, it has been shown that similar to *TGFβ* mutants, mice with reduced *Snai1* function in endothelial-derived cells using the *Tie2cre* model (*Tie2cre;Snai1*$^{fl/+}$) also develop hypoplastic cushions as a result of impaired EMT [29]. As expected, *Snai1* overexpression in avian AVC explants is sufficient to induce EMT at the level of VEC transformation and mesenchyme cell migration and invasion [29]. In other noncardiac systems, Snai1 is known to regulate EMT by directly repressing expression of cell adhesion molecules including *E-cadherin* [33]. The downregulation of cell adhesion genes in *Tie2cre;Snai1*$^{fl/+}$ mice suggests similar mechanisms in the developing heart, therefore attenuating the breakdown of cell–cell contacts between VECs in *Snai1* mutants. It has also shown that *Snail* directly regulates *mmp15:* a matrix metalloproteinase family member important for cell motility. Therefore, in addition to maintained VEC contacts, the hypoplastic endocardial cushion phenotype observed in *Tie2cre;Snail$^{fl/+}$* mice may also be the result of impaired cell invasion into the cardiac jelly [29]. This collection of studies indicate that Tgfβ signaling pathways are important for several biological processes required for endocardial cushion formation including EMT initiation, breakdown of cell–cell contacts, mesenchymal cell motility, and proliferation.

The upstream regulation of Tgfβ during EMT has not been extensively reported; however, miR-23b and miR-199a are predicted to target Tgfβ signaling and overexpression in chick AVC cushion explants impairs EMT [34]. In addition to molecular regulation, shear stress has been shown to activate Tgfβ-Alk5 signaling in endothelial cells overlying the cushion, suggesting that the hemodynamic environment also plays a role in this process [35].

In addition to targeting downstream transcription factors, Tgfβ cross talks with other signaling pathways during endocardial cushion development including β-catenin. *Tie2*-cre deletion of β-catenin in endothelial and endothelial-derived cells inhibits Tgfβ-mediated induction of EMT in mice [36], suggesting a requirement of the canonical Wnt signaling in this process. Tbx20 was later shown to regulate *Lef1*, a key transcriptional mediator of Wnt/β-catenin signaling, during early stages of valve formation [37]. In zebrafish, overexpression of a secreted Wnt inhibitor (*Dickkopf1*) blocks cushion formation, while truncation mutants develop hyperplastic cushions as a result of increased cell proliferation [38]. The notion that Wnt signaling regulates VIC proliferation is continued in the avian system where the Wnt receptor Frzb and Wnt9a promote mesenchyme cell number in the AV cushions [21]. Consistently, expression patterns of other Wnt pathway genes including *Wnt2a, LEF1,* and *Fzd2* in murine mesenchyme cells of the AVC and OFT cushions suggest conserved mechanisms across species.

### 3.2.2 Notch Signaling

Activation of Notch signaling requires interaction of a transmembrane receptor (Notch1–4) to a specific ligand (Jagged (Jag-)1, 2, Delta-like (Dll) 1, 3, 4) presented on neighboring cells. Upon receptor–ligand interaction, Notch receptors undergo proteolytic processing and release the Notch intracellular domain (NICD). NICD is then translocated to the nucleus where it binds the DNA-binding co-repressor RBPJ (recombination signal-binding protein for immunoglobulin κ J region), and recruits the coactivator MAML1 (Mastermind-like 1) to de-repress and activate transcription of downstream target genes including HES and HEY families of transcription factors [39] (Fig. 3). Components of the Notch signaling pathway have been implicated in EMT and many of which are required for murine endocardial cushion formation in vivo (Table 2).

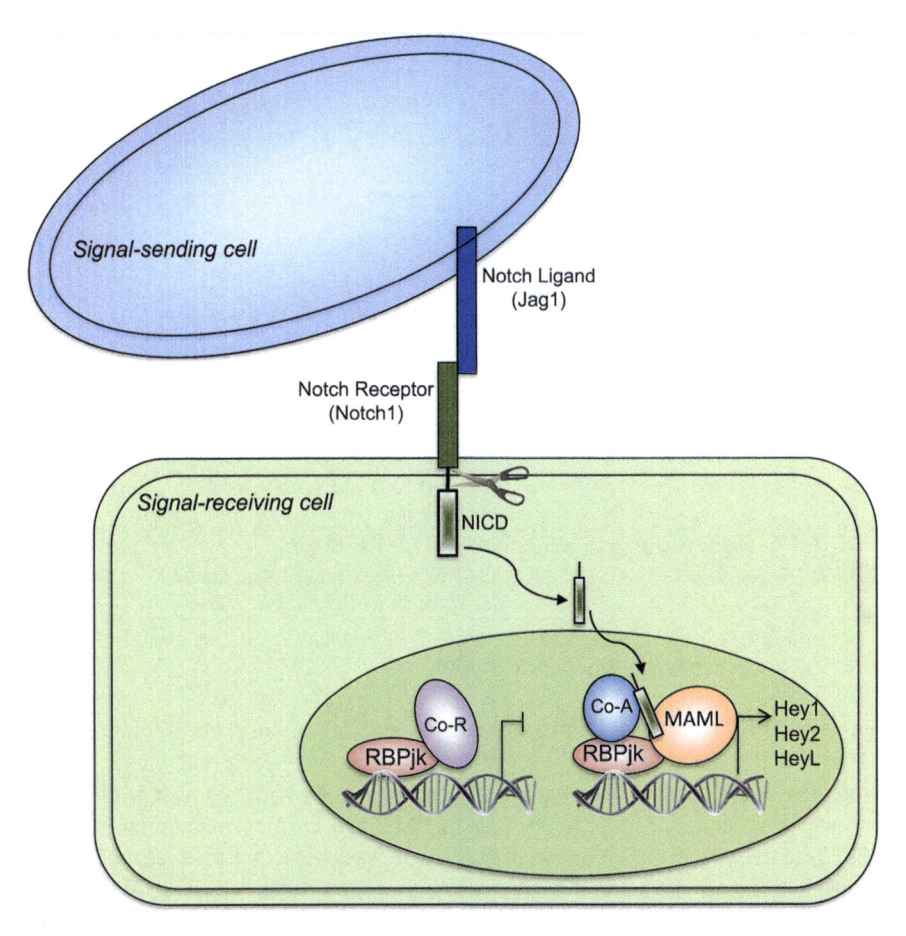

**Fig. 3** Canonical Notch signaling between neighboring cells. Interaction of the Notch ligand (Jagged1, (Jag1)) presented on the surface of the signal-sending cell (blue) with the Notch receptor (Notch1) on the signal-receiving cell (green) leads to cleavage of the Notch intracellular domain (NICD). This cleaved fragment is then relocated to the nucleus of the signal-receiving cell where it binds with cofactors (Co-A, RBPjk, MAML) to largely relieve repression of target genes. Co-R, co-repressor; RBPjk, Recombination Signal Sequence-Binding Protein; Co-A, coactivator; MAML, mastermind-like; NICD, Notch intracellular domain

### 3.2.3 Vascular Endothelial Growth Factor (VEGF)

Vascular endothelial growth factor (VEGF) A is a potent cytokine highly expressed by the myocardium and VECs prior to cushion formation. However, expression of the ligand and its receptors (VEGFR) becomes restricted to the endocardium once cushion formation has been initiated [40, 41]. Given the role that VEGF plays in response to hypoxia in the vascular system, it has been suggested that this temporal and spatial change in VEGF expression during EMT correlates with oxygen levels in

**Table 2** Notch signaling in endocardial cushion development

| Target gene | Model | Endocardial cushion phenotype | References |
|---|---|---|---|
| Hey1, HeyL | Global double knockout | Impaired EMT in E9.5 AV canal explant cultures | [191] |
| Hey2 | Predicted null allele ($Hey2^{lacz}$) | EMT occurs, but cushion maturation is impaired | [192] |
| Jag1 | VE-cadherin-Cre deletion | Hypoplastic endocardial cushions at E10.5, temporarily delayed EMT | [149] |
| MAML | Dominant negative (dn), Isl1-cre deletion | Hypocellular OFT endocardial cushions at E10.5 | [193] |
| MAML | Tet-inducible dn under control of VE-cadherin promoter | Tet-treatment at E10.5; impaired EMT at E11.5 and E12.5 | [194] |
| NICD1 | Mesp1-cre transgenic | Ectopic cell masses in endocardial cushions at E9.5 | [195] |
| Notch1 | Global knockout | Impaired EMT in E9.5 AV canal explant cultures | [196] |
| Notch1 | VE-cadherin-Cre deletion | Delayed EMT at E10.5 | [149] |
| RBPjk | Global knockout | EMT is initiated but endocardial cushions lack mesenchyme cells at E9.5 | [196] |

*EMT* endothelial-to-mesenchymal transformation, *E* embryonic day, *AV* atrioventricular, *OFT* outflow tract, *tet* tetracycline, *dn* dominant negative

the developing embryonic heart [42]. Studies in chick, mouse, and zebrafish demonstrate that endocardial VEGF signaling contributes to early valvulogenesis by promoting the VEC phenotype and proliferation, thereby maintaining a proliferative population of endothelial cells throughout EMT and cushion maturation [41, 43–46]. From these studies, it is apparent that VEGF levels of expression must be tightly controlled during endocardial cushion formation: too much VEGF inhibits EMT, while too little attenuates VEC proliferation, thereby decreasing VEC availability for transformation [41, 45]. Interestingly, it appears that VEGF signaling is required for complete EMT in the OFT, but has less essential roles in AVC endocardial cushion formation [47]. This observation has been attributed to the differential mediation of VEGF signaling through VEGFR1 in the OFT, and VEGFRII in the AVC [47]. Further downstream, VEGF regulates the transcription factor nuclear factor of activated T-cells cytoplasmic 1 (NFATc1) in both developing and mature VECs to promote VEC cell proliferation [46, 48]. While $Nfatc1^{-/-}$ mice undergo successful endocardial cushion EMT [49–51], a more recent role has recently emerged in fate decisions of VECs during transformation stages.

### 3.2.4 Nuclear Factor of Activated T-Cells-1 (Nfatc1)

During a tightly regulated temporal window, VECs break cell–cell contacts, transform, and migrate into the underlying cardiac jelly. However after around E13 in the mouse when cushion formation is complete, VECs no longer transform, despite maintained expression of many EMT-inducers. For many years it has remained a challenge to identify the molecular cues that determine the fate of individual VECs during this window of events: which VECs transform and which VECs remain in the endothelium. In the heart, Nfatc1 is expressed in VECs during endocardial cushion formation [49–51]; however, a role during early stages of valve formation has remained elusive. A study by Wu et al. [52] identified a transcriptional enhancer region (En) within *Nfatc1* that may play an important role in regulating VEC fate during EMT. In contrast to *Nfatc1* expression in all VECs, the study used reporter mice to reveal a small subpopulation of VECs that express the *Nfatc1*-En. VECs positive for *Nfatc1-En* expression do not appear to undergo EMT and remain part of the proliferative endothelial cell population surrounding the developing valve. Further, Nfatc1 inhibits EMT in a cell-autonomous manner by suppressing transcription of *Snail* and *Snai2* [52]. This regulation of VECs by Nfatc1 is the first to provide a mechanism for cell fate decisions of endothelial cells during endocardial cushion EMT.

### 3.2.5 Sox9

The SRY transcription factor Sox9 is activated during initiation of EMT in the AV canal and OFT regions [53]. Targeted deletion to generate *Sox9*-null embryos results in hypoplastic endocardial cushions, and similarly deletion in the *Tie2-cre* lineage leads to premature lethality around E12.5 associated with small endocardial cushions and reduced mesenchyme cell proliferation [54]. These studies identify a critical role for Sox9 in maturation of mesenchyme precursor cells during early valvulogenesis.

## 3.3 The Contribution of Non-endothelial Cell Lineages to the Developing Heart Valve Structures

Cell fate lineage analysis using *Tie2cre;ROSA26R* mice showed that the majority of mesenchymal VICs that form the endocardial cushions in the AV canal and OFT regions are derived from endothelial cells as a result of EMT [25, 26]. However, using additional *Cre*-drivers contribution from additional sources has been demonstrated, some of which are dependent on the valve location. Lineage tracing in mice using the *Wt1/IRES/GFP-Cre* model demonstrates significant contribution of epicardial-derived cells to the parietal leaflets of the AV valves following endocardial cushion formation [55]. However, using a different model (*Wt1CreERT2/+; Rosa26mTmG*) induced at E10.5, the contribution of epicardial-derived cells was scant with enriched contribution to the annulus [56]. Earlier studies using the chick-quail chimera system further support the contribution of these cells to the AV, but not OFT valves [57]. The *Wnt1-cre* transgene fate maps cells originating from the neural crest cell lineage and studies in mice show contribution of these cells to the

OFT endocardial cushions and aorticopulmonary septal structures [26, 58]. While neural crest cell contribution was thought to be restricted to the OFT, later studies reported cells in the AV cushions at E12.5 [59]. Fate mapping of the secondary heart field lineage using the *Mef2c-Cre* model identified an additional cell population contributing to endothelial cells of the outflow tract cushions [60]. These collective studies demonstrate that the valve precursor cell pool is derived from multiple sources, however the role of these differential cell lineages to valve development and disease remain largely elusive.

## 3.4 Molecular Regulation of Embryonic Heart Valve Remodeling

Once endocardial cushion formation is complete, the cushions undergo extensive remodeling as development of the valve primorida is initiated (Fig. 2). In the AVC, remodeling begins with fusion of the inferior and superior endocardial cushions that will give rise to the septal leaflets of the mitral and tricuspid valves [25, 26]. This is in addition to the smaller lateral cushions on the left and right side myocardial walls that serve as precursor to the parietal, or mural leaflets. In the OFT regions, distal cushions formed from EMT and migrating neural crest cells, separate the developing aortic and pulmonary roots, and fused proximal cushions also give rise to non-valvular structures. However, it is fusion and remodeling of the central parts of the developing cushions, along with septation events that create the foundation for developing aortic and pulmonary valve cusps [61]. Remodeling is prominent from ~E14.5 and continues into the postnatal period and characterized by condensation of the mesenchyme cushion cells that remain proliferative at the leading edge to promote elongation. At this time, the mesenchyme cells actively breakdown the primitive ECM within the valve primorida and begin secreting more specialized matrices that will later form the trilaminar structure. Beyond the endocardial cushion stage, the nomenclature of these cells has not been well defined, and their dynamic expression pattern and involvement in matrix remodeling suggest a role beyond mesenchymal cell function, and therefore these cells are referred to as embryonic VICs (eVICs) [17]. While our understanding of the remodeling process is still developing, key regulators beyond endocardial cushion formation are summarized in Table 3. In the adult in the absence of disease, the valve endothelium is maintained, VICs are quiescent (qVICs) and thought to function similar to fibroblasts by mediating ECM turnover to maintain homeostasis and biomechanical function of the aging leaflets (Fig. 2). However, in response to pathological stimuli or changes in the hemodynamic environment, the valve cell populations undergo significant phenotypic changes and mediate pathological ECM remodeling leading to biomechanical failure.

Table 3 Molecular regulators of embryonic heart valve remodeling

| Target gene | Model | Valve remodeling phenotype | References |
|---|---|---|---|
| *Adam17* | *Tie2-cre;Adam17$^{fl/fl}$* | Outflow tract valve thickening at E18.5, increased cell number | [197] |
| *AdamTs5* | *AdamTs5$^{-/-}$* | Enlarged pulmonary valves at E14.5 and E17.5, increased cell number | [198] |
| *Bmp2* | Bmp2 treatment of HH Stage 24 AV cushion mesenchyme cells | Increased versican and hyaluronan expression | [199] |
| *Cadherin-11* | siRNA in porcine aortic valve interstitial cells (pAVICs) | Reduced cell migration, impaired 3D matrix compaction | [200] |
| *CXCR7* | *CXCR7$^{-/-}$* | Semilunar valve thickening at E18.5, increased cell proliferation from E14.5 | [201] |
| *HB-EGF* | *HB-EGF$^{-/-}$* *Tie2-cre;HB-EGF$^{fl/fl}$* | Thickened valve primorida associated with increased proliferation between E14.5-E16.5 | [202–204] |
| *Scleraxis (Scx)* | *Scx$^{-/-}$* | Fragmented collagen fibers, decreased chondroitin sulfate proteoglycans, thickened valve leaflets by E17.5 | [205] |
| *Sox9* | *Col2a1-cre;Sox9$^{fl/fl}$* | Reduced type II collagen, cartilage link protein expression in valve leaflets at E17.5 | [54] |
| *Tbx20* | *Nfatc1Cre/+;Tbx20$^{f/-}$* | Embryonic lethal E14.5-E16.5, valve elongation defects at E15.5 | [37] |
| *Tgfβ2* | *Tgfβ2$^{-/-}$* | Hyperplastic valves, increased hyaluronan and cartilage link protein at E18.5 | [206] |

*E* embryonic day and *OFT* outflow tract

## 4 Heart Valve Disease

In the USA, heart valve disease is an increasing burden with an overall prevalence of 2.5%. In 18–44-year-olds this number is as low as 0.7% and as high as 11.7% in individuals over the age of 75 [1, 62]. In addition to aging, heart valve disease is associated with risk factors similar to coronary heart disease including high blood cholesterol, high blood pressure, smoking, insulin resistance, diabetes, and obesity which can alter the structure and biomechanical properties of the valve [63]. Affected individuals carrying these risks usually develop valve malfunction later in life and therefore their disease is considered acquired. Alternatively, infants can be born with congenital valve malformations as a result of aberrations during embryonic development, and these structural defects often accelerate dysfunction during early postnatal stages of life. There is also emerging evidence to suggest that acquired valve disease may be the result of embryonic defects that become manifested later in life (reviewed [13]). Valve disease affects all sets of valves, although several pathologies are noted more frequently in those positioned on the left side of the heart. The pathogenesis of valve disease has been extensively examined at the histological level and often characterized by disruptions to the endothelial cell layer, "activation" of otherwise quiescent VICs, and changes in the organization and contribution of ECM components. The cause and effect of changes in the cellular components of the valve are not clear, however significant changes in the ECM lead to deformations in the valve leaflets or cusps and alterations in the biomechanical properties. This is largely detected as thickening of the valve leaflets or cusps, and depending on the ECM pathology can lead to biomechanical weakening, or prolapsed valves whereby the leaflets are unable to coapt resulting in regurgitation. Alternatively, valve pathology is presented as stiffened or stenotic valves that fail to fully open. Whether the valve is regurgitant or stenotic the secondary effects on cardiac function can be detrimental to the affected individual.

## 4.1 Bicuspid Aortic Valve

Bicuspid aortic valve (BAV) disease is the most common congenital malformation with an estimated prevalence of 1–2% [64]. Unlike the healthy tricuspid aortic valve, the biscupid valve consists of two leaflets (Fig. 4) and aortic stenosis and insufficiency are common, with over 33% of initially asymptomatic patients experiencing a major cardiovascular event or requiring surgery over the age of 20 [65]. The BAV exists in different morphological phenotypes: the most prevalent (71%) is type-I morphology featuring two cusps of unequal size resulting from the fusion of the right and left coronary cusps, while 15% feature right- and non-coronary cusp fusion and 3% present with non- and left-coronary cusp fusion [66–69]. BAV has strong genetic contributors, as evidenced by reports of familial clustering and calculated heritability [70]. Table 4 summarizes the causative genes linked to BAV in the human population and associated mouse models.

The abnormal formation of two cusps of a BAV is often described as resulting from fusion of two cusps [71]. However, the relationship of the individual valve cusps to specific endocardial cushion progenitors in the outflow tract regions is not known, and BAV could result from failure of primordial cusps to separate as a result of alterations in cell migration, cell survival, and signal transduction. Longitudinal analysis of the developmental origins of a BAV has not been reported, but the evolving development of new animal models (Table 4) will allow for morphologic and mechanistic studies in the embryo and after birth.

In patients with BAV, the structural defect in cusp morphology present at birth (Fig. 4) leads to lifelong dysfunction and abnormal biomechanics, often associated with secondary complications including CAVD. The prevalence of calcification is increased by 20% in BAV patients and may take as little as 10–12 years for the disease to result in severe stenosis [72, 73]. It is not clear why the bicuspid valve is so susceptible to calcification, but studies suggest inheritance with the BAV defect, suggesting a genetic root [74–77]. An alternative theory considers the role of mechanical stresses in a bicuspid setting. Calcification of tricuspid aortic valves

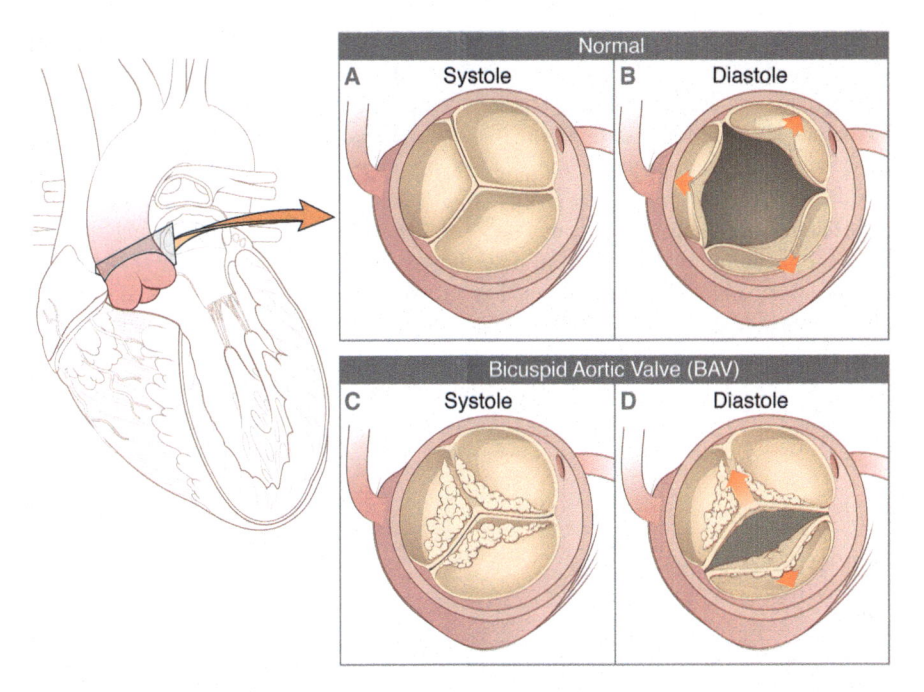

Molecular and Cellular Developments in Heart Valve Development and Disease 221

**Fig. 4** Structural malformations in congenital bicuspid aortic valve disease. (**A, B**) The normal aortic valve consists of three cusps, and during diastole (**A**) and systole (**B**) the cusps sufficiently coapt and fully open respectively to regulate unidirectional blood flow. In contrast, bicuspid aortic valves consist of two cusps that results in narrowing (**D**) and leads to lifelong changes in valve biomechanics, stenosis and increases the risk of calcification (shown as nodules in **C, D**)

**Table 4** Causative genes linked to bicuspid aortic valve

| Target gene | Mouse model | References |
|---|---|---|
| ACTA2 | | [207] |
| GATA5 | $GATA5^{-/-}$<br>$Tie2cre;GATA5^{fl/fl}$ | [77, 208] |
| NOTCH1 | $Notch1^{+/-}$ on Western diet<br>$NOS3^{-/-};Notch1^{+/-}$ | [76, 81, 148] |
| NKX2.5 | $Nkx2.5^{+/-}$ | [209, 210] |
| KCNJ2 | | [211] |
| MATR3 | $Matr3^{GT-ex13}$ | [212] |
| FBN1 | | [213, 214] |
| MADH6 (Smad6) | | [215] |

largely forms on the aortic surface, a region subject to low and oscillatory shear stress, and high diastolic pressure. Studies have demonstrated that the magnitude and nature of shear stresses experienced by the atrial (fibrosa) and ventricular (ventricularis) aortic valve surfaces are very different at the gene expression level

[78]. For example, anti-calcification markers (osteoprotegerin, parathyroid hormone, c-type natriuretic peptide, chordin) are enriched in endothelial cells on the ventricularis surface in the porcine model, compared to the fibrosa [78]. Furthermore, side- and shear-specific expression of pro-osteogenic genes related to canonical BMP signaling was seen on the fibrosa side [79]. In BAV, there is increased turbulence downstream of the BAV and the leaflets, in particular the VECs are exposed to increased fluid shear stress (reviewed [80]). Given the recent role of VECs in mediating calcification in underlying VICs [81] it is plausible that calcification in bicuspid valves is accelerated as a result of altered biomechanics, mechano-responsive signaling pathways in VECs, and cell–cell communications with VICs. The deposition of calcific nodules in addition to remodeling of other ECM components as a result of disturbed blood flow (and strain) will together contribute to biomechanical stiffening of the valve, further limiting leaflet movement and leading to stenosis.

## 4.2 Mitral Valve Prolapse

The syndrome of mitral valve prolapse or myxomatous mitral valve disease is a common disease occurring in 0.6–2.4% of the population, thus being more common than BAV [82]. The disease is usually diagnosed in adulthood, but is heritable and linked to connective tissue disorders and mutations in ECM, signaling and cytoskeletal genes (Table 5). Thus MVP often results from a molecular abnormality present at birth and should be considered a congenital malformation in many cases. In addition, mitral regurgitation and leaflet remodeling occur frequently in adults with ischemic heart disease or dilated cardiomyopathy [83–85]. It seems likely that genetic mutations associated with congenital mitral valve defects also could affect the severity of valvular dysfunction in adult heart disease, but this has not yet been reported. Clinically, MVP is associated with thickened valve leaflets that "bulge" or prolapse into the atrium, thus preventing valve closure (Fig. 5) [86]. The valve thickening is largely due to increased deposition of collagens and proteoglycans, causing pathological changes in valve connective tissue homeostasis and biomechanics [6, 12, 87, 88]. Mouse models (Table 5) demonstrate that end-stage mitral valve disease can result from developmental defects that lead to secondary changes in ECM-related signaling pathways, further promoting pathogenesis in the adult [89, 90].

MVP is observed in patients with Marfan syndrome, caused by mutations in *Fibrillin-1* (*FBN1*), a major component of ECM microfibrils associated with elastin [91–93]. Advances in understanding valve disease pathogenesis in Marfan syndrome patients have been made using mice harboring a cysteine substitution (C1039G) in an epidermal growth factor-like domain in *fibrillin-1* that recapitulate human disease phenotypes [89, 94, 95]. $Fbn1^{C1039G/+}$ heterozygous mice survive until adulthood and have MVP, however $Fbn1^{C1039G/C1039G}$ mice die prematurely (P7-P10) from aortic dissection [89]. The progression of heart valve pathogenesis in mouse models

**Table 5** Genes associated with mitral valve prolapse

| Target gene | Associated syndrome | Mouse model | References |
| --- | --- | --- | --- |
| *Elastin (Eln)* | Williams syndrome | *$Eln^{+/-}$ | [98, 216] |
| *Fibrillin-1 (Fbn1)* | Marfan syndrome | *$Fbn1^{C1039G/+}$ | [89] |
| *Dachsous cadherin-related 1 (Dchs1)* | Non-syndromic mitral valve prolapse | *$Dchs1^{+/-}$ | [108] |
| *Filamin-A (FlnA)* | X-linked myxomatous valvular dystrophy | *$Tie2cre;FlnA^{fl/fl}$ *$Nfatc1cre;FlnA^{fl/fl}$ | [107, 217, 218] |
| *Type I Collagen (Col1)* | Ehlers-Danlos syndrome Osteogenesis imperfecta | $Mov13^{-/-}$ | [219, 220] |
| *Type II Collagen (Col2)* | Sticker syndrome | $Col2a1^{-/-}$ | [221] |
| *Type III Collagen (Col3)* | Ehlers-Danlos syndrome | $Col3a1^{-/-}$ | [222, 223] |
| *Type V collagen (Col5)* | Ehlers-Danlos syndrome | *$ColVa1^{+/-}$ $ColVa1^{-/-}$ | [101, 224, 225] |
| *Type XI collagen (Col11)* | Ehlers-Danlos syndrome Stickler syndrome Marshall syndrome | *$ColXIa1^{-/-}$ | [101, 226] |
| *Tenascin-X (TnX)* | Ehlers-Danlos syndrome | $TnX^{-/-}$ | [227] |
| *TGFβR1/2* | Loeys-Dietz syndrome Marfan syndrome | $Tgf\beta r1$ (p. Y378*) $Tgf\beta r2^{G357W/+}$ | [228, 229] |

*Indicates published heart valve phenotype

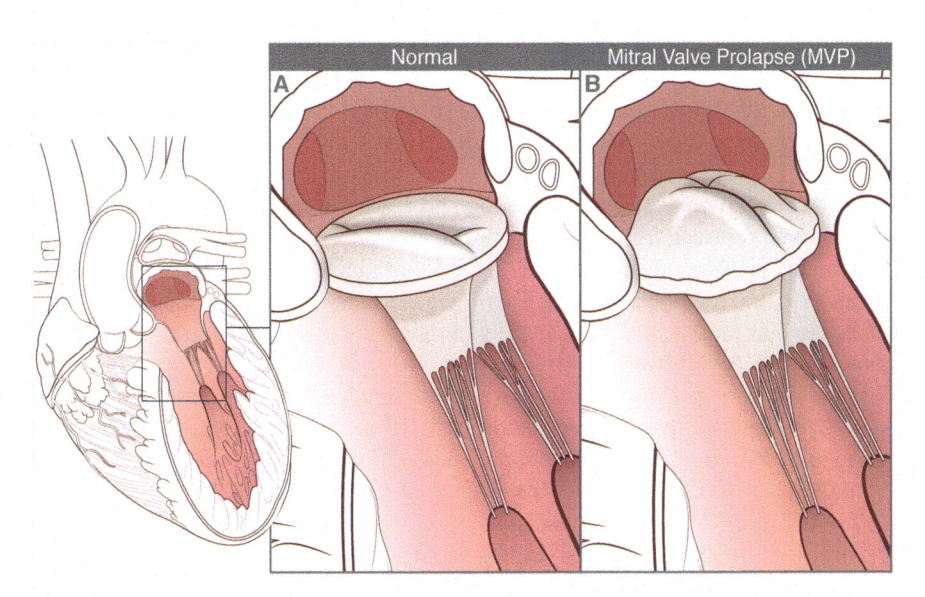

**Fig. 5** Structural alterations in mitral valve prolapse disease. (**A**) In the normal mitral valve, the two valve leaflets open (diastole) and close (systole) to regulate unidirectional blood flow. In mitral valve prolapse (**B**), valve leaflets are often thickened due to changes in extracellular matrix composition and distribution. As a result, leaflets "bulge" or prolapse upward or back into the atria preventing complete closure and leading to blood regurgitation

of MVP has been attributed to increased TGFβ signaling, leading to valve leaflet thickening, ECM disorganization, apoptosis, and cell proliferation [89, 96–98]. Further support for the link between aberrant TGFβ signaling and MVP is the demonstration that Loeys-Dietz syndrome, linked to congenital MVP, is caused by mutations in TGFβ receptor type I [92]. Treatment with TGFβ neutralizing antibodies or with the angiotensin II type 1 receptor blocker Losartan prevents the development of Marfan-related aortic and mitral valve anomalies in mice [99, 100].

Collagen is the major component of the valve ECM, and mutations in collagen and associated glycoprotein genes have been identified in patients with MVP [101–103]. Human gene mutations in collagen types (col) *I, II III, V,* and *XI* and the collagen-associated glycoprotein *tenascin* cause Ehlers-Danlos and Stickler syndromes (reviewed in [104]). *ColVa1*$^{+/-}$ and *ColXIa1*$^{-/-}$ mutant mice recapitulate features of Ehlers-Danlos syndrome (EDS) and display thickened valve leaflets with altered matrix organization, including increased fibrillar collagen protein deposition and gene expression [7]. However, premature lethality of these animals prohibits valve functional studies in newborns or examination of degenerative valve disease in adults [7]. *Tenascin-X*$^{-/-}$ mice also exhibit hyperextensibility of the skin, characteristic of EDS, but are viable [105]. While valve dysfunction or pathogenesis has not been reported in these animals, altered valve biomechanics and MVP might be expected to occur based on the cardiac valve abnormalities associated with other ECM gene mutations that cause EDS.

Increased proteoglycan accumulation is a hallmark of MVP, but mutations in proteoglycans or associated genes have not yet been reported in this patient population [103]. Therefore increased proteoglycan deposition in MVP pathogenesis may be secondary to aberrant valve development and subsequent dysfunction. However, genetic alterations in proteoglycans and their associated molecules in mice can lead to abnormal valve leaflet morphogenesis during development and pathogenesis after birth. Mice deficient for the versican-specific protease *Adamts9* exhibit abnormal valve morphogenesis during development and myxomatous mitral valve leaflets in adults as a result of versican accumulation [106]. As cleaved versican and Adamts9 are both expressed during valvulogenesis, it is likely that the adult mitral valve abnormalities observed in *Adamts9*$^{+/Lacz}$ mice originate in the developmental alterations in matrix remodeling [106]. There is increasing evidence that abnormalities in ECM proteins present at birth can lead to progressive valve dysfunction and disease, including MVP (reviewed [104]).

In addition to the ECM components, more recent studies have identified a role for genes associated with the cytoskeleton in MVP development. Mice with targeted deletion of *Filamin-A* in endothelial-derived cells using *Tie2-cre* develop thickened mitral valves by E17.5, supporting a developmental origin. Affected valves are composed of an abnormal abundance of proteoglycans and collagens, and such phenotypes are thought to be due to decreased serotonylation of Filamin A [107]. Furthermore, mutations in the cell polarity gene *dachsous (DCHS1)* have been identified in non-syndromic MVP, and morphological and functional mitral valve phenotypes have been recapitulated in zebrafish and mice. The mechanisms of

*DCHS1* dysfunction in mitral valve disease are not completely understood, but in vitro manipulation in VICs suggests a role in cell migration and patterning [108].

Together, the combinatory approaches of human genetics with the development of animal models will increase our current understanding of the etiology and pathogenesis of human valve disease.

## 4.3 Pathobiology of Valve Interstitial Cells

In contrast to healthy valves, there is a wealth of evidence to suggest that VICs are no longer quiescent in disease but activated (aVIC) similar to myofibroblasts. Studies in humans [109–111], large animals [112–114], and mice [98, 115–117] have demonstrated that in disease states VICs upregulate the myofibroblast marker SMA, which is thought to recapitulate embryonic phenotypes. However, a recent study did not detect SMA expression in vivo in mesenchyme cells of the cushion until after E12.5 [17]. Nonetheless, SMA is highly expressed in transformed mesenchyme cells in endocardial cushion assays in vitro [36, 118–122]. Notably, VIC activation is not ubiquitous to all resident VICs suggesting heterogeneity among the population and there is a diverse profile of "activated" markers that are increased in subsets of VICs depending on disease, age, sex, and species [123]. In myxomatous mitral valves from human patients and canine models, SMA, Periostin, Vimentin, Desmin and Embryonic Smooth muscle myosin heavy chain (SMemb) protein levels have been observed in subsets of VIC populations, suggestive of activation [17, 109, 112]. Many VICs express multiple activation markers, others do not express any, while SMA (myofibroblast) and Periostin (fibroblast) are detected in diverse populations, highlighting the dynamic and differential phenotypes in mitral valve disease [17, 109].

Histological analysis of human calcific aortic valve disease (CAVD) also report increased SMA in pediatric and adult patients [111], with an additional study by Latif et al., describing aberrant Calponin, SM22, Caldesmon, and Smooth muscle myosin heavy chain expression, reflecting a more diverse VIC population containing myofibroblast- and smooth muscle-like lineages in this disease type [110]. In this later study, changes in smooth muscle and myofibroblast markers were regionalized within the trilaminar layers of the ECM and changes in protein expression were not uniform to all aVICs, highlighting the complexity of differential VIC phenotypes in CAVD. In addition to VIC activation, CAVD has been characterized by ectopic expression of osteoblast markers including Runx2 in affected VIC populations. In vitro, this "osteoVIC" [16] phenotype is thought to precede increased SMA and activation and mediate deposition of mineralized matrix, the contributor of calcification [124, 125]. A comparative post-SMA-positive VIC in myxomatous disease has not been described but warrants further investigation. Worthy of mention, SMA is not always increased in valve pathology and there are reports of unchanged or decreased expression [126–128], suggesting that VIC activation is not always present in disease, or SMA is not a ubiquitous marker of this "activated" phenotype.

The role of aVICs in valve disease is not known and it remains unclear if the changes in VIC phenotypes are cause or effect. Studies by Rabkin et al. support a causative role in myxomatous disease as SMA-positive VICs co-express matrix metalloproteinases (MMPs) that are thought to break down healthy ECM within the valve, and subsequently replace it with an abundance of proteoglycans and fragmented collagen fibers [86, 109]. However, in CAVD it is not known if activation is required for osteoVIC "transdifferentiation" and calcification. While VIC activation is characteristic of several different forms of valve pathology, the mechanisms that underlie activation are not fully understood, but there is increasing evidence from in vitro studies to suggest molecular, cellular, and biomechanical influences. Culture of VICs on stiff substrates significantly increases SMA expression [129–133] more robustly after passage 1 [113], and Xu et al. reported decreased SMA in confluent porcine VICs [134]. Similar to fibroblast activation, Tgfβ1 is a potent inducer of SMA expression in cultured VICs [135–137], which is largely mediated by canonical Smad, and noncanonical MAPK signaling [138, 139]. In addition, increased RhoA and Rho kinase activity is sufficient to promote myofibroblast-related phenotypic markers [140]. In contrast, FGF-2 [138], reduced β-catenin [141], and knockdown of the actin filament stabilizer, cofilin [113] have all been shown to prevent Tgfβ1- and substrate-induced VIC activation. Interestingly, substrate-induced VIC activation can be reversed by transferring cells to a soft substrate [129], or adding polyunsaturated fatty acids to the growth medium [142], highlighting the plasticity of cultured aVICs and potential reversibility of activation. At the cellular level, substrate-induced VIC activation is significantly attenuated when cocultured with VECs both in the presence [143] and in the absence [144] of physical interactions. While the dogma has been that VIC activation marked by SMA expression is a key process in the onset of progression of heart valve pathology, the role of aVICs in disease has not directly been tested and therefore remains unknown. Furthermore it is considered that SMA may not only mark resident aVICs but also represent newly transformed mesenchymal cells recently shown to be present in disease as a result of adult EMT [121, 144, 145].

## 4.4 Pathobiology of Valve Endothelial Cells

Compared to VICs, much less is known about the pathobiology of VECs, although a substantial number of studies within the last decade have given further insight into their role in health and disease. Morphologically, diseased valves are associated with damage to the overlying valve endothelium [146, 147]. Deterioration of the physical barrier likely increases exposure of the underlying VICs and ECM to physical forces evoked by the hemodynamic environment, in addition to circulating risk factors, hence the association of VEC denudation with ECM degradation and altered cell turnover (Han 2103). In addition to physical damage, disruptions in intrinsic and extrinsic signaling pathways in VECs can lead to pathological changes in the VICs, suggesting molecular communication between cell populations.

As discussed previously, Notch signaling is imperative for embryonic valve formation and mutations in the human population cause aortic valve defects [76]. Activated Notch1, determined by nuclear localization of the Notch1 intracellular domain (N1ICD), is enriched in the valve endothelium and haploinsufficient mice develop aortic valve calcification [148]. Furthermore, endothelial-specific deletion of the Notch1 ligand, Jagged 1 (Jag1), also promotes calcific aortic valve disease during adulthood [149], suggesting that Notch1 in postnatal VECs is required to maintain valve homeostasis and prevent calcification. The upstream regulation of Notch1 in VECs after birth is not clear, but shear stress increases endothelial cell signaling via Notch1 [150], and stress induced ostogenic and inflammatory responses requires Notch1 signaling in iPSC-derived endothelial cells [151]. In addition to biomechanics, nitric oxide (NO) signaling has also been implicated as a regulator of Notch signaling in VECs as calcification of porcine aortic valve interstitial cells (pAVICs) following endothelial nitric oxide depletion can be rescued by the addition of activated Notch1 (N1ICD) [81]. Furthermore, compound $Nos3^{-/-}$;$Notch1^{+/-}$ mice develop aortic valve disease at a penetrance of 100% which is significantly greater than ~25–30% in $Nos3^{-/-}$ mice and ~10% in $Notch1^{+/-}$ mice, suggesting genetic interaction of these two pathways in VECs [81].

Decreased NO signaling and increased oxidative stress have been implicated in heart valve disease and thought to be specific to valve endothelial cells similar to the process described in atherosclerosis. In stenotic aortic heart valves, excess production of superoxide in VECs is spatially localized with calcific lesions in humans [152] and hypercholesterolemic mice [153]. Increases in VEC oxidative stress are due to decreased expression and activity of antioxidants as well as the presence of uncoupled NO synthases [152]. Further, it has been shown that VECs actively contribute to calcification through TNF-α induced superoxide formation [154, 155]. Although the mechanisms underlying induction of endothelial oxidative stress in the valve is still largely unknown, these studies provide evidence to suggest that both genetic and environmental factors play a role.

### 4.4.1 VEC-VIC Communications

Studies have shown that damaged or altered signaling within the overlying valve endothelium leads to pathological changes in underlying VICs and ECM. In addition, in vitro studies from several labs have shown that VECs can molecularly communicate with VICs through paracrine signaling to regulate their phenotype [81, 143, 144, 156, 157]. VICs cultured in the absence of VECs are "activated," transdifferentiate towards a myofibroblast-like cell type [144], and undergo premature spontaneous calcification [81]. These pathologic phenotypes can be rescued by coculture with endothelial cells, or treatment with an NO donor [81, 143, 144, 156] further highlighting the importance of this signaling pathway in maintaining function of the endothelium. In addition to NO, a recent in vivo study identifies *Tgfβ1* as a critical growth factor secreted by VECs to VICs to prevent calcification, by promoting nuclear localization of the transcription factor, Sox9 [158]. Collectively, these data show that a disruption in VEC signaling leads to deregulation of VIC homeostasis, resulting in the onset of disease states.

### 4.4.2 VEC Heterogeneity

The spatial localization of VECs on the underside and overlying surface of the valve leaflets allow for differential exposure to mechanical stimuli. In the aortic position, VECs on the underside (ventricularis) experience laminar shear and low stress, while those on the overlying surface (fibrosa) are subjected to oscillatory flow and high pressure in diastole [159]. As VECs serve as mechanosensors, the biomechanical environment can influence their function and the signaling pathways transmitted from the endothelium to underlying VICs and ECM. Shear stress can protect against calcification [160–162] and calcific lesions occur preferentially on the aortic (fibrosa) surface of the valve exposed to oscillatory shear. This difference in disease susceptibility has been attributed to the differential sheer-stress response of VEC gene programs [78, 79, 163]. VECs exposed to oscillatory flow express less anti-calcific and anti-inflammatory genes and more anti-oxidative mRNAs [78], and using induced pluripotent stem cell-derived (iPSC) endothelial cells, this is thought to be mediated by Notch1 [151]. These studies suggest that VECs are heterogeneous and their regulatory function on VICs is influenced by mechanical stimuli.

In addition to being heterogeneous, adult VECs demonstrate degrees of "plasticity." Unlike vascular endothelial cells, VECs have a unique ability to undergo endothelial to mesenchymal transition (EMT), a process well studied during valvulogenesis. However, in vitro studies have shown that subsets of mature ovine VECs have the potential to undergo EMT in response to TGFβ1 treatment [121] and subsequently differentiate into osteogenic and chondrogenic lineages [164]. In addition, exposure of adult porcine VECs to inflammatory cytokines (interleukin-6, tumor necrosis factor-α) also induced EMT in vitro [165]. In vivo, mesenchyme-like cells have been observed in sub-endothelial locations in large animal models of mitral valve regurgitation [166] and a mouse model of calcification ($ApoE^{-/-}$) [157], suggesting EMT. The development of inducible valve endothelial cell-Cre lines will facilitate in the fate mapping of these cells in the future and their contribution to the underlying VIC population in disease. EMT events in healthy adult valves are rare and it is speculated that this is due to the suppressive effects that VICs have on mesenchymal transformation (and calcification) of VECs in vitro [144, 157]. The ability of VECs to undergo EMT in adult valves indicates that there is subpopulation potentially capable of self-renewal that may serve to continually replenish the mature VIC population and whose activity is enhanced during the disease process [145].

## 5  Current Treatments and Clinical Perspectives

Despite the increasing burden of heart valve disease there are currently no pharmacological reagents available to cure valve disease. Drugs such as statins, Angiotensin-converting enzyme (ACE) inhibitors, vasodilators, and β-blockers have shown promising trends in their ability to partially alleviate the effects of either valve stenosis or regurgitation, but solidifying evidence of improved clinical outcomes from these treatments is lacking and therefore the medical consensus on the benefits of these drugs remains blurred [167]. Advancements in the development of mechanistic-based therapies have been hindered by our lack of understanding of the processes that initiate and promote heart valve disease onset and progression. It is also not clear if heart valve disease phenotypes are reversible and so at best, pharmacological therapeutics may only be able to halt progression if the "window of opportunity" has passed.

At present, tissue engineered heart valves (TEHV) are the most promising replacement application due to many factors including their long-term structural integrity and decreased immunogenicity [168, 169]. Several clinical trials involving TEHVs have resulted in improved clinical outcomes [168]; however the ideal valve replacement that can grow, remodel, repair, and withstand the hemodynamic environment upon implantation has not been successfully developed. The application of seeding stem cells onto TEHV scaffolds before implantation has shown encouraging results both in vivo and vitro [168] with focus on bone marrow derived mesenchymal stem cells (BMSCs) and circulating endothelial progenitor cells (EPCs). These populations are not only interesting for their use in TEHVs but also as an endogenous therapeutic target that could be enhanced to support valve maintenance and repair, or serve as a platform to deliver corrected genetic material.

As discussed earlier, some VECs retain their embryonic plasticity into adulthood including their ability to undergo EMT, differentiate, and express progenitor markers [145]. Similarly, many VICs express mesenchymal stem-like cell markers [170] and bone marrow transplant studies have shown that cells of bone marrow origin incorporate into the VIC population in humans [171] and mice [172, 173]; however the contribution of these cells to disease pathogenesis or age-related degeneration are unknown.

Currently, the areas of research discussed here continue to push the field towards possible therapeutics that target specific valve cell populations, mRNAs or proteins and facilitate repair mechanisms to prevent and alleviate both chronic and acute valve dysfunction.

## References

1. d'Arcy JL, et al. Valvular heart disease: the next cardiac epidemic. Heart. 2011;97(2):91–3.
2. Anderson RH. Clinical anatomy of the aortic root. Heart. 2000;84(6):670–3.
3. Tao G, Kotick JD, Lincoln J. Heart valve development, maintenance, and disease: the role of endothelial cells. Curr Top Dev Biol. 2012;100:203–32.

4. Icardo JM, Colvee E. Atrioventricular valves of the mouse: III. Collagenous skeleton and myotendinous junction. Anat Rec. 1995;243(3):367–75.
5. Kunzelman KS, et al. Finite element analysis of the mitral valve. J Heart Valve Dis. 1993;2(3):326–40.
6. Rabkin-Aikawa E, Mayer JE Jr, Schoen FJ. Heart valve regeneration. Adv Biochem Eng Biotechnol. 2005;94:141–79.
7. Lincoln J, Lange AW, Yutzey KE. Hearts and bones: shared regulatory mechanisms in heart valve, cartilage, tendon, and bone development. Dev Biol. 2006;294(2):292–302.
8. Aldous IG, et al. Differences in collagen cross-linking between the four valves of the bovine heart: a possible role in adaptation to mechanical fatigue. Am J Physiol Heart Circ Physiol. 2009;296(6):H1898–906.
9. Balachandran K, Sucosky P, Yoganathan AP. Hemodynamics and mechanobiology of aortic valve inflammation and calcification. Int J Inflamm. 2011;2011:263870.
10. Grande-Allen KJ, Liao J. The heterogeneous biomechanics and mechanobiology of the mitral valve: implications for tissue engineering. Curr Cardiol Rep. 2011;13(2):113–20.
11. Sacks MS, David Merryman W, Schmidt DE. On the biomechanics of heart valve function. J Biomech. 2009;42(12):1804–24.
12. Schoen FJ. Aortic valve structure-function correlations: role of elastic fibers no longer a stretch of the imagination. J Heart Valve Dis. 1997;6(1):1–6.
13. Hinton RB, Yutzey KE. Heart valve structure and function in development and disease. Annu Rev Physiol. 2011;73:29–46.
14. Scott M, Vesely I. Aortic valve cusp microstructure: the role of elastin. Ann Thorac Surg. 1995;60(2 Suppl):S391–4.
15. Misfeld M, Sievers HH. Heart valve macro- and microstructure. Philos Trans R Soc Lond B Biol Sci. 2007;362(1484):1421–36.
16. Liu AC, Joag VR, Gotlieb AI. The emerging role of valve interstitial cell phenotypes in regulating heart valve pathobiology. Am J Pathol. 2007;171(5):1407–18.
17. Horne TE, et al. Dynamic heterogeneity of the heart valve interstitial cell population in mitral valve health and disease. J Cardiovasc Dev Dis. 2015;2(3):214–32.
18. Helmke BP, Davies PF. The cytoskeleton under external fluid mechanical forces: hemodynamic forces acting on the endothelium. Ann Biomed Eng. 2002;30(3):284–96.
19. Miragoli M, et al. Side-specific mechanical properties of valve endothelial cells. Am J Physiol Heart Circ Physiol. 2014;307(1):H15–24.
20. Eisenberg L. Medicine—molecular, monetary, or more than both? JAMA. 1995;274(4):331–4.
21. Person AD, Klewer SE, Runyan RB. Cell biology of cardiac cushion development. Int Rev Cytol. 2005;243:287–335.
22. Lopez-Sanchez C, Garcia-Martinez V. Molecular determinants of cardiac specification. Cardiovasc Res. 2011;91(2):185–95.
23. Markwald RR, et al. Developmental basis of adult cardiovascular diseases: valvular heart diseases. Ann N Y Acad Sci. 2010;1188:177–83.
24. Schroeder JA, et al. Form and function of developing heart valves: coordination by extracellular matrix and growth factor signaling. J Mol Med. 2003;81(7):392–403.
25. Lincoln J, Alfieri CM, Yutzey KE. Development of heart valve leaflets and supporting apparatus in chicken and mouse embryos. Dev Dyn. 2004;230(2):239–50.
26. de Lange FJ, et al. Lineage and morphogenetic analysis of the cardiac valves. Circ Res. 2004;95(6):645–54.
27. Romano LA, Runyan RB. Slug is an essential target of TGFbeta2 signaling in the developing chicken heart. Dev Biol. 2000;223(1):91–102.
28. Cho HJ, et al. Snail is required for transforming growth factor-beta-induced epithelial-mesenchymal transition by activating PI3 kinase/Akt signal pathway. Biochem Biophys Res Commun. 2007;353(2):337–43.

29. Tao G, et al. Mmp15 is a direct target of Snail during endothelial to mesenchymal transformation and endocardial cushion development. Dev Biol. 2011;359(2):209–21.
30. Niessen K, et al. Slug is a direct Notch target required for initiation of cardiac cushion cellularization. J Cell Biol. 2008;182(2):315–25.
31. Oram KF, Carver EA, Gridley T. Slug expression during organogenesis in mice. Anat Rec A Discov Mol Cell Evol Biol. 2003;271(1):189–91.
32. Carver EA, et al. The mouse snail gene encodes a key regulator of the epithelial-mesenchymal transition. Mol Cell Biol. 2001;21(23):8184–8.
33. Cano A, et al. The transcription factor snail controls epithelial-mesenchymal transitions by repressing E-cadherin expression. Nat Cell Biol. 2000;2(2):76–83.
34. Bonet F, et al. MiR-23b and miR-199a impairs epithelial-to-mesenchymal transition during atrioventricular endocardial cushion formation. Dev Dyn. 2015;244:1259.
35. Ten Dijke P, et al. TGF-beta signaling in endothelial-to-mesenchymal transition: the role of shear stress and primary cilia. Sci Signal. 2012;5(212):pt2.
36. Liebner S, et al. Beta-catenin is required for endothelial-mesenchymal transformation during heart cushion development in the mouse. J Cell Biol. 2004;166(3):359–67.
37. Cai X, et al. Tbx20 acts upstream of Wnt signaling to regulate endocardial cushion formation and valve remodeling during mouse cardiogenesis. Development. 2013;140(15):3176–87.
38. Hurlstone AF, et al. The Wnt/beta-catenin pathway regulates cardiac valve formation. Nature. 2003;425(6958):633–7.
39. Iso T, Kedes L, Hamamori Y. HES and HERP families: multiple effectors of the Notch signaling pathway. J Cell Physiol. 2003;194(3):237–55.
40. Miquerol L, et al. Multiple developmental roles of VEGF suggested by a LacZ-tagged allele. Dev Biol. 1999;212(2):307–22.
41. Dor Y, et al. A novel role for VEGF in endocardial cushion formation and its potential contribution to congenital heart defects. Development. 2001;128(9):1531–8.
42. Dor Y, et al. VEGF modulates early heart valve formation. Anat Rec A Discov Mol Cell Evol Biol. 2003;271(1):202–8.
43. Ferrara N, et al. Heterozygous embryonic lethality induced by targeted inactivation of the VEGF gene. Nature. 1996;380(6573):439–42.
44. Carmeliet P, et al. Abnormal blood vessel development and lethality in embryos lacking a single VEGF allele. Nature. 1996;380(6573):435–9.
45. Miquerol L, Langille BL, Nagy A. Embryonic development is disrupted by modest increases in vascular endothelial growth factor gene expression. Development. 2000;127(18):3941–6.
46. Combs MD, Yutzey KE. VEGF and RANKL regulation of NFATc1 in heart valve development. Circ Res. 2009;105(6):565–74.
47. Stankunas K, et al. VEGF signaling has distinct spatiotemporal roles during heart valve development. Dev Biol. 2010;347(2):325–36.
48. Johnson EN, et al. NFATc1 mediates vascular endothelial growth factor-induced proliferation of human pulmonary valve endothelial cells. J Biol Chem. 2003;278(3):1686–92.
49. Ranger AM, et al. The transcription factor NF-ATc is essential for cardiac valve formation. Nature. 1998;392(6672):186–90.
50. de la Pompa JL, et al. Role of the NF-ATc transcription factor in morphogenesis of cardiac valves and septum. Nature. 1998;392(6672):182–6.
51. Lange AW, Yutzey KE. NFATc1 expression in the developing heart valves is responsive to the RANKL pathway and is required for endocardial expression of cathepsin K. Dev Biol. 2006;292(2):407–17.
52. Aad G, et al. Search for the Higgs boson in the $H \rightarrow WW \rightarrow lnujj$ decay channel in pp collisions at radicals = 7 TeV with the ATLAS detector. Phys Rev Lett. 2011;107(23):231801.
53. Akiyama H, et al. Essential role of Sox9 in the pathway that controls formation of cardiac valves and septa. Proc Natl Acad Sci U S A. 2004;101(17):6502–7.
54. Lincoln J, et al. Sox9 is required for precursor cell expansion and extracellular matrix organization during mouse heart valve development. Dev Biol. 2007;305(1):120–32.

55. Wessels A, et al. Epicardially derived fibroblasts preferentially contribute to the parietal leaflets of the atrioventricular valves in the murine heart. Dev Biol. 2012;366(2):111–24.
56. Zhou B, et al. Genetic fate mapping demonstrates contribution of epicardium-derived cells to the annulus fibrosis of the mammalian heart. Dev Biol. 2010;338(2):251–61.
57. Gittenberger-de Groot AC, et al. Epicardium-derived cells contribute a novel population to the myocardial wall and the atrioventricular cushions. Circ Res. 1998;82(10):1043–52.
58. Jiang X, et al. Fate of the mammalian cardiac neural crest. Development. 2000;127(8):1607–16.
59. Nakamura T, Colbert MC, Robbins J. Neural crest cells retain multipotential characteristics in the developing valves and label the cardiac conduction system. Circ Res. 2006;98(12):1547–54.
60. Verzi MP, et al. The right ventricle, outflow tract, and ventricular septum comprise a restricted expression domain within the secondary/anterior heart field. Dev Biol. 2005;287(1):134–45.
61. Spicer DE, et al. The anatomy and development of the cardiac valves. Cardiol Young. 2014;24(6):1008–22.
62. Nkomo VT, et al. Burden of valvular heart diseases: a population-based study. Lancet. 2006;368(9540):1005–11.
63. Stewart BF, et al. Clinical factors associated with calcific aortic valve disease. Cardiovascular Health Study. J Am Coll Cardiol. 1997;29(3):630–4.
64. McBride KL, Garg V. Heredity of bicuspid aortic valve: is family screening indicated? Heart. 2011;97(15):1193–5.
65. Michelena HI, et al. Natural history of asymptomatic patients with normally functioning or minimally dysfunctional bicuspid aortic valve in the community. Circulation. 2008;117(21):2776–84.
66. Sievers HH, Schmidtke C. A classification system for the bicuspid aortic valve from 304 surgical specimens. J Thorac Cardiovasc Surg. 2007;133(5):1226–33.
67. Braverman AC, et al. The bicuspid aortic valve. Curr Probl Cardiol. 2005;30(9):470–522.
68. Sabet HY, et al. Congenitally bicuspid aortic valves: a surgical pathology study of 542 cases (1991 through 1996) and a literature review of 2,715 additional cases. Mayo Clin Proc. 1999;74(1):14–26.
69. Fernandes SM, et al. Morphology of bicuspid aortic valve in children and adolescents. J Am Coll Cardiol. 2004;44(8):1648–51.
70. Lincoln J, Garg V. Etiology of valvular heart disease. Circ J. 2014;78(8):1801–7.
71. Ward C. Clinical significance of the bicuspid aortic valve. Heart. 2000;83(1):81–5.
72. Fedak PW, et al. Clinical and pathophysiological implications of a bicuspid aortic valve. Circulation. 2002;106(8):900–4.
73. Tzemos N, et al. Outcomes in adults with bicuspid aortic valves. JAMA. 2008;300(11):1317–25.
74. Cripe L, et al. Bicuspid aortic valve is heritable. J Am Coll Cardiol. 2004;44(1):138–43.
75. Huntington K, Hunter AGW, Chan KL. A prospective study to assess the frequency of familial clustering of congenital bicuspid aortic valve. J Am Coll Cardiol. 1997;30(7):1809–12.
76. Garg V, et al. Mutations in NOTCH1 cause aortic valve disease. Nature. 2005;437(7056):270–4.
77. Padang R, et al. Rare non-synonymous variations in the transcriptional activation domains of GATA5 in bicuspid aortic valve disease. J Mol Cell Cardiol. 2012;53(2):277–81.
78. Simmons CA, et al. Spatial heterogeneity of endothelial phenotypes correlates with side-specific vulnerability to calcification in normal porcine aortic valves. Circ Res. 2005;96(7):792–9.
79. Holliday CJ, et al. Discovery of shear- and side-specific mRNAs and miRNAs in human aortic valvular endothelial cells. Am J Physiol Heart Circ Physiol. 2011;301(3):H856–67.
80. Saikrishnan N, Mirabella L, Yoganathan AP. Bicuspid aortic valves are associated with increased wall and turbulence shear stress levels compared to trileaflet aortic valves. Biomech Model Mechanobiol. 2015;14(3):577–88.

81. Bosse K, et al. Endothelial nitric oxide signaling regulates Notch1 in aortic valve disease. J Mol Cell Cardiol. 2013;60:27–35.
82. Iung B, Vahanian A. Epidemiology of valvular heart disease in the adult. Nat Rev Cardiol. 2011;8(3):162–72.
83. Timek TA, et al. Mitral leaflet remodeling in dilated cardiomyopathy. Circulation. 2006;114 (1 Suppl):I518–23.
84. Chaput M, et al. Mitral leaflet adaptation to ventricular remodeling: occurrence and adequacy in patients with functional mitral regurgitation. Circulation. 2008;118(8):845–52.
85. Chaput M, et al. Mitral leaflet adaptation to ventricular remodeling: prospective changes in a model of ischemic mitral regurgitation. Circulation. 2009;120(11 Suppl):S99–103.
86. Gupta V, et al. Abundance and location of proteoglycans and hyaluronan within normal and myxomatous mitral valves. Cardiovasc Pathol. 2009;18(4):191–7.
87. Schoen FJ. Evolving concepts of cardiac valve dynamics: the continuum of development, functional structure, pathobiology, and tissue engineering. Circulation. 2008;118 (18):1864–80.
88. Tamura K, Fukuda Y, Ferrans VJ. Elastic fiber abnormalities associated with a leaflet perforation in floppy mitral valve. J Heart Valve Dis. 1998;7(4):460–6.
89. Ng CM, et al. TGF-beta-dependent pathogenesis of mitral valve prolapse in a mouse model of Marfan syndrome. J Clin Invest. 2004;114(11):1586–92.
90. Kern CB, et al. Proteolytic cleavage of versican during cardiac cushion morphogenesis. Dev Dyn. 2006;235(8):2238–47.
91. Dietz HC, et al. Marfan syndrome caused by a recurrent de novo missense mutation in the fibrillin gene. Nature. 1991;352(6333):337–9.
92. Dietz HC, et al. Recent progress towards a molecular understanding of Marfan syndrome. Am J Med Genet C Semin Med Genet. 2005;139C(1):4–9.
93. Robinson PN, et al. Mutations of FBN1 and genotype-phenotype correlations in Marfan syndrome and related fibrillinopathies. Hum Mutat. 2002;20(3):153–61.
94. Dallas SL, et al. Dual role for the latent transforming growth factor-beta binding protein in storage of latent TGF-beta in the extracellular matrix and as a structural matrix protein. J Cell Biol. 1995;131(2):539–49.
95. Isogai Z, et al. Latent transforming growth factor beta-binding protein 1 interacts with fibrillin and is a microfibril-associated protein. J Biol Chem. 2003;278(4):2750–7.
96. Massam-Wu T, et al. Assembly of fibrillin microfibrils governs extracellular deposition of latent TGF beta. J Cell Sci. 2010;123(Pt 17):3006–18.
97. Hanada K, et al. Perturbations of vascular homeostasis and aortic valve abnormalities in fibulin-4 deficient mice. Circ Res. 2007;100(5):738–46.
98. Hinton RB, et al. Elastin haploinsufficiency results in progressive aortic valve malformation and latent valve disease in a mouse model. Circ Res. 2010;107(4):549–57.
99. Habashi JP, et al. Losartan, an AT1 antagonist, prevents aortic aneurysm in a mouse model of Marfan syndrome. Science. 2006;312(5770):117–21.
100. Geirsson A, et al. Modulation of transforming growth factor-beta signaling and extracellular matrix production in myxomatous mitral valves by angiotensin II receptor blockers. Circulation. 2012;126(11 Suppl 1):S189–97.
101. Lincoln J, et al. ColVa1 and ColXIa1 are required for myocardial morphogenesis and heart valve development. Dev Dyn. 2006;235(12):3295–305.
102. Peacock JD, et al. Reduced sox9 function promotes heart valve calcification phenotypes in vivo. Circ Res. 2010;106(4):712–9.
103. Prunotto M, et al. Endocellular polyamine availability modulates epithelial-to-mesenchymal transition and unfolded protein response in MDCK cells. Lab Investig. 2010;90(6):929–39.
104. Lincoln J, Yutzey KE. Molecular and developmental mechanisms of congenital heart valve disease. Birth Defects Res A Clin Mol Teratol. 2011;91(6):526–34.
105. Mao JR, et al. Tenascin-X deficiency mimics Ehlers-Danlos syndrome in mice through alteration of collagen deposition. Nat Genet. 2002;30(4):421–5.

106. Kern CB, et al. Reduced versican cleavage due to Adamts9 haploinsufficiency is associated with cardiac and aortic anomalies. Matrix Biol. 2010;29(4):304–16.
107. Sauls K, et al. Developmental basis for filamin-A-associated myxomatous mitral valve disease. Cardiovasc Res. 2012;96(1):109–19.
108. Durst R, et al. Mutations in DCHS1 cause mitral valve prolapse. Nature. 2015;525:109.
109. Rabkin E, et al. Activated interstitial myofibroblasts express catabolic enzymes and mediate matrix remodeling in myxomatous heart valves. Circulation. 2001;104(21):2525–32.
110. Latif N, et al. Expression of smooth muscle cell markers and co-activators in calcified aortic valves. Eur Heart J. 2015;36(21):1335–45.
111. Wirrig EE, Hinton RB, Yutzey KE. Differential expression of cartilage and bone-related proteins in pediatric and adult diseased aortic valves. J Mol Cell Cardiol. 2011;50(3):561–9.
112. Disatian S, et al. Interstitial cells from dogs with naturally occurring myxomatous mitral valve disease undergo phenotype transformation. J Heart Valve Dis. 2008;17(4):402–11. discussion 412
113. Pho M, et al. Cofilin is a marker of myofibroblast differentiation in cells from porcine aortic cardiac valves. Am J Physiol Heart Circ Physiol. 2008;294(4):H1767–78.
114. Olsson M, Rosenqvist M, Nilsson J. Expression of HLA-DR antigen and smooth muscle cell differentiation markers by valvular fibroblasts in degenerative aortic stenosis. J Am Coll Cardiol. 1994;24(7):1664–71.
115. Munjal C, et al. TGF-beta mediates early angiogenesis and latent fibrosis in an Emilin1-deficient mouse model of aortic valve disease. Dis Model Mech. 2014;7(8):987–96.
116. Matsumoto Y, et al. Regular exercise training prevents aortic valve disease in low-density lipoprotein-receptor-deficient mice. Circulation. 2010;121(6):759–67.
117. Tanaka K, et al. Age-associated aortic stenosis in apolipoprotein E-deficient mice. J Am Coll Cardiol. 2005;46(1):134–41.
118. Sakabe M, et al. Rho kinases regulate endothelial invasion and migration during valvuloseptal endocardial cushion tissue formation. Dev Dyn. 2006;235(1):94–104.
119. Okagawa H, Markwald RR, Sugi Y. Functional BMP receptor in endocardial cells is required in atrioventricular cushion mesenchymal cell formation in chick. Dev Biol. 2007;306 (1):179–92.
120. Sugi Y, et al. Bone morphogenetic protein-2 can mediate myocardial regulation of atrioventricular cushion mesenchymal cell formation in mice. Dev Biol. 2004;269(2):505–18.
121. Paranya G, et al. Aortic valve endothelial cells undergo transforming growth factor-beta-mediated and non-transforming growth factor-beta-mediated transdifferentiation in vitro. Am J Pathol. 2001;159(4):1335–43.
122. Nakajima Y, et al. Expression of smooth muscle alpha-actin in mesenchymal cells during formation of avian endocardial cushion tissue: a role for transforming growth factor beta3. Dev Dyn. 1997;209(3):296–309.
123. McCoy CM, Nicholas DQ, Masters KS. Sex-related differences in gene expression by porcine aortic valvular interstitial cells. PLoS One. 2012;7(7):e39980.
124. Poggio P, et al. Noggin attenuates the osteogenic activation of human valve interstitial cells in aortic valve sclerosis. Cardiovasc Res. 2013;98(3):402–10.
125. Chen J, et al. Notch1 mutation leads to valvular calcification through enhanced myofibroblast mechanotransduction. Arterioscler Thromb Vasc Biol. 2015;35(7):1597–605.
126. Thalji NM, et al. Nonbiased molecular screening identifies novel molecular regulators of fibrogenic and proliferative signaling in myxomatous mitral valve disease. Circ Cardiovasc Genet. 2015;8(3):516–28.
127. Dupuis LE, et al. Insufficient versican cleavage and Smad2 phosphorylation results in bicuspid aortic and pulmonary valves. J Mol Cell Cardiol. 2013;60:50–9.
128. Monzack EL, Masters KS. Can valvular interstitial cells become true osteoblasts? A side-by-side comparison. J Heart Valve Dis. 2011;20(4):449–63.
129. Quinlan AM, Billiar KL. Investigating the role of substrate stiffness in the persistence of valvular interstitial cell activation. J Biomed Mater Res A. 2012;100(9):2474–82.

130. Chen JH, Simmons CA. Cell-matrix interactions in the pathobiology of calcific aortic valve disease: critical roles for matricellular, matricrine, and matrix mechanics cues. Circ Res. 2011;108(12):1510–24.
131. Gould ST, Anseth KS. Role of cell-matrix interactions on VIC phenotype and tissue deposition in 3D PEG hydrogels. J Tissue Eng Regen Med. 2016;10(10):E443–53.
132. Cushing MC, et al. Material-based regulation of the myofibroblast phenotype. Biomaterials. 2007;28(23):3378–87.
133. Duan B, et al. Stiffness and adhesivity control aortic valve interstitial cell behavior within hyaluronic acid based hydrogels. Acta Biomater. 2013;9(8):7640–50.
134. Xu S, et al. Cell density regulates in vitro activation of heart valve interstitial cells. Cardiovasc Pathol. 2012;21(2):65–73.
135. Liu AC, Gotlieb AI. Transforming growth factor-beta regulates in vitro heart valve repair by activated valve interstitial cells. Am J Pathol. 2008;173(5):1275–85.
136. Cushing MC, Liao JT, Anseth KS. Activation of valvular interstitial cells is mediated by transforming growth factor-beta1 interactions with matrix molecules. Matrix Biol. 2005;24 (6):428–37.
137. Walker GA, et al. Valvular myofibroblast activation by transforming growth factor-beta: implications for pathological extracellular matrix remodeling in heart valve disease. Circ Res. 2004;95(3):253–60.
138. Cushing MC, et al. Fibroblast growth factor represses Smad-mediated myofibroblast activation in aortic valvular interstitial cells. FASEB J. 2008;22(6):1769–77.
139. Hutcheson JD, et al. 5-HT(2B) antagonism arrests non-canonical TGF-beta1-induced valvular myofibroblast differentiation. J Mol Cell Cardiol. 2012;53(5):707–14.
140. Gu X, Masters KS. Role of the Rho pathway in regulating valvular interstitial cell phenotype and nodule formation. Am J Physiol Heart Circ Physiol. 2011;300(2):H448–58.
141. Chen JH, et al. Beta-catenin mediates mechanically regulated, transforming growth factor-beta1-induced myofibroblast differentiation of aortic valve interstitial cells. Arterioscler Thromb Vasc Biol. 2011;31(3):590–7.
142. Witt W, et al. Reversal of myofibroblastic activation by polyunsaturated fatty acids in valvular interstitial cells from aortic valves. Role of RhoA/G-actin/MRTF signalling. J Mol Cell Cardiol. 2014;74:127–38.
143. Gould ST, et al. The role of valvular endothelial cell paracrine signaling and matrix elasticity on valvular interstitial cell activation. Biomaterials. 2014;35(11):3596–606.
144. Shapero K, et al. Reciprocal interactions between mitral valve endothelial and interstitial cells reduce endothelial-to-mesenchymal transition and myofibroblastic activation. J Mol Cell Cardiol. 2015;80:175–85.
145. Bischoff J, Aikawa E. Progenitor cells confer plasticity to cardiac valve endothelium. J Cardiovasc Transl Res. 2011;4(6):710–9.
146. Han Y, et al. Reactive oxygen species in paraventricular nucleus modulates cardiac sympathetic afferent reflex in rats. Brain Res. 2005;1058(1–2):82–90.
147. Stein PD, et al. Scanning electron microscopy of operatively excised severely regurgitant floppy mitral valves. Am J Cardiol. 1989;64(5):392–4.
148. Nigam V, Srivastava D. Notch1 represses osteogenic pathways in aortic valve cells. J Mol Cell Cardiol. 2009;47(6):828–34.
149. Hofmann JJ, et al. Endothelial deletion of murine Jag1 leads to valve calcification and congenital heart defects associated with Alagille syndrome. Development. 2012;139(23):4449–60.
150. Masumura T, et al. Shear stress increases expression of the arterial endothelial marker ephrinB2 in murine ES cells via the VEGF-Notch signaling pathways. Arterioscler Thromb Vasc Biol. 2009;29(12):2125–31.
151. Theodoris CV, et al. Human disease modeling reveals integrated transcriptional and epigenetic mechanisms of NOTCH1 haploinsufficiency. Cell. 2015;160(6):1072–86.

152. Miller JD, et al. Dysregulation of antioxidant mechanisms contributes to increased oxidative stress in calcific aortic valvular stenosis in humans. J Am Coll Cardiol. 2008;52(10):843–50.
153. Weiss RM, et al. Calcific aortic valve stenosis in old hypercholesterolemic mice. Circulation. 2006;114(19):2065–9.
154. Farrar EJ, Huntley GD, Butcher J. Correction: endothelial-derived oxidative stress drives myofibroblastic activation and calcification of the aortic valve. PLoS One. 2015;10(5): e0128850.
155. Farrar EJ, Huntley GD, Butcher J. Endothelial-derived oxidative stress drives myofibroblastic activation and calcification of the aortic valve. PLoS One. 2015;10(4):e0123257.
156. Richards J, et al. Side-specific endothelial-dependent regulation of aortic valve calcification: interplay of hemodynamics and nitric oxide signaling. Am J Pathol. 2013;182(5):1922–31.
157. Hjortnaes J, et al. Valvular interstitial cells suppress calcification of valvular endothelial cells. Atherosclerosis. 2015;242(1):251–60.
158. Huk DJ, et al. Valve endothelial cell-derived Tgfβ1 signaling promotes nuclear localization of Sox9 in interstitial cells associated with attenuated calcification. Atheroscler Thromb Vasc Biol. 2016;36(2):328–38.
159. Arjunon S, et al. Aortic valve: mechanical environment and mechanobiology. Ann Biomed Eng. 2013;41(7):1331–46.
160. Braddock M, et al. Fluid shear stress modulation of gene expression in endothelial cells. News Physiol Sci. 1998;13:241–6.
161. Ishida T, et al. MAP kinase activation by flow in endothelial cells. Role of beta 1 integrins and tyrosine kinases. Circ Res. 1996;79(2):310–6.
162. Butcher JT, Simmons CA, Warnock JN. Mechanobiology of the aortic heart valve. J Heart Valve Dis. 2008;17(1):62–73.
163. Butcher JT, et al. Transcriptional profiles of valvular and vascular endothelial cells reveal phenotypic differences: influence of shear stress. Arterioscler Thromb Vasc Biol. 2006;26 (1):69–77.
164. Wylie-Sears J, et al. Mitral valve endothelial cells with osteogenic differentiation potential. Arterioscler Thromb Vasc Biol. 2011;31(3):598–607.
165. Mahler GJ, Farrar EJ, Butcher JT. Inflammatory cytokines promote mesenchymal transformation in embryonic and adult valve endothelial cells. Arterioscler Thromb Vasc Biol. 2013;33(1):121–30.
166. Dal-Bianco JP, et al. Active adaptation of the tethered mitral valve: insights into a compensatory mechanism for functional mitral regurgitation. Circulation. 2009;120(4):334–42.
167. Borer JS, Sharma A. Drug therapy for heart valve diseases. Circulation. 2015;132 (11):1038–45.
168. Sanz-Garcia A, et al. Heart valve tissue engineering: how far is the bedside from the bench? Expert Rev Mol Med. 2015;17:e16.
169. Murphy SV, Atala A. Organ engineering—combining stem cells, biomaterials, and bioreactors to produce bioengineered organs for transplantation. BioEssays. 2013;35(3):163–72.
170. Nomura A, et al. CD34-negative mesenchymal stem-like cells may act as the cellular origin of human aortic valve calcification. Biochem Biophys Res Commun. 2013;440(4):780–5.
171. Deb A, et al. Bone marrow-derived myofibroblasts are present in adult human heart valves. J Heart Valve Dis. 2005;14(5):674–8.
172. Hajdu Z, et al. Recruitment of bone marrow-derived valve interstitial cells is a normal homeostatic process. J Mol Cell Cardiol. 2011;51(6):955–65.
173. Visconti RP, et al. An in vivo analysis of hematopoietic stem cell potential: hematopoietic origin of cardiac valve interstitial cells. Circ Res. 2006;98(5):690–6.
174. Wang J, et al. Atrioventricular cushion transformation is mediated by ALK2 in the developing mouse heart. Dev Biol. 2005;286(1):299–310.
175. Ma L, Martin JF. Generation of a Bmp2 conditional null allele. Genesis. 2005;42(3):203–6.
176. Sridurongrit S, et al. Signaling via the Tgf-beta type I receptor Alk5 in heart development. Dev Biol. 2008;322(1):208–18.

177. Liu W, et al. Bmp4 signaling is required for outflow-tract septation and branchial-arch artery remodeling. Proc Natl Acad Sci U S A. 2004;101(13):4489–94.
178. Solloway MJ, Robertson EJ. Early embryonic lethality in Bmp5;Bmp7 double mutant mice suggests functional redundancy within the 60A subgroup. Development. 1999;126(8):1753–68.
179. Kim RY, Robertson EJ, Solloway MJ. Bmp6 and Bmp7 are required for cushion formation and septation in the developing mouse heart. Dev Biol. 2001;235(2):449–66.
180. Todorovic V, et al. Long form of latent TGF-beta binding protein 1 (Ltbp1L) regulates cardiac valve development. Dev Dyn. 2011;240(1):176–87.
181. Choi M, et al. The bone morphogenetic protein antagonist noggin regulates mammalian cardiac morphogenesis. Circ Res. 2007;100(2):220–8.
182. Snider P, et al. Ectopic noggin in a population of Nfatc1 lineage endocardial progenitors induces embryonic lethality. J Cardiovasc Dev Dis. 2014;1(3):214–36.
183. Moskowitz IP, et al. Transcription factor genes Smad4 and Gata4 cooperatively regulate cardiac valve development. [corrected]. Proc Natl Acad Sci U S A. 2011;108(10):4006–11.
184. Galvin KM, et al. A role for smad6 in development and homeostasis of the cardiovascular system. Nat Genet. 2000;24(2):171–4.
185. Dickson MC, et al. RNA and protein localisations of TGF beta 2 in the early mouse embryo suggest an involvement in cardiac development. Development. 1993;117(2):625–39.
186. Letterio JJ, et al. Maternal rescue of transforming growth factor-beta 1 null mice. Science. 1994;264(5167):1936–8.
187. Bartram U, et al. Double-outlet right ventricle and overriding tricuspid valve reflect disturbances of looping, myocardialization, endocardial cushion differentiation, and apoptosis in TGF-beta(2)-knockout mice. Circulation. 2001;103(22):2745–52.
188. Sanford LP, et al. TGFbeta2 knockout mice have multiple developmental defects that are non-overlapping with other TGFbeta knockout phenotypes. Development. 1997;124(13):2659–70.
189. Jiao K, et al. Tgfbeta signaling is required for atrioventricular cushion mesenchyme remodeling during in vivo cardiac development. Development. 2006;133(22):4585–93.
190. Robson A, et al. The TGFbeta type II receptor plays a critical role in the endothelial cells during cardiac development. Dev Dyn. 2010;239(9):2435–42.
191. Fischer A, et al. Combined loss of Hey1 and HeyL causes congenital heart defects because of impaired epithelial to mesenchymal transition. Circ Res. 2007;100(6):856–63.
192. Donovan J, et al. Tetralogy of fallot and other congenital heart defects in Hey2 mutant mice. Curr Biol. 2002;12(18):1605–10.
193. High FA, et al. Murine Jagged1/Notch signaling in the second heart field orchestrates Fgf8 expression and tissue-tissue interactions during outflow tract development. J Clin Invest. 2009;119(7):1986–96.
194. Chang AC, et al. A Notch-dependent transcriptional hierarchy promotes mesenchymal transdifferentiation in the cardiac cushion. Dev Dyn. 2014;243(7):894–905.
195. Watanabe Y, et al. Activation of Notch1 signaling in cardiogenic mesoderm induces abnormal heart morphogenesis in mouse. Development. 2006;133(9):1625–34.
196. Timmerman LA, et al. Notch promotes epithelial-mesenchymal transition during cardiac development and oncogenic transformation. Genes Dev. 2004;18(1):99–115.
197. Wilson CL, et al. Endothelial deletion of ADAM17 in mice results in defective remodeling of the semilunar valves and cardiac dysfunction in adults. Mech Dev. 2013;130(4–5):272–89.
198. Dupuis LE, et al. Altered versican cleavage in ADAMTS5 deficient mice; a novel etiology of myxomatous valve disease. Dev Biol. 2011;357(1):152–64.
199. Inai K, et al. BMP-2 induces versican and hyaluronan that contribute to post-EMT AV cushion cell migration. PLoS One. 2013;8(10):e77593.
200. Bowen CJ, et al. Cadherin-11 coordinates cellular migration and extracellular matrix remodeling during aortic valve maturation. Dev Biol. 2015;407:145.

201. Yu S, et al. The chemokine receptor CXCR7 functions to regulate cardiac valve remodeling. Dev Dyn. 2011;240(2):384–93.
202. Jackson LF, et al. Defective valvulogenesis in HB-EGF and TACE-null mice is associated with aberrant BMP signaling. EMBO J. 2003;22(11):2704–16.
203. Nanba D, et al. Loss of HB-EGF in smooth muscle or endothelial cell lineages causes heart malformation. Biochem Biophys Res Commun. 2006;350(2):315–21.
204. Yamazaki S, et al. Mice with defects in HB-EGF ectodomain shedding show severe developmental abnormalities. J Cell Biol. 2003;163(3):469–75.
205. Barnette DN, et al. Tgfbeta-Smad and MAPK signaling mediate scleraxis and proteoglycan expression in heart valves. J Mol Cell Cardiol. 2013;65:137–46.
206. Azhar M, et al. Transforming growth factor Beta2 is required for valve remodeling during heart development. Dev Dyn. 2011;240(9):2127–41.
207. Guo DC, et al. Mutations in smooth muscle alpha-actin (ACTA2) lead to thoracic aortic aneurysms and dissections. Nat Genet. 2007;39(12):1488–93.
208. Laforest B, Andelfinger G, Nemer M. Loss of Gata5 in mice leads to bicuspid aortic valve. J Clin Invest. 2011;121(7):2876–87.
209. Biben C, et al. Cardiac septal and valvular dysmorphogenesis in mice heterozygous for mutations in the homeobox gene Nkx2-5. Circ Res. 2000;87(10):888–95.
210. Qu XK, et al. A novel NKX2.5 loss-of-function mutation associated with congenital bicuspid aortic valve. Am J Cardiol. 2014;114(12):1891–5.
211. Andelfinger G, et al. KCNJ2 mutation results in Andersen syndrome with sex-specific cardiac and skeletal muscle phenotypes. Am J Hum Genet. 2002;71(3):663–8.
212. Quintero-Rivera F, et al. MATR3 disruption in human and mouse associated with bicuspid aortic valve, aortic coarctation and patent ductus arteriosus. Hum Mol Genet. 2015;24(8):2375–89.
213. Pepe G, et al. Identification of fibrillin 1 gene mutations in patients with bicuspid aortic valve (BAV) without Marfan syndrome. BMC Med Genet. 2014;15:23.
214. Nataatmadja M, et al. Abnormal extracellular matrix protein transport associated with increased apoptosis of vascular smooth muscle cells in marfan syndrome and bicuspid aortic valve thoracic aortic aneurysm. Circulation. 2003;108(Suppl 1):II329–34.
215. Tan HL, et al. Nonsynonymous variants in the SMAD6 gene predispose to congenital cardiovascular malformation. Hum Mutat. 2012;33(4):720–7.
216. Ewart AK, et al. Hemizygosity at the elastin locus in a developmental disorder, Williams syndrome. Nat Genet. 1993;5(1):11–6.
217. Kyndt F, et al. Mutations in the gene encoding filamin A as a cause for familial cardiac valvular dystrophy. Circulation. 2007;115(1):40–9.
218. Lardeux A, et al. Filamin-a-related myxomatous mitral valve dystrophy: genetic, echocardiographic and functional aspects. J Cardiovasc Transl Res. 2011;4(6):748–56.
219. Nuytinck L, et al. Glycine to tryptophan substitution in type I collagen in a patient with OI type III: a unique collagen mutation. J Med Genet. 2000;37(5):371–5.
220. Lohler J, Timpl R, Jaenisch R. Embryonic lethal mutation in mouse collagen I gene causes rupture of blood vessels and is associated with erythropoietic and mesenchymal cell death. Cell. 1984;38(2):597–607.
221. Li SW, et al. Transgenic mice with targeted inactivation of the Col2 alpha 1 gene for collagen II develop a skeleton with membranous and periosteal bone but no endochondral bone. Genes Dev. 1995;9(22):2821–30.
222. Kuivaniemi H, Tromp G, Prockop DJ. Mutations in fibrillar collagens (types I, II, III, and XI), fibril-associated collagen (type IX), and network-forming collagen (type X) cause a spectrum of diseases of bone, cartilage, and blood vessels. Hum Mutat. 1997;9(4):300–15.
223. Liu X, et al. Type III collagen is crucial for collagen I fibrillogenesis and for normal cardiovascular development. Proc Natl Acad Sci U S A. 1997;94(5):1852–6.
224. Wenstrup RJ, et al. COL5A1 haploinsufficiency is a common molecular mechanism underlying the classical form of EDS. Am J Hum Genet. 2000;66(6):1766–76.

225. Wenstrup RJ, et al. Murine model of the Ehlers-Danlos syndrome. col5a1 haploinsufficiency disrupts collagen fibril assembly at multiple stages. J Biol Chem. 2006;281(18):12888–95.
226. Griffith AJ, et al. Marshall syndrome associated with a splicing defect at the COL11A1 locus. Am J Hum Genet. 1998;62(4):816–23.
227. Bristow J, et al. Tenascin-X, collagen, elastin, and the Ehlers-Danlos syndrome. Am J Med Genet C Semin Med Genet. 2005;139C(1):24–30.
228. Renard M, et al. Absence of cardiovascular manifestations in a haploinsufficient Tgfbr1 mouse model. PLoS One. 2014;9(2):e89749.
229. Gallo EM, et al. Angiotensin II-dependent TGF-beta signaling contributes to Loeys-Dietz syndrome vascular pathogenesis. J Clin Invest. 2014;124(1):448–60.

# Mechanical Mediation of Signaling Pathways in Heart Valve Development and Disease

**Ishita Tandon, Ngoc Thien Lam, and Kartik Balachandran**

**Abstract** Heart valves are elegant, dynamic, pliable structures experiencing complex and varied mechanical forces during development and throughout the course of their postnatal lifetime. Recent work has shown a remarkable link between the mechanical environment of these valve leaflets and their structure, function, and biological behavior. Additionally, these mechanobiological responses, while being regulated by a similar set of signaling pathways, may be very different depending on their occurrence during valvulogenesis or during adulthood. This chapter reviews what is currently known about these differentially mechanoregulated signaling pathways in the heart valve, highlighting any gaps in knowledge.

**Keywords** Heart valve · Valve formation · Valvulogenesis · Valve disease · Mechanobiology · Signaling pathways

## 1 Introduction

Heart valves are elegant, dynamic structures experiencing complex and varied mechanical forces during development and throughout the course of their postnatal lifetime [1, 2]. These forces have a remarkable impact on the structure, function, and homeostatic signaling mechanisms within the valves [1, 3–8]. These signaling mechanisms play a major role in the proper development of heart valves by governing processes like cell turnover and phenotype, among others [1, 7, 9]. At times, mutated genes and misregulated signaling pathways lead to valvulopathies such as bicuspid aortic valve disease, valve calcification, and myxomatous valve disease [7, 10, 11]. Intriguingly, some of these regulatory signaling pathways appear to have very different consequences depending on their occurrence during development or in mature valves [12–14]. Detailed knowledge of these complex signaling

---

I. Tandon · N. T. Lam · K. Balachandran (✉)
Department of Biomedical Engineering, University of Arkansas, Fayetteville, AR, USA
e-mail: kbalacha@uark.edu

mechanisms and in particular their interaction with the dynamic mechanical environment is thus necessary for a complete understanding of the development and pathology of heart valves.

Cardiac morphogenesis begins with a tubular heart structure where a crucial process known as endothelial-mesenchymal transformation (EndoMT) takes place to give rise to cardiac cushions—the early precursors of valve leaflets known as valve primordia. These atrioventricular (AV) and outflow tract (OFT) cushions populated with mesenchymal cells are profoundly visible at just 4 weeks in utero, when the fetal heart has become a looped structure. The tissues from these cushions gradually erode and undergo transformation and maturation to give rise to rudimentary valves and subsequently the AV and semilunar (SL) valves. This transition is not yet clearly understood. These valves are composed of interstitial cells (VICs), endothelial cells (VECs). The tissue extracellular matrix (ECM), which comprises collagen as well as elastin gradually increases in content as well as anisotropy as the valve matures [15–17].

The signaling mechanisms regulating each step in the development of the heart valve, from EndoMT to the synthesis and remodeling of matrix in matured valves, include several widely studied pathways such as VEGF/NFATc1, shp2/ERK1/2, wnt/β-catenin, TGF-β, BMP, and FGF [18–22]. Pathologies arise when one or more of these pathways are misregulated, some pivotal signaling molecules are mutated, or unwanted expression of genes occurs. Intriguingly, many of these pathways have been indicated to be mechanosensitive in a wide range of organ systems, not just in the cardiac valve [20]. In particular, processes such as macromolecule transport, altered gene expression, fibrosis, and tissue calcification show direct dependency on the mechanical forces. It is therefore important to understand the mechanism by which these forces regulate the signaling pathways and how the interaction between biochemical and biomechanics—i.e., mechanobiology, affects the development and pathobiology of the heart valves. This knowledge would enable the development of better therapeutic strategies by improving upon drug designs and refining regenerative medicine approaches.

In this chapter we first describe the mechanical forces acting in the developmental and adult phases of the cardiac valves. We then describe in detail the signaling pathways involved in valvulogenesis and major diseases and how they are affected by these mechanical forces. Both congenital and adult diseases are discussed. The comparison of the pathways involved in developing and diseased conditions have been made to deepen the understanding of how a single pathway behaves differently in healthy versus diseased conditions. This chapter also highlights possible future avenues for research in this field.

## 2 The Mechanical Environment of the Developing and Adult Heart Valve

The four heart valves function to maintain unidirectional blood flow in the heart and experiences complex and varied mechanical forces [23]. This section provides a brief overview of the distinct mechanical forces acting during the developing and adult phase and their overall significance in maintaining the structure and function of the valve.

### 2.1 The Developing Heart Valve (Fig. 1)

The transformation of the fetal tubular heart structure to a developed four-chambered heart with four, fibrous well-defined heart valves is accompanied by varying active mechanical forces such as pressure, strain, shear, and also normal and circumferential forces and frictional fluid shear stress by virtue of blood flow. Initially the heart is a tubular structure which forms an asymmetric loop in three distinct stages [16]. The cardiac jelly forms into endocardial cushions and cushions function as primordial valves facilitating unidirectional flow through the heart [24]. At this stage, the

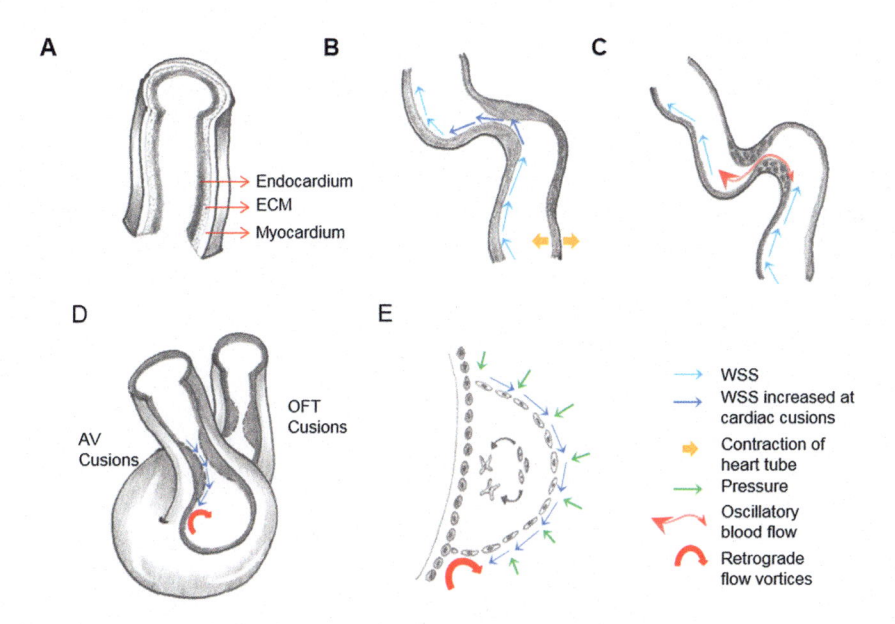

**Fig. 1** Mechanical forces acting on the heart valve during valvulogenesis. (**a–c**) Early heart tube with contractile function. (**b**) Formation of Atrio Ventricular Cushions; (**c**) Myocardial contraction produces oscillatory blood flow creating oscillatory shear stress at the AVC. (**d**) Cardiac looping; (**e**) middle stage of Valve morphogenesis showing EndoMT

pressure and fluid shear stress due to blood flow and circumferential stress due to stretching of the walls begin increasing in prominence. Trabeculation occurs to increase the surface area and compaction within the primordial myocardium, followed by septation and increased contractility [25]. It is at this point that EndoMT, a process via which endocardial endothelial cells delaminate and differentiate into mesenchymal-like cells initiates, resulting in the sprouting and formation of the tricuspid and mitral valve leaflets from AV cushions and SL valve leaflets from the endocardial cushions in the outflow tract [24]. The role of specific mechanical and structural events pre-EndoMT is still poorly understood [16, 24].

Apart from the major extrinsic forces, there are constant cell-ECM and cell-cell tractions acting within the local microenvironment. These tractions are more significant in magnitude in the fetus than in an adult, suggesting their greater importance during the developmental stage [16]. Fluid shear stress increases during the looping, especially in the endocardial cushion region, and is greatest when and where the AV and OT cushions are developing into matured valves and is known to tightly regulate the genes involved in valve remodeling [26]. Aberrant or altered blood flow patterns in these initial stages causes improper cardio- and valvulogenesis [26, 27]. Inability to respond to the stress signal has also been reported to result in malformed mitral and pulmonary valves [16, 26, 28].

Considerable experimental work has reported the importance of the mechanical environment for proper morphogenesis of the AV valves [24, 27, 28]. Studies have shown that during early embryogenesis before the convective transport begins, the tubular embryonic heart had already begun to beat. Evidently this cardiac contraction is important for proper cardiogenesis and blood flow; however, its overall significance is still not clearly understood [26]. It was noted that during the looping, and before the proper development of valve leaflets or cusps, the pressure-volume loops for the heart were reported to be quite similar to that in the fully developed hearts [25]. Indeed, it was shown that altered flow dynamics resulted in functionally and structurally impaired valve development [29–31]. Overall, intracardiac flow and contractile function create the biomechanical factors necessary for impeccable development of heart. A slight deviation from the tightly regulated conditions leads to valve defects. However these mechanisms of mechanoregulation need to be more deeply elucidated [23, 24, 30].

## 2.2 The Mature Heart Valve (Fig. 2)

Cardiac valves were initially thought to function passively under the influence of mechanical forces acting on them by the virtue of flowing blood. Later evidence showed that the pathological conditions building up in the diseased valve were mostly associated with one or more specific regions of altered hemodynamic or mechanical stresses, suggesting an intimate involvement of the external mechanical environment. It was also noted that the different cells and the ECM components in the valve dynamically alter their biological behavior in order to withstand the

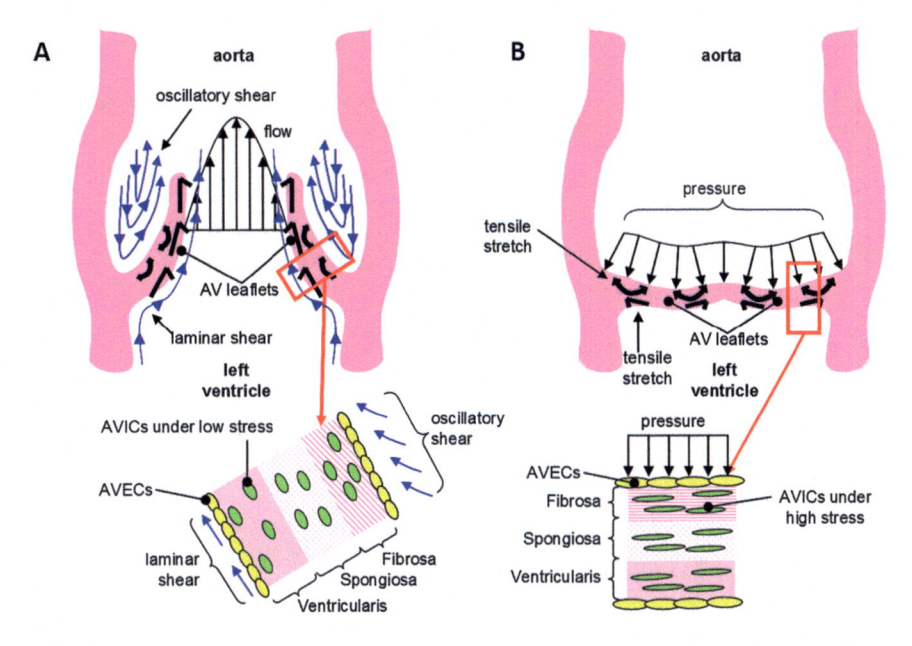

**Fig. 2** Mechanical forces on the adult valve. Mechanical forces acting on the valve in an (**a**) open and (**b**) closed position, using the aortic valve as an example. Figure adapted from [1]

mechanical forces, thereby maintaining proper function [17]. This has led to the exploration of potential link between these forces and the signaling involved in disease initiation of the valves.

The forces experienced in the mature heart valve are well known and have been extensively studied, and are thus not reviewed here in detail. While the magnitudes of the forces differ between the left- and right-sided valves, in general, the types of mechanical forces remain the same. Spatiotemporally varying pressure, membrane stress, fluid shear stress, bending strain and stress, and tensile strain and stress act cyclically for the human life span of approximately 70–90 years [1, 32]. The flow of the blood results in an alternating pulsatile shear stress on the inflow side of the heart valve when the valve is open and blood is flowing in a laminar to turbulent flow regime. The outflow side of the heart is characterized by oscillatory shear stress during valve closure, thereby stopping blood from flowing retrograde [29, 30]. There is a pressure gradient created across the valve which increases in the case of stenosis. The axial shear stress acting on the closed valve also creates a transverse pressure across the valve which is accompanied by shear and bending strain that increases in magnitude in the case of the diseased valve. Studies have shown that changes in shear stress according to evolving anatomy during growth have an impact on the cellular expression [33]. The tension and compression experienced on the convex and concave side of the valve constitute the bending stress during valve opening and closing. Axial stretch and circumferential stretch allows the valve cusps to completely close, forming a coaptive seal without allowing regurgitation. The effectiveness of this coaptation gradually decreases for the valve with aging. Intriguingly, despite the lower pressure environment, the right-sided valves experience higher axial stretches compared to the left-sided valve, likely due to their reduced thicknesses and mechanical strength [1].

## 2.3 Signaling Pathways Involved During Valve Morphogenesis (Fig. 3)

Valvulogenesis occurs during the third week of human pregnancy after the formation of the cardiac jelly. It begins with the formation of endocardial cushions in the AV canal and OFT regions of the developing heart through a process known as EndoMT [34]. It is

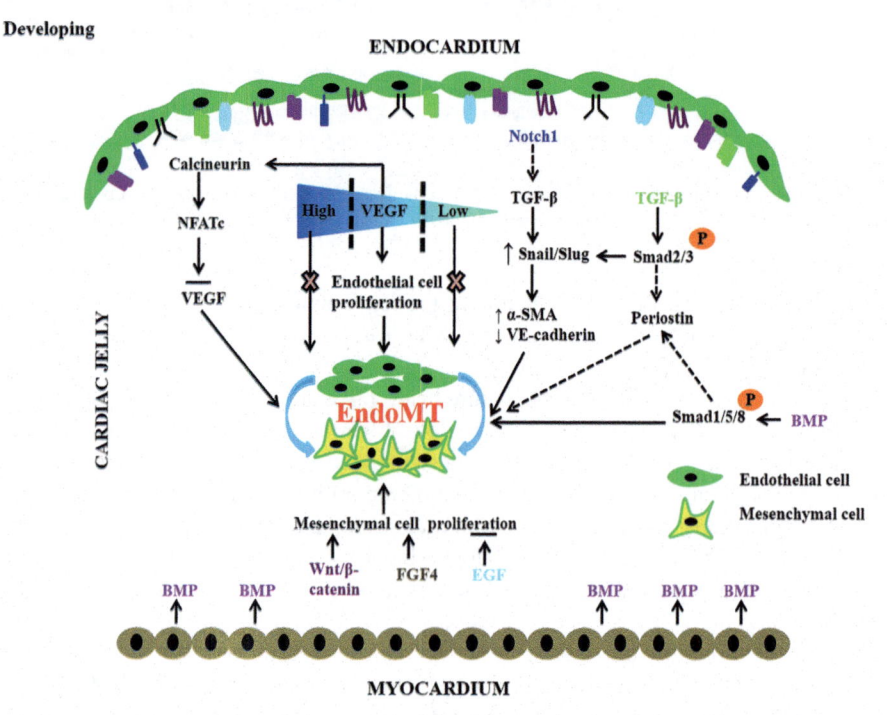

**Fig. 3** Signaling pathways in valvulogenesis. Sufficient expression of VEGF promotes endothelial cell proliferation and differentiation into mesenchymal cells. This also leads to the activation of Calcineurin which in turn activates the NFATc. NFATc suppresses VEGF expression and allows the endocardium to transform into mesenchymal cells. In the developing heart, Notch signaling increases activity of transcription factor Snail/Slug with or without the presence of TGF-β signaling (dashed line). Snail/Slug leads to downregulation of endothelial markers (VE-cadherin, Tie, Tek, and PECAM1) and upregulation of mesenchymal markers (α-SMA, PDGFR, and fibronectin), resulting in EndoMT. Similarly, after Smad2/3 is phosphorylated, TGF-β signaling acts through Snail/Slug to promote EMT while BMP signaling leads to the phosphorylation of Smad1/5/8. TGF-β and BMP2 may induce periostin expression (dashed line) and enhance EndoMT. Wnt/β-catenin and FGF4 enhance mesenchymal cell proliferation while EGF signaling prevents it

reported that the glycosaminoglycan hyaluronan, a major component of the cardiac jelly, causes the activation of ErbB2 and ErbB3 that coordinately provides activation signals for EndoMT. In response to these signals, adjacent endocardial cells become activated, lose cell-cell adhesion, gain migratory properties, invade the cardiac jelly, and transform into mesenchymal cells. These cells start to proliferate, differentiate, and remodel the ECM to further expand and elongate the endocardial cushions [34, 35]. During the last phase of valvular formation, additional ECM components are formed and the rudimentary valves are reshaped into mature leaflets with tri-layered architecture. Obviously, an orchestra of many transcriptional factors and signaling pathways has to be involved and finely tuned to ensure proper valve formation stage after stage. While the molecular regulation of cardiac valvulogenesis is studied elaborately, the role of hemodynamic signaling in valve formation is under-researched. In the context of this section, we will focus on some major signaling pathways that are shared among different stages of valve development including VEGF, NFATc, Notch, Wnt/β-catenin, BMP, TGF-β, FGF, and EGF signaling. We will also summarize some evidence that highlights the relationship between mechanical signaling and valvulogenesis.

*Vascular endothelial growth factor (VEGF)* is an important mediator of EndoMT and its expression must be tightly controlled during the initiation of EndoMT. VEGF is a signal protein that regulates angiogenesis, vasculogenesis, vascular permeability, and endothelial cell proliferation during valve development. VEGF transmits signals upon binding to its receptor complexes VEGF-R1/flt1, VEGF-R2/KDR, and neuropilin-1, which are expressed on the endothelial cell surface exclusively [18]. It was reported that VEGF is highly expressed by the myocardium and endocardium before endocardial cushion formation. However, at the initiation of EndoMT, its expression is downregulated. Surprisingly, high VEGF levels appear to terminate EndoMT [20]. Similarly, mutant mouse embryo displayed multiple cardiac abnormalities when VEGF gene expression increased two to threefold [36]. On the other hand, null mutation of Hhex, the homeobox gene, caused major defects in cardiac development due to decreased apoptosis and dysregulated EndoMT. Interestingly, in these mutated mice, the expression level of VEGF was increased threefold between E9.5 and E11.5. Addition of sFlt-1, a VEGF signaling inhibitor, to the aortic valve explants from mutated mice resulted in the recovery of normal EndoMT [37]. These observations suggest that the level of VEGF expression directly controls the EndoMT process, and only basal level of expression is required for EndoMT initiation. This suggestion is further supported in case of fetal hypoxia or hyperglycemia. Overexpression of VEGF, as seen in hypoxia, causes endothelial cells to maintain their endothelial phenotype, thus preventing cells from undergoing EndoMT. Similarly, under-expression of VEGF, as seen in hyperglycemia, results in the depletion of endothelial cells and leads to shortage of cell source for the cardiac cushions to undergo EndoMT [18]. The overall outcome if EndoMT cannot occur is that the cardiac cushions will not form. As a result, normal heart valves cannot be developed and the embryo will suffer some form of congenital valvular heart disease, underlining the importance of the strict control of VEGF signaling.

Post-EndoMT and primordial valve growth require high level of endothelial cell proliferation and ECM synthesis. This correlates to high level of expression of VEGF after the onset of EndoMT but its expression is restricted to endocardium only [38]. Studies on human pulmonary valve endothelial cells treated with VEGF show that VEGF promotes VEC proliferation through NFATc1 activation. Increased endothelial cell proliferation is not only necessary to populate the growing leaflet surface area but also to prevent endothelial cells from differentiating into mesenchymal cells [38, 39].

*NFATc1* signaling functions spatially from myocardium to endocardium to initiate valve formation and maintain embryonic valve maturation [40]. NFATc1 belongs to the NFAT family of calcium-sensitive transcription factors expressed by AV and SL endocardial cells throughout growth and remodeling [20]. Immunofluorescent staining of embryonic sections show that endocardial cells express NFATs before EndoMT of the OFTs and also during the valve and septal primordial growth. NFAT expression appears to be dispensable in EndoMT but becomes more critical during valve formation [41]. For instance, NFATc1-deficient mice have normal initiation of endocardial cushion formation and EndoMT but suffer severe ventricular septal and valve defects (i.e., valvular incompetence, stenosis, and conotruncal abnormalities) which results in death by E14.5 [42]. Another study reported that endocardial endothelial cells that have already undergone EndoMT do not stain for NFATc1 [18]. Chang et al. proposed a model to explain the role of NFATc and calcineurin signaling in valve formation. First, NFATc2, c3, and c4 in the myocardium are activated by calcineurin B which in turn causes the suppression of VEGF expression, thus allows the overlaying endocardium to transform into mesenchymal cells. CnB then activates NFATc1 in the endocardium, adjacent to the earlier myocardial site of NFAT action, to provide signals for valve elongation and refinement [43]. This model is supported by several lines of evidence. For instance, mutant embryos lacking NFATc2/c3/c4 had significantly reduced numbers of mesenchymal cells in the endocardial cushion together with aberrantly thinned myocardium, or loss of CnB in the cushion only damaged valve elongation but not EndoMT [41].

Overall, experimental results from VEGF and NFATc signaling imply that VEGF is a downstream signal of NFAT in developing and postnatal valve endothelium. Cells undergo EndoMT when low level of VEGF expression together with the activation of NFATc2, c3, and c4 occur. After that, VEGF expression is upregulated to terminate the EndoMT while NFATc1 in endocardium directs valve maturation.

*Notch signaling* is a critical regulator of cardiac development, especially in the control of endocardial cushion EndoMT. Notch1, 2, and 4 receptors and their ligands (Jag1/2 and Dll4) are expressed by endothelial cells in the AV canal and OFT before and throughout EndoMT [40]. Notch is a transmembrane receptor protein that is composed of a short extracellular region, a single transmembrane pass, and a small intracellular region. Upon binding to its ligands, Notch intracellular domain is translocated to the nucleus and modifies gene expression [44]. In the developing cardiac cushion, Notch signaling increases expression of the TGF-β pathway which in turn increases activity of transcription factor Snail. Snail causes downregulation of endothelial markers (VE-cadherin, TIE, TEK, and PECAM1) and upregulation of mesenchymal markers (α-SMA, PDGFR, and fibronectin) which initiates EndoMT overall [45]. Disruption of Notch signaling specifically reduced TGF-β2 expression in the myocardium as well its transcription factor Snail [18]. In mouse embryos that lacked expression of Notch1, failure of EndoMT was manifested by histological analysis of E9.5 AV canal endocardium which showed that cells did not lose their cell adhesion molecules and did not invade the cardiac jelly [45]. Mutations in Notch1 were shown to cause early developmental defects in the human aortic valve and lead to high chances of developing aortic valve disease [46]. These observations suggest that the Notch signaling is required for endocardial cushion endothelial cell EndoMT and it can function independently or in combination with TGF-β signaling to regulate EndoMT. Notch signaling may also play a role in valve leaflet maturation by initiating valve polarity and stratification in response to shear stress on the flow side of the valve. However, more evidence is needed to confirm this last hypothesis [20].

*Wnt/β-catenin* signaling is essential for normal heart valve development as demonstrated in multiple animal models [47–49]. During valvulogenesis, Wnt/β-catenin regulates endocardial cushion as well as OFT development and promotes valve progenitor cell proliferation. Wnt signaling acts by the binding of a Wnt-protein ligand to a frizzled family receptor which results in the translocation of β-catenin followed by the activation of β-catenin responsive genes. Zebrafish with APC mutation, a negative regulator that phosphorylates β-catenin, was able to complete gastrulation but its embryonic heart displayed multiple defects with excessive endocardial cushions. Conversely, overexpression of APC or Dickkopf 1, a Wnt inhibitor, stopped EndoMT, thus halting cushion formation. Injection of APC RNA in wild-type embryos increased the number of embryos lacking endocardial cushions [48]. Analysis of valve markers from in situ hybridization confirmed that Wnt/β-catenin signaling is active in myocardial and endocardial cells that are located in the valve-forming region [50]. In line with these observations, it is reported that Wnt signaling plays a role in promoting mesenchymal cell proliferation in the AV cushions. Wnt-9a overexpression increases cell proliferation and mesenchymal cell number in the AV canal and results in enlarged, hyperplastic, multiple AV cushion projections into the lumen. Wnt-9a appears to be an important regulator of the expansion of endocardial cushion endothelial and mesenchymal cell populations that are critical for appropriate growth and remodeling of the AV valves [51]. In late stages of valvulogenesis, Wnt signaling involves in regulating fibrosa layer maturation as well as directing cell lineages. Few initial studies include the observatory expression of Wnt3a, Wnt7b (involved in bone development), and the Wnt pathway reporter TOPGAL in remodeling mouse AV and SL valve leaflets, or the expression of periostin, an ECM protein that is expressed in cells from mesenchymal lineage, when treated avian embryo VICs with Wnt [20]. More studies are needed to elucidate the role of Wnt signaling in valve stratification.

*Bone morphogenetic proteins* act as a myocardial-derived signal required for the initiation of endocardial cushion formation and EndoMT. BMPs are members of the TGF-β superfamily, and initiate their signaling effects through cell surface receptors (Alk2, Alk3, and BMPRII) with serine/threonine kinase activity, which in turn phosphorylates downstream substrate proteins known as R-Smads. R-Smads associating with the co-mediator Smads and enter the nucleus where it functions as transcription factors to regulate gene expression [18, 20]. Several BMPs are expressed in the heart throughout development. BMP2 and BMP4 are expressed in both the AV and OFT myocardium with higher expression of BMP2 in the AV and higher BMP4 expression in the OFT [52–54]. BMP6 is expressed in the endocardium, myocardium, and AV mesenchyme while BMP7 is expressed in the OFT myocardium only [55]. Although being expressed in different locations, disruption in BMPs signaling all associates with defects in heart development. BMP-2 deficient mice had severe defects in AV cushion morphogenesis by 9.5 dpc and suffered heart failure at 10.5 dpc [52]. BMP2 is not present in healthy human valves but is significantly detectable in calcified stenotic valves [56]. Mice with targeted deletion of BMP4 showed no mesoderm formation and died between 6.5 and 9.5 dpc [57]. BMP5; BMP7 double mutants displayed some developmental defects but had normal embryos. However, double mutants missed endocardial cushions due to significant delays in heart development which was also responsible for embryonic lethality around 10.5 dpc [58]. These results highlight the important role of BMP signaling in initiation of cardiac cushion EndoMT. At early valvulogenesis, BMPs are secreted from the myocardium and signals the EndoMT through their receptors. BMP ligands interact with BMPR2 and Alks to activate Smad1/5/8. These activated Smads promote endothelial cell migration, proliferation, and differentiation into mesenchymal cells [18]. BMP signaling also plays a role in post-EndoMT AV cushion mesenchyme differentiation and maturation. Inai et al. reported that BMP-2 induced migration of mesenchymal cells as well as increased expression of an ECM protein, periostin at both mRNA and protein levels [59]. Indeed, several studies showed that BMPs are capable of promoting growth of endocardial cushions, proliferation and differentiation of valve primordial mesenchymal cell as well as maintenance of ECM architecture during valve growth via a complex network of transcription factors. For instance, Tbx20 expression in chicken embryo results in increased expression of MMP9/MMP13 while Twist1 expression causes increased expression of cell migration markers genes including periostin, cadherin 11, and MMP2 [60, 61]. BMP signaling may also contribute to the AV stratification especially at the spongiosa layer. During endocardial cushion culture, treatment of valve progenitor cells with BMP2 or FGF4 induced expression of aggrecan or tenascin proteins characteristic of cartilage or tendon cell lineage. This suggests the ability of cardiac progenitor valve cells to differentiate into different cell fates depending on exposure to BMP or FGF signaling [62, 63].

*TGF-βs* are among the first signaling molecules that are involved in heart valve development. Several TGF-β ligands and receptors expressed in the AVC and OFT suggest the importance of this family during endocardial cushion formation and EMT [20, 64]. TGF-β2 is highly expressed in mouse and chick embryos, during EndoMT within the cushion myocardium and invading mesenchymal cells [65, 66]. TGF-β1 is expressed in AV endocardium during and post-EndoMT. TGF-β3 is expressed only in cushion mesenchymal at later stages of cushion remodeling [67]. Accumulation of evidence points to the role of TGF-β2 in the initiation and cessation of endocardial cushion EndoMT and remodeling while TGF-β3 may participate in later stages of establishing valve structure and function [38, 65, 68, 69]. TGF-β signals through the binding event with their transmembrane receptors which leads to the phosphorylation and activation of cytosolic Smad2/3 proteins. Smads induce expression of the transcription factor Slug/Snail causing upregulation of mesenchymal specific genes and downregulation of endothelial specific genes, thereby promotes AV canal endothelial cell activation and invasion during EndoMT [18, 20, 38]. At later stages of valvulogenesis, TGF-β2 expression appears to be required to ensure proper reduction in mesenchymal proliferation as the valve primordia starts to mature and stratify its morphology with more diverse and specified valve cell profiles [20]. Few studies also demonstrate a synergistic cross talk between periostin, TGF-β, and BMP signaling that overall goal was to enhance EndoMT [70].

*FGF and EGF signaling* take part in later stages of valvulogenesis and they are important mediators of valve maturation. During growth of endocardial cushions and valve primordia, FGF4 stimulates proliferation of mesenchymal cells while EGF signaling inhibits this process to prevent hyperplasia [66]. EGF binds to its receptor containing EGF and ErbB to activate the tyrosine phosphatase shp2, which in turn signals ERK-mediated inhibition of proliferation by a Ras-specific pathway. Disruption of EGFR, ErbB2, or ErbB3 results in hyperplastic semilunar valves in which mesenchymal cells are present excessively. Mutant mice also show premature death [71, 72]. In addition to promoting cellular proliferation, FGF4 signaling also contributes to the maturation of valve supporting structures (chordae tendineae) by inducing expression of the tendon-related transcription factor scleraxis and tenascin [20].

## 2.4 Signaling Pathways During Valve Disease (Fig. 4)

The heart is the first organ to form and function in the embryo. Any defect in heart formation results in cardiac malformations which are characteristic of congenital heart defects (CHD). Deficits in the valvulogenesis (formation of the AV and OT valves) account for 20–25% of CHD and they can be caused by genetic or environmental factors [73, 74]. As we discussed in the previous section, many signaling pathways are involved in initiating valve formation and promoting valve maturation with precise leaflet morphology, ECM layers, and supporting structures. Thus, one could expect that disruption to any of these pathways will lead to disease, either in the form of congenital valve disease or acquired valve disease.

*Notch signaling* plays an important role in valve development, as outlined earlier. Notch1-deficient mice suffer early developmental defects in the aortic valve and progressive aortic valve disease through repression of Runx2 activity [46]. Notch2 and JAG1 mutations are associated with a spectrum of cardiovascular anomalies and they are also primary causes of Alagille syndromes. These patients exhibit pulmonary artery stenosis and hypoplasia, pulmonic valve stenosis and Tetralogy of Fallot.

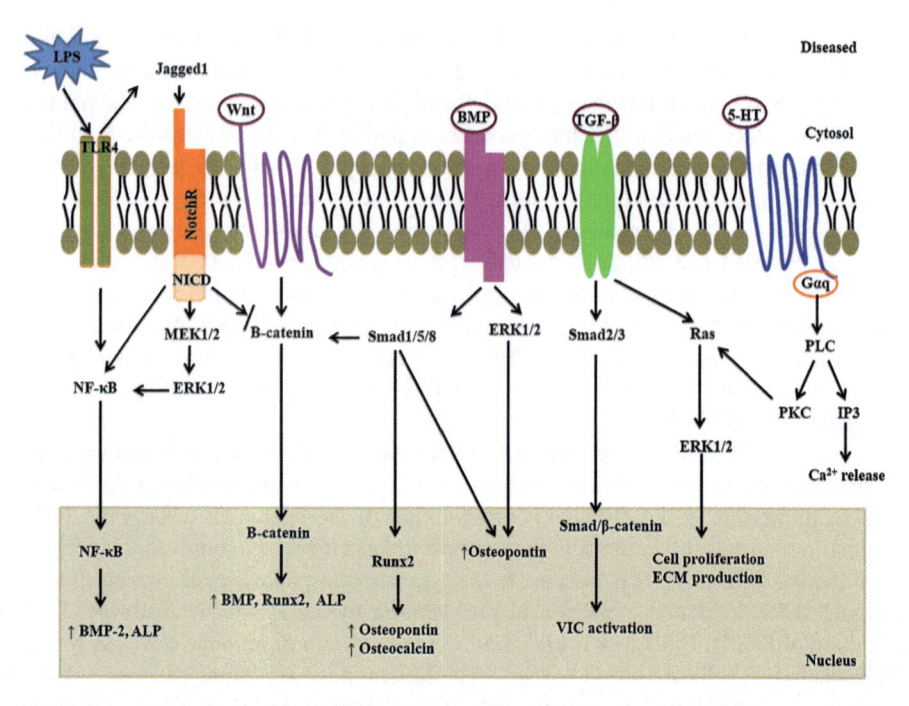

**Fig. 4** Signaling pathways in valvular heart disease. Notch1 interacts with Toll-like receptor 4 in human VIC and mediate the pro-osteogenic response to LPS stimulation through ERK1/2 and NF-κB activation. Notch1 signaling also can interfere Wnt/β-catenin stabilization and signaling, thus preventing BMP2 and Runx2 expression. BMP-2 causes upregulation of Runx2, osteopontin, and osteocalcin through activation of ERK1/2 and Smad1/5/8 pathways. TGF-β signaling induces VIC activation via Smad or Ras-pathway which results in increased ALP activity, cell proliferation, ECM biosynthesis that eventually contributes to valve calcification process. TGF-β also involves in the 5-HT-mediated signaling. Binding of 5HT to the receptor activates the Gαq-PLC-PKC pathway which leads to upregulation of TGF-β1. They act together to regulate cell proliferation and ECM production

Various studies where genes in Notch signaling pathway (Jagged1, Hesr1/Hey1, Hesr2/Hey2) were mutated demonstrated in similar cardiac effects [75]. It was found that Notch1 interacts with Toll-like receptor 4 (TLR4) in human AVICs and mediates the pro-osteogenic response to TLR4 stimulation through modulation of ERK1/2 and NF-κB activation [76]. Blockade of Dll4-Notch signaling using anti-Dll4 monoclonal antibody caused decreased atherosclerosis, plaque calcification, and fat accumulation [77]. These results demonstrate the functions of Notch signaling during valve development and that disruption of Notch signaling components can cause severe heart valve anomalies.

*BMP signaling* functions in multiple aspects of cardiogenesis from cardiogenic mesoderm specification to second heart field, heart chamber and endocardial cushion morphogenesis, epicardium development and differentiation [78]. In the previous section, we already discussed its role in valvulogenesis and how defective BMP signals leads to impaired embryonic valve formation. Such disruption will cause

CHD in most cases due to the intrinsically malformed valves. Apart from its role in heart development, BMP signaling is thought to be highly involved in calcific aortic valve disease (CAVD). Cells isolated from human stenotic aortic valves exhibited markedly increased BMP2 protein expression. Cultured cells treated with BMP2 cause upregulation of early osteogenic markers via the activation of the Smad1 and ERK1/2 pathways [79]. In fact, the expression of potent osteogenic morphogens, BMP2/BMP4 is seen in myofibroblasts and pre-osteoblasts in areas of the valves where ossification occurred [80]. In another study, the researchers observed strong expression of phospho-Smad1/5/8 in the calcified fibrosa endothelium together with low level of Smad6 expression compared to noncalcified valves [81]. These studies clearly demonstrate the role of the BMP pathway in the pathogenesis of CAVD and suggest that inhibitor of BMP signaling can be a potential treatment for CAVD.

*TGF-β signaling* regulates various biological processes, specifically in embryogenesis, development, and tissue homeostasis. Disruption of TGF-β leads to a spectrum of diseases, including hypertension, aneurysms, atherosclerosis, and cardiovascular disease [82–84]. Elevated TGF-β levels are thought to cause microfibrillar deficiency and inadequate matrix sequestration which are characteristics of genetic disorder Marfan syndrome [85]. Mutations in TGF-βR2 are also associated with Marfan syndrome type 2. TGF-β also appears to involve in myocardial infarction (MI). TGF-β-509C/T and T29C polymorphisms, for instance, are coupled with increased risk of MI in males [86, 87]. Many studies have showed that TGF-β1 promotes calcification of aortic smooth muscle cells and aortic valve interstitial cells in animal models. Cultured cells in calcified medium treated with TGF-β1 showed nodule formation coupled with significantly increased alkaline phosphatase activity and MMP2/9 expression [88]. Addition of TGF-β to culture medium induced VIC activation with significantly increased α-SMA expression as well as contractility [89]. It is well-known that myofibroblastic activation of VICs manifests as increased ECM biosynthesis, MMPs and TIMPs expression as well as increased proliferation and migration [90]. Interestingly, TGF-β1 is showed to be able to induce the activation of Wnt/β-catenin signaling through TGF-βR1 kinase activity. Without β-catenin expression, the effect of TGF-β1 on myofibroblast differentiation is inhibited [91]. Obviously, activation of TGF-β1 signaling in valve cells is associated with disruption in ECM homeostasis that eventually impairs valve structure and function. These results support the idea that TGF-β can be used as a future therapeutic target for treatment of valve diseases.

*Serotonin (5HT) signaling* has been proposed to be responsible for the valvulopathy that is observed after prolonged administration of serotonergic drugs. Elevated 5HT levels as seen in carcinoid heart disease are thought to cause excessive ECM secretion and thicken the valve leaflets due to activation of G-protein-coupled receptors [92]. Among all 5HT subtype receptors, 5HTR2A and 5HTR2B are the most frequently pharmaceutical targeted, suggesting their role in valvular pathological processes. Addition of 5HT to the aortic valve interstitial cell culture causes upregulation of TGF-β1 as well as increased collagen synthesis. Administration of 5HTR2A antagonist significantly terminated the effect of 5HT on TGF-β1 activity in a dose-dependent manner. It was also reported that 5HT increased PLC and PKC activity coupled with Gαq overexpression. These results suggest that 5HT mediates valvulopathy through a Gαq-PCL-PKC signaling pathway which leads to the upregulation of TGF-β1 [93, 94]. Other studies, in contrast, propose the important role of 5HTR2B in the maintenance of ECM homeostasis in cardiac tissues. Gq-coupled 5HTR2B receptor-null mice show incomplete cardiac development characterized by ventricular dilation, decreased systolic function, and lack of tissue integrity [95]. Treatment with 5HTR2B antagonist significantly promoted an antifibrotic environment in a mouse model by decreasing mRNA levels of TGF-β1, connective growth factor, and plasminogen activator inhibitor-1 [96]. Disruption of 5HTR2B is thought to be responsible for preventing the proliferation of VICs, EMC accumulation as well as inhibiting TGF-β1 signaling, thus preventing VICs from differentiation to an activated phenotype [92]. In conclusion, 5HTR2A/5HTR2B can be a promising strategy for future treatment of heart disease.

## 3 Mechanobiology of Valvulogenesis and Disease

Heart valves are inexorably linked to their surrounding mechanical and hemodynamic environment and vice versa. Mechanical signaling regulates valve development, tissue homeostasis, and disease, among other processes, while the valves open and close continuously to maintain their proper mechanical environment. Despite this vital connection, it is still unclear how mechanical forces modulate the multiple signaling pathways that ultimately result in homeostasis and/or disease during development and in the adult heart valve. The spatiotemporally complex combination of mechanical forces that act on and surround the heart valves makes it extremely hard for researchers to be able to simulate these conditions in vitro or in vivo. However, efforts have been undertaken to understand these mechanoresponses in simplified model benchtop systems. Here we will summarize some of the previous work that have been done to highlight the importance of mechanical signaling in contributing toward heart valve development and disease.

The most dominant mechanical stimuli during heart valve development are oscillatory flow and shear stresses generated by the blood flow [25, 28, 34]. During valvulogenesis, the retrograde flow fraction (RFF is the degree of reversing flow) is highest in the AV canal where it also accompanied with the expression of shear-related genes, including Notch1b, KLF2A, and BMP4. Klf2a-deficient zebrafish shows decreased endothelial number, abnormal cell shape as the effect of decreased RFF [97]. In fact, it has been reported that Notch1 expression only occurs in the presence of blood flow in the mouse embryo and that without Notch1, all Notch ligands and effectors cannot be upregulated by flow [98].

Many studies have indicated that primordial VECs are sensitive to shear stress which is probably due to their constant conditioning with blood flow. In response to shear stress, VECs showed decreased cell number but increased protein content [8]. Mutant mice for shear-responsive genes, including ET-1 and NOS3, displayed various cardiac defects and also suffered embryonic lethality [99, 100]. In addition, the blood flow and pressure gradient can affect the magnitude of shear stress during valvulogenesis. At the same time, valve formation is influenced by differences in wall shear stress [24]. Using a computational approach, Yalcin et al. suggested that low wall shear stress promoted the formation of myocardial trabeculation in contrast to high unidirectional shear stress, which enhances cardiac jelly expansion and EndoMT [101].

As we mentioned previously, BMP signaling is involved in aortic valve calcification and accumulation of evidence suggests that this signaling is affected by mechanical forces. 10% and 15% cyclic stretch applied to aortic valve leaflets causes significant increased BMP-2/4 expression as well as apoptosis compared to fresh samples. Addition of BMP antagonist reduced calcium content, ALP activity, Runx2, and osteocalcin expression in the valve cusps [7]. Tissue exposed to abnormal fluid shear stresses increased TGF-β1, BMP-4 expression as well as ECM degradation [102]. The addition of BMP-2 to isolated cells from stenotic human leaflets increased the expression of osteogenic markers, including Runx2 and osteopontin through either ERK1/2 or Smad1 pathway [79]. Cyclic stretch also influenced TGF-β signaling. TGF-β1 treatment in the presence of applied cyclic stretch caused increased expression of contractile and biosynthetic proteins in VICs. This phenotypic shift might result in excessive ECM synthesis that alters ECM structure, impair valve function, and eventually cause degenerative valvular disease [4]. Similarly, exposing VICs to various strain levels after treatment with TGF-β1 accelerates the nodule formation and applied strain regulates maturation of calcific nodule via apoptosis [103]. Mechanosensitivity of 5-HT signaling has also been observed in several studies. The combination of 5-HT and cyclic stretch applied to aortic valve cusps resulted in increased collagen biosynthesis, cell proliferation, and tissue stiffness. These responses are terminated by 5-HTR2A antagonist but not 5-HTR2B antagonist [104], although both receptor subtypes are significantly upregulated in response to cyclic stretch [105]. 5-HTR2B expression was also upregulated in response to mechanical stretch in pressure-induced cardiomyopathy rat model [106].

It has recently been noted that several pathways such as BMP, TGF-β, Wnt/β-catenin, and Notch1 among others play a regulatory role during development as well as disease [12]. The reasons for such differential regulation of these signaling pathways remain unknown, but could possibly involve subtle spatiotemporal variations in the mechanical environment. Indeed, a recent study reported that different EndoMT signaling pathways are switched on and off depending on the precise magnitude of applied cyclic stretching on cells [12]. A table (Table 1) has been compiled comparing the different effects of mechanosensitive signaling pathways in development and disease. Future work in this area of valve mechanobiology will greatly improve our understanding and aid in the push toward developing therapies for treating valvular heart diseases.

**Table 1** Compilation of mechanosensitive signaling pathways that are common to development and disease, their differential function and regulation, and potential causes

| | | Development | | | Disease | | | | |
|---|---|---|---|---|---|---|---|---|---|
| | | Function | Target molecules | Mechanosensitivity | Cause for disregulation/mutation | Function | Target molecules | Mechanosensitivity | Disease caused |
| 1 | VEGF | Regulate endothelial cell proliferation | IP3, ERK, NFATc1 | – Expressed only in the endocardium under the effect of unidirectional flow. | | | | | |
| 2 | NFATc | Transcriptional regulator of endothelial program | AP-1, Calcineurin | | | | | | |
| 3 | Notch | Provide initial event in the activation of EndoMT | Snail | – Notch1 active in areas with high reversing flow in the AV canal | – Mutated—Hesr1/Hey1, Hesr2/Hey2, Notch2 and JAG1 | | ERK1/2, NF-κB | | – Alagille syndrome<br>– Pulmonary artery stenosis and hypoplasia<br>– Pulmonic valve stenosis and tetralogy of Fallot |
| 4 | Wnt/β-catenin | Activate mesenchymal program | | | | | | | |
| 5 | BMP | Myocardial-derived signal for EndoMT | Smad, periostin | – BMP4 active in areas with high reversing flow in the AV canal.<br>– BMP expression only seen in the midzone between the AV cushion and myocardium. | | – Upregulation of early osteogenic markers | Smad1/5/8, ERK1/2 | Increased by abnormal fluid shear stress | – Intrinsically malformed valves, CAVD |

| 6 | TGF-β | Initiate EndoMT | Snail/Slug, Periostin | – Increased expression in the endocardium of the AV valves on the atrial inflow side. | – Upregulated TGFβ mutated TGFβR2 | – Myofibroblastic activation of VICs | Smad2/3, Ras | Increased by abnormal fluid shear stress | – Hypertension<br>– Aneurysms<br>– Atherosclerosis<br>– Marfan syndrome<br>– Myocardial infarction |
|---|---|---|---|---|---|---|---|---|---|
| 7 | FGF | Stimulate mesenchymal cell proliferation | | | | | | | |
| 8 | EGF | Inhibit mesenchymal cell proliferation | Shp2, Ras | | | | | | |
| 9 | Serotonin (5HT) | Maintenance of ECM homeostasis in cardiac tissues | | | – Prolonged administration of serotonergic drugs | – Excessive ECM secretion, thickening of the valve leaflets, increased collagen synthesis | Gαq, PKC, Ras | – 5-HTR2A and 5-HTR2B upregulated in response to cyclic stretch.<br>– 5-HTR2B upregulated in response to mechanical stretch. | |

Empty cells indicate lack of knowledge in that specific area

## 4 Conclusion

The previous two decades have witnessed tremendous growth in the subject of heart valve mechanobiology. While this growth has been driven primarily by the push toward developing a tissue-engineered heart valve, there is also considerable interest in understanding the differential role mechanical-mediated signaling plays in valve development and disease. Pushing this field forward will require novel benchtop bioreactor systems that can recapitulate the complex mechanical environment of the heart valves and also systems to prolong these cultures in the order of months and years. It is hoped that future research in this area will illuminate the myriad of ways in which the structure, biology, and function of heart valves are regulated by mechanical forces, and subsequently use this knowledge to design novel treatments.

## References

1. Balachandran K, Sucosky P, Yoganathan AP. Hemodynamics and mechanobiology of aortic valve inflammation and calcification. Int J Inflam. 2011;2011:263870.
2. Sacks MS, David Merryman W, Schmidt DE. On the biomechanics of heart valve function. J Biomech. 2009;42(12):1804–24.
3. Sacks MS, Yoganathan AP. Heart valve function: a biomechanical perspective. Philos Trans R Soc Lond B Biol Sci. 2007;362(1484):1369–91.
4. Merryman WD, et al. Synergistic effects of cyclic tension and transforming growth factor-beta1 on the aortic valve myofibroblast. Cardiovasc Pathol. 2007;16(5):268–76.
5. Merryman WD, et al. Differences in tissue-remodeling potential of aortic and pulmonary heart valve interstitial cells. Tissue Eng. 2007;13:2281.
6. Huang HY, Liao J, Sacks MS. In-situ deformation of the aortic valve interstitial cell nucleus under diastolic loading. J Biomech Eng. 2007;129(6):880–9.
7. Balachandran K, et al. Elevated cyclic stretch induces aortic valve calcification in a bone morphogenic protein-dependent manner. Am J Pathol. 2010;177(1):49–57.
8. Butcher JT, Nerem RM. Valvular endothelial cells regulate the phenotype of interstitial cells in co-culture: effects of steady shear stress. Tissue Eng. 2006;12(4):905–15.
9. Mahler GJ, Butcher JT. Inflammatory regulation of valvular remodeling: the good(?), the bad, and the ugly. Int J Inflam. 2011;2011:721419.
10. Atkins SK, et al. Bicuspid aortic valve hemodynamics induces abnormal medial remodeling in the convexity of porcine ascending aortas. Biomech Model Mechanobiol. 2014;13:1209.
11. Richards JM, et al. The mechanobiology of mitral valve function, degeneration, and repair. J Vet Cardiol. 2012;14(1):47–58.
12. Balachandran K, et al. Cyclic strain induces dual-mode endothelial-mesenchymal transformation of the cardiac valve. Proc Natl Acad Sci U S A. 2011;108(50):19943–8.
13. Markwald RR, Butcher JT. The next frontier in cardiovascular developmental biology—an integrated approach to adult disease? Nat Clin Pract Cardiovasc Med. 2007;4(2):60–1.
14. Wylie-Sears J, et al. Mitral valve endothelial cells with osteogenic differentiation potential. Arterioscler Thromb Vasc Biol. 2011;31(3):598–607.
15. Goodwin RL, et al. Three-dimensional model system of valvulogenesis. Dev Dyn. 2005;233 (1):122–9.
16. Lindsey SE, Butcher JT, Yalcin HC. Mechanical regulation of cardiac development. Front Physiol. 2014;5:318.

17. Butcher JT, Simmons CA, Warnock JN. Mechanobiology of the aortic heart valve. J Heart Valve Dis. 2008;17(1):62–73.
18. Armstrong EJ, Bischoff J. Heart valve development: endothelial cell signaling and differentiation. Circ Res. 2004;95(5):459–70.
19. Combs MD, Yutzey KE. VEGF and RANKL regulation of NFATc1 in heart valve development. Circ Res. 2009;105(6):565–74.
20. Combs MD, Yutzey KE. Heart valve development: regulatory networks in development and disease. Circ Res. 2009;105(5):408–21.
21. Zhou J, et al. Cadherin-11 expression patterns in heart valves associate with key functions during embryonic cushion formation, valve maturation and calcification. Cells Tissues Organs. 2013;198(4):300–10.
22. Moskowitz IP, et al. Transcription factor genes Smad4 and Gata4 cooperatively regulate cardiac valve development. Proc Natl Acad Sci U S A. 2011;108(10):4006–11.
23. Hinton RB, Yutzey KE. Heart valve structure and function in development and disease. Annu Rev Physiol. 2011;73:29–46.
24. Lindsey SE, Butcher JT. The cycle of form and function in cardiac valvulogenesis. Aswan Heart Cent Sci Pract Ser. 2011;2011:10.
25. Goenezen S, Rennie MY, Rugonyi S. Biomechanics of early cardiac development. Biomech Model Mechanobiol. 2012;11(8):1187–204.
26. Granados-Riveron JT, Brook JD. The impact of mechanical forces in heart morphogenesis. Circ Cardiovasc Genet. 2012;5(1):132–42.
27. Boselli F, Freund JB, Vermot J. Blood flow mechanics in cardiovascular development. Cell Mol Life Sci. 2015;72(13):2545–59.
28. Steed E, Boselli F, Vermot J. Hemodynamics driven cardiac valve morphogenesis. Biochim Biophys Acta. 2016;1863(7B):1760–6.
29. Johnson B, et al. Altered mechanical state in the embryonic heart results in time-dependent decreases in cardiac function. Biomech Model Mechanobiol. 2015;14(6):1379–89.
30. Kalogirou S, et al. Intracardiac flow dynamics regulate atrioventricular valve morphogenesis. Cardiovasc Res. 2014;104(1):49–60.
31. Tao G, Kotick JD, Lincoln J. Heart valve development, maintenance, and disease: the role of endothelial cells. Curr Top Dev Biol. 2012;100:203–32.
32. Chester AH, et al. The living aortic valve: from molecules to function. Global Cardiol Sci Pract. 2014;2014(1):52–77.
33. Back M, et al. Biomechanical factors in the biology of aortic wall and aortic valve diseases. Cardiovasc Res. 2013;99(2):232–41.
34. Riem Vis PW, et al. Environmental regulation of valvulogenesis: implications for tissue engineering. Eur J Cardiothorac Surg. 2011;39(1):8–17.
35. Camenisch TD, et al. Heart-valve mesenchyme formation is dependent on hyaluronan-augmented activation of ErbB2-ErbB3 receptors. Nat Med. 2002;8(8):850–5.
36. Miquerol L, Langille BL, Nagy A. Embryonic development is disrupted by modest increases in vascular endothelial growth factor gene expression. Development. 2000;127(18):3941–6.
37. Hallaq H, et al. A null mutation of Hhex results in abnormal cardiac development, defective vasculogenesis and elevated Vegfa levels. Development. 2004;131(20):5197–209.
38. Butcher JT, Markwald RR. Valvulogenesis: the moving target. Philos Trans R Soc Lond B Biol Sci. 2007;362(1484):1489–503.
39. Johnson EN, et al. NFATc1 mediates vascular endothelial growth factor-induced proliferation of human pulmonary valve endothelial cells. J Biol Chem. 2003;278(3):1686–92.
40. Rabkin E, et al. Activated interstitial myofibroblasts express catabolic enzymes and mediate matrix remodeling in myxomatous heart valves. Circulation. 2001;104(21):2525–32.
41. Chang CP, et al. A field of myocardial-endocardial NFAT signaling underlies heart valve morphogenesis. Cell. 2004;118(5):649–63.
42. de la Pompa JL, et al. Role of the NF-ATc transcription factor in morphogenesis of cardiac valves and septum. Nature. 1998;392(6672):182–6.

43. Lambrechts D, Carmeliet P. Sculpting heart valves with NFATc and VEGF. Cell. 2004;118 (5):532–4.
44. Andersson ER, Sandberg R, Lendahl U. Notch signaling: simplicity in design, versatility in function. Development. 2011;138(17):3593–612.
45. Timmerman LA, et al. Notch promotes epithelial-mesenchymal transition during cardiac development and oncogenic transformation. Genes Dev. 2004;18(1):99–115.
46. Garg V, et al. Mutations in NOTCH1 cause aortic valve disease. Nature. 2005;437 (7056):270–4.
47. Marvin MJ, et al. Inhibition of Wnt activity induces heart formation from posterior mesoderm. Genes Dev. 2001;15(3):316–27.
48. Hurlstone AF, et al. The Wnt/beta-catenin pathway regulates cardiac valve formation. Nature. 2003;425(6958):633–7.
49. Liebner S, et al. Beta-catenin is required for endothelial-mesenchymal transformation during heart cushion development in the mouse. J Cell Biol. 2004;166(3):359–67.
50. Gitler AD, et al. Molecular markers of cardiac endocardial cushion development. Dev Dyn. 2003;228(4):643–50.
51. Person AD, et al. Frzb modulates Wnt-9a-mediated beta-catenin signaling during avian atrioventricular cardiac cushion development. Dev Biol. 2005;278(1):35–48.
52. Ma L, et al. Bmp2 is essential for cardiac cushion epithelial-mesenchymal transition and myocardial patterning. Development. 2005;132(24):5601–11.
53. Delot EC. Control of endocardial cushion and cardiac valve maturation by BMP signaling pathways. Mol Genet Metab. 2003;80(1–2):27–35.
54. Liu W, et al. Bmp4 signaling is required for outflow-tract septation and branchial-arch artery remodeling. Proc Natl Acad Sci U S A. 2004;101(13):4489–94.
55. Kim RY, Robertson EJ, Solloway MJ. Bmp6 and Bmp7 are required for cushion formation and septation in the developing mouse heart. Dev Biol. 2001;235(2):449–66.
56. Kaden JJ, et al. Expression of bone sialoprotein and bone morphogenetic protein-2 in calcific aortic stenosis. J Heart Valve Dis. 2004;13(4):560–6.
57. Winnier G, et al. Bone morphogenetic protein-4 is required for mesoderm formation and patterning in the mouse. Genes Dev. 1995;9(17):2105–16.
58. Solloway MJ, Robertson EJ. Early embryonic lethality in Bmp5;Bmp7 double mutant mice suggests functional redundancy within the 60A subgroup. Development. 1999;126 (8):1753–68.
59. Inai K, et al. BMP-2 induces cell migration and periostin expression during atrioventricular valvulogenesis. Dev Biol. 2008;315(2):383–96.
60. Shelton EL, Yutzey KE. Tbx20 regulation of endocardial cushion cell proliferation and extracellular matrix gene expression. Dev Biol. 2007;302(2):376–88.
61. Shelton EL, Yutzey KE. Twist1 function in endocardial cushion cell proliferation, migration, and differentiation during heart valve development. Dev Biol. 2008;317(1):282–95.
62. Lincoln J, Alfieri CM, Yutzey KE. BMP and FGF regulatory pathways control cell lineage diversification of heart valve precursor cells. Dev Biol. 2006;292(2):292–302.
63. Zhao B, et al. BMP and FGF regulatory pathways in semilunar valve precursor cells. Dev Dyn. 2007;236(4):971–80.
64. Brown CB, et al. Antibodies to the type II TGFbeta receptor block cell activation and migration during atrioventricular cushion transformation in the heart. Dev Biol. 1996;174 (2):248–57.
65. Azhar M, et al. Ligand-specific function of transforming growth factor beta in epithelial-mesenchymal transition in heart development. Dev Dyn. 2009;238(2):431–42.
66. Person AD, Klewer SE, Runyan RB. Cell biology of cardiac cushion development. Int Rev Cytol. 2005;243:287–335.
67. Molin DG, et al. Expression patterns of Tgfbeta1-3 associate with myocardialisation of the outflow tract and the development of the epicardium and the fibrous heart skeleton. Dev Dyn. 2003;227(3):431–44.

68. Beffagna G, et al. Regulatory mutations in transforming growth factor-beta3 gene cause arrhythmogenic right ventricular cardiomyopathy type 1. Cardiovasc Res. 2005;65(2):366–73.
69. Hu BC, et al. The association between transforming growth factor beta3 polymorphisms and left ventricular structure in hypertensive subjects. Clin Chim Acta. 2010;411(7–8):558–62.
70. Conway SJ, Doetschman T, Azhar M. The inter-relationship of periostin, TGF beta, and BMP in heart valve development and valvular heart diseases. ScientificWorldJournal. 2011;11:1509–24.
71. Chen B, et al. Mice mutant for Egfr and Shp2 have defective cardiac semilunar valvulogenesis. Nat Genet. 2000;24(3):296–9.
72. Erickson SL, et al. ErbB3 is required for normal cerebellar and cardiac development: a comparison with ErbB2-and heregulin-deficient mice. Development. 1997;124 (24):4999–5011.
73. Hoffman JI, Kaplan S. The incidence of congenital heart disease. J Am Coll Cardiol. 2002;39 (12):1890–900.
74. Loffredo CA. Epidemiology of cardiovascular malformations: prevalence and risk factors. Am J Med Genet. 2000;97(4):319–25.
75. Huang JB, et al. Molecular mechanisms of congenital heart disease. Cardiovasc Pathol. 2010;19(5):e183–93.
76. Zeng Q, et al. Notch1 promotes the pro-osteogenic response of human aortic valve interstitial cells via modulation of ERK1/2 and nuclear factor-kappaB activation. Arterioscler Thromb Vasc Biol. 2013;33(7):1580–90.
77. Fukuda D, et al. Notch ligand delta-like 4 blockade attenuates atherosclerosis and metabolic disorders. Proc Natl Acad Sci U S A. 2012;109(27):E1868–77.
78. Wang J, Greene SB, Martin JF. BMP signaling in congenital heart disease: new developments and future directions. Birth Defects Res A Clin Mol Teratol. 2011;91(6):441–8.
79. Yang X, et al. Bone morphogenic protein 2 induces Runx2 and osteopontin expression in human aortic valve interstitial cells: role of Smad1 and extracellular signal-regulated kinase 1/2. J Thorac Cardiovasc Surg. 2009;138(4):1008–15.
80. Mohler ER 3rd, et al. Bone formation and inflammation in cardiac valves. Circulation. 2001;103(11):1522–8.
81. Ankeny RF, et al. Preferential activation of SMAD1/5/8 on the fibrosa endothelium in calcified human aortic valves—association with low BMP antagonists and SMAD6. PLoS One. 2011;6 (6):e20969.
82. ten Dijke P, Arthur HM. Extracellular control of TGFbeta signalling in vascular development and disease. Nat Rev Mol Cell Biol. 2007;8(11):857–69.
83. Pardali E, Goumans MJ, ten Dijke P. Signaling by members of the TGF-beta family in vascular morphogenesis and disease. Trends Cell Biol. 2010;20(9):556–67.
84. Doetschman T, et al. Transforming growth factor beta signaling in adult cardiovascular diseases and repair. Cell Tissue Res. 2012;347(1):203–23.
85. Benke K, et al. The role of transforming growth factor-beta in Marfan syndrome. Cardiol J. 2013;20(3):227–34.
86. Koch W, et al. Association of transforming growth factor-beta1 gene polymorphisms with myocardial infarction in patients with angiographically proven coronary heart disease. Arterioscler Thromb Vasc Biol. 2006;26(5):1114–9.
87. Yokota M, et al. Association of a T29-->C polymorphism of the transforming growth factor-beta1 gene with genetic susceptibility to myocardial infarction in Japanese. Circulation. 2000;101(24):2783–7.
88. Clark-Greuel JN, et al. Transforming growth factor-beta1 mechanisms in aortic valve calcification: increased alkaline phosphatase and related events. Ann Thorac Surg. 2007;83 (3):946–53.
89. Walker GA, et al. Valvular myofibroblast activation by transforming growth factor-beta: implications for pathological extracellular matrix remodeling in heart valve disease. Circ Res. 2004;95(3):253–60.

90. Liu AC, Joag VR, Gotlieb AI. The emerging role of valve interstitial cell phenotypes in regulating heart valve pathobiology. Am J Pathol. 2007;171(5):1407–18.
91. Chen JH, et al. Beta-catenin mediates mechanically regulated, transforming growth factor-beta1-induced myofibroblast differentiation of aortic valve interstitial cells. Arterioscler Thromb Vasc Biol. 2011;31(3):590–7.
92. Hutcheson JD, et al. Serotonin receptors and heart valve disease—it was meant 2B. Pharmacol Ther. 2011;132(2):146–57.
93. Jian B, et al. Serotonin mechanisms in heart valve disease I: serotonin-induced up-regulation of transforming growth factor-beta1 via G-protein signal transduction in aortic valve interstitial cells. Am J Pathol. 2002;161(6):2111–21.
94. Xu J, et al. Serotonin mechanisms in heart valve disease II: the 5-HT2 receptor and its signaling pathway in aortic valve interstitial cells. Am J Pathol. 2002;161(6):2209–18.
95. Nebigil CG, et al. Ablation of serotonin 5-HT(2B) receptors in mice leads to abnormal cardiac structure and function. Circulation. 2001;103(24):2973–9.
96. Fabre A, et al. Modulation of bleomycin-induced lung fibrosis by serotonin receptor antagonists in mice. Eur Respir J. 2008;32(2):426–36.
97. Vermot J, et al. Reversing blood flows act through klf2a to ensure normal valvulogenesis in the developing heart. PLoS Biol. 2009;7(11):e1000246.
98. Gordon WR, et al. Mechanical Allostery: evidence for a force requirement in the proteolytic activation of notch. Dev Cell. 2015;33(6):729–36.
99. Yanagisawa H, et al. Dual genetic pathways of endothelin-mediated intercellular signaling revealed by targeted disruption of endothelin converting enzyme-1 gene. Development. 1998;125(5):825–36.
100. Kurihara Y, et al. Aortic arch malformations and ventricular septal defect in mice deficient in endothelin-1. J Clin Invest. 1995;96(1):293–300.
101. Yalcin HC, et al. Hemodynamic patterning of the avian atrioventricular valve. Dev Dyn. 2011;240(1):23–35.
102. Sun L, Sucosky P. Bone morphogenetic protein-4 and transforming growth factor-beta1 mechanisms in acute valvular response to supra-physiologic hemodynamic stresses. World J Cardiol. 2015;7(6):331–43.
103. Fisher CI, Chen J, Merryman WD. Calcific nodule morphogenesis by heart valve interstitial cells is strain dependent. Biomech Model Mechanobiol. 2013;12(1):5–17.
104. Balachandran K, et al. Elevated cyclic stretch and serotonin result in altered aortic valve remodeling via a mechanosensitive 5-HT(2A) receptor-dependent pathway. Cardiovasc Pathol. 2012;21(3):206–13.
105. Balachandran K, et al. Aortic valve cyclic stretch causes increased remodeling activity and enhanced serotonin receptor responsiveness. Ann Thorac Surg. 2011;92(1):147–53.
106. Liang YJ, et al. Mechanical stress enhances serotonin 2B receptor modulating brain natriuretic peptide through nuclear factor-kappaB in cardiomyocytes. Cardiovasc Res. 2006;72(2):303–12.

# Tissue Engineered Heart Valves

Jay M. Reimer and Robert T. Tranquillo

**Abstract** Tissue engineered heart valves are being developed in order to provide an alternative prosthetic valve to patients suffering from valvular heart disease. They aim to address the limitations of currently existing bioprosthetic and mechanical heart valves, which have a limited functional life or require lifelong anticoagulation, respectively. Tissue engineered valves generally consist of three parts: a biodegradable polymeric scaffold for initial structural integrity and cell attachment sites, entrapped or seeded cells that remodel that biodegradable scaffold, and stimulation paradigms to direct cellular activity (especially production of a functional extracellular matrix). In vitro functional testing is useful to assess valve designs based on their hydrodynamic performance under physiologic pressure and flow conditions. However, in vivo testing is crucial since tissue engineered heart valves aim to provide a living valve capable of cell-mediated repair, remodeling, and growth. The aforementioned considerations comprise the focus of this chapter.

**Keywords** Heart valve engineering · Tissue engineered heart valves · Biopolymer scaffold · Synthetic polymer scaffold · Animal model · Pulse duplicator

## 1 Background

Valvular heart disease, a widespread disease that has been estimated to afflict ~2.5% of the US population, limits the efficient flow of blood through the heart due to improper valve opening or closing [1]. Incomplete valve opening due to reduced leaflet mobility is referred to as valve stenosis. Valve regurgitation occurs when the leaflets fail to coapt, such as leaflet prolapse, during diastole; this allows blood to flow backward through the valve. Valve dysfunction (regurgitation and/or stenosis) can occur progressively or result from a specific event. Most commonly, patients

---

J. M. Reimer · R. T. Tranquillo (✉)
Department of Biomedical Engineering, University of Minnesota, Minneapolis, MN, USA
e-mail: tranquillo@umn.edu

|  |  | All Valves | | Aortic Valve | | Mitral Valve | | Pulmonary Valve | | Tricuspid Valve | |
|---|---|---|---|---|---|---|---|---|---|---|---|
|  |  | Replace | Repair | Replace | Repairs | Replace | Repairs | Replace | Repairs | Replace | Repairs |
| All ages | | 91,803 | 18,689 | 73,679 | 2,279 | 15,981 | 14,646 | 1,418 | 375 | 735 | 1,389 |
|  |  | 83.1% | 16.9% | 66.7% | 2.1% | 14.5% | 13.3% | 1.3% | .3% | .7% | 1.3% |
| Age group | "Pediatric" <1 | 57 | 681 | * | 85 | * | 75 | 57 | 251 | * | 270 |
|  |  | 0.1% | 0.6% |  | 0.1% |  | 0.1% | 0.1% | 0.2% |  | 0.2% |
|  | 1-17 | 982 | 901 | 352 | 322 | 118 | 260 | 512 | 96 | * | 223 |
|  |  | 0.9% | 0.8% | 0.3% | 0.3% | 0.1% | 0.2% | 0.5% | 0.1% |  | 0.2% |
|  | "Adult" 18-44 | 5,715 | 1,662 | 3,629 | 373 | 1,204 | 1,078 | 624 | * | 258 | 211 |
|  |  | 5.2% | 1.5% | 3.3% | 0.3% | 1.1% | 1.0% | 0.6% |  | 0.2% | 0.2% |
|  | 45-64 | 24,879 | 7,099 | 18,694 | 740 | 5,818 | 6,062 | 157 | * | 210 | 297 |
|  |  | 22.5% | 6.4% | 16.9% | 0.7% | 5.3% | 5.5% | 0.1% |  | 0.2% | 0.3% |
|  | 65-84 | 53,279 | 7,871 | 44,702 | 706 | 8,390 | 6,808 | * | * | 205 | 357 |
|  |  | 58.1% | 7.1% | 40.5% | 0.6% | 7.6% | 6.2% |  |  | 0.2% | 0.3% |
|  | 85+ | 6,694 | 411 | 6,253 | 53 | 441 | 358 | * | * | * | * |
|  |  | 6.1% | 0.4% | 5.7% | 0.0% | 0.4% | 0.3% |  |  |  |  |

| ICD-9-CM Codes Used | Aortic Valve | Mitral Valve | Pulmonary Valve | Tricuspid Valve |
|---|---|---|---|---|
| Replacement | 35.21, 35.22,35.05, 35.06 | 35.23, 35.24 | 35.25, 35.26 | 35.27, 35.28 |
| Repair | 35.01, 35.11 | 35.02, 35.12 | 35.03, 35.13 | 35.04, 35.14 |

**Fig. 1** 2011 Valvular heart disease hospital discharges stratified by procedure type (repair or replacement) and patient age. Bold numbers = Total number of patients for a given subset. Percentages = Number of given subset to the total number of valve procedures (repair + replacement) (Adapted from 2011 HCUPnet data). Note: sum of all age groups does not necessarily equal "all ages" group due to the associated standard errors

undergo progressive valve deterioration as they age due to leaflet calcification, hypertension, and/or atherosclerosis. A subset of patients suffers from congenital cardiovascular defects, which affect valve performance at a much younger age.

Congenital cardiovascular defects are estimated to affect at least 40,000 infants each year in the USA alone and ~25% of these patients (240/100,000 live births [LBs]) require an invasive treatment within their first year of life [1, 2]. Defects affecting heart valves include bicuspid aortic valves (1370/100,000 LBs), Tetralogy of Fallot (40/100,000 LBs), pulmonary stenosis (60/100,000 LBs), pulmonary atresia (7–8/100,000 LBs), truncus arteriosus (7/100,000 LBs), and tricuspid atresia/stenosis (6.7/100,000 LBs) [1, 3, 4]. Tetralogy of Fallot encompasses multiple defects including a ventricular septal defect, pulmonary stenosis, an overriding aortic valve to the right, and right ventricular hypertrophy [5]. Atresia is a more extreme version of stenosis in which the valve opening is abnormally closed or absent.

Dysfunctional valves can sometimes be repaired (nearly 19,000 in the USA in 2011 alone), but are more commonly replaced, ~83% of the time (nearly 92,000 in the USA in 2011), with prosthetic valves to restore heart function. The aortic valve most commonly requires intervention, followed by the mitral, pulmonary, and tricuspid valves. The total number of heart valve procedures—stratified by procedure type, age, and valve position—can be seen in Fig. 1. For younger patients, congenital defects (such as Tetralogy of Fallot) can sometimes be repaired, but pulmonary regurgitation may persist. This is typically well tolerated initially, but late pulmonary valve replacement is increasingly used due to the effects of prolonged pulmonary insufficiency [6–9]. These patients present a unique challenge for replacement valves as the child undergoes anatomical growth during their maturation.

## 1.1 Existing Therapies

Two general heart valve replacement options exist currently: bioprosthetic and mechanical heart valves [10, 11]. These replacement valves are differentiated by their materials and design. There are two standard replacement procedures for prosthetic valves: surgical and transcatheter. Each valve and procedure type has advantages and disadvantages, which affect their selection for use in diseased patients [12–18].

### 1.1.1 Replacement Procedures

Surgical valve replacement, first developed in the early 1960s, traditionally includes opening the patient's chest cavity, putting the patient on bypass, and replacing the afflicted valve. More recently, transcatheter valve replacement has been developed, which utilizes specially designed, expandable bioprosthetic valves. This technique is much less invasive than traditional surgical valve replacement. Access to the defective valve is most commonly made through the femoral artery, but transapical and transaortic approaches have also been described. Since transcatheter valve replacement is relatively new, it is not yet indicated for all patient subsets despite some promising outcomes. Longer term patient monitoring to collect additional data will be needed in order to better understand the safety and efficacy of transcatheter versus surgical valve replacement.

### 1.1.2 Existing Replacement Valves

Bioprosthetic heart valves are derived from biological tissues whereas mechanical valves utilize inert materials such as pyrolytic carbon or titanium. Mechanical heart valves are thus typically very durable, and can maintain function after 30 years, but require aggressive anticoagulation to remain patent and to prevent thromboembolism [19]. Current anticoagulation regimens require patient lifestyle changes—such as routine blood tests and dietary restrictions (avoiding vitamin K)—to ensure that therapeutic levels are maintained [14, 19, 20]. Furthermore, anticoagulation is associated with numerous side effects, including bleeding or hemorrhaging [13, 16, 19].

Bioprosthetic heart valves are generally more hemocompatible, but tend to perform adequately for only 10–20 years [13, 21]. Tissue sources for bioprosthetic heart valves include bovine tissue, porcine tissue, and a limited supply of human cadaveric valves [22]. Tissues obtained for bioprosthetic heart valves are typically rendered inert by chemical fixation (nonhuman tissues) or cryopreservation, which can negatively affect the tissue's microstructure and in vivo durability [23–25]. Specifically, these treatments can limit host cell invasion, which is necessary for matrix repair and regeneration [23]. This deficiency limits the functional life of bioprosthetic heart valves and makes it more likely that patients younger than 65 will likely require a reoperation to replace a worn out prosthetic heart valve.

The tissues for bioprosthetic valves are often sewn onto a rigid frame or an expandable stent (transcatheter valves), but can also be implanted without either. The tissues for transcatheter valves are the same as those used for traditional bioprosthetic valves and are thus not expected to have a longer functional lifetime. Given that transcatheter valves are relatively new, they have been more extensively used in patients who are considered at higher risk for traditional surgical valve replacement. However, the use of transcatheter valves is expected to grow in the future, pending continued demonstration of their safety and efficacy.

Reoperations to replace defective prosthetic valves are associated with a number of risks, including mortality [13, 17, 26]. Addressing bioprosthetic valve failure is challenging, although valve-in-valve procedures have been reported in which a second valve is inserted within the defective bioprosthetic valve [27–29]. However, concerns remain regarding the valve positioning, coronary obstruction, and higher pressure gradients due to a narrowed valve orifice [29].

### 1.1.3 Replacement Heart Valve Selection

The particular prosthetic valve and replacement procedure used depends on factors such as current disease status, expected patient lifespan, comorbidities, patient preferences, and associated costs [13, 19]. In general, mechanical valves are selected when the patient is expected to outlive the functional life of bioprosthetic valves or unable to withstand the anticoagulation regimen required for mechanical heart valve replacement. Valve selection for pediatric patients is especially problematic since neither mechanical nor bioprosthetic valves are capable of growth. Thus, these patients often must undergo one or more reoperations to replace outsized prosthetic valves. Additionally, the chemically treated and cryopreserved tissues used are prone to calcification in children, which can result in valve failure.

### 1.1.4 Emerging Replacement Valves

Decellularization has also been extensively explored as a way to reduce the immunogenicity of allogeneic (human) and xenogeneic (nonhuman) tissues without cryopreservation or chemical fixation [24, 25, 30–33]. There are several published clinical trials that evaluated the use of decellularized, non-cryopreserved human heart valves in the pulmonary or aortic position [24, 34–36]. Although the initial results have been promising, their long-term function remains unproven and they are limited in supply due to donor shortage [22, 25, 31, 32, 35, 37, 38]. Decellularized xenogeneic valves have also been investigated extensively in animal models [31, 33, 39]. Although some human data has been reported, the study noted a high incidence of conduit failure within 2 years of implantation [40]. Furthermore, explant histology showed minimal host cell invasion, apart from the presence of inflammatory giant-type cells on the lumenal surface. Providing various sizes of decellularized native valves can be challenging, particularly for allogeneic valves where there are a limited number of donors.

A more detailed description of decellularized xenogeneic or allogeneic valves will not be discussed in this chapter, despite considering them to be tissue engineered heart valves (TEHVs). Additional information on these approaches can be found elsewhere and in the citations provided above. Instead, the remainder of this chapter will focus on TEHVs that are not derived from native valves.

## 2 Tissue Engineered Heart Valves

Tissue engineered heart valves are being developed to address the shortcomings of current and emerging replacement options. In order to be a viable replacement option, TEHVs must first demonstrate excellent valve function (i.e., minimal regurgitation and low systolic pressure drops) under physiological pressures and flowrates. Additional design criteria include high durability, hemocompatibility, immunocompatibility, and capacity for growth (for pediatric applications). To date, most TEHV designs have focused on imitating the function rather than the form of native heart valves. For example, most tissues used for TEHVs consist of a single layer as opposed to the tri-layer tissue found in native heart valve leaflets.

Various approaches have been described to generate an engineered tissue suitable for a heart valve application. TEHVs typically consist of a degradable scaffold and cells, which are often exposed to chemical and/or mechanical stimulation to produce a desired response, such as collagen secretion. Although differing in various aspects, these approaches all aim to generate a "living" tissue capable of repairing and remodeling itself in vivo. The hope for such TEHVs is a longer functional life and/or somatic growth (pediatric patients). Reported studies have utilized various cell types, scaffold materials, mechanical/chemical stimuli, and valve designs [41]. Components reported in the development of engineered tissues for TEHVs are discussed below.

### 2.1 Cell Sources

The primary responsibility for cells in heart valve tissue engineering is to remodel the starting polymer scaffold and secrete extracellular matrix components such as collagen, elastin, and/or proteoglycans. The cells' specific roles within the engineered tissue guide their selection. A universal cell type has yet to be identified or adopted for heart valve tissue engineering applications, though most researchers use a cell type that is easily sourced and can be easily expanded in vitro. Commonly used cell types include smooth muscle cells (SMCs), myofibroblasts, dermal fibroblasts, and mesenchymal stem cells (MSCs) [42, 43].

The cells generally fall into one of three categories depending on how they are obtained: autologous (from the same patient), allogeneic (from the same species), or xenogeneic (from a different species). Another method avoids isolating cells altogether and instead relies on in situ cell invasion and extracellular matrix deposition [44, 45]. In one specific approach, valve molds were implanted subdermally for several weeks, during which time host cells invaded and secreted extracellular matrix proteins [46]. Another group implanted an electrospun scaffold into the designated valve position and allowed host cells to repopulate this matrix [47]. This nontraditional method avoids prolonged in vitro culture but is dependent on appropriate cell recruitment and tissue formation, which may vary between individuals. Other challenges include recruiting circulating cells, guiding their cell fate and behavior, and controlling tissue formation [45].

Historically for cardiovascular tissue engineering, and heart valve tissue engineering in particular, it has been important to consider the thrombogenicity and immunogenicity of the engineered tissue. These risks can be attenuated by using autologous cell sources. However, obtaining and expanding autologous cells can be a lengthy and arduous task, especially considering that these cells could behave differently depending on the individual and they may express a diseased phenotype. This process can be further complicated if a valve replacement is needed quickly, as can be the case for pediatric patients suffering from congenital defects.

### 2.1.1 Decellularization

The advent of decellularization has had an impact on heart valve tissue engineering. This process enables researchers to effectively remove the cellular components from a tissue, while leaving the extracellular matrix intact [48]. Acellular, non-fixed TEHVs offer several advantages over their cellularized counterparts. First, decellularization allows TEHVs to be stored for longer periods of time prior to use, which is important for commercialization. Additionally, decellularization, if performed properly, removes the antigenic cellular components and would enable researchers to use allogeneic and possibly xenogeneic cells to batch-produce heart valve tissues. An in-depth description of decellularization protocols is outside the scope of this report, but there are several literature reviews that can provide more detail [49]. In general, these protocols use a combination of zwitterionic, ionic, or nonionic detergents in addition to extensive rinsing or perfusion.

## 2.1.2 Recellularization

TEHVs rely on cells to remodel the matrix and respond to physiological growth cues over time. One factor to consider is that decellularized TEHVs must rely on host cell invasion and proliferation for tissue regeneration and growth. This recellularization process takes time and thus the acellular engineered tissue must also be able to withstand the in vivo environment without cells until sufficient cell invasion occurs. The invaded cells also must not grossly remodel the tissue such that the valve geometry and/or its function are compromised. The recruitment of endothelial cells to cover the blood-contacting surfaces is particularly important, as this will reduce the thrombogenicity of the TEHVs.

## 2.2 Scaffold Materials

Polymeric scaffolds are an integral part of engineered tissues and provide initial mechanical support, an initial geometry, and attachment sites/cues for cells. These scaffolds are typically temporary and degrade as the cells secrete their own extracellular matrix. It is important to consider the initial mechanical strength of the scaffold and its degradation rate so that physical support for the cellular component of the engineered tissue is maintained during maturation. Additionally, the scaffold should be non-immunogenic and non-thrombogenic. It must also promote cellular attachment and interaction, and its degradation by-products should not elicit an inflammatory response [42, 50]. The two most common sources of scaffold materials for TEHVs are grouped into two categories: synthetic polymers and biopolymers. Recently, a hybrid scaffold has also been described that is unique from the two traditional categories.

### 2.2.1 Synthetic Polymer Scaffolds

Synthetic polymers are frequently used for heart valve tissue engineering. Primary advantages of this scaffold type are that they are widely available and can be configured to nearly any microscopic morphology and macroscopic geometry. Additionally, synthetic polymers can have a large range of mechanical and chemical properties based on how they are synthesized. Fabrication methods include salt-leaching, rapid prototyping, electrospinning, and phase-separation [42].

Synthetic polymers do not inherently contain biological signaling components like those found in biological polymers, although researchers have the ability to incorporate bioactive components [50]. Cell engraftment can be achieved by directly seeding cells onto the synthetic scaffold or using a cell carrier, such as fibrin. Controlling the polymer degradation rate is also important since it provides mechanical support and attachment sites for cells. Accelerated or slow degradation can affect the in vitro remodeling and maturation of the tissue. Additionally, if polymer degradation isn't completed prior to implantation, its degradation by-products must not cause deleterious effects on adjacent cells [51].

One of the first reports of a TEHV made from a synthetic scaffold used a polyglycolic acid (PGA)–polylactic acid (PLA) copolymer layered between nonwoven PGA [52]. Later, the use of polyhydroxyoctanoate (PHO) was explored to remedy problems associated with high scaffold stiffness, but ultimately proved to be less than ideal because of its slow degradation profile [53, 54]. PGA coated with poly-4-hydroxybutyrate (P4HB) has a more favorable degradation profile and has seen more extensive use [55, 56]. In general, most synthetic scaffolds for heart valve tissue engineering have been derived from aliphatic polyesters, polyhydroxyalkanoates, or combinations of the two [42, 57].

### 2.2.2 Biopolymer Scaffolds

Biopolymer scaffolds naturally incorporate biological signaling components, which are important to modulate cell attachment and activity [42, 50]. These materials can be obtained autologously to minimize immunocompatibility concerns, but this can introduce variability among the scaffolds, including purity and degradation rate, and it could limit commercialization and clinical use. For these reasons, xenogeneic sources of biopolymers are more commonly used in cardiovascular tissue engineering approaches. Another challenge associated with biopolymers is that they lack the initial mechanical strength required to withstand physiologic conditions. This has prompted researchers to explore various culture conditions and bioreactors (which will be discussed in more detail later), to improve the mechanical properties of the TEHVs.

Several biopolymer sources, particularly collagen and fibrin, have been explored for use in heart valve tissue engineering [42, 50, 57]. Unlike with synthetic polymers, cells can be directly entrapped into these scaffolds because polymerization occurs in an aqueous solution at physiological pH and temperature. The cell and monomer solutions are mixed together prior to polymerization to facilitate cell distribution throughout the resulting hydrogel. This method of cell seeding is advantageous because it enables researchers to incorporate cells directly, and homogeneously, into their nascent engineered tissue.

Type I collagen is a natural choice as a biological scaffold because it is a major extracellular matrix protein found in cardiovascular tissues, including heart valves. Collagens are homologous across species and thus are biocompatible and weakly immunogenic [58, 59]. However, cells cultured in type I collagen gels exhibited lower collagen synthesis and cell proliferation compared to fibrin gels [59–61].

Fibrin has been an attractive biopolymer for heart valve tissue engineering because cells are able to proliferate more and produce more collagen and total protein than in collagen gels [62–64]. One challenge for the use of fibrin is controlling its degradation rate, as it can be enzymatically degraded rapidly depending on the cell type and in vitro conditions [62]. This degradation must be matched with the deposition of cell-produced collagen to provide sufficient mechanical properties to the maturing engineered tissue.

### 2.2.3 Hybrid Scaffolds

Hybrid scaffolds are an alternative source to synthetic and biopolymeric scaffolds. These scaffolds have been recently described and are a combination of nitinol and cells seeded with [65] or without [66] a biological material (collagen). Other reports have previously demonstrated the capability of fabricating nitinol-based heart valves [67, 68].

The hybrid scaffolds are fabricated by coating the metal meshes with cell-seeded collagen gels, which produce matrix proteins over a period of multiple weeks. These scaffolds are inherently strong and are capable of withstanding physiological pressure gradients due to the incorporation of the metal mesh [67, 69]. Although the biological scaffold material or the cells are not the primary load-bearing component, they serve to improve the biocompatibility of the TEHV [70].

One of the major issues using this type of scaffold is cell–metal interface and ensuring that there is proper attachment and cellular responses to these materials. Hybrid scaffolds are also not optimal for pediatric patients since they are inert and lack the ability to grow with the patient. Apart from these concerns, the in vivo function of TEHVs with hybrid scaffolds has yet to be assessed [41].

## 2.3 Stimulation Paradigms

The conditions imposed upon the nascent tissue have a significant impact on cellular function and ultimately tissue development. Most engineered tissues undergo some sort of mechanical conditioning to generate a mechanically robust and mature tissue in order to withstand physiologic conditions. This maturation process, particularly for approaches that use biopolymer scaffolds to grow connective tissues, often focuses on biochemical and mechanical stimulation paradigms to enhance cellular collagen synthesis. Biochemical agents are most commonly added to the culture medium periodically and have demonstrated the ability to stimulate collagen synthesis in vitro. These agents include fibroblast growth factor (FGF), transforming growth factor (TGF), plasmin, insulin, and ascorbic acid [71–73].

Cells are able to sense mechanical stimuli via mechanotransduction, which can be utilized in order to induce a desired cellular response [74, 75]. Most commonly, researchers mechanically stretch the engineered tissue to stimulate cellular collagen production. These systems are often capable of applying complex strain and shear stress patterns in the constructs. Regardless of the design, bioreactors must meet certain requirements such as adequate gas exchange, nutrient delivery, and sterility. For heart valve tissue engineering, bioreactors that replicate physiological pulsatile pressures or utilize controlled cyclical stretch have been described [76–88]. The

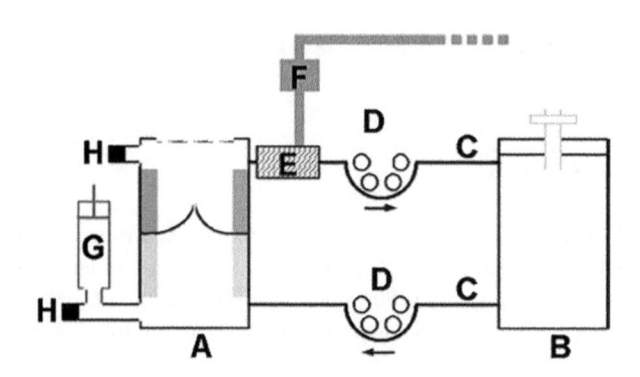

Fig. 2 Schematic of a diastolic pulse duplicator system consisting of a (A) bioreactor and a (B) medium chamber. The medium is pumped through the (C) tubing connecting the two chambers using (D) roller pumps. Compressed air is introduced into a (E) polycarbonate cylinder, which encases part of the tubing, using a (F) magnet valve. Compliance is incorporated into the system using a (G) syringe and pressure is measured using (H) sensors on either side of the TEHV. With kind permission from Springer Science+Business Media: Annals of Biomedical Engineering, Tissue engineering of human heart valve leaflets: a novel bioreactor for a strain-based conditioning approach, 33, 2005, 1778–1788, A. Mol, N.J. Driessen, M.C. Rutten, S.P. Hoerstrup, C.V. Bouten, and F.P. Baaijens, Fig. 2 [81]

most suitable system depends on a number of factors such as the desired level of control on the regimen, bioreactor complexity, and valve design.

Bioreactors require a number of components in order to replicate physiologic pulsatile pressure waveforms. Common features among the different designs include a pump to induce fluid motion, a mounting chamber for the TEHV, a culture medium reservoir, compliance chambers for energy dissipation, and a tunable element to control system pressure [43]. These systems condition the whole valve during the entire cardiac cycle (systole and diastole). Another iteration of these systems is to simulate only diastolic pressure conditions by exposing the leaflets to cyclic back pressure [81]. An example system and its components are described in more detail in Fig. 2.

Although the aforementioned systems can replicate physiological pressure conditions, the strain magnitude depends on the tissue properties and is not controlled. There have been published reports demonstrating the advantages of using incremental strain regimens and/or transmural flow [76–78]. Syedain and colleagues designed a cyclic stretch bioreactor capable that had a defined strain magnitude by mounting a TEHV within a latex tube [77]. Since the latex tube was much stiffer than the engineered tissue, strain magnitudes could be prescribed throughout the culture period independent of the tissue properties.

## 3 Valve Design

Traditional TEHVs, those using synthetic or biopolymeric scaffolds, must be fabricated into the heart valve geometry. The inherent properties of synthetic and biopolymeric scaffolds allow researchers to fabricate TEHVs in ways not possible with traditional bioprosthetic valves or decellularized native tissues. Almost all TEHV designs incorporate distinct leaflets that open and close in response to pressure gradients. Valvular function also can be achieved by attaching an engineered tissue tube to a three-pronged frame or to create a tubular valve [89] that is the functional equivalent of a trileaflet valve [90–92]. Some valve designs for pediatric patients also incorporate a root, or flow conduit, since cardiovascular congenital defects can affect the outflow tract as well as the leaflets [88, 93].

Biopolymeric and synthetic scaffolds are easily moldable since they often begin as an aqueous solution or typically incorporate a thermoplastic polyhydroxyalkanoate (such as P4HB), respectively [42, 55, 94, 95]. This has enabled researchers to use casting molds (as shown in Fig. 3) that integrate the valve root and leaflets together. This allowed researchers to create valves with leaflets directly attached to the root from the onset of valve fabrication [73, 83].

While these molds were innovative and provided a proof of concept, there are several challenges associated with this approach. First, machining the molds is challenging given the unique geometries utilized. Additionally, the presence of sharp corners and high stress points in these molds can lead to tissue thinning and tearing. Another major challenge is ensuring that valve geometry and coaptation is maintained throughout the culture period. Although cell induced compaction can be exploited to generate anisotropic tissues [73, 96], over-compaction can result in

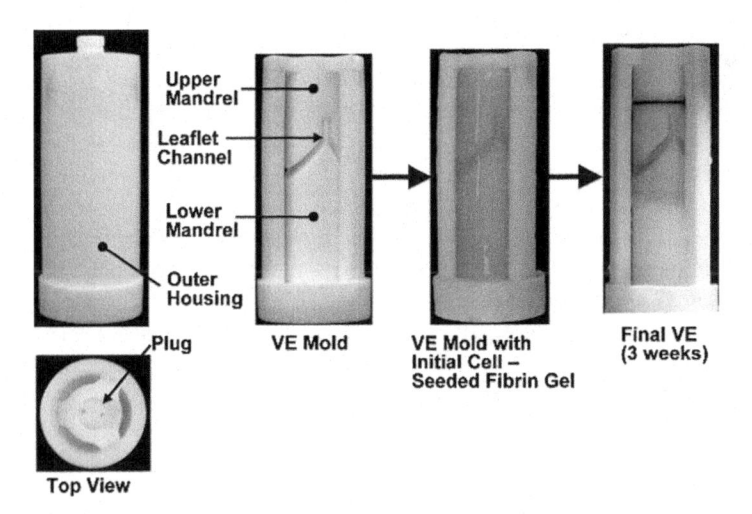

**Fig. 3** Valve equivalent mold and fabrication procedure using a biopolymeric (fibrin) scaffold [73]. Reprinted with permission from (Tissue Engineering Part A, Volume 14 & Issue 1, pages 83–95), published by Mary Ann Liebert, Inc., New Rochelle, NY

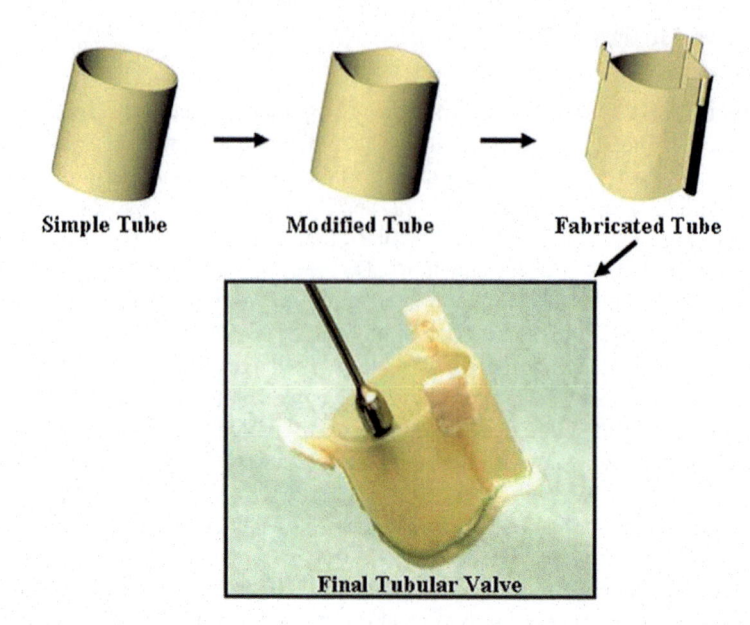

**Fig. 4** Schematic showing the design and modification of the Medtronic 3F bioprosthetic valve, which utilizes tubular leaflet design. During implantation, the fabricated tube is attached around the base and at the three commissural tabs. Exposure to back pressure collapses the tube in the regions in between the commissural tabs, resulting in the formation of leaflets [89]. Reprinted from Journal of Thoracic and Cardiovascular Surgery, Volume 130/edition 2, James Cox, Niv Ad, Keith Myers, Mortiz Gharib, and R.C. Quijano, Tubular heart valves: a new tissue prosthesis design—preclinical evaluation of the 3F aortic bioprosthesis, 520–527, Copyright (2005), with permission from Elsevier

geometrical changes that compromise the valve. For example, over-compaction in the leaflet channels results in loss of coaptation.

A simpler approach is based on a tubular valve design [89]. Instead of using a complex mold to define the heart valve root and leaflet geometries, tissue tubes are constrained such that they collapse inward under back pressure at three equi-spaced positions (commissures) around the circumference that anchor the tube (Fig. 4). This creates the equivalent of three leaflets, which respond to dynamic pressure gradients across a tubular heart valve. These principles have been applied to heart valve tissue engineering as researchers have reported tubular TEHVs using tubular tissues based on synthetic and biopolymer scaffolds [88, 90, 92, 93]. Most tubular TEHVs utilize a single tube that is attached to a frame, an expandable stent, or designed to be implanted within the native vasculature.

The use of inert frames, stents, chemically fixed tissues, or embedded synthetic materials preclude them from growing and are thus suboptimal for pediatric patients. One new approach avoids these components and might be more suitable for pediatric valve replacements [93]. It consists of two completely biological, decellularized engineered tubes (primarily cell-produced collagen) that are attached using a degradable suture line and does not incorporate any inert materials, frames, or stents. It relies on host cell invasion and matrix production to fuse the two engineered tubes along the degradable suture line and subsequently respond to physiologic growth cues in vivo. However, it remains to be seen whether this or other TEHVs are amenable to long-term function and growth.

## 4 In Vitro Functional Testing

Functional testing systems in the laboratory are invaluable in validating and improving valve designs for TEHVs prior to expensive animal studies. These pulsatile flow testing systems are similar to some of the aforementioned bioreactor systems in terms of components required. They are not exclusive to TEHVs, but instead are useful for testing all types of valves (including mechanical and bioprosthetic valves). They typically include pressure and flow probes, a pump to induce fluid motion, compliance chambers, a valve housing chamber, and a fluid reservoir [85].

The primary goal of functional testing is to elucidate the TEHV's function under physiologic pressure and flow conditions [97]. Important hydrodynamics include systolic pressure drop, regurgitation, and effective orifice area (EOA). Detailed descriptions of these metrics and other valve testing requirements can be found in ISO 5840. A high-speed camera is typically placed end-on to visualize and record leaflet motion during the cardiac cycle. This allows researchers to visualize any fluttering, prolapse, or asymmetry in the valve leaflets. An additional camera can be used to visualize root motion if it is incorporated into the prosthetic heart valve and the test system allows for it.

Accelerated wear testing (AWT) is a type of functional testing designed to evaluate the fatigue properties of prosthetic heart valves under pulsatile flow and physiologic loading. Complete valve opening and physiologic diastolic pressure gradients are the only requirements for AWT. This allows researchers to utilize higher frequency pumps to evaluate a heart valve's durability after a large number of cycles. Traditional mechanical and bioprosthetic heart valves must withstand 600 million or 200 million cycles per ISO 5840, respectively. Often valves undergo full functional testing (physiologic flow and pressure conditions) periodically during AWT to assess their hydrodynamic performance.

The appropriate number of test cycles for TEHVs is unclear since engineered valves are designed on the premise that cells will repair and remodel the extracellular matrix following implantation [42]. These cells can be either transplanted with the valve or host cells that repopulate the acellular valve, or a combination of both. The complex in vivo environment cannot be completely mimicked with in vitro systems. The ideal number of cycles depends on the timeline necessary for cellular production of matrix proteins needed to sustain valve function following implantation. Discussions with the FDA on these points will be crucial to ensure that the protocols for appropriate in vitro functional tests and their duration are established as TEHVs become closer to commercialization.

## 5 In Vivo Testing

Large animal testing is a requirement for the preclinical assessment of prosthetic heart valves. These tests allow researchers to assess the performance of the surgical procedure and valve's hydrodynamic performance in a beating heart over an extended period of time. These studies are highly regulated and require approval by a standard governing body known as the Institutional Animal Care and Use Committee, or IACUC. The following sections highlight the animal models and results with TEHVs to date.

### 5.1 Animal Models for TEHVs

Chronic in vivo TEHV studies are conducted primarily in sheep, which are the current "gold standard" for preclinical prosthetic heart valve replacement studies [98]. Studies with TEHVs are often performed in the pulmonary position of sheep due to the lower pressure gradients associated with this position compared to the aortic valve. Congenital defects also often affect this valve. Sheep are advantageous because they have similar normal cardiovascular physiological parameters (blood pressure, cardiac output, heart rate, and intracardiac pressure) and valve diameters that are similar to humans. The ovine model is also an aggressive calcification model, particularly in juvenile animals, compared to other large animal models.

Canine models were common historically since they have good temperaments and can be trained easily. However, they are not as common now due to the occurrence of leaflet fusion and high collateral coronary circulation [98]. Swine models are advantageous since they have very similar cardiac anatomy to humans in terms of their valves, coronary arteries, and conduction system. However, they have a high incidence of postoperative mortality and arrhythmias which limits their widespread use. There has also been a report of a nonhuman primate model being used, but this is currently not very common [99].

Heart valve replacement studies, particularly those with TEHVs, are conducted using healthy animals and primarily aim to assess valve function. Naturally occurring and reproducible valvular defects have been described in rats and mice, but are much less common in larger animal models [100]. Researchers have described large animal models with iatrogenic valve stenosis, regurgitation, and anatomical abnormalities [98, 100]. However, these are not commonly used, particularly for TEHV studies.

The animal model's growth rate is also important to consider when conducting chronic animal studies, particularly if the desired valve orifice sizes and study aims necessitate using immature animals. Growing animal models are intriguing for TEHV studies since they aim to assess somatic growth of the TEHV, which is an unmet clinical need for pediatric valve replacement patients. Accelerated growth rates compared to human can result in "prosthesis-patient mismatch" and valve stenosis if the animal outgrows its prosthetic heart valve [18]. The aforementioned animal models all reach full maturity within 12–18 months of age, which is substantially faster than in humans [98, 101–103]. This timescale can be challenging for TEHVs, especially pediatric valves, which rely on transplanted or invaded cells to dictate growth in response to physiological cues.

## 5.2 Preclinical Testing Results with TEHVs

Shinoka et al. [104] was one of the first to demonstrate the feasibility of TEHVs for valve replacement. A single sheep pulmonary valve leaflet was replaced with autologous or allogeneic tissue engineered leaflets, which were fabricated by seeding a combination of fibroblasts, smooth muscle cells, and endothelial cells seeded onto polyglactin/PGA mesh sheets. Lambs with one of these allogeneic leaflets experienced leaflet retraction and an acute inflammatory response despite receiving immunosuppression therapy, whereas those with an autologous leaflet did not. While the single leaflet approach has limited applications, this study demonstrated the feasibility of tissue engineering in heart valve replacements.

The same research group soon developed a complete trileaflet TEHV by seeding autologous (ovine) myofibroblasts and endothelial cells onto a PGA/P4HB scaffold [55]. Following pulsatile bioreactor culture, the TEHVs were implanted into the pulmonary position of sheep for up to 20 weeks and valve function was monitored via echocardiography. The tensile strength of the engineered leaflets were initially higher than those of the native leaflets, but was comparable after 20 weeks in vivo. Histology showed increased organization and layering similar to native leaflets after 20 weeks with the presence of collagen, glycosaminoglycans, and elastin. Although the TEHV leaflets remained functional throughout the duration of the study, significant regurgitation was observed after 16 and 20 weeks. Authors attributed this loss of valve function to the enlarging flow conduit and/or shrinkage of the TEHV leaflets.

A minimally invasive TEHV was fabricated by seeding myofibroblasts or stem cells onto a PGA/P4HB scaffold using fibrin as a cell carrier [105]. Following bioreactor culture, the TEHVs were transapically implanted into the pulmonary position of sheep for either 4 or 8 weeks. Examination following explantation revealed thickened and non-coapting leaflets. A similar approach, but without bioreactor culture, demonstrated valve function up to 4 weeks in the pulmonary position of primates [106]. Substantial cellular remodeling and endothelial cell coverage was observed without tissue thickening. However, structural shortening of the TEHV leaflets was detected after explantation.

Autologous pulmonary and aortic TEHVs have been explored in canine and goat models. Using a custom mold and in situ fabrication, these "Biovalves" were highly collagenous with a small amount of elastin [107]. Early generations of this TEHV allowed substantial regurgitation in dogs [108], which ultimately led to design changes in the valve mold. The most recent Biovalve incorporated a 3D printed mold and were implanted in the aortic position for 1 month [107]. Another group implanted an electrospun supramolecular polymer scaffold that was functionalized with biological components to recruit circulating host cells [47]. They reported sustained mechanical and biological function of their TEHV up to 6 month in their preliminary report.

Two different groups have described the fabrication of TEHVs using fibrin scaffolds with complex molds. Flanagan et al. used autologous cell populations (smooth muscle cells and fibroblasts) and bioreactor culture for 4 weeks to generate TEHVs sufficient for implantation [109]. The mature TEHVs were seeded with autologous endothelial cells prior to implantation into the pulmonary position of mature sheep for 3 months. Extensive in vivo remodeling resulted in the replacement of the fibrin scaffold with cell-produced extracellular matrix. An endothelium was observed on the valvular surfaces and there was no evidence of thrombi, calcification, stenosis, or aneurysms. Despite these promising results, valvular function was not maintained due to cell-mediated leaflet shortening.

Another fibrin-based TEHV was fabricated by using human dermal fibroblasts and a cyclic stretch bioreactor [73, 110]. Cell-mediated fibrin gel contraction was exploited by this group to dictate collagen fiber alignment and mechanical anisotropy similar to the native heart valve root and leaflets [73, 96]. These bileaflet TEHVs were implanted interpositionally into the pulmonary position of sheep after comprising the native pulmonary valve leaflets [110]. A concentric sleeve was placed along the entire root length to mitigate the risk of suture pullout or root rupture.

Echocardiography immediately following implantation showed coapting leaflets with regurgitation, orifice area, and pressure gradients comparable to the native pulmonary valve. However, echocardiography at 4 weeks revealed moderate regurgitation due to significant leaflet shrinkage. Only one shortened leaflet was observed in each of the two valves explanted after 8 weeks. Extensive tissue remodeling and endothelialization was observed following implantation. Elevated collagen and elastin concentrations and minimal calcification were observed.

These studies demonstrated the feasibility of TEHVs, but ultimately failed due to progressive leaflet contraction in vivo. Syedain et al. demonstrated that the transplanted fibroblasts maintained a contractile phenotype following implantation. Flanagan et al. also reported that cells in the explanted leaflets expressed alpha smooth muscle actin. Proposed solutions included fabricating TEHVs with additional coaptation area, utilizing stiffer polymers to prevent over-compaction, or cell removal following in vitro tissue remodeling [55, 109, 110].

Decellularization has been explored recently as a means to eliminate in vivo leaflet shortening by removing the cellular components. It also reduces the immunogenicity of the TEHVs. However, this approach relies on rapid host cell invasion to repopulate the TEHVs so that cell-mediated repair and remodeling (and potentially growth) can occur. It is imperative that the acellular TEHVs can withstand this initial phase prior to recellularization in vivo. Decellularization was first performed on TEHVs fabricated using synthetic polymer scaffolds attached to expandable nitinol stents [94], and subsequently in TEHVs fabricated using a fibrin gel [90].

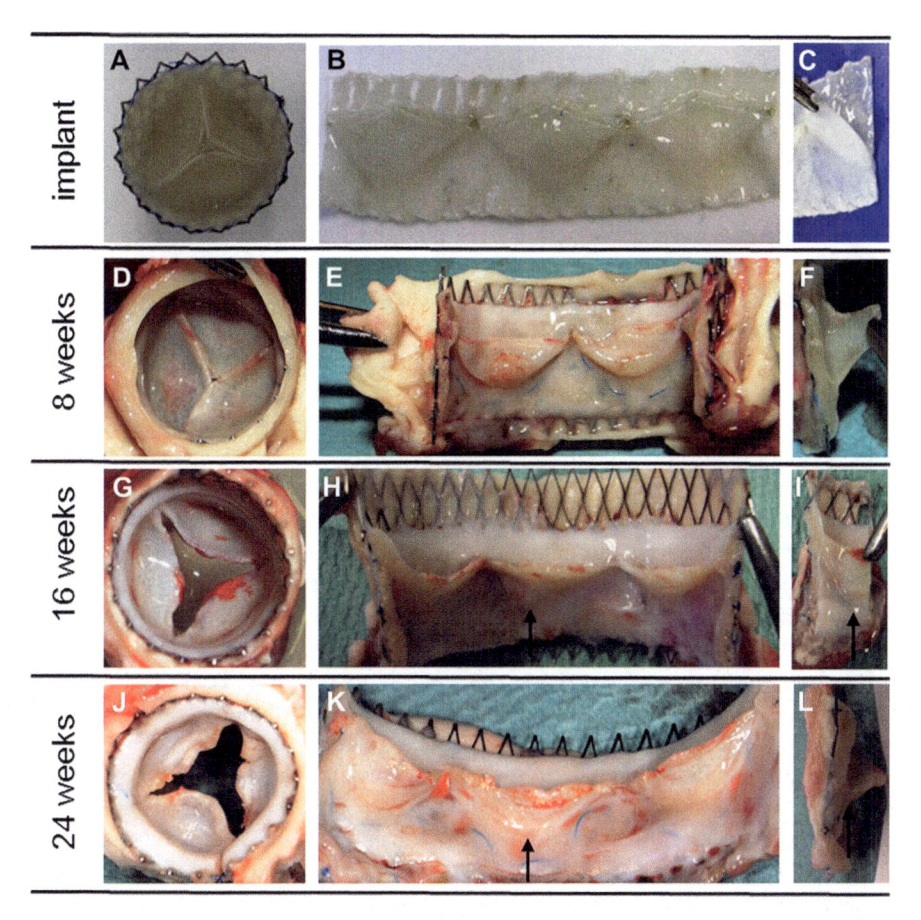

**Fig. 5** Macroscopic appearance of a decellularized TEHV (derived using synthetic polymers), (**a–c**) before and after implantation for (**d–f**) 8, (**g–i**) 16, and (**j–l**) 24 weeks. Leaflet coaptation was (**d**) maintained for 8 weeks, but (**g, j**) was not evident thereafter. (**e, f, h, i, k, l**) Leaflet fusion to the root (arrows) progressed upward over time, resulting in smaller leaflets. Reprinted from the Journal of American College of Cardiology, Volume 63/Issue 13, A. Driessen-Mol, M.Y. Emmert, P.E. Dijkman, L. Frese, B. Sanders, B. Weber, N. Cesarovic, M. Sidler, J. Leenders, R. Jenni, J. Grunenfelder, V. Falk, F.P. Baaijens, and S.P. Hoerstrup, Transcatheter implantation of homologous "off-the-shelf" tissue-engineered heart valves with self-repair capacity: long-term functionality and rapid in vivo remodeling in sheep. Pages 1320–1329, 2014, with permission from Elsevier [91]

Decellularized TEHVs implanted into the pulmonary position of sheep exhibited extensive host cell invasion and matrix remodeling. Recellularization was fastest in the valve root, followed by the root/leaflet attachment area and the leaflets. However, mild central regurgitation appeared after 8 weeks and progressed to moderate after 24 weeks [91]. This loss of function was attributed to fusion of the leaflets to the valve root at the leaflet base, as shown in Fig. 5. Similar results were reported in a senescent nonhuman primate model [99]. However, the appearance of mild to moderate regurgitation after 8 weeks was attributed to the passive retraction of the extracellular matrix in the primate study.

Another decellularized TEHV has also been reported recently using engineered tubes derived from a sacrificial fibrin scaffold [90]. Following in vitro maturation and decellularization, the engineered tubes consisted primarily of cell-produced collagen, which was preferentially aligned in the circumferential direction. The engineered tubes were sewn onto a three-pronged frame to create a tubular heart valve and tested under aortic conditions in an in vitro pulse duplicator system [90]. The TEHVs were subsequently implanted into the aortic position of adult sheep for up to 24 weeks [111]. Although there has been one other aortic TEHV study in sheep, it was designed to assess valve delivery, position, and function immediately after implantation [112]. The study by Syedain et al. was the first long-term aortic valve study with a traditional TEHV (derived using synthetic or biopolymeric scaffolds) reported.

Full leaflet motion, laminar flow, and maintained effective orifice area were observed throughout the entirety of the implant period [111]. Two of the four valves implanted exhibited trivial to mild aortic insufficiency immediately after implantation. The insufficiency grade in three of the valves increased after 12 weeks, which was attributed to small matrix tears near the top of the frame struts. However, the aortic insufficiency did not progress further between 12 and 24 weeks. No evidence of calcification was reported in any of the explanted TEHVs. Extensive recellularization was observed near the base of the leaflets after 12 weeks. After 24 weeks, the cells in these regions expressed vimentin and lacked alpha smooth muscle actin, similar to native valve interstitial cells. Increased cell invasion was observed near the leaflet free edge at this time point as well. The invaded cells deposited proteoglycans, elastin, collagen IV, and laminin. An endothelium was similarly forming from the base of the leaflets towards the free edge. Similar results regarding cell infiltration and matrix deposition were observed when a 'pediatric' heart valve flow was implanted into a growing sheep model. This valve was fabricated using the same starting tissue, degradable sutures, and lacked a frame [113].

## 6 Remaining Challenges

The prospect of using tissue engineering principles to generate a living heart valve capable of repairing and remodeling itself is promising and exciting. However, there are numerous challenges remaining before a tissue engineered heart valve can be safely translated to the clinic. These heart valves must demonstrate both short- and long-term function in dynamic mechanical and flow environments. In order to respond to these environments and other biological cues, the correct cells must reside in the TEHV and maintain a homeostatic functional state over an extended period of time [114]. This challenge is exacerbated in pediatric patients who are also undergoing somatic growth.

Due to the breadth of valve fabrication methods (scaffolds, cells, cell stimulation, etc.) continued research is needed to identify the optimal TEHV. The use of animal models will be crucial in order to elucidate the host response to TEHVs and their various components in regards to thrombosis, inflammation, infection, and calcification. Demonstrating repeatable and long-term valve function has been a challenge for the field to date.

Since TEHVs depend on resident cells to confer remodeling and somatic growth in response to physiologic cues in vivo, it will be crucial to assess their presence and phenotype following implantation. Ensuring cellularity is essential for decellularized valves, which are reliant on host cell invasion to provide durability (via secretion of new matrix proteins) and hemocompatibility. Demonstrating success in these areas will be critical to the success of the heart valve tissue engineering field.

## 7  Summary

There are numerous approaches to developing a TEHV, each with their own advantages and disadvantages. Various cell sources, scaffold materials, and cell stimulation paradigms have been explored with varying levels of success. Ensuring the safety and efficacy of a tissue engineered valve prosthesis will require well-designed in vitro and in vivo studies. Translating tissue engineered heart valves into the clinic is promising, but also challenging, as evidenced by outcomes of in vivo TEHV studies to date.

## References

1. Mozaffarian D, Benjamin EJ, Go AS, Arnett DK, Blaha MJ, Cushman M, et al. Heart disease and stroke statistics—2015 update: a report from the American Heart Association. Circulation. 2015;131(4):e29–322. https://doi.org/10.1161/CIR.0000000000000152.
2. Moller J. Prevalence and incidence of cardiac malformations. In: Moller J, editor. Surgery of congenital heart disease: pediatric care consortium 1984–1995. Armonk, NY: Futura Publishing Company Inc.; 1998. p. 19–26.
3. Minnesota Department of Health. Diseases and conditions identified in children. [Webpage] 2015. Available from: http://www.health.state.mn.us/divs/cfh/topic/diseasesconds/. Cited 24 Aug 2015.
4. Centers for Disease Control and Prevention. Data & statistics for congenital defects in the US. 2015. Available from: http://www.cdc.gov/ncbddd/birthdefects/data.html. Cited 24 Aug 2015.
5. Liao K, John R. Handbook of cardiac anatomy, physiology, and devices. 2nd ed. Minneapolis, MN: Spring Science & Business Media; 2009.
6. Waterbolk TW, Hoendermis ES, den Hamer IJ, Ebels T. Pulmonary valve replacement with a mechanical prosthesis. Promising results of 28 procedures in patients with congenital heart disease. Eur J Cardiothorac Surg. 2006;30(1):28–32. https://doi.org/10.1016/j.ejcts.2006.02.069.

7. Bouzas B, Kilner PJ, Gatzoulis MA. Pulmonary regurgitation: not a benign lesion. Eur Heart J. 2005;26(5):433–9. https://doi.org/10.1093/eurheartj/ehi091.
8. Yemets IM, Williams WG, Webb GD, Harrison DA, McLaughlin PR, Trusler GA, et al. Pulmonary valve replacement late after repair of tetralogy of Fallot. Ann Thorac Surg. 1997;64 (2):526–30. https://doi.org/10.1016/S0003-4975(97)00577-8.
9. Tweddell JS, Simpson P, Li SH, Dunham-Ingle J, Bartz PJ, Earing MG, et al. Timing and technique of pulmonary valve replacement in the patient with tetralogy of Fallot. Semin Thorac Cardiovasc Surg Pediatr Card Surg Annu. 2012;15(1):27–33. https://doi.org/10.1053/j.pcsu.2012.01.007.
10. Kheradvar A, Groves EM, Goergen CJ, Alavi SH, Tranquillo R, Simmons CA, et al. Emerging trends in heart valve engineering: part II. Novel and standard technologies for aortic valve replacement. Ann Biomed Eng. 2015;43(4):844–57. https://doi.org/10.1007/s10439-014-1191-5.
11. Kheradvar A, Groves EM, Simmons CA, Griffith B, Alavi SH, Tranquillo R, et al. Emerging trends in heart valve engineering: part III. Novel technologies for mitral valve repair and replacement. Ann Biomed Eng. 2015;43(4):858–70. https://doi.org/10.1007/s10439-014-1129-y.
12. Rahimtoola SH. Choice of prosthetic heart valve for adult patients. J Am Coll Cardiol. 2003;41 (6):893–904. https://doi.org/10.1016/S0735-1097(02)02965-0.
13. Rahimtoola SH. Choice of prosthetic heart valve in adults an update. J Am Coll Cardiol. 2010;55(22):2413–26. https://doi.org/10.1016/j.jacc.2009.10.085.
14. Birkmeyer NJ, Birkmeyer JD, Tosteson AN, Grunkemeier GL, Marrin CA, O'Connor GT. Prosthetic valve type for patients undergoing aortic valve replacement: a decision analysis. Ann Thorac Surg. 2000;70(6):1946–52.
15. El Oakley R, Kleine P, Bach DS. Choice of prosthetic heart valve in today's practice. Circulation. 2008;117(2):253–6. https://doi.org/10.1161/CIRCULATIONAHA.107.736819.
16. Kaneko T, Cohn LH, Aranki SF. Tissue valve is the preferred option for patients aged 60 and older. Circulation. 2013;128:1365–71. https://doi.org/10.1161/CIRCULATIONAHA.113.002584.
17. van Geldorp MWA, Eric Jamieson WR, Kappetein AP, Ye J, Fradet GJ, Eijkemans MJC, et al. Patient outcome after aortic valve replacement with a mechanical or biological prosthesis: weighing lifetime anticoagulant-related event risk against reoperation risk. J Thorac Cardiovasc Surg. 2009;137(4):881–6, 886e1–5. https://doi.org/10.1016/j.jtcvs.2008.09.028.
18. Mankad S. Management of prosthetic heart valve complications. Curr Treat Options Cardiovasc Med. 2012;14(6):608–21. https://doi.org/10.1007/s11936-012-0212-7.
19. Tillquist MN, Maddox TM. Cardiac crossroads: deciding between mechanical or bioprosthetic heart valve replacement. Patient Prefer Adherence. 2011;5:91–9. https://doi.org/10.2147/PPA.S16420.
20. Oterhals K, Fridlund B, Nordrehaug JE, Haaverstad R, Norekvål TM. Adapting to living with a mechanical aortic heart valve: a phenomenographic study. J Adv Nurs. 2013;69(9):2088–98. https://doi.org/10.1111/jan.12076.
21. Hammermeister K, Sethi GK, Henderson WG, Grover FL, Oprian C, Rahimtoola SH. Outcomes 15 years after valve replacement with a mechanical versus a bioprosthetic valve: final report of the Veterans Affairs randomized trial. J Am Coll Cardiol. 2000;36:1152–8.
22. Delmo Walter EM, de By TMMH, Meyer R, Hetzer R. The future of heart valve banking and of homografts: perspective from the Deutsches Herzzentrum Berlin. HSR Proc Intensive Care Cardiovasc Anesth. 2012;4(2):97–108.
23. Elkins RC, Goldstein S, Hewitt CW, Walsh SP, Dawson PE, Ollerenshaw JD, et al. Recellularization of heart valve grafts by a process of adaptive remodeling. Semin Thorac Cardiovasc Surg. 2001;13(4 Suppl 1):87–92.

24. Neumann A, Cebotari S, Tudorache I, Haverich A, Sarikouch S. Heart valve engineering: decellularized allograft matrices in clinical practice. Biomed Tech (Berl). 2013;58(5):453–6. https://doi.org/10.1515/bmt-2012-0115.
25. Elkins R, Dawson P, Goldstein S, Walsh S, Black K. Decellularized human valve allografts. Ann Thorac Surg. 2001;71(5):S428–32. https://doi.org/10.1016/S0003-4975(01)02503-6.
26. Jones JM, O'kane H, Gladstone DJ, Sarsam MA, Campalani G, MacGowan SW, et al. Repeat heart valve surgery: risk factors for operative mortality. J Thorac Cardiovasc Surg. 2001;122 (5):913–8. https://doi.org/10.1067/mtc.2001.116470.
27. Gurvitch R, Cheung A, Ye J, Wood DA, Willson AB, Toggweiler S, et al. Transcatheter valve-in-valve implantation for failed surgical bioprosthetic valves. J Am Coll Cardiol. 2011;58 (21):2196–209. https://doi.org/10.1016/j.jacc.2011.09.009.
28. Gurvitch R, Cheung A, Bedogni F, Webb JG. Coronary obstruction following transcatheter aortic valve-in-valve implantation for failed surgical bioprostheses. Catheter Cardiovasc Interv. 2011;77(3):439–44. https://doi.org/10.1002/ccd.22861.
29. Dvir D, Webb J, Brecker S, Bleiziffer S, Hildick-Smith D, Colombo A, et al. Transcatheter aortic valve replacement for degenerative bioprosthetic surgical valves: results from the global valve-in-valve registry. Circulation. 2012;126(19):2335–44. https://doi.org/10.1161/ CIRCULATIONAHA.112.104505.
30. Cebotari S, Tudorache I, Ciubotaru A, Boethig D, Sarikouch S, Goerler A, et al. Use of fresh decellularized allografts for pulmonary valve replacement may reduce the reoperation rate in children and young adults early report. Circulation. 2011;124(11):S115–23. https://doi.org/10. 1161/Circulationaha.110.012161.
31. Erdbrügger W, Konertz W, Dohmen PM, Posner S, Ellerbrok H, Brodde O-EE, et al. Decellularized xenogenic heart valves reveal remodeling and growth potential in vivo. Tissue Eng. 2006;12(8):2059–68. https://doi.org/10.1089/ten.2006.12.2059.
32. Dohmen PM, Lembcke A, Holinski S, Pruss A, Konertz W. Ten years of clinical results with a tissue-engineered pulmonary valve. Ann Thorac Surg. 2011;92(4):1308–14. https://doi.org/ 10.1016/j.athoracsur.2011.06.009.
33. Konertz W, Angeli E, Tarusinov G, Christ T, Kroll J, Dohmen PM, et al. Right ventricular outflow tract reconstruction with decellularized porcine xenografts in patients with congenital heart disease. J Heart Valve Dis. 2011;20(3):341–7.
34. Ruzmetov M, Shah JJ, Geiss DM, Fortuna RS. Decellularized versus standard cryopreserved valve allografts for right ventricular outflow tract reconstruction: a single-institution comparison. J Thorac Cardiovasc Surg. 2012;143(3):543–9. https://doi.org/10.1016/j.jtcvs.2011.12. 032.
35. Brown JW, Elkins RC, Clarke DR, Tweddell JS, Huddleston CB, Doty JR, et al. Performance of the CryoValve SG human decellularized pulmonary valve in 342 patients relative to the conventional CryoValve at a mean follow-up of four years. J Thorac Cardiovasc Surg. 2010;139:339–48. https://doi.org/10.1016/j.jtcvs.2009.04.065.
36. Tudorache I, Theodoridis K, Baraki H, Sarikouch S, Bara C, Meyer T, et al. Decellularized aortic allografts versus pulmonary autografts for aortic valve replacement in the growing sheep model: haemodynamic and morphological results at 20 months after implantation. Eur J Cardiothorac Surg. 2015;49:1228–38. https://doi.org/10.1093/ejcts/ezv362.
37. Baraki H, Tudorache I, Braun M, Hoffler K, Gorler A, Lichtenberg A, et al. Orthotopic replacement of the aortic valve with decellularized allograft in a sheep model. Biomaterials. 2009;30(31):6240–6. https://doi.org/10.1016/j.biomaterials.2009.07.068.
38. Bibevski S, Wilkinson D, Ruzmetov M, Fortuna R, Turrentine M, Brown JW, et al., cartographers. Performance of synergraft decellularized pulmonary allografts compared with standard cryopreserved allografts: results from multi-institutional data [Abstract]. Circulation. 2012;126 (supplement 21).
39. Tudorache I, Calistru A, Baraki H, Meyer T, Hoffler K, Sarikouch S, et al. Orthotopic replacement of aortic heart valves with tissue-engineered grafts. Tissue Eng Part A. 2013;19 (15–16):1686–94. https://doi.org/10.1089/ten.TEA.2012.0074.

40. Perri G, Polito A, Esposito C, Albanese SB, Francalanci P, Pongiglione G, et al. Early and late failure of tissue-engineered pulmonary valve conduits used for right ventricular outflow tract reconstruction in patients with congenital heart disease. Eur J Cardiothorac Surg. 2012;41 (6):1320–5. https://doi.org/10.1093/ejcts/ezr221.
41. Kheradvar A, Groves EM, Dasi LP, Alavi SH, Tranquillo R, Grande-Allen KJ, et al. Emerging trends in heart valve engineering: part I. Solutions for future. Ann Biomed Eng. 2015;43 (4):833–43. https://doi.org/10.1007/s10439-014-1209-z.
42. Chester A, Yacoub MH, Taylor PM. Heart valve tissue engineering. In: Boccaccini AR, Harding SE, editors. Myocardial tissue engineering. Berlin: Springer; 2011. p. 243–66.
43. Schmidt JB, Tranquillo RT. Tissue-engineered heart valves. In: Iaizzo PA, Bianco R, Hill AJ, St. Louis JD, editors. Heart valves: from design to clinical implantation. Springer: New York; 2013. p. 261–80.
44. van Loon SLM, Smits AIPM, Driessen-Mol A, Baaijens F, Bouten CV. The immune response in in situ tissue engineering of aortic heart valves. In: Aikawa E, editor. Calcific aortic valve disease. Rijeka: InTech; 2013. p. 207–45.
45. Bouten CV, Driessen-Mol A, Baaijens FP. In situ heart valve tissue engineering: simple devices, smart materials, complex knowledge. Expert Rev Med Devices. 2012;9(5):453–5. https://doi.org/10.1586/erd.12.43.
46. Hayashida K, Kanda K, Yaku H, Ando J, Nakayama Y. Development of an in vivo tissue-engineered, autologous heart valve (the biovalve): preparation of a prototype model. J Thorac Cardiovasc Surg. 2007;134(1):152–9. https://doi.org/10.1016/j.jtcvs.2007.01.087.
47. Bouten CV, Smits AI, Talacua H, Muylaert DE, Janssen HM, Bosman A, et al., editors. Biomaterial-based in situ tissue engineering of heart valves. 2015 4th TERMIS world congress, Sept 2015, Boston, MA, Tissue Engineering Part A.
48. Ott HC, Matthiesen TS, Goh SK, Black LD, Kren SM, Netoff TI, et al. Perfusion-decellularized matrix: using nature's platform to engineer a bioartificial heart. Nat Med. 2008;14(2):213–21. https://doi.org/10.1038/nm1684.
49. Gilbert TW, Sellaro TL, Badylak SF. Decellularization of tissues and organs. Biomaterials. 2006;27(19):3675–83. https://doi.org/10.1016/j.biomaterials.2006.02.014.
50. Taylor PM. Biological matrices and bionanotechnology. Philos Trans R Soc Lond Ser B Biol Sci. 2007;362(1484):1313–20. https://doi.org/10.1098/rstb.2007.2117.
51. Sung HJ, Meredith C, Johnson C, Galis ZS. The effect of scaffold degradation rate on three-dimensional cell growth and angiogenesis. Biomaterials. 2004;25(26):5735–42. https://doi.org/10.1016/j.biomaterials.2004.01.066.
52. Zund G, Breuer CK, Shinoka T, Ma PX, Langer R, Mayer JE, et al. The in vitro construction of a tissue engineered bioprosthetic heart valve. Eur J Cardiothorac Surg. 1997;11(3):493–7. https://doi.org/10.1016/S1010-7940(96)01005-6.
53. Stock UA, Nagashima M, Khalil PN, Nollert GD, Herden T, Sperling JS, et al. Tissue-engineered valved conduits in the pulmonary circulation. J Thorac Cardiovasc Surg. 2000;119(4 Pt 1):732–40. https://doi.org/10.1067/mtc.2000.104584.
54. Knight R, Wilcox HE, Korossis SA, Ingham E. The use of acellular matrices for tissue engineering of cardiac valves. Proc Inst Mech Eng H. 2008;222(1):129–43. https://doi.org/10.1243/09544119JEIM230.
55. Hoerstrup SP, Sodian R, Daebritz S, Wang J, Bacha EA, Martin DP, et al. Functional living trileaflet heart valves grown in vitro. Circulation. 2000;102(19 Suppl 3):III44–9. https://doi.org/10.1161/01.CIR.102.suppl_3.III-44.
56. Knight RL, Booth C, Wilcox HE, Fisher J, Ingham E. Tissue engineering of cardiac valves: re-seeding of acellular porcine aortic valve matrices with human mesenchymal progenitor cells. J Heart Valve Dis. 2005;14(6):806–13.
57. Rippel RA, Ghanbari H, Seifalian AM. Tissue-engineered heart valve: future of cardiac surgery. World J Surg. 2012;36(7):1581–91. https://doi.org/10.1007/s00268-012-1535-y.
58. Glowacki J, Mizuno S. Collagen scaffolds for tissue engineering. Biopolymers. 2008;89 (5):338–44. https://doi.org/10.1002/bip.20871.

59. Chevallay B, Herbage D. Collagen-based biomaterials as 3D scaffold for cell cultures: applications for tissue engineering and gene therapy. Med Biol Eng Comput. 2000;38 (2):211–8. https://doi.org/10.1007/BF02344779.
60. Thie M, Schlumberger W, Semich R, Rauterberg J, Robenek H. Aortic smooth muscle cells in collagen lattice culture: effects on ultrastructure, proliferation and collagen synthesis. Eur J Cell Biol. 1991;55(2):295–304.
61. Clark RA, Nielsen LD, Welch MP, McPherson JM. Collagen matrices attenuate the collagen-synthetic response of cultured fibroblasts to TGF-beta. J Cell Sci. 1995;108(Pt 3):1251–61.
62. Ma P. Biomaterials and regenerative medicine. Cambridge: Cambridge University Press; 2012.
63. Coustry F, Gillery P, Maquart FX, Borel JP. Effect of transforming growth factor beta on fibroblasts in three-dimensional lattice cultures. FEBS Lett. 1990;262(2):339–41.
64. Grassl ED, Oegema TR, Tranquillo RT. Fibrin as an alternative biopolymer to type-I collagen for the fabrication of a media equivalent. J Biomed Mater Res. 2002;60(4):607–12. https://doi.org/10.1002/jbm.10107.
65. Alavi SH, Kheradvar A. Metal mesh scaffold for tissue engineering of membranes. Tissue Eng Part C Methods. 2012;18(4):293–301. https://doi.org/10.1089/ten.TEC.2011.0531.
66. Loger K, Engel A, Haupt J, Li Q, Lima de Miranda R, Quandt E, et al. Cell adhesion on NiTi thin film sputter-deposited meshes. Mater Sci Eng C. 2016;59:611–6. https://doi.org/10.1016/j.msec.2015.10.008.
67. Loger K, de Miranda RL, Engel A, Marczynski-Bühlow M, Lutter G, Quandt E. Fabrication and evaluation of nitinol thin film heart valves. Cardiovasc Eng Technol. 2014;5(4):308–16. https://doi.org/10.1007/s13239-014-0194-6.
68. Stepan LL, Levi DS, Carman GP. A thin film nitinol heart valve. J Biomech Eng. 2005;127 (6):915–8.
69. Alavi SH, Kheradvar A. A hybrid tissue-engineered heart valve. Ann Thorac Surg. 2015;99 (6):2183–7. https://doi.org/10.1016/j.athoracsur.2015.02.058.
70. Alavi SH, Liu WF, Kheradvar A. Inflammatory response assessment of a hybrid tissue-engineered heart valve leaflet. Ann Biomed Eng. 2013;41(2):316–26. https://doi.org/10.1007/s10439-012-0664-7.
71. Neidert MR, Lee ES, Oegema TR, Tranquillo RT. Enhanced fibrin remodeling in vitro with TGF-beta1, insulin and plasmin for improved tissue-equivalents. Biomaterials. 2002;23 (17):3717–31.
72. Ramaswamy S, Gottlieb D, Engelmayr GC, Aikawa E, Schmidt DE, Gaitan-Leon DM, et al. The role of organ level conditioning on the promotion of engineered heart valve tissue development in-vitro using mesenchymal stem cells. Biomaterials. 2010;31(6):1114–25. https://doi.org/10.1016/j.biomaterials.2009.10.019.
73. Robinson P, Johnson S, Evans M, Barocas V, Tranquillo R. Functional tissue-engineered valves from cell-remodeled fibrin with commissural alignment of cell-produced collagen. Tissue Eng Part A. 2008;14(1):83–95. https://doi.org/10.1089/ten.a.2007.0148.
74. Alenghat FJ, Ingber DE. Mechanotransduction: all signals point to cytoskeleton, matrix, and integrins. Sci STKE. 2002;2002(119):pe6. https://doi.org/10.1126/stke.2002.119.pe6.
75. Chiquet M, Renedo AS, Huber F, Fluck M. How do fibroblasts translate mechanical signals into changes in extracellular matrix production? Matrix Biol. 2003;22(1):73–80. https://doi.org/10.1016/S0945-053X(03)00004-0.
76. Syedain ZH, Weinberg JS, Tranquillo RT. Cyclic distension of fibrin-based tissue constructs: evidence of adaptation during growth of engineered connective tissue. Proc Natl Acad Sci U S A. 2008;105(18):6537–42. https://doi.org/10.1073/pnas.0711217105.
77. Syedain ZH, Tranquillo RT. Controlled cyclic stretch bioreactor for tissue-engineered heart valves. Biomaterials. 2009;30(25):4078–84. https://doi.org/10.1016/j.biomaterials.2009.04.027.

78. Schmidt JB, Tranquillo RT. Cyclic stretch and perfusion bioreactor for conditioning large diameter engineered tissue tubes. Ann Biomed Eng. 2015;44:1785. https://doi.org/10.1007/s10439-015-1437-x.
79. Dumont K, Yperman J, Verbeken E, Segers P, Meuris B, Vandenberghe S, et al. Design of a new pulsatile bioreactor for tissue engineered aortic heart valve formation. Artif Organs. 2002;26(8):710–4. https://doi.org/10.1046/j.1525-1594.2002.06931_3.x.
80. Hildebrand DK, Wu ZJ, Mayer JE Jr, Sacks MS. Design and hydrodynamic evaluation of a novel pulsatile bioreactor for biologically active heart valves. Ann Biomed Eng. 2004;32(8):1039–49. https://doi.org/10.1114/B:ABME.0000036640.11387.4b.
81. Mol A, Driessen NJ, Rutten MC, Hoerstrup SP, Bouten CV, Baaijens FP. Tissue engineering of human heart valve leaflets: a novel bioreactor for a strain-based conditioning approach. Ann Biomed Eng. 2005;33(12):1778–88. https://doi.org/10.1007/s10439-005-8025-4.
82. Ruel J, Lachance G. A new bioreactor for the development of tissue-engineered heart valves. Ann Biomed Eng. 2009;37(4):674–81. https://doi.org/10.1007/s10439-009-9646-9.
83. Flanagan T, Cornelissen C, Koch S, Tschoeke B, Sachweh J, Schmitz-Rode T, et al. The in vitro development of autologous fibrin-based tissue-engineered heart valves through optimized dynamic conditioning. Biomaterials. 2007;28(23):3388–97. https://doi.org/10.1016/j.biomaterials.2007.04.012.
84. Hoerstrup SP, Sodian R, Sperling JS, Vacanti JP, Mayer JE Jr. New pulsatile bioreactor for in vitro formation of tissue engineered heart valves. Tissue Eng. 2000;6(1):75–9. https://doi.org/10.1089/107632700320919.
85. Berry JL, Steen JA, Koudy Williams J, Jordan JE, Atala A, Yoo JJ. Bioreactors for development of tissue engineered heart valves. Ann Biomed Eng. 2010;38(11):3272–9. https://doi.org/10.1007/s10439-010-0148-6.
86. Engelmayr GC Jr, Soletti L, Vigmostad SC, Budilarto SG, Federspiel WJ, Chandran KB, et al. A novel flex-stretch-flow bioreactor for the study of engineered heart valve tissue mechanobiology. Ann Biomed Eng. 2008;36(5):700–12. https://doi.org/10.1007/s10439-008-9447-6.
87. Ramaswamy S, Boronyak SM, Le T, Holmes A, Sotiropoulos F, Sacks MS. A novel bioreactor for mechanobiological studies of engineered heart valve tissue formation under pulmonary arterial physiological flow conditions. J Biomech Eng. 2014;136(12):1210091–12100914. https://doi.org/10.1115/1.4028815.
88. Weber M, Heta E, Moreira R, Gesche VN, Schermer T, Frese J, et al. Tissue-engineered fibrin-based heart valve with a tubular leaflet design. Tissue Eng Part C Methods. 2014;20(4):265–75. https://doi.org/10.1089/ten.TEC.2013.0258.
89. Cox JL, Ad N, Myers K, Gharib M, Quijano RC. Tubular heart valves: a new tissue prosthesis design—preclinical evaluation of the 3F aortic bioprosthesis. J Thorac Cardiovasc Surg. 2005;130(2):520–7. https://doi.org/10.1016/j.jtcvs.2004.12.054.
90. Syedain ZH, Meier LA, Reimer JM, Tranquillo RT. Tubular heart valves from decellularized engineered tissue. Ann Biomed Eng. 2013;41(12):2645–54. https://doi.org/10.1007/s10439-013-0872-9.
91. Driessen-Mol A, Emmert MY, Dijkman PE, Frese L, Sanders B, Weber B, et al. Transcatheter implantation of homologous "off-the-shelf" tissue-engineered heart valves with self-repair capacity: long-term functionality and rapid in vivo remodeling in sheep. J Am Coll Cardiol. 2014;63(13):1320–9. https://doi.org/10.1016/j.jacc.2013.09.082.
92. Moreira R, Velz T, Alves N, Gesche VN, Malischewski A, Schmitz-Rode T, et al. Tissue-engineered heart valve with a tubular leaflet design for minimally invasive transcatheter implantation. Tissue Eng Part C Methods. 2015;21(6):530–40. https://doi.org/10.1089/ten.TEC.2014.0214.
93. Reimer JM, Syedain ZH, Haynie BH, Tranquillo RT. Pediatric tubular pulmonary heart valve from decellularized engineered tissue tubes. Biomaterials. 2015;62:88–94. https://doi.org/10.1016/j.biomaterials.2015.05.009.

94. Dijkman PE, Driessen-Mol A, Frese L, Hoerstrup SP, Baaijens FP. Decellularized homologous tissue-engineered heart valves as off-the-shelf alternatives to xeno- and homografts. Biomaterials. 2012;33(18):4545–54. https://doi.org/10.1016/j.biomaterials.2012.03.015.
95. Sodian R, Lueders C, Kraemer L, Kuebler W, Shakibaei M, Reichart B, et al. Tissue engineering of autologous human heart valves using cryopreserved vascular umbilical cord cells. Ann Thorac Surg. 2006;81(6):2207–16. https://doi.org/10.1016/j.athoracsur.2005.12.073.
96. Robinson PS, Tranquillo RT. Planar biaxial behavior of fibrin-based tissue-engineered heart valve leaflets. Tissue Eng Part A. 2009;15(10):2763–72. https://doi.org/10.1089/ten.tea.2008.0426.
97. Kelley TA, Marquez S, Popelar CF. In vitro testing of heart valve substitutes. In: Iaizzo PA, Bianco R, Hill AJ, St. Louis JD, editors. Heart valves: from design to clinical implantation. New York: Springer; 2013. p. 283–320.
98. Ahlberg SE, Bateman MG, Eggen MD, Quill JL, Richardson ES, Iaizzo PA. Animal models for cardiac valve research. In: Iaizzo PA, Bianco R, Hill AJ, St. Louis JD, editors. Heart valves: from design to clinical implantation. New York: Springer; 2013. p. 343–57.
99. Weber B, Dijkman PE, Scherman J, Sanders B, Emmert MY, Grunenfelder J, et al. Off-the-shelf human decellularized tissue-engineered heart valves in a non-human primate model. Biomaterials. 2013;34(30):7269–80. https://doi.org/10.1016/j.biomaterials.2013.04.059.
100. Roosens B, Bala G, Droogmans S, Van Camp G, Breyne J, Cosyns B. Animal models of organic heart valve disease. Int J Cardiol. 2013;165(3):398–409. https://doi.org/10.1016/j.ijcard.2012.03.065.
101. Gottlieb D, Fata B, Powell AJ, Cois CA, Annese D, Tandon K, et al. Pulmonary artery conduit in vivo dimensional requirements in a growing ovine model: comparisons with the ascending aorta. J Heart Valve Dis. 2013;22(2):195–203.
102. Molina JE, Edwards JE, Bianco RW, Clack RW, Lang G, Molina JR. Composite and plain tubular synthetic graft conduits in right ventricle-pulmonary artery position: fate in growing lambs. J Thorac Cardiovasc Surg. 1995;110(2):427–35.
103. Akay HO, Ozmen CA, Bayrak AH, Senturk S, Katar S, Nazaroglu H, et al. Diameters of normal thoracic vascular structures in pediatric patients. Surg Radiol Anat. 2009;31(10):801–7. https://doi.org/10.1007/s00276-009-0525-8.
104. Shinoka T, Breuer CK, Tanel RE, Zund G, Miura T, Ma PX, et al. Tissue engineering heart valves: valve leaflet replacement study in a lamb model. Ann Thorac Surg. 1995;60(6 Suppl): S513–6.
105. Schmidt D, Dijkman PE, Driessen-Mol A, Stenger R, Mariani C, Puolakka A, et al. Minimally-invasive implantation of living tissue engineered heart valves. J Am Coll Cardiol. 2010;56(6):510–20. https://doi.org/10.1016/j.jacc.2010.04.024.
106. Weber B, Scherman J, Emmert MY, Gruenenfelder J, Verbeek R, Bracher M, et al. Injectable living marrow stromal cell-based autologous tissue engineered heart valves: first experiences with a one-step intervention in primates. Eur Heart J. 2011;32(22):2830–40. https://doi.org/10.1093/eurheartj/ehr059.
107. Nakayama Y, Takewa Y, Sumikura H, Yamanami M, Matsui Y, Oie T, et al. In-body tissue-engineered aortic valve (Biovalve type VII) architecture based on 3D printer molding. J Biomed Mater Res B Appl Biomater. 2015;103(1):1–11. https://doi.org/10.1002/jbm.b.33186.
108. Yamanami M, Yahata Y, Uechi M, Fujiwara M, Ishibashi-Ueda H, Kanda K, et al. Development of a completely autologous valved conduit with the sinus of Valsalva using in-body tissue architecture technology: a pilot study in pulmonary valve replacement in a beagle model. Circulation. 2010;122(11 Suppl):S100–6. https://doi.org/10.1161/CIRCULATIONAHA.109.922211.
109. Flanagan TC, Sachweh JS, Frese J, Schnoring H, Gronloh N, Koch S, et al. In vivo remodeling and structural characterization of fibrin-based tissue-engineered heart valves in the adult sheep model. Tissue Eng Part A. 2009;15(10):2965–76. https://doi.org/10.1089/ten.TEA.2009.0018.

110. Syedain ZH, Lahti MT, Johnson SL, Robinson PS, Ruth GR, Bianco RW, et al. Implantation of a tissue-engineered heart valve from human fibroblasts exhibiting short term function in the sheep pulmonary artery. Cardiovasc Eng Technol. 2011;2(2):101–12. https://doi.org/10.1007/s13239-011-0039-5.
111. Syedain Z, Reimer J, Schmidt J, Lahti M, Berry J, Bianco R, et al. 6-Month aortic valve implantation of an off-the-shelf tissue-engineered valve in sheep. Biomaterials. 2015;73:175–84. https://doi.org/10.1016/j.biomaterials.2015.09.016.
112. Emmert MY, Weber B, Behr L, Sammut S, Frauenfelder T, Wolint P, et al. Transcatheter aortic valve implantation using anatomically oriented, marrow stromal cell-based, stented, tissue-engineered heart valves: technical considerations and implications for translational cell-based heart valve concepts. Eur J Cardiothorac Surg. 2014;45(1):61–8. https://doi.org/10.1093/ejcts/ezt243.
113. Reimer J, Syedain Z, Haynie B, Lahti M, Berry J, Tranquillo R. Implantation of a tissue-engineered tubular heart valve in growing lambs. Ann Biomed Eng. 2017;45(2):439–451.
114. Schoen FJ. Heart valve tissue engineering: quo vadis? Curr Opin Biotechnol. 2011;22(5):698–705. https://doi.org/10.1016/j.copbio.2011.01.004.

# Decellularization in Heart Valve Tissue Engineering

Katherine M. Copeland, Bo Wang, Xiaodan Shi, Dan T. Simionescu,
Yi Hong, Pietro Bajona, Michael S. Sacks, and Jun Liao

**Abstract** Annually, over 50,000 deaths are attributed to heart valve disease (HVD) in the United States. The most common treatment for HVD, such as stenosis and regurgitation, is total valve replacement using mechanical and bioprosthetic heart valves, which often results in subsequent surgery to replace failed implants. For the pediatric population especially, a viable valve implant with the potential to repair, remodel, and grow within the patient is a great clinical need. Recent research has demonstrated that tissue-engineered heart valves (TEHV) have the potential to deliver a viable valve replacement, which is constructed with scaffolds and functional cells. Acellular heart valve (HV) scaffolds obtained by decellularization, biological polymer scaffolds, and synthetic polymer scaffolds have been widely used for TEHV fabrication, each having advantages and disadvantages. In this chapter, we focus on TEHV via a decellularization approach and provide a systematic review covering (1) the concepts of decellularization and current decellularization methods, (2) a comparative study showing how the ultrastructure and biomechanics of acellular HV scaffolds were affected by different

K. M. Copeland · X. Shi · Y. Hong
Department of Bioengineering, University of Texas at Arlington, Arlington, TX, USA

B. Wang
College of Science, Mathematics and Technology, Alabama State University, Montgomery, AL, USA

D. T. Simionescu
Department of Bioengineering, Clemson University, Clemson, SC, USA

P. Bajona
Department of Cardiovascular and Thoracic Surgery, University of Texas Southwestern Medical Center, Dallas, TX, USA

M. S. Sacks
The Oden Institute and the Department of Biomedical Engineering, The University of Texas at Austin, Austin, TX, USA

J. Liao (✉)
Tissue Biomechanics and Bioengineering Laboratory, The Department of Bioengineering, The University of Texas at Arlington, Arlington, TX, USA
e-mail: jun.liao@uta.edu

decellularization methods, (3) the cell sources and bioreactor systems for TEHV reseeding, conditioning, and integrity testing, and (4) the current accomplishments of HV decellularization in animal studies and clinical trials and the challenges in moving this approach toward clinical applications.

**Keywords** Heart valves · Decellularization · Aortic valve leaflets · Acellular valve scaffolds · Extracellular matrix · Trilayered structure · Collagen · Elastin · Proteoglycans · Tissue engineering · Bioreactors · Valve replacements

# 1 Introduction

Heart valve diseases comprise a considerable portion of cardiovascular diseases, a major cause of morbidity and mortality. Heart valve disease occurs when one or more of the four heart valves no longer functions as it should, commonly due to stenosis, regurgitation, or congenital defects. According to the American Heart Association (AHA), 50,222 deaths in the USA were related to heart valve diseases in 2013 [1, 2]. The tricuspid valve, pulmonary valve, mitral valve, and aortic valve are the four types of valves in the heart that ensure unidirectional blood flow beginning from the right atrium, followed by the right ventricle, pulmonary artery, left atrium, left ventricle, and lastly the aorta. The aortic valve, located between the aorta and the left ventricle, is most prone to disease among the four types of heart valves due to the high working pressure and complicated stresses involved in its role in governing blood flow from the heart to the rest of the body [3]. In fact, aortic valve disease (AVD) led to approximately 39,000 of reported deaths in 2013 [1]. For patients with end-stage heart valve disease, complete heart valve replacement is currently the standard treatment option [2, 4].

Mechanical and bioprosthetic heart valves are the two primary options for heart valve substitutes. Although these substitutes can be successful in clinical applications, each of these options has major shortcomings [5]. Mechanical heart valves are made of synthetic materials, which give them excellent long-term durability of ~20–30 years [6]; however, patients with these valves face the risk of thromboembolic complications and require lifelong anticoagulation therapy [5]. On the other hand, bioprosthetic heart valves are made from chemically crosslinked (e.g., glutaraldehyde) porcine aortic valves or bovine pericardium, which give excellent hemodynamic performance and do not require lifelong anticoagulation therapy. However, bioprosthetic heart valves lack long-term durability and, therefore, have a limited lifespan, up to ~15–20 years [7, 8], due to progressive leaflet structural deterioration and calcification [9, 10]. Cryopreserved homografts have also been employed, but are limited by donor shortage and long-term degeneration [9–11]. A recent option, used in patients who are considered inoperable, is transcatheter aortic valve replacement (TAVR), which works by inserting a collapsible bioprosthetic valve through a catheter and placing and expanding it into the aortic root; nevertheless, TAVR experiences durability limitation similar to bioprosthetic heart valves, as well as its own issues such as paravalvular leak [12].

For the above-mentioned options, it is important to note that all the valve substitutes lack the capability to grow and remodel within the patient and hence present a major challenge in applications in the pediatric population [11]. Thus, pediatric recipients of all types of heart valve prostheses are subjected to subsequent surgeries to replace the valves that they have outgrown [13]. Another major concern with pediatric bioprosthetic heart valves is the acceleration of heart valve calcification [14]. Calcification is commonly caused by cellular xenograft rejection and glutaraldehyde fixation pretreatment, especially in younger, growing recipients [14, 15]. Therefore, there is a great need for tissue-engineered heart valves (TEHV) that experience no deterioration/calcification and can grow and remodel in response to the pediatric patient's somatic growth.

Although diseased aortic valves are generally most common and, therefore, introduce the need for tissue-engineered aortic valves (TEAV), there is also a need for tissue-engineered pulmonary valves (TEPV). For instance, in the Ross procedure for pediatric patients, the diseased aortic heart valves are replaced by autografts of the patient's own pulmonary valve. The pulmonary valve is then replaced by a donor homograft [16]. This homograft requires a need for re-intervention more commonly than the pulmonary autograft. Depending on the age of the patient and the patient's response to the initial homograft, the homograft may need to be replaced by another homograft or even a mechanical or biosynthetic valve [17], thus arises the need for a TEPV.

In general, researchers hope to fabricate these TEHVs by recellularizing three-dimensional (3D) scaffolds to achieve cell-scaffold integration, appropriate mechanical properties, and robust functional performance. Recellularization can be achieved via either in vitro bioreactor conditioning or surgical insertion of a cell-free construct that encourages host cell migration and repopulation in vivo [18]. Both approaches, in vitro recellularization and in vivo cell repopulation, rely on the design and fabrication of scaffold materials.

The most commonly researched scaffold materials for TEHV include synthetic polymer scaffolds, biological polymer scaffolds, and acellular HV scaffolds. Synthetic sources such as poly(glycolic) acid (PGA), poly(lactic) acid (PLA), poly (caprolactone) (PCL), and poly(hydroxyalkanoate) (PHA) offer uniform mechanical and structural control; however, degradation rates and cell compatibility of these scaffolds are not easily predicted [19]. In vitro cellular seeding of polymeric scaffolds has shown to increase cell viability, leading to the production of extracellular matrix (ECM) that is important for the continuance of cellular integration and maintenance of mechanical properties; however, cell proliferation and ECM production rates in current studies have not been optimal [19–21]. On the other hand, biological polymer scaffolds, such as collagen, fibrin, and hyaluronic acid, offer the inherent benefit of cell compatibility but are not as mechanically sound as their synthetic counterparts [19]. It was reported that biological polymer scaffolds are currently mechanically unsuitable for aortic valve replacement since their weak mechanical strength cannot match the high working pressure required of the aortic environment [19, 21, 22].

In this chapter, we will focus on acellular HV scaffolds and provide a systematic review covering (1) the concepts of decellularization and current decellularization methods, (2) a comparative study showing how the ultrastructure and biomechanics of acellular HV scaffolds were affected by different decellularization methods, (3) the cell sources and bioreactor systems for TEHV reseeding, conditioning, and integrity testing, and (4) the current accomplishments of HV decellularization in animal studies and clinical trials and the challenges in moving this approach toward clinical applications.

## 2 Fundamentals of Tissue-Engineered Heart Valves

The goal of tissue engineering is to create a functional and living organ that can repair, remodel, and grow within the patient just as the native organ would in response to injury, disease, and normal physiological forces [11, 23]. To achieve this goal, three requirements must be met: (1) proper 3D scaffolds; (2) isolation of appropriate cell sources (unless using a cell-free approach); and (3) in vitro culture conditions, such as bioreactor conditioning, that can produce the final tissue-engineered product prior to implantation [18, 23].

The ideal TEHV should be widely available, allow cell attachment, proliferation, and migration, provide structural and biomechanical cues and biochemical signals in favor of valvular interstitial cell and valvular endothelial cell behavior, and grow as the patient develops [24, 25]. Additionally, TEHV should have regenerative potential, optimal mechanical properties, low risk of matrix degeneration and thromboembolism, and should be free of adverse immune responses [24, 25]. Therefore, the first challenge in creating a TEHV is constructing 3D scaffolds with proper ECM components, ultrastructure, and mechanical properties. Moreover, it is important to implement the proper cell source and type to achieve successful recellularization of heart valve scaffolds [26]. The two cell types required for TEHV are myofibroblast-like valve interstitial cells (VICs) and valve endothelial cells (VECs) [23]. In the context of TEHV, valve interstitial cells produce ECM and are, therefore, responsible for the development of tissue to match the mechanical properties of native heart valves [23]. Endothelial cells act as a covering for the surface of TEHV and prevent thrombotic complications from occurring [23]. Previous studies show that both types of cells can be harvested or derived from vascular vessels, bone marrow, umbilical cord blood (UCB), and amniotic fluid [23]. Lastly, HV bioreactors, in vitro systems that are designed to mimic the pressure profile of the left ventricle, provide physiological stimulations allowing the cells to thoroughly reseed the HV scaffolds, proliferate, differentiate into or maintain the right phenotypes, and synthesize ECM.

## 3 Components and Ultrastructure of Aortic Valves

As mentioned above, the aortic valve is the most commonly diseased of the four heart valves. Consequently, the aortic valve has been studied most in the field of HV tissue engineering. It is thus important to understand its constitutive ECM components, ultrastructure, and cellular constituents. With the knowledge of the natural design of aortic valve scaffolds and cells, bioengineers can better develop a tissue engineered aortic valve (TEAV).

### *3.1 Structure of Aortic Valve*

The aortic valve consists of three flexible tissue leaflets (or cusps) and a circular root complex, and is categorized as a semilunar valve due to its half-moon shaped leaflet [27]. The aortic valve leaflet is a thin, flexible trilayered structure composed of the fibrosa, spongiosa, and ventricularis layers, which contribute to its complicated and efficient biomechanical functions [28, 29]. The fibrosa is the thickest and main load-bearing layer of the three. It is located at the aortic side and is responsible for tissue strength and durability [28]. The fibrosa is comprised of a dense collagen network, primarily type I collagen. Type I collagen forms thick fiber bundles that are primarily oriented in the circumferential direction of the valve leaflet. Type III collagen is also present and forms finer, more reticulate fibers. In the fibrosa, elastin fibers are sparsely meshed with the collagen fiber network [28]. The middle layer, the spongiosa, consists mainly of glycosaminoglycans (GAGs) [30] and a relatively sparse collagen fiber network that spans across the spongiosa and connects the fibrosa and ventricularis layers [28, 31]. The spongiosa operates as a cushion between the fibrosa layer and ventricularis layer and reduces the shear stresses due to valve flexure, enabling long-term durability of the valves. Moreover, the interconnection of collagen fibers across the spongiosa allows a trilayered structure with less risk of delamination. The ventricularis, the thinnest of the three layers, mainly consists of radially oriented elastin fibers, which support large deformation along the radial direction and minimize energy dissipation during valve loading and unloading [32, 33].

### *3.2 Cellular Components of Aortic Valve*

Cellular components of the aortic valve include VICs and VECs that account for nearly 30% of the volumetric density of the valve [34]. VICs are the principal cell population and are essential to valve function. Structurally and mechanically, VICs are an intermediate cell-type between fibroblasts and smooth muscle cells, but also take on characteristics of contractile myofibroblasts in response to injury or disease

[34, 35]. VICs are responsible for maintaining the structural and mechanical properties of tissues, and also play a role in remodeling and repairing damaged valve tissue by synthesizing important ECM components such as type I collagen and glycosaminoglycans (GAGs) [36, 37]. VECs are oriented orthogonally to the direction of blood flow and form a continuous layer. In response to hemodynamic flow, VECs regulate the phenotype of VICs to maintain the overall leaflet tissue integrity [38]. During the developmental, mature, and aging stages, the cell components, cell density, and cell phenotype of the heart valves change accordingly [39]. Because of the dynamic cellular changes that occur throughout people's lives, it is important to achieve the proper cellular components, cell densities, and cell phenotypes prior to implantation of a tissue-engineered construct. This is especially true when deciding between a cell seeded construct or an acellular construct.

## 4  Decellularization and Applications in TEHV

### 4.1  Concept of Decellularization and Acellular HV Scaffolds

Decellularization is the removal of all cellular materials from a native tissue via chemical, enzymatic, and/or physical means [40]. To obtain acellular HV scaffolds, it is important to identify the optimal decellularization protocols that can thoroughly remove all cellular contents, including the cell membrane and chromosomal debris, while introducing minimal structural disruption and preserving the mechanical properties.

The first successful decellularization protocol was performed on small intestine submucosa in 1995 by Badylak et al. using a chemical detergent treatment [41]. Since then, common decellularization methods, including chemical, enzymatic, and/or physical treatments, have been developed. The combination of these treatments disrupts the cell membrane, releases cell contents, and facilitates removal of cellular components from the ECM. Badylak et al. also notes that the degree to which decellularization occurs depends upon the tissue and the specific decellularization protocol used [41]. It is thus important to utilize the optimal decellularization protocols for each specific tissue to achieve the desired acellular scaffold properties and performance.

Decellularization techniques have been applied to heart valve tissue engineering and have shown promising results in large animal studies, preclinical testing, and clinical trials [26, 42–44]. The advantages of the decellularized heart valves as scaffolds include preservation of leaflet morphology and geometry, native ECM molecules, overall ultrastructure and trilayered configuration, and similar mechanical and functional behavior to that of the native heart valve leaflet.

The trilayered structure design, the distribution and configuration of collagen, elastin, and GAGs in individual layers, along with valve tissue viscoelasticity are thought to contribute to the overall excellent performance and durability of heart valve leaflets [11]. Decellularized heart valves preserve most of the above traits and are, therefore, attractive scaffolds for TEHV [45]. However, there are still concerns about acellular HV scaffolds, such as disruption of certain ECM architecture, challenges in thorough reseeding, immunogenicity, and toxicity by possible residual chemicals used in the decellularization process [19, 46]. In the following section, we briefly summarize various decellularization methods and protocols and discuss their advantages and disadvantages.

## 4.2 Decellularization Techniques Available for HV Decellularization

Currently, valve tissues of human origin or xenogeneic origin have been decellularized with various protocols to obtain acellular HV scaffolds for valvular tissue engineering [47, 48]. Different decellularization protocols involve physical, chemical, or enzymatic treatments, or a combination of those techniques [40, 49].

*Decellularization Using Physical Forces.* In decellularization, the most commonly used physical methods include application of mechanical forces, snap-freezing, sonication, and mechanical agitation [40]. Mechanical forces can cause cell lysis; however, only the cells on the surface of the tissues, the cells near the surface, and the cells associated with sparsely organized ECM structures can be effectively removed with this method [40]. Snap-freezing disrupts cellular membranes by forming intracellular ice crystals that cause cell lysis; however, the intracellular contents still need subsequent processing for removal [40, 50]. Mechanical agitation and sonication have often been applied along with chemical treatments to facilitate cell lysis and cellular debris removal. Overall, physical treatment is insufficient to achieve thorough decellularization when applied alone, and must be combined with other chemical and/or enzymatic treatments to remove cellular remnants [40].

*Decellularization via Chemical Agents.* Chemical agents for decellularization include acids and bases, hypotonic and hypertonic solutions, and detergents [41]. Acids and bases not only cause hydrolytic degradation of biomolecules, but also damage ECM components and decrease mechanical properties [51–54]. Hypertonic saline can cause DNA to be dissociated from proteins [55]; hypotonic solutions can cause cell lysis by an osmotic effect and better preserve the ECM architecture [56]. Detergents, such as Triton X-100, sodium deoxycholate acid, sodium dodecyl sulfate (SDS), deoxycholic acid, and ethylenediaminetetraacetic acid (EDTA), are most commonly used in decellularization. These detergents have the capability to preserve, at various degrees, overall tissue dimension, ECM ultrastructure, and mechanical behaviors, and simultaneously disrupt lipid–lipid bonds/lipid–protein bonds and remove cellular nuclei/DNA from dense tissues and organs [40]. It was reported that the combination of multiple detergents is more effective for cell removal [57] and ultrastructure preservation [58, 59] than the single-detergent treatments; however, detergent treatments alone cannot completely dissolve and remove the cellular fragments [60, 61].

*Decellularization via Enzymatic Treatments.* Enzymatic treatments, including the use of trypsin, endonucleases, exonucleases, collagenase, lipase, dispase, thermolysin, and α-galactosidase, can specifically remove cell residues, leaving behind the overall structural and functional ECM constituents. Nonetheless, enzyme-treated scaffolds showed fragmentation and distortion of the ECM components [62]. Furthermore, enzymatic agents used for decellularization can be difficult to remove from the acellular tissue scaffolds and may subsequently result in an immune response [40]. In decellularization applications, enzymes are usually added into the decellularization solution along with chemical detergents since enzymatic treatment alone is inadequate to achieve complete cell removal [41].

## 5 Effects of Decellularization Protocols on Leaflet Biomechanics and a Study on Fatigue Mechanism

In general, decellularization maintains the overall structural and mechanical characteristics of the native tissues; however, specific decellularization protocols can result in different levels of preservation of those properties. Liao et al. performed a comparative study to determine the effects of three major decellularization protocols on the structural and mechanical properties of aortic valve leaflets [45]. The main decellularization agents of the three major protocols were (1) an anionic detergent—sodium dodecyl sulfate (SDS), (2) an enzymatic agent—Trypsin, and (3) a nonionic detergent—Triton X-100. The various outcomes of the acellular leaflets from those three decellularization protocols were reported [45].

*Effects of Decellularization Protocols on Leaflet Mechanical Behavior.* It was found that under 60 N/m equibiaxial tension, the three protocols all increased the extensibility of the acellular leaflets, with Triton X-100 exhibiting the greatest increase (Fig. 1a) [45]. Analysis of maximum stretch ratios along the circumferential direction ($\lambda_c$) and the radial direction ($\lambda_r$) revealed that Triton X-100 had the greatest effect on extensibility in the radial direction (Fig. 1b). The three protocols did not affect the maximum tensile modulus (MTM) of the radial direction, but increased the MTM of the circumferential direction, with the Triton X-100 protocol causing the largest MTM increase (Fig. 1c). Axial coupling, a parameter reflecting the inter-fiber mechanical interactions and fiber rotations, revealed that all three decellularization protocols induced larger changes in stretch along the opposite stretch axis, which meant a higher degree of axial coupling (Fig. 2a, b). This observation could be explained by the increased fiber mobility after decellularization, i.e., a collagen network less bounded in the acellular leaflets. Bending tests of the leaflets showed that all three decellularization protocols induced a large decrease in flexural rigidity (Fig. 2c). Furthermore, the momentum-curvature curves of the acellular leaflets all showed a nonlinear response in both "with curvature" (WC) and "against curvature"

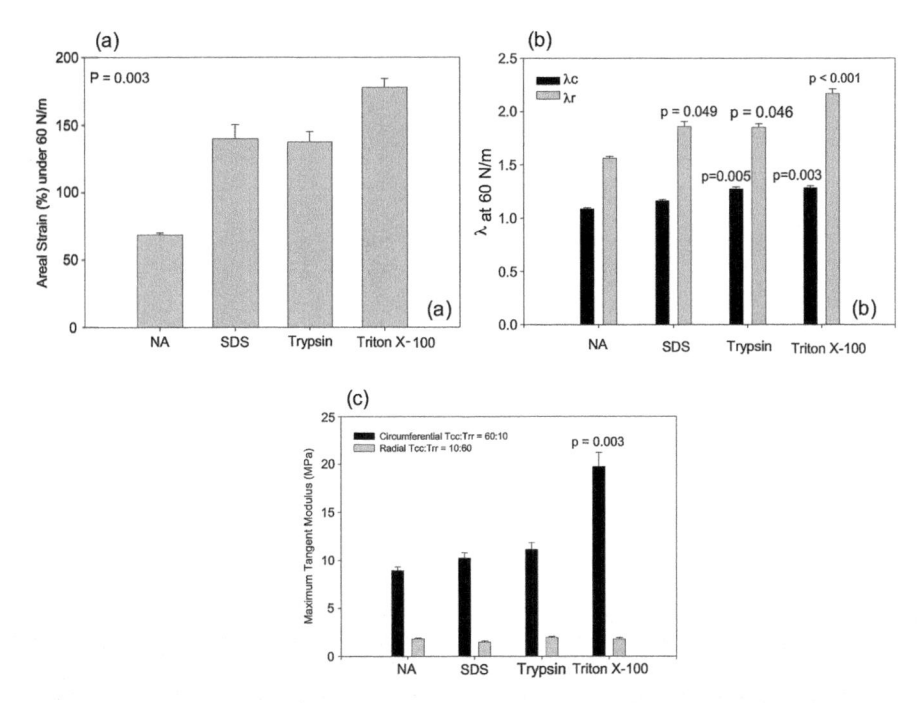

**Fig. 1** Parameters for biaxial behavior of aortic valve leaflets obtained by different decellularization protocols. Net extensibility was represented by areal strain (**a**) and maximum stretch ratios (**b**) in the circumferential and radial directions at 60 N/m equibiaxial tension. Net extensibility increased for all decellularization protocols. (**c**) Maximum tangent modulus (MTM) was increased in the circumferential direction with no effect in the radial direction under 60 N/m equibiaxial tension. Reprinted with permission from *Effects of decellularization on the mechanical and structural properties of the porcine aortic valve leaflet*, by Liao et al., March 2008

(AC) bending while the native leaflets had a much stiffer, near linear moment–curvature relationship (Fig. 2c, d). The decreased flexural rigidity and nonlinear response of the acellular leaflets were likely associated with loss of tissue stiffness and ECM disruption after decellularization [45].

*Effects of Decellularization Protocols on Leaflet Ultrastructure.* Digital images of the intact leaflets were taken before and after decellularization with a bright light shining from the back and through leaflet. It was found that the gross collagen fiber architecture, including the branching of collagen fiber bundles and the thinning regions with finer network features, were preserved after decellularization (Fig. 3a, b). Different decellularization protocols have different effects on the dimensions of the acellular leaflets. Evaluation of area and thickness before and immediately after decellularization showed that (1) the SDS protocol did not cause any statistical differences in area or thickness of the acellular leaflets; (2) the Trypsin protocol increased both the area and thickness of the acellular leaflets ($p = 0.011$ and $p = 0.002$, respectively); and (3) the Triton-X protocol decreased both the area

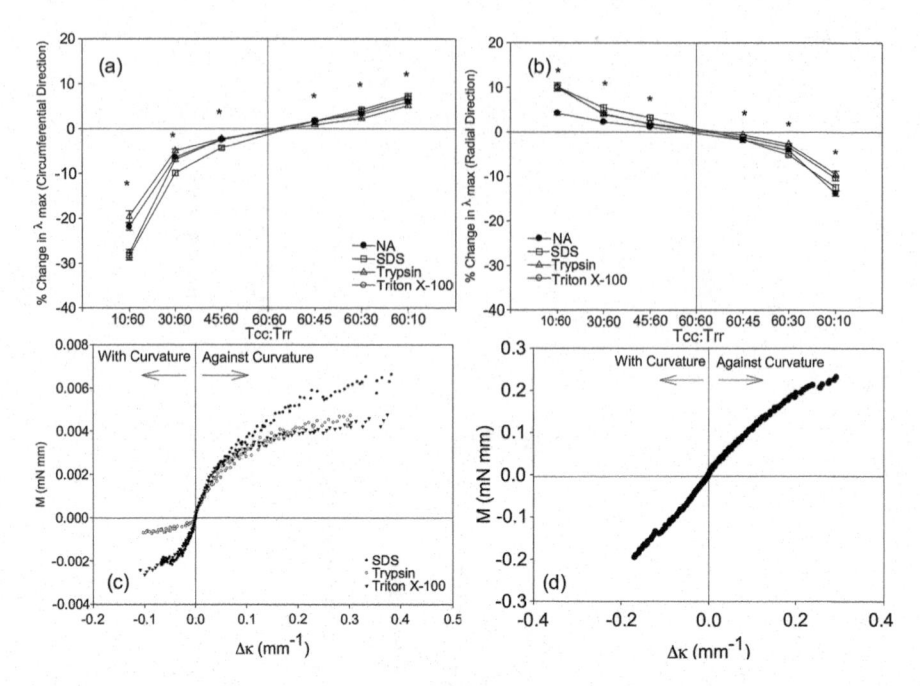

**Fig. 2** Axial cross-coupling for biaxial protocols for the (**a**) circumferential and (**b**) radial directions demonstrated that all decellularization protocols caused changes when compared to the native valve (NA), which is likely caused by the structural changes in the tissue following decellularization. The flexural behavior of (**c**) the decellularized valve leaflets exhibited a nonlinear response both with and against the curvature, while (**d**) the native valve leaflets had an overall linear response. The nonlinear behavior is likely due to the decrease in flexural rigidity and ECM disruption. Reprinted with permission from *Effects of decellularization on the mechanical and structural properties of the porcine aortic valve leaflet*, by Liao et al., March 2008

and thickness of the acellular leaflets ($p < 0.001$ and $p < 0.001$, respectively) (Table 1).

Small angle light scattering measurements were performed over the entire leaflet to reveal the local and gross fiber structure, including the preferred fiber orientation (short dark line) and degree of fiber alignment (color code) in local regions (Fig. 3c–f). Native leaflets showed significant regional variations with various degrees of alignment, in which the coaptation region and the mid-belly region aligned relatively well, but other regions, such as the regions near the annulus and the regions containing the Nodulus of Aranti, showed poorer alignment (Fig. 3c). The SDS and Trypsin protocols were able to preserve the regional variations, while causing an overall decrease in the degree of alignment, especially in the belly region (Fig. 3d, e). The acellular leaflets obtained with the Triton X-100 protocol showed a loss of typical regional variation and displayed more homogenous fiber structure (Fig. 3f).

Histological images of hematoxylin and eosin (H&E) stained heart valve leaflets demonstrated that the three protocols thoroughly removed all cell nuclei (Fig. 4b–d). By comparing the Movat's Pentachrome histological images, it is noticeable that the

# Heart Valve Decellularization

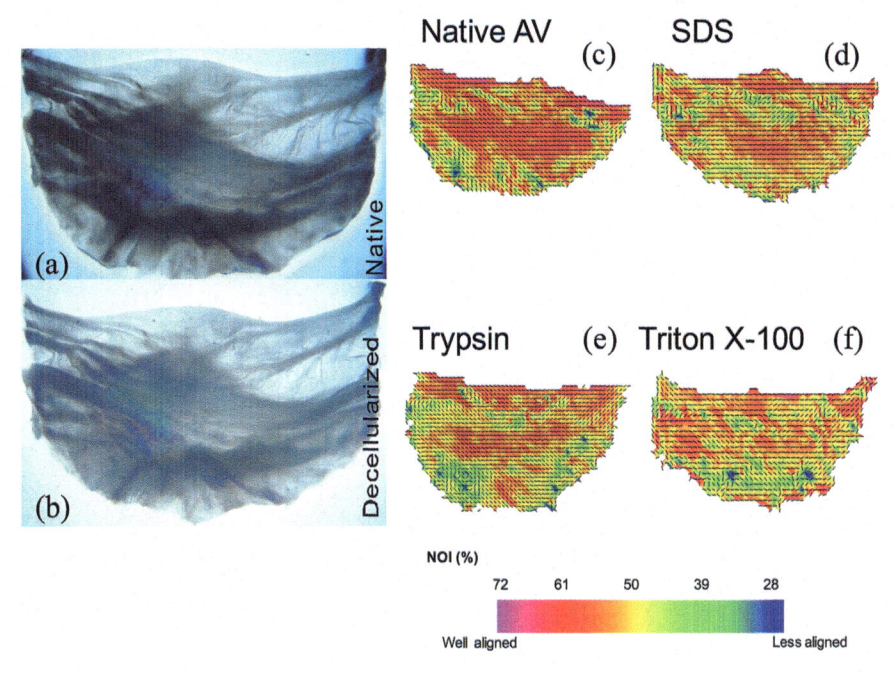

**Fig. 3** Morphology of the native aortic valve leaflet (**a**) and the same leaflet decellularized with the SDS protocol (**b**), showing the overall collagen fiber architecture was preserved after decellularization. Fiber architecture mapped by small angle light scattering (SALS) demonstrated that, when compared to the native valve leaflet (**c**), the SDS-treated valve leaflet (**d**) had better alignment than the trypsin-treated valve leaflet (**e**) and the Triton X-100-treated valve leaflets (**f**) (larger NOI corresponds to better alignment). Reprinted with permission from *Effects of decellularization on the mechanical and structural properties of the porcine aortic valve leaflet*, by Liao et al., March 2008

**Table 1** Different decellularization protocols caused different changes in leaflet dimensions

|  | Area of valve leaflets (mm$^2$) |  | Thickness of valve leaflets (mm) |  |
| --- | --- | --- | --- | --- |
|  | Before decellularization | After decellularization | Before decellularization | After decellularization |
| SDS | 264.5 ± 36.8 | 268.0 ± 35.9 | 0.48 ± 0.07 | 0.43 ± 0.08 |
| Trypsin | 250.5 ± 42.4 | 302.8 ± 49.6* | 0.43 ± 0.04 | 0.48 ± 0.04* |
| Triton X-100 | 260.9 ± 36.3 | 213.2 ± 26.6* | 0.31 ± 0.05 | 0.16 ± 0.04* |

The SDS protocol did not cause any statistical differences in area or thickness of the acellular leaflets. The Trypsin protocol increased both the area and thickness of the acellular leaflets ($p = 0.011$ and $p = 0.002$, respectively). The Triton-X protocol decreased both the area and thickness of the acellular leaflets ($p < 0.001$ and $p < 0.001$, respectively). Values are given as Mean ± SEM. * denotes statistical significance when compared with the area and thickness of the native valve leaflet. Reprinted with permission from *Effects of decellularization on the mechanical and structural properties of the porcine aortic valve leaflet*, by Liao et al., March 2008

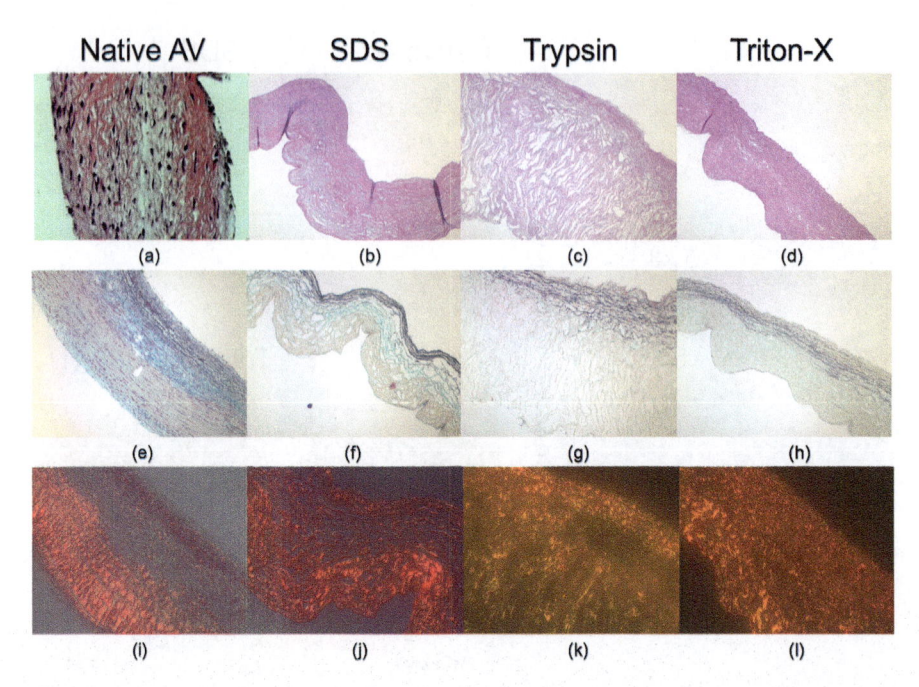

**Fig. 4** Histology for (**a–d**) hematoxylin and eosin (H&E), (**e–h**) Movat's Pentachrome, and (**i–l**) picrosirius red stains. H&E stains verified that no cell nuclei were present after decellularization with each protocol. The trilayered structure of the leaflet decellularized with (**f**) SDS (anionic detergent) was very similar to the trilayered structure of the (**e**) native aortic valve leaflet. However, the spongiosa was depleted with decellularization with (**g**) Trypsin (enzymatic agent) and (**h**) Triton X-100 (nonionic detergent). (**i–l**) The macroscopically well-organized collagen crimp structure was disrupted by all three decellularization protocols as compared to the native aortic valve leaflet. Reprinted with permission from *Effects of decellularization on the mechanical and structural properties of the porcine aortic valve leaflet*, by Liao et al., March 2008

SDS protocol preserved the trilayered structure very well (Fig. 4e, f), while the Trypsin protocol and Triton X-100 protocol caused depletion of GAGs in the spongiosa after decellularization (Fig. 4g, h). Moreover, the Trypsin protocol and the Triton X-100 protocol resulted in significant changes in ECM ultrastructure. The ECM in Trypsin-decellularized leaflets showed a swollen and loose collagen fiber structure, as well as a disrupted elastin network (Fig. 4g). The swelling of the ECM network and loss of elastin constriction could possibly explain the increased leaflet area and thickness in Trypsin-decellularized leaflets (Fig. 4g and Table 1). On the other hand, the acellular leaflets obtained with Triton X-100 showed a condensed collagen fiber morphology and a relatively well-preserved elastin fiber network (Fig. 4d, h), which is consistent with the observed overall tissue shrinkage shown by the decreased leaflet area and thickness (Table 1) [45].

Polarized light microscopy was applied to picrosirius red-stained slices to reveal the collagen crimp structure in the leaflets. The fibrosa layer in the native leaflets exhibited macroscopically well-organized light distinguishing bands, which revealed the well-coordinated collagen crimp structure spanning the fibrosa (wavy collagen fibers) (Fig. 4i). Noticeably, when compared to the native leaflets, the acellular leaflets obtained from all three decellularization protocols showed that the collagen fibers were locally crimped (Fig. 4j–l), but no longer had macroscopic distinguishing bands (Fig. 4i) [45]. Although collagen fibers in the acellular leaflets retained a degree of local crimp structure, their macroscopically disrupted alignment was understandably an important contributor to the altered leaflet mechanical behavior, such as increased tissue extensibility. Moreover, loss of the well-organized collagen crimp structure indicated a loss of certain protective mechanism associated with this natural design [63], which might affect collagen fiber loading behavior and long-term tissue durability [45].

*Fatigue Mechanism of the Acellular Leaflets.* It is important to understand the fundamental fatigue mechanism of the acellular leaflets, so researchers can better predict the fate of implants and envision strategies to prevent structural degeneration and increase valve durability. Liao et al. developed a pulsatile flow loop equipped with a silicone root featuring the Valsalva sinus design to subject the decellularized aortic valve conduits to in vitro cardiac exercising (2 weeks: ~1.0 million cycles; 4 weeks: ~2.0 million cycles) [64]. The exercised leaflets were characterized to assess the structural and mechanical alterations [64]. Although this in vitro cardiac exerciser still represented a harsh environment, in which the effects of solution storage could not be fully excluded and the observed fatigue took place in an accelerated fashion, the fatigue pattern of the acellular leaflets did reveal the trend of leaflet ECM disruption without cellular maintenance.

The observations can be summarized as follows: (1) Although the overall morphology was maintained and the leaflets were able to coapt and support pressure load, the fatigued leaflets exhibited an unfolded and thinned morphology (Fig. 5); (2) the thinned morphology was a result of the straightening of the locally wavy collagen fibers and the disruption of the elastin network; (3) the unfolding of wavy

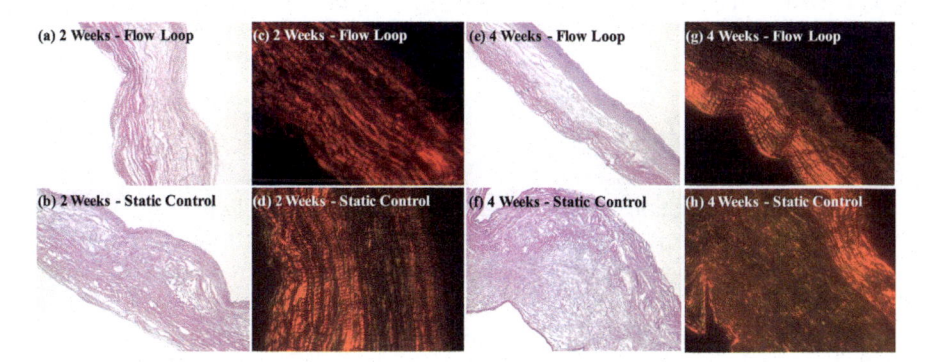

**Fig. 5** H&E staining (**a, b, e, f**) and polarized light (picrosirius red staining) (**c, d, g, h**) showed the straightening of collagen fibers in the exercised leaflets. (**a, c**) 2-week exercising and (**b, d**) 2-week static control; (**e, g**) 4-week exercising and (**f, h**) 4-week static control. Reprinted with permission from *The Intrinsic Fatigue Mechanism of the Porcine Aortic Valve Extracellular Matrix*, by Liao et al., January 2012

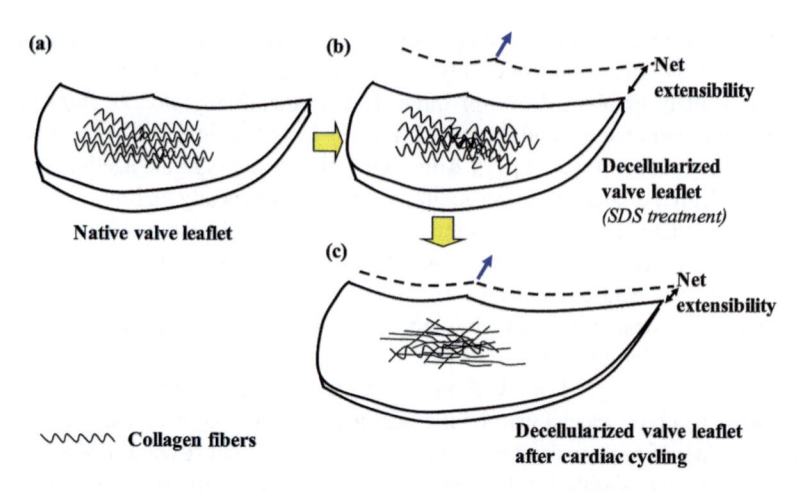

**Fig. 6** A schematic illustration of the underlying structural mechanism for net extensibility changes. (**a**) Native aortic valve leaflet, (**b**) Decellularized leaflet right after SDS treatment, and (**c**) Decellularized valve leaflet after cardiac cycling. Note that the sketches of collagen fibers were simplified for illustrative purposes. Refer to the histology in this study and our previous study for the details of collagen fiber organization. Reprinted with permission from *The Intrinsic Fatigue Mechanism of the Porcine Aortic Valve Extracellular Matrix*, by Liao et al., January 2012

collagen fibers and structural thinning caused a changed leaflet reference status and resulted in lower net tissue extensibility of the fatigued leaflets (Fig. 6) [64]. In short, due to lack of either exogenous stabilizing crosslinks and/or cellular maintenance, the decellularized valve leaflets experienced structural fatigue and the structural disruption was irreversible and cumulative [64]. This study points out two approaches that can help resist fatigue caused by cyclic loading, i.e., exogenous stabilizing crosslinking and recellularization of the acellular leaflets (either in vitro or in vivo), in order to achieve tissue maintenance and remodeling.

## 6 Cell Sources for TEHV

With regard to applications of TEHV, several types of human cells have been studied including vascular-derived cells, bone marrow-derived cells, UCB-derived cells, and amniotic fluid-derived stem cells. Vascular endothelial cells are commonly used in recellularization and can be harvested from the vasculature and proliferated in vitro [44, 65]. Vascular endothelial cells have proven to be effective to reconstruct the right ventricular outflow tract (RVOT) during the Ross operation. In one study, after Dohmen et al. replaced the diseased aortic valve with the autologous pulmonary valve, an acellular pulmonary valve autograft seeded with autologous vascular endothelial cells was used to replace the pulmonary valve and reestablish the RVOT [44]. After 1 year follow-up, the reconstructed RVOT continued to demonstrate excellent hemodynamic function [44]. It is believed that seeding the decellularized heart valves with vascular endothelial cells may increase valve hemocompatibility and decrease the chance of valve degeneration, and hence increase valve durability.

Recently, new cell sources with improved growth and differentiation potential have been evaluated as reseeding cells in HV tissue engineering. Bone marrow-derived cells, UCB-derived cells, and amniotic fluid stem cells are attractive candidates for reseeding because of their proliferative potential. Bone marrow-derived cells are easily isolated and demonstrate high volume cellular expansion in vitro [66]. Due to the excellent proliferation and differentiation potential, bone marrow-derived cells have been used to regenerate heart valves in vitro [21, 67]. It is worth mentioning that bone marrow-derived cells are easily converted into endothelial cells, fibroblasts, myofibroblasts, and smooth muscle cells that bear a resemblance to valvular cells [68]. However, these cells still meet the challenge of inflammatory and thrombotic reactions [69].

The umbilical cord contains various types of progenitor cells including mesenchymal cells and endothelial progenitor cells (EPC) from Wharton's jelly tissue [70]. Umbilical cord- or cord blood-derived cells can be used as an alternative to endothelial cells and fibroblasts, and provide the high number of cells required in regenerative medicine [71]. The extraordinary proliferative capability of UCB-derived cells and the efficiency in providing a large number of endothelial-like cells for HV tissue engineering make them an attractive autologous cell source, particularly for congenital defects [72–75]. The disadvantages of UCB-derived cells is their low adhesive ability after cryopreservation and subsequent thawing, and the long-term influence of storage is still unknown [76]. Amniotic fluid-derived stem cells are unique fetal cell sources that contain multipotent stem cells with potential for tissue engineering applications [77]. Schmidt et al. created a viable autologous heart valve leaflet ex vivo with a single cell source isolated from human amniotic fluid [78]. Both UCB-derived cells and amniotic fluid-derived cells serve as a multipotent stem cell source without raising the ethical concerns involved in the use of embryonic stem cells [79, 80].

Since numerous practical cell sources (vascular-derived cells, bone marrow-derived cells, UCB-derived cells, and amniotic fluid stem cells) exist, no further attention has been given to embryonic stem cells, despite the success of in vitro experiments. This is largely due to the ethical concerns surrounding the use of embryonic stem cells, as well as the uncertainty of long-term biological safety and efficacy following implantation [81]. However, induced pluripotent stem cells (iPS cells) have a promising future in the field of TEHV. Although a current protocol does not exist, efforts are being made to generate engineered VICs from iPS cells [82].

## 7 Bioreactors for TEHV Development

Bioreactor systems are often used in various tissue engineering applications to provide tissue constructs with an environment that mimics the in vivo environment of the target tissues/organs, as well as provide precise physical stimulations, such as mechanical loading, pressure, fluid flow, and electrical stimulation [83–85]. In HV tissue engineering, the goal of a bioreactor is to create a mechanical and biochemical environment that closely mimics the physiological condition experienced by native HV, and is hence designed to control parameters such as temperature, oxygen diffusion, biofactors, flow rate, and pressure gradients that open and close the heart valves [86]. Previous studies have demonstrated that these physical and biochemical cues together are able to facilitate cell migration, proliferation, alignment, and ECM formation and remodeling [19, 21, 87–89]. The bioreactor system functions to condition TEHV prior to surgical implantation, as well as observe and address any structural irregularities [19, 86].

For example, mechanical stimuli such as cyclic stretch, cyclic flexure, and shear stress induced by flow have been included as loading modes in a flex–stretch–flow (FSF) bioreactor [88, 89]. By employing neodymium magnets embedded in a paddlewheel, culture medium was recirculated within the bioreactor system. Using this two-chambered FSF bioreactor, Engelmayr et al. applied stretch, flexure, and shear independently or in combination, and examined their effects on cell differentiation and proliferation, ECM remodeling, and the maturation of the TEHV [88, 89]. Their study concluded a synergistic effect of combined mechanical stimulations on TEHV [88, 89].

Pulsatile flow bioreactors have been designed to condition intact TEHV, aiming to accurately mimic the tissue's physiological environment by providing mechanical forces that stimulate cellular migration and attachment to the acellular constructs [21, 90–92]. By using a bioreactor that could apply pulsatile fluid flow stimulations, Zeltinger et al. developed a TEHV from decellularized porcine aortic valves seeded with human neonatal fibroblasts [91]. This pneumatic bioreactor apparatus consisted of a valve holder assembly that stabilized and submerged the aortic valves in a chamber filled with culture medium. A return flow loop was used to circulate the medium through the valves at a rate of 25 mL/min [91]. This study demonstrated that, after 8 weeks in vitro culture, the TEHV was viable and could secrete several extracellular matrix proteins [91]. Mol et al. seeded a tri-leaflet polyglycolic acid (PGA) heart valve scaffold with human saphenous vein cells in a six-part diastolic pulse duplicator (DPD) bioreactor system, which provided dynamic strains to the leaflets [92]. The six parts of the DPD worked together simultaneously to culture the cells, as well as monitor the pressures upstream and downstream of the leaflets [92]. Their results showed that a large amount of ECM was formed in the conditioned TEHV, and the TEHV exhibited nonlinear mechanical properties [92].

Recently, Sierad et al. designed a state-of-the-art bioreactor to condition and analyze the function of an acellular porcine aortic valve treated with PGG for stabilization purposes [86]. In addition to the three chambers (air, ventricular, and aortic chamber), this pneumatic bioreactor system consisted of a pressurized compliance tank, a reservoir tank, one-way valves, pressure transducers, and a ventilator pump that all fit within a standard incubator (Fig. 7) [86]. Aortic valves were

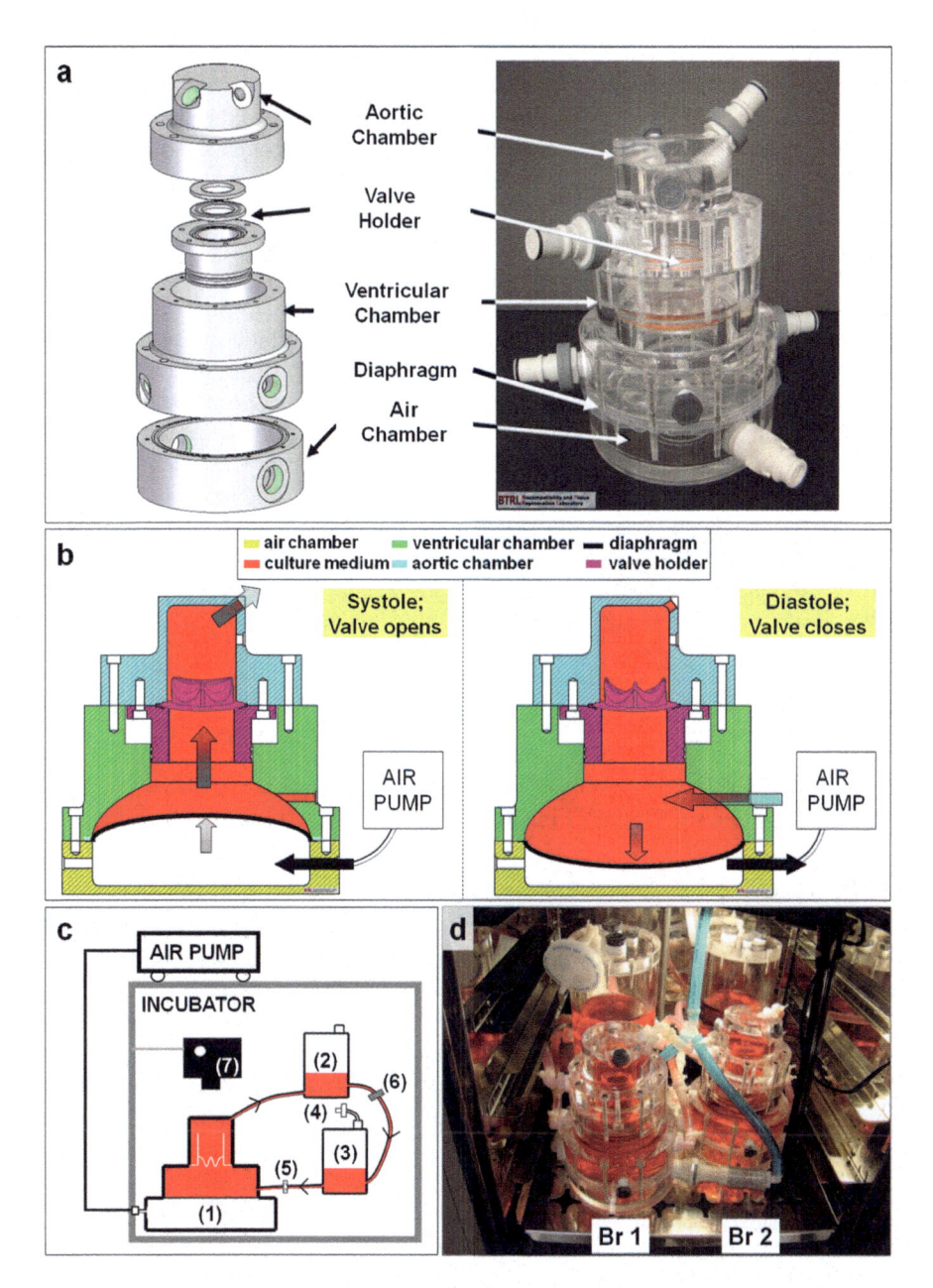

Fig. 7 Heart valve bioreactor and conditioning system. (a) Computer-aided design (left) and manufactured heart valve bioreactor (right) showing the four main components. A transparent silicone membrane diaphragm is mounted between the air chamber and the ventricular chamber. (b) Air and media flow through the system during systole (left) and diastole (right). Color coding aids identification of the components, and black arrows indicate direction of air and media flow. (c) Schematic overview of the entire conditioning system: a three-chambered heart valve bioreactor (1),

mounted between two o-rings on the valve-holder. During inspiration, the ventilator pushed air into the air chamber, which was connected to the ventricular chamber via a silicone diaphragm. The influx of air in the air chamber pushed the diaphragm into the ventricular chamber, resulting in culture media flow through the aortic valve, hence opening the valve. This was followed by expiration, where the ventilator released pressure in the air chamber, resulting in a vacuum pressure inside the ventricular chamber and hence closing the valve via a pressure gradient [86].

Sierad et al. noted that both the noncoronary and left coronary leaflets were functional, noting that both exhibited good apposition and coaptation. On the other hand, the right coronary leaflet did not function perfectly well during systole due to the thick muscular shelf found in the porcine heart. However, since this thick muscular shelf is not present in the human heart, the above observation will not apply to human aortic valve, and the right coronary leaflet of the human heart should behave similarly to that of the porcine noncoronary and left coronary leaflets. Additionally, SEM revealed that after 17 days in the bioreactor, endothelial cells adhered well to the scaffold surface and acquired a morphology similar to that of the native VECs. The endothelial cells were aligned parallel with the circumferential direction along the collagen fibers. Additionally, greater than 90% of the cells were viable, suggesting that the biological scaffold was not toxic to cells.

## 8 TEHV via Decellularization Approach: Large Animal Studies and Clinical Trials

TEHV have not yet been approved for widespread clinical applications due to issues with immunogenicity, calcification, fibrosis, etc.; however, several large animal studies and clinical trials, of which decellularized aortic valves were used as scaffolds, have been performed, and the results demonstrated the potential of TEHV via a decellularization approach. Baraki et al. implanted acellular aortic valve allografts orthotopically in sheep and compared the implants to native allografts [26]. Up to 9 months, the decellularized aortic valves functioned well, exhibited normal morphology and some cellular repopulation, and experienced only trivial regurgitation

**Fig. 7** (continued) an optional pressurized compliance chamber (2), a reservoir tank (3) with sterile filter (4) for gas exchange, one-way valve (5), pressure-retaining valve (6), a webcam (7), and a ventilator pump (air pump). The entire setup fits within a standard cell culture incubator. (**d**) Two identical bioreactor systems (Br 1 and Br 2) with endothelial cell seeded, functioning valves inside an incubator; the bioreactors are in the front row while their corresponding reservoirs are in the back row. The webcams normally mounted onto the top viewing windows of the aortic chamber have been removed to reveal bioreactor details. Reprinted with permission from *Form Follows Function: Advances in Trilayered Structure Replication for Aortic Heart Valve Tissue Engineering*, by Simionescu et al., December 2011

and microthrombi formation [26]. In another study, Lichtenberg et al. reseeded acellular aortic valves using endothelial cells and conditioned the valves with a pulsatile flow bioreactor; they reported that the endothelial cell coverage significantly increased after the reseeded valves were implanted in lambs for 3 months [43, 93]. Additionally, their animal study showed improved function of the reseeded valves when compared to the acellular valves without in vitro endothelial cell reseeding [43, 93]. However, there were also observations that found sustained tissue contraction and valve-thickening after implantation, resulting in less pliable valve tissue [43, 93–95].

Preliminary clinical results demonstrated that decellularized homografts, either reseeded in vitro or without any reseeding, are an appropriate replacement for the diseased valves. Dohmen et al. recellularized acellular pulmonary valve (PV) allografts with autologous vascular endothelial cells in vitro and implanted them into adult patients, yielding successful results with no calcification evident over a period of 10 years [96]. Cebotari et al. created tissue-engineered heart valves with decellularized PV allografts and autologous endothelial progenitor cells (EPCs). Three and a half years after implantation into two pediatric patients (ages 11 and 13) with congenital PV failure, an increase of the PV annulus diameter and decrease of valve regurgitation were observed, showing the TEHVs were able to grow with the somatic development of pediatric patients [93].

In another large-scale decellularized heart valve study, da Costa et al. implanted decellularized (0.1% SDS) aortic valve homografts as root replacements for 41 patients. After 3 years, the implanted aortic valves showed structural integrity, low incidence of calcification, and improved hemodynamic performance and function with trivial regurgitation, and no additional operations were required [19, 97]. The success of da Costa's clinical trials might be due to the less aggressive decellularization approach using 0.1% SDS, which greatly helped preserve the overall morphology, size, structure, and function of the acellular aortic valves with few microstructural disruptions [19, 45, 97]. However, it is important to note that the dense ECM environment still prevented adequate cellular infiltration, which might lead to complications such as cusp shrinkage, fibrosis, or stenosis [19, 43, 49].

## 9 Current Challenges in TEHV via Decellularization

*Immunogenic Responses.* Although there were successful applications of TEHV via decellularization as summarized in the above section [26, 43, 93, 96, 97], a few clinical studies reported failures of the implanted acellular heart valves, which might be related to remaining DNA fragments inside the decellularized xenogenic valve matrices that elicited an immune response [98–100]. For instance, SynerGrafts (Cryolife, Inc., USA), an acellular (nonglutaraldehyde-fixed) porcine aortic prosthetic valve, has been tested in four children; all four grafts failed due to inflammatory response, valve rupture, and degeneration [98]. The inflammatory response, likely caused by the xenogenic valvular collagen matrix and the remaining immunogenic cell fragments, led to thickening of the valves, which likely led to the lack of cellular infiltration. One year postimplantation, the valves experienced rupture and degeneration, and there remained no signs of cell repopulation or endothelialization [98].

Leyh et al. effectively eliminated endothelial and interstitial cells from porcine and ovine pulmonary valve conduits; however, implants in a sheep model showed severe calcification and only partial repopulation with endothelial cells [100]. In an in vitro human peripheral blood mononuclear cell model, Bayrak et al. also demonstrated the strong immune response induced by native porcine matrix scaffolds; on the other hand, glutaraldehyde-treated scaffolds did not induce a strong immune response, but proved nonviable, resulting in residual inflammatory responses [99]. Caution should thus be taken if the decellularized xenogenic heart valves are used, and tests need to be performed to verify the scaffold's minimal immune response [101].

*Recellularization and Conditioning.* Other challenges lay in the recellularization of the decellularized HV scaffolds, which include selecting the right cell sources, promoting the desired cell phenotypes, facilitating cell infiltration and proliferation, and stimulating ECM secretion/development during in vitro culture [27, 102–104]. As mentioned in Sect. 6, various types of cells have been studied and have shown some promising outcomes. However, the acellular leaflets, whose structural and biomechanical properties were preserved relatively well, often have a dense collagen network [19, 45]. As an example, the acellular leaflets obtained from 0.1% SDS treatment preserved the trilayered structure and biomechanical properties well, but resulted in a dense collagen network with small pore size, which represents a big challenge for cell infiltration and proliferation [45]. Moreover, the long wait time for cell expansion, construct seeding, and TEHV conditioning and maturation during in vitro tissue culture is another practical issue for patients who are in urgent need of heart valve replacement [105].

*Disruption of ECM Components and Microstructures.* During decellularization, loss of ECM components, microstructural alterations, and physical property changes all present challenges [40, 41, 45, 49, 59, 64]. One comparative experiment that investigated the flexural responses at the microstructural level in native, decellularized, and glutaraldehyde-fixed porcine aortic valve leaflets can help us better understand the challenges of ECM disruption [19]. In the experiment, the native leaflet, acellular leaflet (0.1% SDS treatment), and glutaraldehyde-fixed leaflet (0.625% glutaraldehyde treatment for bioprosthetic heart valve) were clamped into a "U" shape toward the fibrosa direction in a phosphate-buffered saline (PBS) bath. The "U" shaped leaflet strips were then fixed with 4% glutaraldehyde (fixation for histology), followed by histological analyses.

This study showed that, with the coordination of the three layers, the native leaflet was able to cope well with the extreme flexure. As shown by Fig. 8a, b, the fibrosa layer readjusted the collagen fiber crimps in response to compressive stress without causing fiber buckling; the ventricularis layer showed the uncrimping and recruitment of the fibers due to tensile stress, and the spongiosa appeared to smoothly mediate the transition from tension in the ventricularis to compression in the fibrosa.

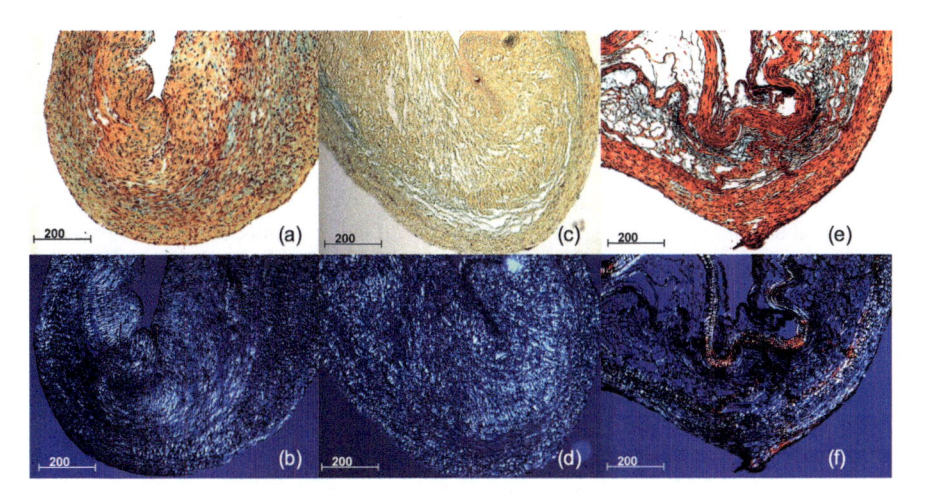

**Fig. 8** (a) Movat's pentachrome staining and (b) polarized light imaging of the native porcine aortic valve leaflet; (c) Movat's pentachrome staining and (d) polarized light imaging of the decellularized (0.1% SDS treatment) aortic valve leaflet; (e) Movat's pentachrome staining and (f) polarized light imaging of the glutaraldehyde fixed (0.625% Glut) porcine aortic valve leaflet. Sample preparation: The leaflet strips were folded into a "U" shape toward the fibrosa side in a PBS bath; the "U" shape leaflet strips were then clamped at two edges and fixed with 4% glutaraldehyde for histology. Scale bar unit: μm. Reprinted with permission from *Form Follows Function: Advances in Trilayered Structure Replication for Aortic Heart Valve Tissue Engineering*, by Simionescu et al., December 2011

Although the acellular leaflet showed the preservation of the overall leaflet structural integrity and trilayered structure (Fig. 8c, d), histology did reflect loss of GAGs in the spongiosa (Fig. 8c) and a large-scale loss of the collagen crimp pattern (Fig. 8d), all similar to what have been observed in our previous study (Fig. 4a, e). It was hence understandable that the acellular leaflet could cope with the extreme flexure relatively well, but with a degree of imperfection. As demonstrated by the histological images, no noticeable fiber buckling was observed in the fibrosa layer that experienced compression, but overstretching of fibers was observed in the ventricularis that experienced tension (Fig. 8c, d). As a sharp comparison, the effectiveness of the acellular leaflet in coping with valvular flexure could be readily appreciated by examining the extreme bending morphology of the glutaraldehyde-treated leaflet (Fig. 8e, f). Glutaraldehyde crosslinking stiffened all fiber networks, and the spongiosa layer lost its cushion functionality. Without a functioning trilayered structure, the chemically fixed leaflet experienced difficulty in coping with extreme flexure, showing serious fiber buckling in the fibrosa that was under compression (Fig. 8e, f).

As demonstrated by the above study [19], one can envision that some decellularization protocols might be more effective than others. Recently, Haupt et al. performed a study investigating how a combined detergent-based decellularization strategy preserves the macro- and micro-structure of heart valves [106]. In this study, porcine pulmonary heart valves were decellularized using ionic and nonionic detergents with low concentrations of enzymes. Valves treated with DET (SDS and Triton X-100 detergents) and ENZ (low concentrations of enzymes) exhibited an intact morphology lacking cellular content. The overall collagen and elastic fiber architecture remained similar to that of native valves, with a loss of glycosaminoglycans.

*ECM Stabilization with Crosslinking and Coating Reagents.* Since TEHV will be subjected to complex, strong mechanical forces, it is important that decellularization protocols do not interrupt the structural integrity and weaken the mechanical properties of valvular tissues. To prevent in vivo structural degeneration and fatigue, stabilize the ECM structure, increase durability, and decrease antigenicity of the decellularized heart valves, researchers proposed to treat the decellularized heart valves with crosslinking reagents, such as pentagalloyl glucose (PGG) and nordihydroguaiaretic, which are mild and nontoxic to cells [107, 108]. For instance, PGG has a high affinity for collagen and elastin and is, therefore, useful for preserving the ECM components, which is needed to maintain valvular structure and function. Compared to glutaraldehyde that was used for bioprosthetic HVs, PGG-treated HVs showed better mechanical behavior, biocompatibility, and biological properties. Especially, PGG-treated pericardium-based valve scaffolds presented no calcification and were supportive of host cell infiltration and matrix remodeling after in vivo implantation [108]. Recently, acellular mitral valves were also treated with PGG to stabilize the scaffolds and retard degradation [109]. Implantation studies showed that these PGG-treated acellular valves did not elicit an immune or inflammatory response, nor did it cause calcification, thus resulting in biocompatible mitral valve scaffolds.

To modify acellular HVs, researchers also coated the decellularized heart valves with biological or biodegradable polymer composites to further promote microenvironments that are favorable for recellularization and HV construct remodeling [110, 111]. For instance, a polyethylene glycol (PEG) hydrogel was used to coat acellular heart valves to improve cell attachment to the scaffold and assist in cell retention [112]. In another instance, researchers noticed that, after scaffold crosslinking, the size and number of pores in the treated scaffold were reduced. To facilitate better cell infiltration and proliferation, a high concentration of acetic acid solution was used to increase the porosity of the scaffolds; the resulting scaffolds were then coated with biocompatible RGD polypeptides to increase cell adhesion [51].

## 10  Final Remarks

To achieve a successful TEHV, an optimal combination of proper 3D scaffolds, cell sources, and culture conditions must be determined. As one of the approaches of TEHV, decellularized heart valve scaffolds have shown great potential in large animal studies, preclinical tests, and clinical trials. The notable benefits of the decellularization approach are the removal of the immunogenic cellular contents, the preservation of leaflet morphology and geometry, native ECM molecules, overall ultrastructure and trilayered configuration, and mechanical and functional behavior similar to the native heart valves. As demonstrated by the previous studies, the optimal performance and durability of valve leaflets greatly benefit from the trilayered structure design being capable of mitigating flexural fatigue, along with the nonlinear, anisotropic, viscoelastic tissue behavior [11, 19, 27, 64, 67, 103, 104, 113].

Due to the high working pressure and complicated stresses/strains that heart valves undergo, TEHV must have "ongoing strength, flexibility, and durability, beginning at the time of implantation and continuing indefinitely thereafter" [113]. It is thus essential to fully understand, from the beginning, how the various decellularization protocols affect the mechanical and structural properties of heart valve tissues. We assessed the effects of SDS, Trypsin, and Triton X-100 on leaflet biomechanics, reporting that (1) all three protocols increased the extensibility of the acellular leaflets, with Triton X-100 exhibiting the greatest increase; (2) all three decellularization protocols significantly decreased leaflet flexural rigidity; (3) all three protocols were able to preserve the gross collagen fiber architecture, including the branching of collagen fiber bundles; (4) the SDS protocol preserved leaflet dimensions well, while the Trypsin protocol increased both the area and thickness (swelling), and the Triton X-100 protocol decreased both the area and thickness of the acellular leaflets (shrinking); (5) the SDS protocol preserved the trilayered structure well when compared with the Trypsin protocol and Triton X-100 protocol; moreover, the Trypsin protocol resulted in a swollen and loose collagen fiber structure and a disrupted elastin network, while the Triton X-100 protocol resulted in a condensed collagen fiber morphology and a relatively well-preserved elastin fiber network; (6) lastly, the acellular leaflets obtained from all three decellularization protocols showed that the collagen fibers were locally crimped but no longer had macroscopically organized wavy patterns as shown in the native leaflets.

The above observations indicate that various decellularization protocols result in different initial biomechanical states of the leaflets, likely presenting various challenges for later leaflet recellularization, performance, and durability. In an in vitro fatigue mechanism study, we found that the fatigued leaflets exhibited an unfolded and thinned morphology, which resulted from the straightening of locally wavy collagen fibers and the disruption of the elastin network; moreover, the unfolding of collagen fibers and structural thinning caused a changed leaflet reference status and hence resulted in lower net tissue extensibility of the fatigued leaflets. We speculate that, in order to resist cyclic fatigue and irreversible and cumulative structural disruption, exogenous stabilizing crosslinking and recellularization of acellular leaflets can be used to slow down structural degeneration and achieve tissue maintenance and remodeling, respectively.

TEHV incorporates two types of cells: myofibroblast-like VICs that mediate valvular ECM for optimal biomechanical properties, and VECs that cover the valve surface for hemocompatibility and durability. Up to now, vascular-derived cells, bone marrow-derived cells, UCB-derived cells, and amniotic fluid-derived stem cells have been investigated as reseeding sources for acellular valve scaffolds. Each cell type has its advantages and disadvantages. In general, they all face challenges such as cell quantity, thorough infiltration of cells into dense valve interstices, defective endothelial coverage of valve surface, and uncertainty to achieve proper cell phenotypes after reseeding. To facilitate recellularization and maturation, TEHV often requires in vitro conditioning in a bioreactor designed to closely mimic the physiological conditions of heart valves. Many studies have shown that, by delivering stimulations such as pressure gradients to open and close the heart valves, shear flow to mediate VECs, and appropriate biofactors, the TEHV exhibits either better recellularization/maturation results or desirable functional performance with remodeling potential [11, 19, 27, 43, 64, 67, 86, 92, 103, 104, 113].

Although there has not been widespread clinical applications for TEHV yet, positive results have been reported in large animal experiments and clinical trials, in which decellularized aortic/pulmonary valve scaffolds were utilized [26, 42–44]. Among them, reiterating two notable clinical trials would help us appreciate the potential of TEHV via the decellularization approach (with or without in vitro reseeding)—(1) in vitro recellularized acellular PV allografts function well without calcification in adult patients over a 10 year period [96]; (2) acellular PV allografts reseeded with autologous EPCs were found to grow in size with the somatic growth of pediatric patients [93].

Many challenges still exist in TEHV via the decellularization approach. Concerns include inflammatory responses, unpredictable degeneration, and lack of cellular repopulation, as observed from the catastrophic failure of early clinical trials of the acellular porcine aortic prosthetic valve (SynerGrafts) [98], as well as other failures of the implanted acellular heart valves, which might be related to the remaining immunogenic cell and DNA fragments inside the decellularized xenogenic valve matrices [98–100]. Other subtle ECM disruptions, such as loss of GAGs, microstructural alterations, and loss of spongiosa mediation, might also present challenges to fatigue resistance and long-term durability of the leaflets. One solution that has been investigated is to treat the acellular heart valves with mild, nontoxic crosslinking reagents (e.g., PGG or nordihydroguaiaretic), with a goal to stabilize ECM structure, decrease antigenicity, and prevent in vivo structural degeneration and fatigue, while not inhibiting cell repopulation [102, 103].

Lastly, TEHV should join the efforts that seek to establish standardization for tissue-engineered medical products (TEMPs). The goal of TEMP standardization is to ensure the repeatability, quality, and reliability of final TE products and hence facilitate both FDA approval and translational applications. We summarize the standards and evaluation criteria for TEHV via the decellularization approach as follows: (1) structural integrity and biomechanical properties, (2) free of residual cellular/DNA debris, (3) cell isolation/expansion for reseeding, (4) recellularization and construct maturation, (5) biocompatibility, hemocompatibility, and immunological characteristics, (6) functional and hemodynamic performance, (7) durability, remodeling, and growth potential, and (8) manufacture and delivery protocols for the final products [114, 115]. This is a long, challenging list and represents the need for systematic, reliable, and repeatable procedures to move TEHV to clinical applications.

**Acknowledgments** This work is supported in part by AHA BGIA-0565346U, GRNT17150041, NIH 1R01EB022018-01, 1R56HL130950-01, 1R15HL140503, T32HL134613, and NSF CAREER #1554835. The authors also thank the support from the Competitiveness Operational Programme 2014–2020, ID P_37_673, MySMIS code: 103431, contract 50/05.09.2016 and the Harriet and Jerry Dempsey Bioengineering Professorship.

# References

1. Mozaffarian D, et al. Executive summary: heart disease and stroke statistics—2016 update: a report from the American Heart Association. Circulation. 2016;133(4):447–54.
2. Davlouros PA, et al. Transcatheter aortic valve replacement and stroke: a comprehensive review. J Geriatr Cardiol. 2018;15(1):95.
3. Stewart BF, et al. Clinical factors associated with calcific aortic valve disease. J Am Coll Cardiol. 1997;29(3):630–4.
4. Mozaffarian D, et al. Heart disease and stroke statistics—2016 update: a report from the American Heart Association. Circulation. 2016;133(4):e38–60.
5. Hammermeister K, et al. Outcomes 15 years after valve replacement with a mechanical versus a bioprosthetic valve: final report of the Veterans Affairs randomized trial. J Am Coll Cardiol. 2000;36(4):1152–8.
6. Tillquist MN, Maddox TM. Cardiac crossroads: deciding between mechanical or bioprosthetic heart valve replacement. Patient Prefer Adherence. 2011;5:91–9.
7. Frater R, et al. Long-term durability and patient functional status of the Carpentier-Edwards Perimount pericardial bioprosthesis in the aortic position. J Heart Valve Dis. 1998;7(1):48–53.
8. Marchand MA, et al. Fifteen-year experience with the mitral Carpentier-Edwards PERIMOUNT pericardial bioprosthesis. Ann Thorac Surg. 2001;71(5):S236–9.
9. Bloomfield P, et al. Twelve-year comparison of a Bjork-Shiley mechanical heart valve with porcine bioprostheses. N Engl J Med. 1991;324(9):573–9.
10. Kaneko T, Cohn LH, Aranki SF. Tissue valve is the preferred option for patients aged 60 and older. Circulation. 2013;128(12):1365–71.
11. Vesely I. Heart valve tissue engineering. Circ Res. 2005;97(8):743–55.
12. Delgado V, et al. Successful deployment of a transcatheter aortic valve in bicuspid aortic stenosis. Circ Cardiovasc Imaging. 2009;2(2):e12–3.
13. Henaine R, et al. Valve replacement in children: a challenge for a whole life. Arch Cardiovasc Dis. 2012;105(10):517–28.
14. Schoen FJ, Levy RJ. Calcification of tissue heart valve substitutes: progress toward understanding and prevention. Ann Thorac Surg. 2005;79(3):1072–80.
15. Manji RA, et al. Glutaraldehyde-fixed bioprosthetic heart valve conduits calcify and fail from xenograft rejection. Circulation. 2006;114(4):318–27.
16. Ross DN. Replacement of aortic and mitral valves with a pulmonary autograft. Lancet. 1967;2 (7523):956–8.
17. Charitos EI, et al. Reoperations on the pulmonary autograft and pulmonary homograft after the Ross procedure: an update on the German Dutch Ross Registry. J Thorac Cardiovasc Surg. 2012;144(4):813–23.
18. Lam MT, Wu JC. Biomaterial applications in cardiovascular tissue repair and regeneration. Expert Rev Cardiovasc Ther. 2012;10(8):1039–49.
19. Simionescu D, et al. Form follows function: advances in trilayered structure replication for aortic heart valve tissue engineering. J Healthcare Eng. 2012;3(2):179–202.

20. Engelmayr GC, et al. The independent role of cyclic flexure in the early in vitro development of an engineered heart valve tissue. Biomaterials. 2005;26(2):175–87.
21. Hoerstrup SP, et al. New pulsatile bioreactor for in vitro formation of tissue engineered heart valves. Tissue Eng. 2000;6(1):75–9.
22. Ramamurthi A, Vesely I. Evaluation of the matrix-synthesis potential of crosslinked hyaluronan gels for tissue engineering of aortic heart valves. Biomaterials. 2005;26(9):999–1010.
23. Meyer U, et al. Fundamentals of tissue engineering and regenerative medicine. Berlin: Springer; 2009.
24. Bouten CV, et al. Substrates for cardiovascular tissue engineering. Adv Drug Deliv Rev. 2011;63(4–5):221–41.
25. Rippel RA, Ghanbari H, Seifalian AM. Tissue-engineered heart valve: future of cardiac surgery. World J Surg. 2012;36(7):1581–91.
26. Baraki H, et al. Orthotopic replacement of the aortic valve with decellularized allograft in a sheep model. Biomaterials. 2009;30(31):6240–6.
27. Sacks MS, Schoen FJ, Mayer JE Jr. Bioengineering challenges for heart valve tissue engineering. Annu Rev Biomed Eng. 2009;11:289–313.
28. Vesely I, Noseworthy R. Micromechanics of the fibrosa and the ventricularis in aortic valve leaflets. J Biomech. 1992;25(1):101–13.
29. Brazile B, et al. On the bending properties of porcine mitral, tricuspid, aortic, and pulmonary valve leaflets. J Long-Term Eff Med Implants. 2015;25(1–2):41–53.
30. Tomasek JJ, et al. Myofibroblasts and mechano-regulation of connective tissue remodelling. Nat Rev Mol Cell Biol. 2002;3(5):349–63.
31. Stella JA, Sacks MS. The digital leaflet: quantitative image analysis and 3-D digital reconstruction of the aortic valve leaflet. In: ASME 2007 summer bioengineering conference. American Society of Mechanical Engineers; 2007.
32. Adham M, et al. Mechanical characteristics of fresh and frozen human descending thoracic aorta. J Surg Res. 1996;64(1):32–4.
33. Isenberg BC, Williams C, Tranquillo RT. Small-diameter artificial arteries engineered in vitro. Circ Res. 2006;98(1):25–35.
34. Filip DA, Radu A, Simionescu M. Interstitial cells of the heart valves possess characteristics similar to smooth muscle cells. Circ Res. 1986;59(3):310–20.
35. Liu AC, Joag VR, Gotlieb AI. The emerging role of valve interstitial cell phenotypes in regulating heart valve pathobiology. Am J Pathol. 2007;171(5):1407–18.
36. Merryman WD, et al. Viscoelastic properties of the aortic valve interstitial cell. J Biomech Eng. 2009;131(4):041005.
37. Serini G, Gabbiani G. Mechanisms of myofibroblast activity and phenotypic modulation. Exp Cell Res. 1999;250(2):273–83.
38. Butcher JT, Nerem RM. Valvular endothelial cells regulate the phenotype of interstitial cells in co-culture: effects of steady shear stress. Tissue Eng. 2006;12(4):905–15.
39. Aikawa E, et al. Human semilunar cardiac valve remodeling by activated cells from fetus to adult: implications for postnatal adaptation, pathology, and tissue engineering. Circulation. 2006;113(10):1344–52.
40. Gilbert TW, Sellaro TL, Badylak SF. Decellularization of tissues and organs. Biomaterials. 2006;27(19):3675–83.
41. Crapo PM, Gilbert TW, Badylak SF. An overview of tissue and whole organ decellularization processes. Biomaterials. 2011;32(12):3233–43.
42. Neumann A, et al. Early systemic cellular immune response in children and young adults receiving decellularized fresh allografts for pulmonary valve replacement. Tissue Eng Part A. 2014;20(5–6):1003–11.
43. Lichtenberg A, et al. Preclinical testing of tissue-engineered heart valves re-endothelialized under simulated physiological conditions. Circulation. 2006;114(1 Suppl):I559–65.

44. Dohmen PM, et al. Ross operation with a tissue-engineered heart valve. Ann Thorac Surg. 2002;74(5):1438–42.
45. Liao J, Joyce EM, Sacks MS. Effects of decellularization on the mechanical and structural properties of the porcine aortic valve leaflet. Biomaterials. 2008;29(8):1065–74.
46. Zhou J, et al. Impact of heart valve decellularization on 3-D ultrastructure, immunogenicity and thrombogenicity. Biomaterials. 2010;31(9):2549–54.
47. Dainese L, et al. Heart valve engineering: decellularized aortic homograft seeded with human cardiac stromal cells. J Heart Valve Dis. 2012;21(1):125–34.
48. Cigliano A, et al. Fine structure of glycosaminoglycans from fresh and decellularized porcine cardiac valves and pericardium. Biochem Res Int. 2012;2012:979351.
49. Rieder E, et al. Decellularization protocols of porcine heart valves differ importantly in efficiency of cell removal and susceptibility of the matrix to recellularization with human vascular cells. J Thorac Cardiovasc Surg. 2004;127(2):399–405.
50. Lehr EJ, et al. Decellularization reduces immunogenicity of sheep pulmonary artery vascular patches. J Thorac Cardiovasc Surg. 2011;141(4):1056–62.
51. Dong X, et al. RGD-modified acellular bovine pericardium as a bioprosthetic scaffold for tissue engineering. J Mater Sci Mater Med. 2009;20(11):2327–36.
52. Prasertsung I, et al. Development of acellular dermis from porcine skin using periodic pressurized technique. J Biomed Mater Res B Appl Biomater. 2008;85(1):210–9.
53. Reing JE, et al. The effects of processing methods upon mechanical and biologic properties of porcine dermal extracellular matrix scaffolds. Biomaterials. 2010;31(33):8626–33.
54. Gorschewsky O, et al. Quantitative analysis of biochemical characteristics of bone-patellar tendon-bone allografts. Biomed Mater Eng. 2005;15(6):403–11.
55. Cox B, Emili A. Tissue subcellular fractionation and protein extraction for use in mass-spectrometry-based proteomics. Nat Protoc. 2006;1(4):1872–8.
56. Xu CC, Chan RW, Tirunagari N. A biodegradable, acellular xenogeneic scaffold for regeneration of the vocal fold lamina propria. Tissue Eng. 2007;13(3):551–66.
57. Petersen TH, et al. Tissue-engineered lungs for in vivo implantation. Science. 2010;329 (5991):538–41.
58. Hudson TW, Liu SY, Schmidt CE. Engineering an improved acellular nerve graft via optimized chemical processing. Tissue Eng. 2004;10(9–10):1346–58.
59. Cebotari S, et al. Detergent decellularization of heart valves for tissue engineering: toxicological effects of residual detergents on human endothelial cells. Artif Organs. 2010;34 (3):206–10.
60. Meyer SR, et al. Comparison of aortic valve allograft decellularization techniques in the rat. J Biomed Mater Res A. 2006;79(2):254–62.
61. Lumpkins SB, Pierre N, McFetridge PS. A mechanical evaluation of three decellularization methods in the design of a xenogeneic scaffold for tissue engineering the temporomandibular joint disc. Acta Biomater. 2008;4(4):808–16.
62. Grauss RW, et al. Histological evaluation of decellularised porcine aortic valves: matrix changes due to different decellularisation methods. Eur J Cardiothorac Surg. 2005;27 (4):566–71.
63. Fung Y-C. Biomechanics: mechanical properties of living tissues. New York: Springer Science & Business Media; 2013.
64. Liao J, et al. The intrinsic fatigue mechanism of the porcine aortic valve extracellular matrix. Cardiovasc Eng Technol. 2012;3(1):62–72.
65. Dohmen PM, et al. Mid-term clinical results using a tissue-engineered pulmonary valve to reconstruct the right ventricular outflow tract during the Ross procedure. Ann Thorac Surg. 2007;84(3):729–36.
66. Pittenger MF, et al. Multilineage potential of adult human mesenchymal stem cells. Science. 1999;284(5411):143–7.
67. Hoerstrup SP, et al. Functional living trileaflet heart valves grown in vitro. Circulation. 2000;102(19 Suppl 3):III44–9.

68. Latif N, et al. Characterization of structural and signaling molecules by human valve interstitial cells and comparison to human mesenchymal stem cells. J Heart Valve Dis. 2007;16(1):56–66.
69. Rotmans JI, et al. In vivo cell seeding with anti-CD34 antibodies successfully accelerates endothelialization but stimulates intimal hyperplasia in porcine arteriovenous expanded polytetrafluoroethylene grafts. Circulation. 2005;112(1):12–8.
70. Zhang S, et al. Ovarian cancer stem cells express ROR1, which can be targeted for anti-cancer-stem-cell therapy. Proc Natl Acad Sci U S A. 2014;111(48):17266–71.
71. Sodian R, et al. Use of human umbilical cord blood-derived progenitor cells for tissue-engineered heart valves. Ann Thorac Surg. 2010;89(3):819–28.
72. Schmidt D, et al. Umbilical cord blood derived endothelial progenitor cells for tissue engineering of vascular grafts. Ann Thorac Surg. 2004;78(6):2094–8.
73. Schmidt D, et al. Living patches engineered from human umbilical cord derived fibroblasts and endothelial progenitor cells. Eur J Cardiothorac Surg. 2005;27(5):795–800.
74. Schmidt D, et al. Engineering of biologically active living heart valve leaflets using human umbilical cord-derived progenitor cells. Tissue Eng. 2006;12(11):3223–32.
75. Schmidt D, et al. Living autologous heart valves engineered from human prenatally harvested progenitors. Circulation. 2006;114(1 Suppl):I125–31.
76. Corselli M, et al. Clinical scale ex vivo expansion of cord blood-derived outgrowth endothelial progenitor cells is associated with high incidence of karyotype aberrations. Exp Hematol. 2008;36(3):340–9.
77. De Coppi P, et al. Isolation of amniotic stem cell lines with potential for therapy. Nat Biotechnol. 2007;25(1):100–6.
78. Schmidt D, et al. Cryopreserved amniotic fluid-derived cells: a lifelong autologous fetal stem cell source for heart valve tissue engineering. J Heart Valve Dis. 2008;17(4):446–55; discussion 455.
79. Prusa A-R, Hengstschlager M. Amniotic fluid cells and human stem cell research: a new connection. Med Sci Monit. 2002;8(11):RA253–7.
80. Tsai MS, et al. Isolation of human multipotent mesenchymal stem cells from second-trimester amniotic fluid using a novel two-stage culture protocol. Hum Reprod. 2004;19(6):1450–6.
81. Siepe M, et al. Stem cells used for cardiovascular tissue engineering. Eur J Cardiothorac Surg. 2008;34(2):242–7.
82. Vander Roest MJ, Merryman WD. A developmental approach to induced pluripotent stem cells-based tissue engineered heart valves. Future Medicine. 2017;13:1–4.
83. Martin I, Wendt D, Heberer M. The role of bioreactors in tissue engineering. Trends Biotechnol. 2004;22(2):80–6.
84. Bancroft GN, Sikavitsas VI, Mikos AG. Design of a flow perfusion bioreactor system for bone tissue-engineering applications. Tissue Eng. 2003;9(3):549–54.
85. Pörtner R, et al. Bioreactor design for tissue engineering. J Biosci Bioeng. 2005;100(3):235–45.
86. Sierad LN, et al. Design and testing of a pulsatile conditioning system for dynamic endothelialization of polyphenol-stabilized tissue engineered heart valves. Cardiovasc Eng Technol. 2010;1(2):138–53.
87. Mol A, et al. The relevance of large strains in functional tissue engineering of heart valves. Thorac Cardiovasc Surg. 2003;51(2):78–83.
88. Engelmayr GC Jr, et al. A novel bioreactor for the dynamic flexural stimulation of tissue engineered heart valve biomaterials. Biomaterials. 2003;24(14):2523–32.
89. Engelmayr GC Jr, et al. A novel flex-stretch-flow bioreactor for the study of engineered heart valve tissue mechanobiology. Ann Biomed Eng. 2008;36(5):700–12.
90. Weston MW, Yoganathan AP. Biosynthetic activity in heart valve leaflets in response to in vitro flow environments. Ann Biomed Eng. 2001;29(9):752–63.
91. Zeltinger J, et al. Development and characterization of tissue-engineered aortic valves. Tissue Eng. 2001;7(1):9–22.

92. Mol A, et al. Tissue engineering of human heart valve leaflets: a novel bioreactor for a strain-based conditioning approach. Ann Biomed Eng. 2005;33(12):1778–88.
93. Cebotari S, et al. Clinical application of tissue engineered human heart valves using autologous progenitor cells. Circulation. 2006;114(1 Suppl):I132–7.
94. Schmidt D, et al. Minimally-invasive implantation of living tissue engineered heart valves: a comprehensive approach from autologous vascular cells to stem cells. J Am Coll Cardiol. 2010;56(6):510–20.
95. Syedain ZH, et al. Implantation of a tissue-engineered heart valve from human fibroblasts exhibiting short term function in the sheep pulmonary artery. Cardiovasc Eng Technol. 2011;2(2):101–12.
96. Dohmen PM, et al. Ten years of clinical results with a tissue-engineered pulmonary valve. Ann Thorac Surg. 2011;92(4):1308–14.
97. da Costa FD, et al. The early and midterm function of decellularized aortic valve allografts. Ann Thorac Surg. 2010;90(6):1854–60.
98. Simon P, et al. Early failure of the tissue engineered porcine heart valve SYNERGRAFT® in pediatric patients. Eur J Cardiothorac Surg. 2003;23(6):1002–6.
99. Bayrak A, et al. Human immune responses to porcine xenogeneic matrices and their extracellular matrix constituents in vitro. Biomaterials. 2010;31(14):3793–803.
100. Leyh RG, et al. In vivo repopulation of xenogeneic and allogeneic acellular valve matrix conduits in the pulmonary circulation. Ann Thorac Surg. 2003;75(5):1457–63.
101. Dohmen PM, Konertz W. Results with decellularized xenografts. Circ Res. 2006;99(4):e10.
102. Mol A, et al. Tissue engineering of heart valves: advances and current challenges. Expert Rev Med Devices. 2009;6(3):259–75.
103. Schoen FJ, Levy RJ. Tissue heart valves: current challenges and future research perspectives. J Biomed Mater Res A. 1999;47(4):439–65.
104. Mendelson K, Schoen FJ. Heart valve tissue engineering: concepts, approaches, progress, and challenges. Ann Biomed Eng. 2006;34(12):1799–819.
105. Song JJ, Ott HC. Organ engineering based on decellularized matrix scaffolds. Trends Mol Med. 2011;17(8):424–32.
106. Haupt J, et al. Detergent-based decellularization strategy preserves macro- and microstructure of heart valves. Interact Cardiovasc Thorac Surg. 2018;26(2):230–6.
107. Lu X, et al. Crosslinking effect of nordihydroguaiaretic acid (NDGA) on decellularized heart valve scaffold for tissue engineering. J Mater Sci Mater Med. 2010;21(2):473–80.
108. Tedder ME, et al. Stabilized collagen scaffolds for heart valve tissue engineering. Tissue Eng Part A. 2009;15(6):1257–68.
109. Deborde C, et al. Stabilized collagen and elastin-based scaffolds for mitral valve tissue engineering. Tissue Eng A. 2016;22(21–22):1241–51.
110. Stamm C, et al. Biomatrix/polymer composite material for heart valve tissue engineering. Ann Thorac Surg. 2004;78(6):2084–92; discussion 2092–3.
111. Lutolf MP, Hubbell JA. Synthetic biomaterials as instructive extracellular microenvironments for morphogenesis in tissue engineering. Nat Biotechnol. 2005;23(1):47–55.
112. Ouyang H, et al. [Research on application of modified polyethylene glycol hydrogels in the construction of tissue engineered heart valve]. Zhonghua wai ke za zhi [Chin J Surg]. 2008;46(22):1723–6.
113. Schoen FJ. Heart valve tissue engineering: quo vadis? Curr Opin Biotechnol. 2011;22(5):698–705.
114. Hjortnaes J, et al. Translating autologous heart valve tissue engineering from bench to bed. Tissue Eng Part B Rev. 2009;15(3):307–17.
115. Bouten CV, Driessen-Mol A, Baaijens FP. In situ heart valve tissue engineering: simple devices, smart materials, complex knowledge. Expert Rev Med Devices. 2012;9(5):453–5.

# Novel Bioreactors for Mechanistic Studies of Engineered Heart Valves

**Kristin Comella and Sharan Ramaswamy**

**Abstract** Beyond the identification of the most appropriate cells and scaffold materials, translation of cardiovascular tissue engineering structures requires optimization of construct biological and mechanical properties in order to permit their long-term functionality in the native hemodynamic environment. Unfortunately, rudimentary tissue growth technologies such as plate or rotisserie culture do not lead to the generation of functional tissues, thereby limiting their usefulness. Dynamic culture systems or bioreactors with true construct mechanical conditioning capabilities thus form an essential part of the research and development pathway in cardiovascular regenerative medicine. This is because engineered tissues cultured under specific bioreactor mechanical environments enhance their biological properties such as functional stem cell differentiation to complex tissue phenotypes and also augment construct structural properties as a result of accelerated tissue formation. In the heart valve tissue engineering arena, based on at least a decade of scientific results, there is now general acceptance that bioreactor usage is a critical preclinical step. In this book chapter, we describe important considerations in bioreactor design as well as focus on the critical scientific findings in the development of de novo valvular tissues, including our own experience in this area. We subsequently detail scaffold and cell sources that have been used in conjunction with these devices and finally conclude by addressing the key challenges that still remain.

**Keywords** Tissue Engineering · Heart valve disease · Somatic growth · Extracellular matrix · Bioreactors · Flow · Stretch · Flexure · Scaffolds · Stem cells

K. Comella · S. Ramaswamy (✉)
Tissue Engineered Mechanics, Imaging and Materials Laboratory (TEMIM Lab),
Department of Biomedical Engineering, College of Engineering and Computing,
Florida International University, Miami, FL, USA
e-mail: sramaswa@fiu.edu

# 1 Introduction

Millions of people worldwide are suffering from valvular heart disease and the rate of incidence is expected to significantly increase in the future [1]. Current techniques for prosthetic valve replacement are reasonable in some cases but have distinct limitations in others. In particular, no robust treatment option is available for the treatment of critical congenital valve diseases in children. An inability of prosthetic valves to accommodate somatic growth and practical size limitations of current commercial valves represent specific technical challenges. The field of tissue engineered heart valves holds promise of filling this unmet need. The living, biological heart valve would function in a similar manner as native tissue with normal hemodynamic performance and minimal degeneration. To achieve this task, current consensus in the scientific community is that a bioreactor will be necessary to help create a tissue engineered valve with ideal function. In this context, the bioreactor is a device that will help to promote engineered tissue properties via the delivery of appropriate biomechanical and biochemical stimuli during tissue culture. There are many variables to consider when designing a bioreactor to condition a tissue engineered heart valve (TEHV). From a biomimetic standpoint, this design needs to consider the hemodynamic environment of the native, healthy heart valve. In addition, the type of scaffold material, the choice of cell source(s), and the duration of culture will affect the de novo tissue properties.

# 2 Bioreactors

## 2.1 Design

The basic components of bioreactors as described by Berry et al. to design engineered heart valves are [2]:

1. A driving force or pump for fluid movement

    Typical pumps include pneumatically controlled diaphragms, reciprocating pistons, peristaltic pumps, or meshing gears teeth. The pumps are typically controlled by a computer system as it is desirable to have complete control over the fluid movement.
2. A reservoir (for cell culture media)

    Culture media should support the growth of the cell type and can contain a variety of different growth factors. Minimizing the overall volume of media with help to reduce costs.
3. A test section containing the heart valve

    Location of the test sections can disturb or change the flow patterns. Many groups are using a spiral-bound section to allow for minimal flow disturbances.
4. A fluid capacitance to store and release energy at every pump cycle

As system pressure builds due to forward or reverse fluid flow, a fluid capacitance can help to release this energy.
5. A resistance element
The overall system pressure can be controlled with the resistance element.
6. Allows for Gas exchange
Gas exchange by bubbling of the media or diffusion at the reservoir surface can keep the culture media oxygenated to support cell growth.

### 2.1.1 Static Designs

A simple example of a static bioreactor is a petri dish or culture flask. These typically include a non-stirred environment or may include packed-bed reactors with porous substrate which allows for culture perfusion [3]. Even though rotisserie culture systems involve rotational movement of tubes containing the engineered tissues, they are still considered static in the context of the negligible mechanical stresses that are imparted onto the developing tissue constructs. Seliktar et al. [4] described an initial static bioreactor that was subsequently modified to permit dynamic mechanical conditioning of tissue engineered constructs composed of cells in a scaffold. The response to this conditioning included increased contraction and mechanical strength as compared to traditional static cultures.

### 2.1.2 Dynamic Designs

Dynamic bioreactors typically include pulse duplicators and devices for organ tissue culture. This environment allows for the maturation and physical conditioning of engineered tissue. Gandaglia et al. [5] provided a description for a dynamic bioreactor that is similar to the defined components described by Berry et al. [2] above. The dynamic bioreactor for engineered heart valves consist of the following components: (1) a ventricle, driven by a motor (mechanical, pneumatic, or electric) which creates specific pressure and flow patterns, (2) a system to adjust resistance, (3) a chamber to shelter the heart valve, (4) a hydraulic circuit for the different unit joint, (5) probes and sensors for measuring and adjusting flow and pressure, temperature, pH, $pO_2$, $pCO_2$, and metabolite concentration, and (6) hardware and software to control the system.

Berry et al. [2] made a schematic of the typical pulsatile bioreactor designs. Small differences in design can lead to varied environments and therefore major differences in the tissue engineered heart valve properties (Fig. 1).

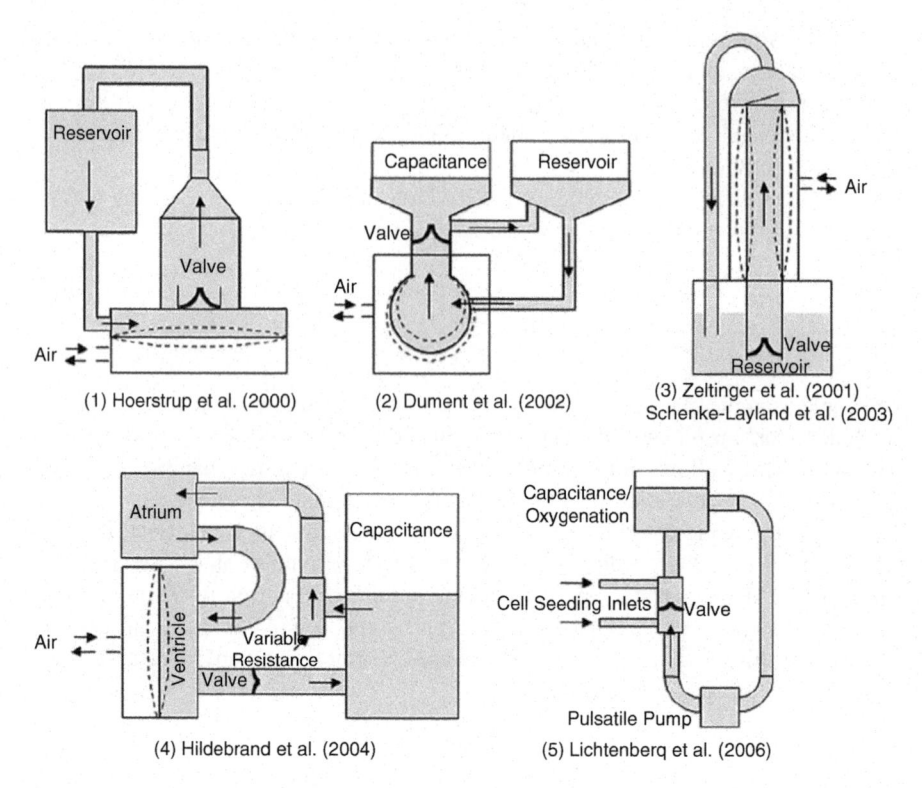

**Fig. 1** Schematic diagrams of five contemporary heart valve bioreactor designs. These designs have been reported to deliver physiologic pulsatile flow and pressure while maintaining the proper temperature, gas exchange, and sterility [2]

## 2.2 Mechanical Environment

The mechanical environment for normal heart valve tissue can be broadly described as a combination of cyclic stretch, forward and oscillating blood-induced shear stresses as well as cyclic flexure. The hemodynamics is time-dependent and is based on heart function over the cardiac cycle as well as the opening and closing events of the valve. Therefore, designing engineered valves that can withstand this environment is challenging. Many investigators have focused on creating specialized bioreactors that divide the environment into components of stretch, flex or flow [2].

### 2.2.1 Stretch

The commercially available Flexcell system (Flexcell International, Hillsborough, USA) provides an environment for studying cyclic stretch. The Flexcell System is a computerized, pressure-operated instrument that applies a defined controlled, static or variable duration cyclic tension, to cells growing in vitro. This system utilizes regulated vacuum pressure to deform flexible bottomed culture plates. Utilizing the Flexcell system, Carrion et al. [6] demonstrated that bicuspid aortic valve (BAV) leaflets experience increased biomechanical strain as compared to tricuspid aortic valves in response to mechanical stretch. Human aortic valve interstitial cells were exposed to cyclic stretch of 14% at 1 Hz or static condition, the control condition, for 24 h. They demonstrated the molecular pathogenesis involved in the calcification of BAVs especially when exposed to mechanical stretch. Cyclic stretch has been shown to affect cytoskeletal organization as it is sensed by cardiomyocytes. The stretch signal sensing is mediated via focal adhesion kinase and leads to intracellular reorganization and orientation. The oblique orientation of the cell with regard to the direction of stretch may define a directed force vector which could allow the cell to orientate [7]. In one study, human adipose stem cells exposed to mechanical stretch at 12% at 1 Hz frequency for up to 96 h did not differentiate to cardiovascular cell phenotypes [8]. The ability of cells to secrete an array of cytokines, however, remained unchanged by mechanical stretch [8]. In another study, activating transcription factor 3 which is known to have a cardioprotective activity was shown to be activated by mechanical stretch [9].

Ku et al. [10] demonstrated that collagen synthesis by valve interstitial cells as well as mesenchymal stem cells is a function of both the degree and duration of stretch. Depending on the stretching profile that is applied, the properties could be varied to make an appropriate tissue engineered heart valve.

Syedain and Tranquillo [11] considered a controlled cyclic stretch bioreactor as opposed to pulsed flow. This led to improved tensile and compositional properties. They demonstrated that the TEHV could withstand the cyclic pressures of a pulmonary artery in the Ovine model. Another study used aortic valve interstitial cells (VIC) on a porcine aortic valve in a cyclic stretch and pressure bioreactor [12]. The leaflets were stretched at a frequency of 1.167 Hz (equivalent to 70 bpm) at 10% or 15% stretch and a cyclic pressure of either 120/80 or 140/100 mm Hg was applied. The combination of cyclic stretch and pressure was shown to help modulate the VIC phenotype.

### 2.2.2 Flexure

Dynamic flexure or bending effects on valve leaflets is an important variable to consider when developing TEHVs. Engelmayr et al. [13] developed a bioreactor that considered dynamic flexure. The linear actuator was used to simulate the cardiac cycle at 1 Hz (60 bpm). The scaffolds were displaced 6.35 mm with a flexure angle of 62°. They assessed unidirectional cyclic flexural properties of a nonwoven mesh of polyglycolic acid (PGA) fiber as well as a nonwoven mesh of PGA and poly L-lactic acid (PLLA). Both samples were coated with poly 4-hydroxybutyrate (P4HB). The PGA/PLLA/P4HB scaffolds that were subjected to dynamic flexure were significantly less stiff than the static controls and they developed directional anisotropy. The PGA/P4HB scaffolds showed less stiffness after dynamic flexure but did not develop directional anisotropy. These experiments indicated that dynamic flexure does produce changes in the scaffolds and that this should be considered when developing engineered heart valve tissues.

In another set of experiments, Engelmayr et al. [14] considered cell-seeded scaffolds. Ovine vascular smooth muscle cells (SMC) were seeded on PGA/PLLA scaffolds and subjected to either static conditions or unidirectional cyclic three-point flexure at a frequency of 1 Hz and a central displacement of 6.35 mm. The flexure group showed an increase in stiffness, collagen, and vimentin expression as well as more homogeneous cell distribution as compared to the static group. Similar results were obtained using scaffolds seeded with ovine bone marrow-derived mesenchymal cells (BMSCs) [15].

Mirnajafi et al. [16] designed a device to study controlled cyclic flexural loading on bioprosthetic heart valves. They were able to study layer-specific structural damage induced by cyclic flexural tensile at a frequency of 1 Hz and a central displacement of 6.35 mm and compressive stresses alone. Results indicated that rigidity was reduced after only 10 million cycles.

### 2.2.3 Flow

Some bioreactors focus on steady fluid flow mechanical conditioning. One system examined the effects of laminar flow (7 days exposure to ~0.066 $dyn/cm^2$, followed by a step change to ~0.13 $dyn/cm^2$ for an additional 7 days) on tissue induced by dialysis roller pump (500 mL/min), and nutrition flow by a perfusion pump (3 mL/h) [17]. Mahler et al. [18] considered a microfluidic bioreactor that applies oscillatory shear stresses. The system was used to study the effects of flow on the transformation of endothelial to mesenchymal cells. For the oscillatory shear experiments, a square wave flow pattern at 1 Hz was used, in which the flow was fully forward at 2, 10, or 20 $dyn/cm^2$ for one-half of the cycle and fully reversed at $-2$, $-10$ or $-20$ $dyn/cm^2$ respectively for the remainder of the cycle. In previous reports, steady shear flows led to differences in morphology and transcriptional profiles [19].

Ramaswamy et al. [20] considered design of a bioreactor that combined all three physiological characteristics of flexure, stretch, and flow in combination or independently. The device consists of four identical chambers and up to three specimens can be placed into each. Specimens can be subjected to changes in curvature, uniaxial tensile strains, and laminar flow. This system provides a useful tool to study the effect of mechanical stimuli on engineered heart valve tissue formation. The flow chamber takes a U-shaped configuration (see Fig. 2). The device is connected to a peristaltic pump and an actuator. There are four chambers each containing three samples. Samples are fixed with a pin at one end and a circular moving post on the other end. In a combination of cyclic flexure and steady flow (flex-flow) experiments for example, the moving post moves linearly using the linear actuator in the axial direction to initiate cyclic flexure. A peristaltic pump is used to create a continuous flow rate (e.g. 850 mL/min).

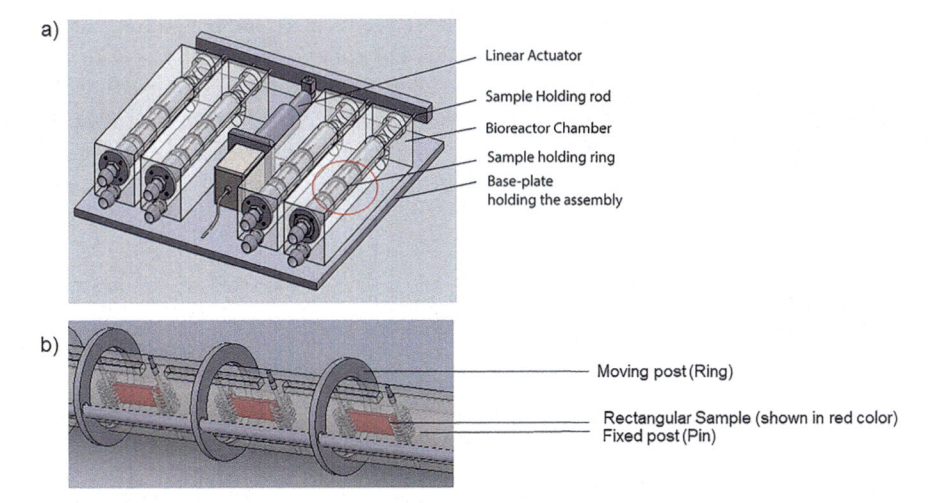

**Fig. 2** (a) Schematic diagram of the custom-built U-shaped bioreactor connected to a linear actuator which guides the rods that threads through samples, permitting them to bend and straighten. (b) Inset shows three samples inside the conditioning chamber that can be moved one end (ring) and is fixed on the other (pin). In the current study, the actuator was set to a 1 Hz frequency to permit cyclic flexure while the pump operated at a steady flow rate of 850 mL/min [21]

Using this bioreactor under flex-flow conditions to engineer aortic valvular tissues (1 Hz cyclic flexure and shear stress of ~4.73 dyn/cm$^2$ on the "ventricular" side of the specimen), the Ramaswamy group successfully differentiated BMSCs simultaneously in vitro, to both endothelial cells and activated interstitial cells (i.e., alpha-smooth muscle actin positive) which distributed preferentially to the engineered tissue surfaces and interstitium, respectively, similar to native heart valve morphology [21].

In addition to the physiological ranges of flexure, flow, and stretch as well as the intrinsic biochemistry of cell culture media being used, other parameters need to also be considered when conditioning engineered heart valves. Some of these parameters include fluid pressure, temperature, and oxygen diffusion which can all be regulated to promote extracellular matrix (ECM) formation, cellular proliferation and/or cell organization. As one example, pressures in some bioreactors have been typically set to 80–120 mm Hg to simulate systemic circulation or 20–40 mm Hg for the pulmonary circulation [22].

# 3 Scaffolds

An important factor in the tissue engineering of heart valves is naturally the choice of scaffold material. The type of material utilized in tissue engineering can provide a natural environment which mimics the physiological conditions in the body. The native ECM acts as the backbone for cells to organize and assemble into functioning tissue. An engineered ECM must mimic this environment in order to create and maintain tissues for functional use as an implant. Most cell types utilized in bioreactors are adherent cells and therefore a cell adhesion substrate is critical for success. Cells can be embedded in a complex microenvironment that may include various growth factors and/or cytokines. Scaffolds can be made of synthetic materials or natural materials.

## 3.1 Synthetic

The most common synthetic materials are polyhydroxyalkanoate (PHA), PGA, poly lactic acid (PLA), and PLLA/PLA blends. One of the major concerns when developing an appropriate scaffold material is the vascularization to support proliferation of cells. Another critical component is to design the appropriate degradation time schedule. The material must allow for cell adhesion and replacement of scaffold material with the development of de novo ECM. If the scaffold degrades too early, the entire construct may fail mechanically.

Shinoka et al. [23] described a PLA scaffold seeded with fibroblasts which formed a tissue-like sheet. The sheet was then seeded with endothelial cells to form a cellular monolayer. The implantation of the seeded scaffold as leaflet replacements was completed in seven animals and all survived the procedure. There was no evidence of stenosis and histology demonstrated appropriate cellular architecture and extracellular matrix. Sodian et al. [24] have used cell-seeded PLA scaffold as well as PHA and P4HB synthetic polymers for heart valve tissue engineering purposes. After comparing PGA, PHA, and P4HB, there was significantly higher cell attachment and collagen content on PGA samples. A variety of different combinations of scaffold material have been reported; however, most have described short-term success in vitro which may not translate into long-term success with in vivo possibility [5, 25–27].

## 3.2 Natural

Natural scaffolds are typically obtained from decellularized animal valves or made from collagen and fibrin-based scaffolds. When obtaining animal valves the processing must be done carefully to preserve the structure of the ECM to allow for cell seeding. Several studies have been described where detergents do not damage the ECM matrix structure and integrity. Some studies have focused on low concentration of detergents for scaffold purification [28, 29] or crosslinking of collagen with nordihydroguaiaretic acid (NDGA) [30]. These valves have several drawbacks which include inflammation, fibrosis, degradation, damage, and possible death of patients. In addition, these designs may be cost prohibitive. Gerdisch et al. [31] have performed studies for mitral valve repair using decellularized ECM from porcine small intestinal submucosa (CorMatrix Cardiovascular Inc., Roswell, GA). A total of 19 patients underwent the surgery and this preliminary study showed promised with good valvular function and no calcification on echocardiographic images.

Another method of natural designs involves collagen and fibrinogen gels to help entrap the seeded cells. By mixing the cells in the gel, the structure contracts and the cells remove water. Unfortunately the final construct remains unpredictable due to large differences in cell density [32, 33]. Moreira et al. [34] discussed the use of a living textile reinforced mitral valve to overcome the limitation of fibrin gel.

Hong et al. [35] fabricated a novel scaffold using an electrospinning technique. Cells were seeded onto decellularized pig valves that had been coated with poly (3-hydroxybutyrate-co-4-hydroxybutyrate). The valves were placed in a pulsatile flow bioreactor and exposed to defined physical signals with a fixed frequency (1 Hz) and pressure conditions (3 L/min and 60/40 mm Hg) over a time period of either 9 or 16 days. As compared to decellularized valves without the coating, the hybrid scaffold showed a significant increase in mechanical strength. Both groups showed similar proliferation of cells and comparable cell mass.

## 4 Cell Types

After considering the type of scaffold and bioreactor that should be utilized, the cell source(s) to be seeded on the scaffold will need to be chosen. Most groups have focused on using adherent cell types and in many cases stem cells.

### 4.1 Mesenchymal Stem Cells (MSCs)

Adult stem cells are found in every part of the body and their primary role is to heal and maintain the tissue with which they reside. In addition, they have the ability to differentiate into specialized cell types. Adult stem cells can be harvested from a patient's own tissue, such as adipose (fat) tissue, bone marrow, teeth, and more. Adult stem cells have a limited potential for growth when placed into culture and do not grow as rapidly as embryonic stem cells.

There is a population of cells known as MSCs which are different from blood cells which may give rise to bone, cartilage, and even fat cells. Caplan [36] first described MSCs in 1990 and the ability of these cells to go down different pathways of differentiation.

Several groups have described the use of MSCs in experiments with bioreactors. MSCs could be used in an autologous (from same patient) or allogeneic (non-matched donor) manner because they usually will not illicit an immune response, i.e., they are considered immunomodulatory [37]. In addition to the immune privileged nature of MSCs, they are a preferable source in that they will respond to mechanical stimuli [38–41] and allow natural remodeling to heart valve structure [42].

## 4.2 Fibroblasts/Smooth Muscle Cells/Valvular Interstitial Cells

Fibroblasts are a cell that are responsible for synthesizing the ECM and creating the framework of tissue. Smooth muscle cells (SMCs) or myocytes are responsible for creating muscle tissue in a process called myogenesis. Valvular interstitial cells (VICs) are the cells that make up the interstitial spaces of aortic valves. They exhibit an intermediate phenotype between fibroblasts and SMCs. They are responsible for creating the ECM in the valve and maintaining its mechanical characteristics [43].

Appleton et al. [44] combined vascular SMCs with a mixture of growth factors to create a surrogate cell source for developing a valve graft. They demonstrated that by combining SMCs with transforming growth factor beta-1, epidermal growth factor, and platelet-derived growth factor, they could increase proliferation and expression of ECM constituents found in the aortic valve. There are reports of calcification and other potential drawbacks to cells in the heart valve lineages such as a low rate of matrix remodeling and lack of elastin production. In addition, scaffolds seeded with these cells displayed low mechanical strength with a disarray of fibrous components. Ultimately even if these obstacles are overcome, the reality of being able to incorporate somatic cell sources such as SMCs for heart valve tissue engineering is likely to be impractical due to distinct supply limitations.

## 4.3 Endothelial Progenitor Cells

Endothelial progenitor cells (EPCs) are a mixture of cells that are responsible for regeneration of the endothelial lining of blood vessels. Because of their ability to create structured ECM, EPCs represent a viable cell source for TEHVs. In addition, EPCs can be used as a cell source for engineered valves because they have characteristics of both interstitial and endothelial tissues. Some researchers have combined EPCs with smooth muscle cells and fibroblasts to help improve the ECM formation [45]. Valves that had been conditioned in a bioreactor demonstrated unobstructed opening and closing. In addition, the tissue presented with collagen types I and II with a homogeneous cell distribution.

## 4.4 Embryonic/IPS/Amniotic/Cord

Embryonic stem cells (EScs) are the most primitive stem cell. They are derived from the inner cell mass of the blastocyst and are present in the embryo before implantation occurs in the uterus. EScs are capable of dividing rapidly in culture and can be grown indefinitely in an undifferentiated state. They are capable of differentiating into all cells of the adult body. EScs were originally discovered and derived from a mouse in 1981 by Evans and Kaufman [46] and later from a human by Thomson in 1998. Thomson [47] first developed a technique to isolate and grow human embryonic stems cells in culture. The inner cell mass (ICM) of the blastocyst is a pluripotent cell that can become specialized into any cell of the body. Cells from the ICM may maintain the potential to form any cell in the body even after extensive proliferation in culture.

Induced pluripotent stem cells (iPS) are progenitors that have been derived from an adult somatic cell via transcription factor encoding by genes. The adult cell is re-programmed to behave very similar to an ESc. The cells have similar expression of genes and proteins, potency and differentiability. Because iPS cells can be derived from the patient as an autologous product, there is less risk of rejection and no need for immunosuppression. In addition, the production of iPS cells does not require the destruction of an embryo, therefore eliminating some of the ethical debate surrounding embryonic stem cells. These cells are typically derived using transfection techniques to make the non-pluripotent cells express proteins or transcription factors that are typically expressed by pluripotent cells. Takahashi and Yamanaka [48] pioneered these techniques in 2006.

Both EScs and iPS cells represent a novel approach to tissue engineering because of their capacity to differentiate into any cell or tissue. Some groups have explored the use of these types of cells for heart valve tissue formation. As one example, Ghodsizad et al. [49] described the use of embryonic cells seeded in a decellularized valve in a magnetically guided multifunctional bioreactor. Valves were prepared using decellularized porcine pericardium. Half of the samples were prepared using a magnetic field to prevent the loss of cells. This allowed for homogeneous and efficient cell seeding of the decellularized matrix. An absence in cell seeding was observed in the nonmagnetic group. It is important to note that iPS is a new technology and issues relating to repeatability and avoidance of teratomas are likely to delay its applicability for use in translational heart valve tissue engineering investigations. Furthermore, ethical and legal concerns with embryonic cell sources are likely to result in translational studies focused primarily on adult stem cell progenitors for at least the foreseeable future.

Schmidt et al. [50] studied the use of amniotic fluid-derived cells as a potential source for heart valve tissue engineering. They demonstrated that the cells could maintain their characteristics after cryopreservation. In particular, cells expressing CD133 showed characteristics similar to endothelial-like cells and were able to produce ECM elements on a leaflet. Periodontal ligament cells (PDLs) have demonstrated potential in heart valve tissue engineering after flow-based, bioreactor mechanical conditioning [51]. PDLs can differentiate down pathways of endothelial cell and smooth muscle cell phenotypes by applying steady shear stress of 1 $dyn/cm^2$.

## 5 Preclinical/Clinic Studies

A few studies have been completed where TEHVs were implanted in animals. Many preclinical animal studies were completed with the valve in the pulmonary position [52]. The benefit to testing the valve in the pulmonary position is that the conditions on the right side of the heart are less demanding than those of the systemic circulation. Dohmen et al. [52] described the use of tissue engineered valves in a juvenile sheep model. Decellularized porcine matrices were seeded with vascular endothelial cells. The valves were implanted into the right ventricular outflow tract. The valves exhibited endothelial cells along the inner surface of the valve as well as fibroblast growth on the decellularized matrix. In addition, the calcium contents were very low on the TEHV.

In another study, synthetic scaffolds were seeded with autologous bone marrow cells and implanted into nonhuman primates [53]. Echocardiography showed preserved valve function at 4 weeks post implant. Substantial cellular remodeling and ingrowth on the scaffold was visualized via histology and immunohistochemistry. In addition, there was evidence of new living tissue replacing the biodegradable synthetic matrix. Bert et al. [54] also demonstrated successful implantation of TEHV in young healthy baboons.

Hoerstrup et al. [55] preconditioned leaflet heart valves in pulse duplicators for 14 days prior to implanting in lambs. The valves grown in the bioreactor were under gradually increasing nutrient medium flow and pressure conditions (125 mL/min at 30 mm Hg to 750 mL/min at 55 mm Hg). They were able to function for up to 5 months when they were explanted and resembled normal heart valves in properties and structure. Converse et al. [56] have recently reported on a prototype of a full-scale system that produces pulsatile conditioning using cyclic waveforms. After 72 h of conditioning, full tissue recellularization was not achieved but a subsurface pilot population of cells resulted.

Preconditioning valves prior to implantation has been challenging mostly due to the lack of a disposable system that ensures sterility and can create an appropriate mechanical environment. When moving TEHV therapies to the clinic, it is important to develop a construct that is non-thrombogenic, biocompatible, capable of cell growth, have long-term efficacy, and result in clinical improvement for the patient. Dohmen et al. [57] reported on 63 patients with calcific aortic valve disease who underwent a valve replacement using a TEHV. A biopsy sample of the TEHV indicated the presence of endothelial cells and recellularization.

In 2006, researchers evaluated TEHVs seeded with autologous progenitor cells implanted into pediatric patients [58]. The cells were isolated from peripheral mononuclear cells isolated from human blood. No signs of valve degeneration were observed in either patient. Two clinical trials have been reported using decellularized xenograph heart valves [59]. Several adverse events were reported highlighting the importance of rigorous models prior to human studies. Three of the four children that received the implants died and therefore the fourth valve was removed prophylactically. Two of the three deaths were sudden with severely degenerated valves and the third had rupture after 7 days.

CorMatrix ECM Technology has been used at more than 825 hospitals across the USA and has been implanted in nearly 95,000 cardiac procedures. The FDA issued clearance of an Investigational Device Exemption (IDE) for an Early Feasibility Study of its CorMatrix® ECM® Tricuspid Heart Valve. Gerdisch et al. [31] demonstrated early feasibility on 19 patients for mitral valve repair. The ECM scaffold appeared to resist calcification and infection and may be a satisfactory material for mitral valve surgical situations; however, additional studies are warranted to determine long-term feasibility.

## 6 Conclusions

Tissue engineered valves provide new hope in the development of cardiac valve replacement. The main considerations when designing a TEHV include the scaffold and cell types as well as mechanical/physiological forces to condition the growing construct. The ability to develop a TEHV to replace damaged valve tissue decreases the risk of complications and improves valve performance. Tissue engineering presents the opportunity to introduce a natural tissue that may have long-term viability and fewer side effects. There are many hurdles still to overcome but TEHVs represent an opportunity to revolutionize the cardiac field. For example, valves which are exposed to dynamic and complex forces must integrate with the native tissue. Hence, designing a TEHV that is capable of integrating with native tissues and functioning in the high throughput of the heart is a challenge. Establishing appropriate in vitro models for optimizing mechanical cues for de novo valvular tissue growth and appropriate in vivo preclinical models can help to advance the field. Major research goals include understanding mechanisms of action, developing biomarkers, assays, and tools, defining appropriate endpoints and establishing reliable animal models. Preliminary work in the clinic is promising but many questions remain unanswered. The major clinical goals include characterizing and designing quality tissue constructs for implant, sterility concerns, addressing rejection complications, and early prediction of clinical outcomes. Thus, there are indeed many variables to be considered and further analyzed in TEHV design and approaches. Nonetheless, TEHVs represent an area in tissue engineering where clinical translation in the foreseeable future is likely, based on the state of the field today. The use of bioreactors as part of this process, in terms of a preconditioning tool, has proven to be essential, thus augmenting the role of the Bioengineer in the overall TEHV treatment strategy.

# References

1. Nkomo VT, Gardin JM, Skelton TN, et al. Burden of valvular heart diseases: a population-based study. Lancet. 2006;368:1005–11. https://doi.org/10.1016/S0140-6736(06)69208-8.
2. Berry JL, Steen JA, Koudy Williams J, Jordan JE, Atala A, Yoo JJ. Bioreactors for development of tissue engineered heart valves. Ann Biomed Eng. 2010;38(11):3272–9. https://doi.org/10.1007/s10439-010-0148-6.
3. Ohshima N, Yanagi K, Miyoshi H. Packed-bed type reactor to attain high density culture of hepatocytes for use as a bioartificial liver. Artif Organs. 1997;21:1169–76.
4. Seliktar D, Black RA, Vito RP, Nerem RM. Dynamic mechanical conditioning of collagen-gel blood vessel constructs induces remodeling in vitro. Ann Biomed Eng. 2000;28:351–62.
5. Gandaglia A, Bagno A, Naso F, Spina M, Gerosa G. Cells, scaffolds and bioreactors for tissue-engineered heart valves: a journey from basic concepts to contemporary developmental innovations. Eur J Cardiothorac Surg. 2011;39(4):523–31.
6. Carrion K, Dyo J, Patel V, Sasik R, Mohamed SA, Hardiman G, Nigam V. The long non-coding HOTAIR is modulated by cyclic stretch and WNT/ß-CATENIN in human aortic valve cells and is a novel repressor of calcification genes. PLoS One. 2014;9(5):e96577. https://doi.org/10.1371/journal.pone.0096577.
7. Dhein S, Schreiber A, Steinbach S, Apel D, Salameh A, Schlegel F, Kostelka M, Dohmen PM, Mohr FW. Mechanical control of cell biology. Effects of cyclic mechanical stretch on cardiomyocyte cellular organization. Prog Biophys Mol Biol. 2014;115(2–3):93–102. https://doi.org/10.1016/j.pbiomolbio.2014.06.006.
8. Girão-Silva T, Bassaneze V, Campos LC, Barauna VG, Dallan LA, Krieger JE, Miyakawa AA. Short-term mechanical stretch fails to differentiate human adipose-derived stem cells into cardiovascular cell phenotypes. Biomed Eng Online. 2014;13:54. https://doi.org/10.1186/1475-925X-13-54.
9. Koivisto E, Jurado Acosta A, Moilanen AM, Tokola H, Aro J, Pennanen H, Säkkinen H, Kaikkonen L, Ruskoaho H, Rysä J. Characterization of the regulatory mechanisms of activating transcription factor 3 by hypertrophic stimuli in rat cardiomyocytes. PLoS One. 2014;9(8):e105168. https://doi.org/10.1371/journal.pone.0105168.
10. Ku CH, Johnson PH, Batten P, Sarathchandra P, Chambers RC, Taylor PM, Yacoub MH, Chester AH. Collagen synthesis by mesenchymal stem cells and aortic valve interstitial cells in response to mechanical stretch. Cardiovasc Res. 2006;71(3):548–56.
11. Syedain ZH, Tranquillo RT. Controlled cyclic stretch bioreactor for tissue-engineered heart valves. Biomaterials. 2009;30(25):4078–84. https://doi.org/10.1016/j.biomaterials.2009.04.027.
12. Thayer P, Balachandran K, Rathan S, Yap CH, Arjunon S, et al. The effects of combined cyclic stretch and pressure on the aortic valve interstitial cell phenotype. Ann Biomed Eng. 2011;39:1654–67. https://doi.org/10.1007/s10439-011-0273-x.
13. Engelmayr GC, Hildebrand DK, Sutherland FW, Mayer JE, Sacks MS. A novel bioreactor for the dynamic flexural stimulation of tissue engineered heart valve biomaterials. Biomaterials. 2003;24:2523–32.
14. Engelmayr GC, Engelmayr GC Jr, Rabkin E, Sutherland FW, Schoen FJ, Mayer JE Jr, Sacks MS. The independent role of cyclic flexure in the early in vitro development of an engineered heart valve tissue. Biomaterials. 2005;26:175–87.
15. Engelmayr GC Jr, Sales VL, Mayer JE Jr, Sacks MS. Cyclic flexure and laminar flow synergistically accelerate mesenchymal stem cell-mediated engineered tissue formation: implications for engineered heart valve tissues. Biomaterials. 2006;27:6083–95.
16. Mirnajafi A, Zubiate B, Sacks MS. Effects of cyclic flexural fatigue on porcine bioprosthetic heart valve heterograft biomaterials. J Biomed Mater Res A. 2010;94(1):205–13.
17. Jockenhoevel S, Zund G, Hoerstrup SP, Schnell A, Turina M. Cardiovascular tissue engineering: a new laminar flow chamber for in vitro improvement of mechanical tissue properties. ASAIO J. 2002;48:8–11.

18. Mahler GJ, Frendl CM, Cao Q, Butcher JT. Effects of shear stress pattern and magnitude on mesenchymal transformation and invasion of aortic valve endothelial cells. Biotechnol Bioeng. 2014;111(11):2326–37.
19. Butcher JT, Nerem R. Valvular endothelial cells regulate the phenotype of interstitial cells in co-culture: effects of steady shear stress. Tissue Eng. 2006;12:905–15.
20. Ramaswamy S, Boronyak SM, Le T, Holmes A, Sotiropoulos F, Sacks MS. A novel bioreactor for mechanobiological studies of engineered heart valve tissue formation under pulmonary arterial physiological flow conditions. J Biomech Eng. 2014;136(12):121009.
21. Rath S, Salinas M, Villegas AG, Ramaswamy S. Differentiation and distribution of marrow stem cells in flex-flow environments demonstrate support of the valvular phenotype. PLoS One. 2015;10(11):e0141802. https://doi.org/10.1371/journal.pone.0141802.
22. Xing Y, He Z, Warnock JN, Hilbert SL, Yoganathan AP. Effects of constant static pressure on the biological properties of porcine aortic valve leaflets. Ann Biomed Eng. 2004;32:555–62.
23. Shinoka T, Breuer CK, Tanel RE, Zund G, Miura T, Ma PX, Langer R, Vacanti JP, Mayer JEJ. Tissue engineering heart valves: valve leaflet replacement study in a lamb model. Ann Thorac Surg. 1995;60:S513.
24. Sodian R, Hoerstrup SP, Sperling JS, Daebritz S, Martin DP, Moran AM, Kim BS, Schoen FJ, Vacanti JP, Mayer JE Jr. Early in vivo experience with tissue-engineered trileaflet heart valves. Circulation. 2000;102(19 Suppl 3):III22–9.
25. Filova E, Straka F, Mirejovsky T, et al. Tissue-engineered heart valves. Physiol Res. 2009;58 (Suppl 2):S141–58.
26. Freed LE, Guilak F, Guo XE, et al. Advanced tools for tissue engineering: scaffolds, bioreactors, and signaling. Tissue Eng. 2006;12(12):3285–305.
27. Ghanbari H, Kidane AG, Burriesci G, et al. The anti-calcification potential of a silsesquioxane nanocomposite polymer under in vitro conditions: potential material for synthetic leaflet heart valve. Acta Biomater. 2010;6(11):4249–60.
28. Dumont K, Yperman J, Verbeken E, et al. Design of a new pulsatile bioreactor for tissue engineered aortic heart valve formation. Artif Organs. 2002;26(8):710–4.
29. Eaglstein WH, Falanga V. Tissue engineering and the development of Apligraf, a human skin equivalent. Clin Ther. 1997;19(5):894–905.
30. Cebotari S, Tudorache I, Jaekel T, et al. Detergent decellularization of heart valves for tissue engineering: toxicological effects of residual detergents on human endothelial cells. Artif Organs. 2010;34(3):206–10.
31. Gerdisch MW, Shea RJ, Barron MD. Clinical experience with CorMatrix extracellular matrix in the surgical treatment of mitral valve disease. J Thorac Cardiovasc Surg. 2014;148:1370–8.
32. Flanagan TC, Sachweh JS, Frese J, Schnoring H, Gronloh N, Koch S, Tolba RH, Schmitz-Rode T, Jockenhoevel S. *In vivo* remodeling and structural characterization of fibrin-based-tissue-engineered heart valves in the adult sheep model. Tissue Eng Part A. 2009;15:2965.
33. Syedain Z, Lathi M, Johnson SL, Robinson P, Ruth GR, Bianco R, Tranquillo RT. Implantation of a tissue-engineered heart valve from human fibroblasts exhibiting short term function in the sheep pulmonary artery. Cardiovasc Eng Technol. 2011;2:101.
34. Moreira R, Gesche VN, Hurtado-Aguilar LG, Schmitz-Rode T, Frese J, Jockenhoevel S, Mela P. TexMi: development of tissue-engineered textile-reinforced mitral valve prosthesis. Tissue Eng Part C Methods. 2014;20(9):741–8. https://doi.org/10.1089/ten.tec.2013.0426.
35. Hong H, Dong N, Shi J, et al. Fabrication of a novel hybrid scaffold for tissue engineered heart valve. J Huazhong Univ Sci Technol Med Sci. 2009;29(5):599–603.
36. Caplan AI. Mesenchymal stem cells. J Orthopaed Res. 1991;9(5):641–50.
37. Rubtsov YP, Suzdaltseva YG, Goryunov KV, Kalinina NI, Sysoeva VY, Tkachuk VA. Regulation of immunity via multipotent mesenchymal stromal cells. Acta Nat. 2012;4(1):23–31.
38. Lu X, Zhai W, Zhou Y, et al. Crosslinking effect of nordihydroguaiaretic acid (NDGA) on decellularized heart valve scaffold for tissue engineering. J Mater Sci Mater Med. 2010;21(2):473–80.

39. Lutter G, Metzner A, Jahnke T, et al. Percutaneous tissue engineered pulmonary valved stent implantation. Ann Thorac Surg. 2010;89(1):259–63.
40. Martin DJ, Warren LA, Gunatillake PA, et al. Polydimethylsiloxane/polyether-mixed macrodiol-based polyurethane elastomers: biostability. Biomaterials. 2000;21(10):1021–9.
41. Martin I, Wendt D, Heberer M. The role of bioreactors in tissue engineering. Trends Biotechnol. 2004;22(2):80–6.
42. Mathur AB, Collier TO, Kao WJ, et al. In vivo biocompatibility and biostability of modified polyurethanes. J Biomed Mater Res. 1997;36(2):246–57.
43. Taylor PM, Batten P, Brand NJ, Thomas PS, Yacoub MH. The cardiac valve interstitial cell. Int J Biochem Cell Biol. 2003 Feb;35(2):113–8.
44. Appleton AJ, Appleton CT, Boughner DR, Rogers KA. Vascular smooth muscle cells as a valvular interstitial cell surrogate in heart valve tissue engineering. Tissue Eng Part A. 2009;15 (12):3889–97.
45. Moreira R, Velz T, Alves N, et al. Tissue-engineered heart valve with a tubular leaflet design for minimally invasive transcatheter implantation. Tissue Eng Part C Methods. 2015;21(6):530–40. https://doi.org/10.1089/ten.tec.2014.0214.
46. Evans MJ, Kaufman MH. Establishment in culture of pluripotential cells from mouse embryos. Nature. 1981;292:154–6. https://doi.org/10.1038/292154a0.
47. Thomson JA. Embryonic stem cell lines derived from human blastocysts. Science. 1998;282:1145–7. https://doi.org/10.1126/science.282.5391.1145.
48. Takahashi K, Yamanaka S. Induction of pluripotent stem cells from mouse embryonic and adult fibroblast cultures by defined factors. Cell. 2006;126:663–76.
49. Ghodsizad A, Bordel V, Wiedensohler H, Elbanayosy A, Koerner MM, Gonzalez Berjon JM, Barrios R, Farag M, Zeriouh M, Loebe M, Noon GP, Koegler G, Karck M, Ruhparwar A. Magnetically guided recellularization of decellularized stented porcine pericardium-derived aortic valve for TAVI. ASAIO J. 2014;60(5):582–6. https://doi.org/10.1097/MAT. 0000000000000110.
50. Schmidt D, Achermann J, Odermatt B, Genoni M, Zund G, Hoerstrup SP. Cryopreserved amniotic fluid-derived cells: a lifelong autologous fetal stem cell source for heart valve tissue engineering. J Heart Valve Dis. 2008;17(4):446–55.
51. Martinez C, Rath S, Van Gulden S, Pelaez D, Alfonso A, et al. Periodontal ligament cells cultured under steady-flow environments demonstrate potential for use in heart valve tissue engineering. Tissue Eng A. 2012;19:458–66. https://doi.org/10.1089/ten.TEA.2012.0149.
52. Dohmen PM, Ozaki S, Nitsch R, Yperman J, Flameng W, Konertz W. A tissue engineered heart valve implanted in a juvenile sheep model. Med Sci Monit. 2003;9:BR137–44.
53. Weber B, Scherman J, Emmert MY, Gruenenfelder J, Verbeek R, Bracher M, Black M, Kortsmit J, Franz T, Schoenauer R, Baumgartner L, Brokopp C, Agarkova I, Wolint P, Zund G, Falk V, Zilla P, Hoerstrup SP. Injectable living marrow stromal cell-based autologous tissue engineered heart valves: first experiences with a one-step intervention in primates. Eur Heart J. 2011;32:2830.
54. Bert AA, Drake WB, Quinn RW, Brasky KM, O'Brien JE Jr, Lofland GK, Hopkins RA. Transesophageal echocardiography in healthy young adult male baboons (Papio hamadryas anubis): normal cardiac anatomy and function in subhuman primates compared to humans. Prog Pediatr Cardiol. 2013;35(2):109–20.
55. Hoerstrup SP, Sodian R, Daebritz S, Wang J, Bacha EA, Martin DP, Moran AM, Guleserian KJ, Sperling JS, Kaushal S, et al. Functional living trileaflet heart valves grown in vitro. Circulation. 2000;102(19 Suppl 3):III44–9.
56. Converse GL, Buse EE, Neill KR, McFall CR, Lewis HN, VeDepo MC, Quinn RW, Hopkins RA. Design and efficacy of a single-use bioreactor for heart valve tissue engineering. J Biomed Mater Res B Appl Biomater. 2017;105:249–59. https://doi.org/10.1002/jbm.b.33552.
57. Dohmen PM, Hauptmann S, Terytze A, Konertz WF. In-vivo repopularization of a tissue-engineered heart valve in a human subject. J Heart Valve Dis. 2007;16(4):447–9.

58. Cebotari S, Lichtenberg A, Tudorache I, Hilfiker A, Mertsching H, Leyh R, Breymann T, Kallenbach K, Maniuc L, Batrinac A, Repin O, Maliga O, Ciubotaru A, Haverich A. Clinical application of tissue engineered human heart valves using autologous progenitor cells. Circulation. 2006;114(1 Suppl):I132–7.
59. Simon P, Kasimir MT, Seebacher G, et al. Early failure of the tissue engineered porcine heart valve SYNERGRAFT in pediatric patients. Eur J Cardiothorac Surg. 2003;23(6):1002–6; discussion 1006.

# Bioprosthetic Heart Valves: From a Biomaterials Perspective

**Naren Vyavahare and Hobey Tam**

**Abstract** For the past 50+ years, the heart valve replacement (HVR) industry has gone through multiple growth phases in terms of innovation and market growth, and is now a formidable submarket of the cardiovascular medical device space. HVRs started off as small, compact simplistic devices made from synthetic parts. Now, these archaic prototypes have evolved into a diverse product offering from a multitude of small to large healthcare firms that range from completely synthetic materials to crosslinked tissue-based HVRs to new research being performed to investigate engineered tissue valves. These innovative leaps in technology and growth in commerce would not have been possible without the collaboration of multidisciplinary investigators unified through their passion in pursuing one goal—a superior healthcare option that resulted in a better quality of life for individuals throughout the world needing a new heart valve. Everything from biomaterial design to micro/macro-biomechanics to build computational modeling to optimize valve design has been utilized to create the current product line and are currently still in use as our metrics and methods get more advanced to innovate the future of heart valves as well as cardiovascular device technology. The future of HVRs rests on academia and industry coming together to move technology forward to provide patients in dire need of a more durable HVR device. Therefore, the rest of the content covered in this book is a comprehensive review of current art and models in existence to design an effective HVR in the efforts of empowering individuals wishing to bring healthcare options to a patient segment in dire need of change. The following chapter will cover past and current approaches in designing and fabricating HVR materials, current performance of HVRs, material design considerations of next generation materials, and major research interests in the next generation of HVR materials. The rest of this publication will cover approaches in properly leveraging this base biomaterial in valve-specific design to innovate a more effective HVR.

N. Vyavahare (✉) · H. Tam
Clemson University, Clemson, SC, USA
e-mail: narenv@clemson.edu

**Key words** Xenografts · Heart valve replacement · Tissue-based materials · Tissue crosslinking · Heart valve design

# 1 Introduction

For the past 50+ years, the heart valve replacement (HVR) industry has gone through multiple growth phases in terms of innovation and market growth, and is now a formidable submarket of the cardiovascular medical device space. HVRs started off as small, compact simplistic devices made from synthetic parts. Now, these archaic prototypes have evolved into a diverse product offering from a multitude of small to large healthcare firms that range from completely synthetic materials to crosslinked tissue-based HVRs to new research being performed to investigate engineered tissue valves. These innovative leaps in technology and growth in commerce would not have been possible without the collaboration of multidisciplinary investigators unified through their passion in pursuing one goal—a superior healthcare option that resulted in a better quality of life for individuals throughout the world needing a new heart valve. Everything from biomaterial design to micro/macro-biomechanics to build computational modeling has been utilized to create the current optimal valve designs. The future of HVRs also rests on academia and industry coming together to move technology forward to provide patients in dire need of a HVR device that is durable. The following chapter will cover past and current approaches in designing and fabricating HVR materials, current performance of HVRs, material design considerations of next generation materials, and major research interests in the next generation of HVR materials. The rest of this publication will cover approaches in properly leveraging this base biomaterial in valve-specific design to innovate a more effective HVR.

# 2 Growing the Heart Valve Market Space

The cardiovascular medical device and therapeutics space as an aggregate market is valued at $92BIL with a 7% CAGR across the industry [1]. Heart valve replacements represent a $1.25BIL segment of this market [1] with over 300,000 heart valve replacements per year globally [2, 3]. However, in emerging markets such as the Chinese, Brazilian, and Indian healthcare markets, the cardiovascular space is experiencing upwards of 10–15% CAGR annually. With 85% of developed industry concentrated in developed economies (only 18% of the human population resides in these regions), the cardiovascular market (including the heart valve replacement market) is poised to continue large growth in the next 50 years if suppliers of the cardiovascular space can meet patient needs in these countries. Market growth will concentrate on penetrating emerging economies with patient segments that have not

## 3 Current Indications for Use

Between 275,000 and 370,000 heart valve replacements occur annually [2, 5–8]. This number is a small fraction of the 15 million people suffering from aortic or mitral valve disease. Over 20,000 cases of valve disease result in death annually in the United States, and contribute to an additional 42,000 deaths [9]. Valve disease can be broken into two categories: valvular stenosis and valvular insufficiency. Stenotic valves are characterized by narrowing of the valve opening, typically caused by stiffening of the leaflets. Valvular insufficiency, or regurgitation, is characterized by incomplete coaptation of the leaflets, which can also be caused by stiffening and by structural degradation of the leaflet tissue [10]. One of the most common causes of stenotic valves is calcific degeneration of the leaflets [11–15]. Calcific nodules form within the leaflet, creating a thick fibrotic, stiffer tissue (Fig. 1). Calcific degeneration progresses similarly to atheroma in the vasculature and typically occurs in adults over 65 years of age. Because of the stiff leaflets and smaller opening, the heart must pump harder to move the same amount of blood. This puts increased strain on the heart and reduces the flow of blood to the body. An increased pressure gradient across the valve can put greater stress on the stiffened leaflets and this can cause calcified fragments to break off and embolize.

Stenosis and regurgitation are two ways to characterize valve disease and failure, but there are further contributing causes and disorders to valve disease. These include rheumatic fever, infective endocarditis, myxomatous degeneration, calcific degeneration, and congenital anomalies. Each of these disorders contributes to valve disease, predominantly affecting the highly stressed mitral and aortic valves.

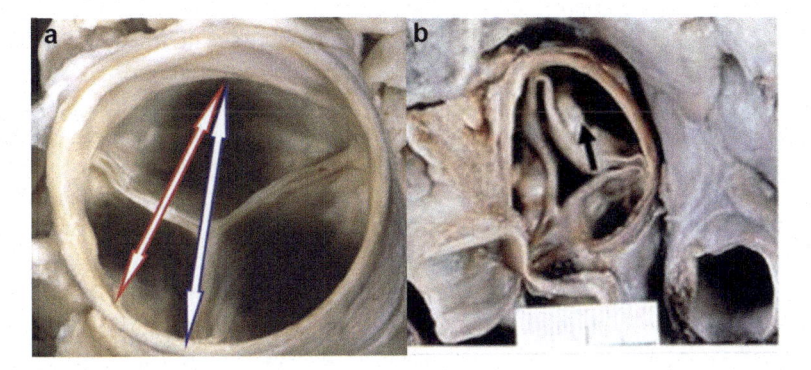

**Fig. 1** (a) Healthy human aortic valve [16] and (b) explanted calcified human aortic valve with arrow pointing to calcific nodule [17]

Incidence of problems with these valves arises from the higher pressures and stress placed on the right side of the heart. Valvular diseases can be alternatively classified as congenital (preexisting at birth) heart defects, or acquired diseases [3, 10, 18]. Congenital defects are associated with inherited disorders, and they affect approximately 3% of the population. However, they are usually asymptomatic until discovered in a routine examination. Two common congenital defects are myxomatous mitral valves and bicuspid aortic valves [3, 18]. Further disorders affecting the valves are autoimmune and connective tissue disorders. Myxomatous mitral valves, commonly called "floppy valves," present as prolapsed valves, extending back into the left atrium during contraction. The leaflets contain thickened spongiosa caused by an accumulation of dermatin sulfate [3, 10]. In fact, the term myxoma is derived from the Greek word for mucus, which contains a mixture of proteoglycans. The increase of the spongiosa causes a thicker, more rubbery tissue, and weakens the fibrosa as collagen bundles are pushed farther apart [10]. Mitral valve prolapse can result in valve insufficiency, but can also be asymptomatic and cause heart failure [3, 10].

Bicuspid aortic valves appear to have only two leaflets, and can appear as one normal leaflet and two leaflets fused together, or two symmetric leaflets. Similar to myxomatous degeneration, bicuspid aortic valves are asymptomatic in early life unless observed under examination. The presence of a bicuspid aortic valve can lead to aortic root dilation or cusp prolapse later in life [10, 18]. Aside from congenital disorders, several acquired diseases affect the heart valves. Incidence is affected by environmental factors such as lifestyle choices and infection. The most common acquired diseases affecting heart valves are infective endocarditis, rheumatic heart disease, and degenerative diseases such as calcification and aortic root dilation.

Infective endocarditis is defined as the infection and vegetation of the endocardium or heart valves [19–22]. The majority of infections are caused by gram-positive cocci such as coagulase-positive *Staphylococcus aureus* or coagulase-negative *Staphylococcus epidermis*. Valve leaflets have no vascular supply; consequently, infection must come from the blood itself. Valve damage is a predisposing factor for infective endocarditis due to compromised native barriers to infection [19]. Bacteria or fungi can become trapped during the repair process within a network of immune cells, platelets, and fibrin. These organisms cause rapid tissue destruction, leading to large vegetations or growths, which cause incomplete coaptation or stenosis of the valve [23]. Larger vegetations can embolize into the bloodstream, leading to side effects and infection progression downstream. Since valve damage promotes bacterial attachment, conditions such as congenital heart defects, degenerative disease, rheumatic disease, and mechanical wear can be predisposing factors for endocarditis [10, 19, 23]. Rheumatic fever can develop in children following contraction of strep throat, caused by Streptococci bacteria, and Rheumatic fever is believed to be caused by antibody cross-reactivity [24–26]. In rare occurrences, antibodies are released which can react with proteins in the tissue. Rheumatic fever can lead to rheumatic heart disease, collectively causing damage to all parts of the heart and valvular defects [10]. However, incidences affecting the joints are more common

[27]. Rheumatic fever has largely been eliminated in first and second world countries, mainly remaining a problem in developing countries [28].

Degenerative diseases are passive disorders that cause wear, tear, and calcification of the valve [29]. They are more common in older patients due to tissue fatigue, and present as floppy valves, aortic dilation, or valve calcification. Preexisting conditions such as congenital defects or other diseases increase the propensity of the valve to wear, causing quicker degeneration of the tissue [3, 10, 14, 15, 29]. Over 25% of patients over 65 suffer from aortic calcific sclerosis [30], and roughly 43% of all heart valve diseases present as aortic sclerosis [9]. Aortic stenosis causes include calcification and thickening of the aortic valve, and progresses similarly to atherosclerosis [3, 14, 15]. Lifestyle choices remain a significant contribution to the development of this disorder. Though mostly a degenerative disease, calcification is also an active process influenced by phenotypic shifts of cells [3, 14, 15, 31, 32]. Increased macrophage activity occurs, and valvular interstitial cells (VICs) shift toward osteoblast-like properties, which produce bone mineralization. Degenerative diseases greatly interest the cardiac community, since they are both prevalent in native heart valve disease, and remain the number one mode of failure in bioprosthetic valves.

## 4 Current Treatments for Valvular Disease

Early intervention can help reduce the severity of valvular disease before it necessitates surgery or replacement. Noninvasive visualization techniques such as echocardiography or coronary angiography can correctly identify valve disease. Echocardiography can show the size of the valve openings. Doppler mode uses ultrasound to determine regurgitation or stenosis severity. Coronary angiography utilizes an injectable contrast agent to visualize blood flow and identify a narrowed valve or backflow of blood. Blood cultures and echocardiography are used to identify infective endocarditis. However, if heart valve disease is not caught early enough or prevetive measures are not taken, structural degeneration or calcification can result in valve insufficiencies caused by valve stenosis (closing or reduction of effective orifice area of valve opening) or regurgitation (reduced ability for valve leaflets to come back together for coaptation). Figure 2 illustrates a normally functioning (a) opened (top) and closed (bottom) valve. Figure 2b illustrates a stenotic valve (top) and a regurgitant valve (bottom). Both can compromise systemic blood flow and result in fatal consequences in the patient.

Upon identification of valvular disease, corrective actions for treatment can be taken. Lifestyle changes or drug therapy affect progression of acquired diseases through mediating and reducing the severity of symptoms. A variety of percutaneous interventional methods exist. One such example is balloon valvuloplasty, which works similarly to angioplasty to help reopen stenotic valves. This treatment is most commonly used in the tricuspid, pulmonary, and mitral valves. Risk for embolism exists in valves with vegetations or calcific deposits. Surgical repair includes the use

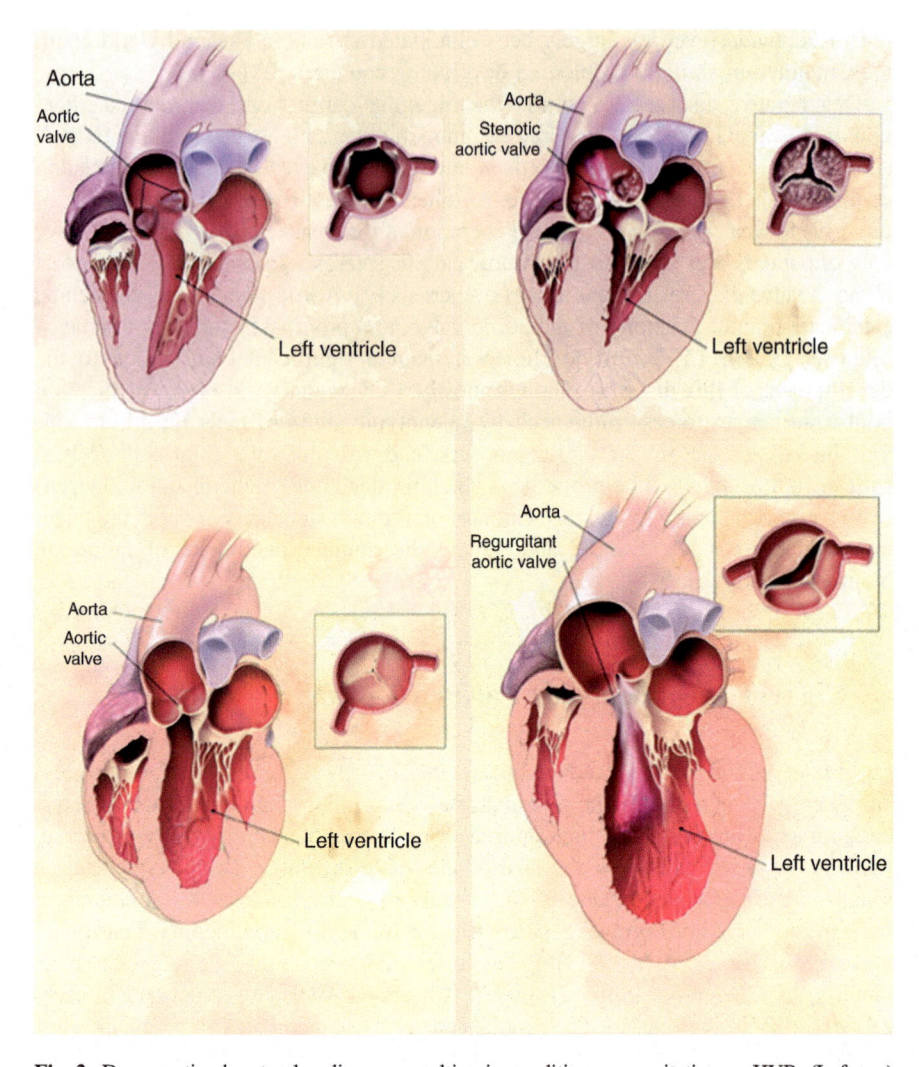

**Fig. 2** Degenerative heart valve diseases resulting in conditions necessitating an HVR. (Left top) Normal effective orifice area. (Left bottom) Normal coaptation of heart valve leaflets during ventricular diastole. (Right Top) Stenotic valve that cannot open into optimal effective orifice area. (Right Bottom) Regurgitant heart valve from heart valve leaflets not being able to accomplish coaptation

of autologous or xenographic patches, removal of calcific deposits, or reinforcement of the valve. Surgery can also repair congenital defects of heart valves [33, 34]. If repair cannot sufficiently treat a diseased valve, the next treatment option is usually valve replacement. This therapy typically consists of an open-heart surgery, including a heart lung bypass machine. Recent attempts to create minimally invasive treatments have resulted in percutaneous valves, which utilize catheterization techniques to introduce the valve [5, 35, 36]. Replacement valve options include mechanical heart valves or bioprosthetic tissue valves. Some investigation has been made in polymeric valves, but they were unsuccessful due to inadequate combination of mechanical properties and non-thrombogenicity [6, 37]. Attempts mainly included polyurethanes, polyesters, and polyethers [6, 37]. Polyesters and polyethers degrade under hydrolysis and oxidative deamination; polyurethanes were used for their excellent hemocompatibility and unique two phase structure. A large push in research is towards the development of tissue engineered heart valves derived from a patient's own cells [38].

Preexisting conditions and age play a role when optimal valve replacement therapy is chosen for each patient. Mechanical heart valves require lifelong systemic anticoagulant therapy while bioprosthetic valves do not [33, 39]. Patients who cannot tolerate this therapy due to blood disorders, pregnancy, or other contraindications should not receive MHVs [40–44]. Bioprosthetic tissue valves possess a short implant life, so they are typically implanted in older patients [3, 45]. More about these valve types, their selection factors, and shortcomings are discussed in the following section. Clinically used prosthetic valves can be broken into two categories: mechanical heart valves (MHVs) and bioprosthetic heart valves (BHVs). Mechanical heart valves primarily consist of metal, and have various mechanical designs to improve valve performance. Bioprosthetic heart valves are tissue based and made of either human (allograft) or animal-based (xenograft) tissues. Another treatment option typically used in young children is the Ross Procedure [6, 46]. The patient's aortic valve is replaced by the patient's pulmonary valve which in turn is replaced by an allograft. This allows the valve in the aortic position to remodel and grow as the patient grows [47]. Recent pushes in the field have investigated the use of polymeric heart valves, as well as tissue engineering approaches for heart valves [3]. Each type of heart valve has various designs, failure modes, advantages, and disadvantages, which will be discussed later in this section.

A set of design considerations was originally presented by Harken in 1962 [48], and has been modernized by Schoen and Levy (Table 1) [3]. Ideally, a replacement valve would perform as closely as possible to a native valve. Design considerations include mechanical factors, and implant-host factors. Further design factors are ease of surgery and patient comfort. Each valve type tackles some of the design requirements in different ways. For example, mechanical heart valves possess good

Table 1 Desirable characteristics of a heart valve substitute (adapted from [3])

| |
|---|
| Nonobstructive |
| Prompt and complete closure |
| Nonthrombogenic |
| Infection resistant |
| Chemically inert and nonhemolytic |
| Long-term durability |
| Easily and permanently implanted |
| Patient-prosthesis interface heals appropriately |
| Not annoying to patient |

durability due to the use of metal, so altered flow patterns are the design factor that represents the highest capacity to improve. Similarly, tissue valves possess good native hemocompatibility, so design changes focus on improving durability and reducing implant calcification.

## 5 Current Gold Standards in Heart Valve Replacements

There are currently three large companies that comprise the majority of the aggregate value of the heart valve replacement market. These companies are Edwards Lifesciences, St. Jude Medical, and Medtronic [1]. Smaller companies that make up the rest of the aggregate value of the heart valve replacement market are companies like Sorin, Onyx, and other medium to smaller producers [1].

Current heart valve replacement technology offerings available on the market can be divided into two main product offering segments: mechanical heart valves (MHVs) and bioprosthetic heart valves (BHVs). Each type addresses different patient segments with different value proposition offerings to fit those patient segment needs. It should be noted that when the heart valve market is segmented by different product offerings, BHVs account for the majority of the market share, and the rest of the market is in MHVs. This means that BHVs are the predominant and preferred product offering in the heart valve replacement space. Edwards Lifesciences leads this product offering segment, followed by Medtronic, then St. Jude [1]. However, despite over 40 years of innovation, both modes of replacing heart valve function have failure rates at 10–15 years needing replacement or compromising of patient [2, 3, 5, 7, 49]. Also, Edwards Lifesciences' design and utility patents on their heart valve replacements are due to expire soon or have already expired thereby creating a potential power vacuum in the heart valve replacement space. Without these market barriers on specific methods of fabricating the valve and valve design, other players may mimic their designs at will. The following sections delve deeper into the technology developed in each product offering segment.

## 6 Mechanical Heart Valves (MHVs)

The core concept behind MHVs is to completely replace mechanical function of the heart valve with synthetic materials. MHVs usually incorporate a bileaflet design often made from biologically inert materials such as polycarbonate. The major value proposition in using these materials and design is that (1) the materials are engineered significantly past the lifespan of a human and (2) the geometry, dynamics, and overall design of the heart valve provide adequate pulsatile system blood flow to the body. MHVs are composed entirely of synthetic materials, typically composed of metal or both metal and polymer. Figure 3 below depicts several

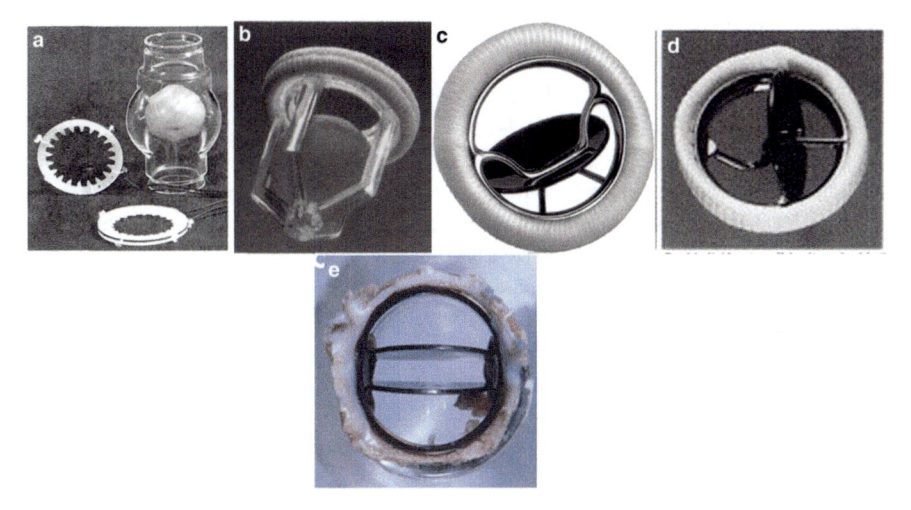

**Fig. 3** Mechanical Heart Valves. (**a**) Hufnagel [33], (**b**) Starr-Edwards [51], (**c**) Bjork-Shiley [52], (**d**) Medtronic-Hall [33], (**e**) St. Jude Bileaflet [53]

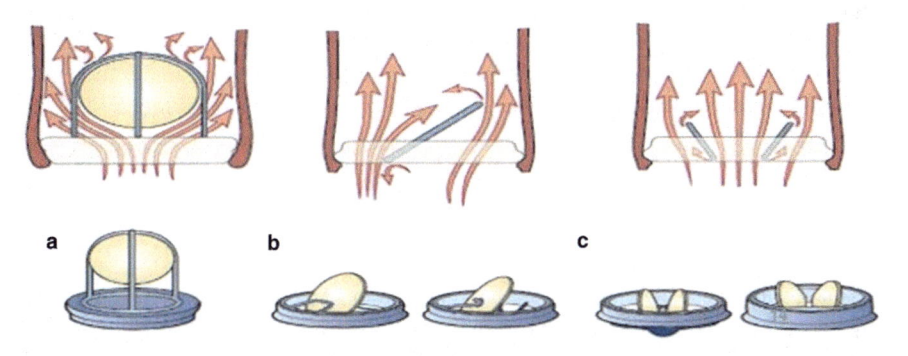

**Fig. 4** Contrasting MHV designs resulting in differing hemodynamics. (Top) Side view with arrows indicating different laminar flow of blood shearing against surfaces. (Bottom) Isometric view of MHV designs during ventricular systole. (**a**) Ball and cage model. (**b**) Single leaflet model. (**c**) Bileaflet model

models of MHVs. All MHVs consist of three major parts: sewing ring to attach to the tissue, housing, and occluder. Flexible fabric made of Dacron (polyethylene terephthalate) or ePTFE (expanded polytetrafluoroethylene) typically composes the sewing ring to allow attachment to the body via sutures. The housing consists of a base for attachment of the sewing ring, and guides the motion of the occluder. The occluder is the only moving part of the valve and it moves to alternately block and allow blood flow. Figure 4 depicts different designs of MHV leaflets to allow for pulsatile blood flow. MHVs exhibit decreased hemocompatibility due to their materials and geometry and thus they require lifelong anticoagulant therapy. However these valves also exhibit high durability due to their materials properties, which contributes to their low material failure rates. Therefore MHVs are typically used for younger patients with long life expectancy. These patients must also not have preexisting conditions such as blood disorders or pregnancy risk which contraindicate anticoagulant therapy [40, 41, 44, 50].

Drawbacks of this MHV design are that any occluder design, although optimized to reduce irregular flow patterns to mimic native blood flow as much as possible, still forces blood past a very non-native junction at the strut where the polycarbonate occlude(s) is attached. This causes irregular shearing patterns in the blood that activate the coagulation cascade making MHVs highly thrombogenic. It should be noted that MHV thrombogenicity has no correlation with the selection of material — the inherent geometry of the MHV causes irregular shear patterns within the blood that elicit a coagulation response. Therefore, MHV recipients must be put on anticoagulant therapy such as warfarin, coumadin, or other blood-thinning anticoagulants for the duration of implant use. Usually, patients are compromised due to this anticoagulant therapy because these anticoagulation agents are highly toxic and can lead to other lethal complications. This design is valuable in the small patient segment in developed economies that fall between the ages of 45 and 55 years and cannot afford more expensive BHVs. This patient demographic is still very mobile and immune competent. Implanting the alternative, BHVs, in this patient segment would be catastrophic as BHVs fail more rapidly in younger patients [54, 55]; therefore, surgeons opt for MHVs for their durability and patients' ability to handle the anticoagulation therapy. It ultimately becomes a lifestyle choice [56, 57].

## 7 Bioprosthetic Heart Valves Overview (BHVs)

The majority of the HVR patient segment receives a BHV. BHVs are fabricated from chemically treated tissues derived from connective tissue in animals. Two biologically based materials have been approved for clinical use as xenogenic tissue in BHVs: whole porcine aortic valves and bovine pericardium-based valves [3, 58–60]. The heart valve tissue bares all the loads and fatigue cycling of the system; therefore, biomaterial engineering is of utmost importance in BHV engineering. Currently, the standard practice in industry is to chemically tan bovine pericardium or porcine aortic heart valve leaflets with glutaraldehyde to produce a durable yet viscoelastic material that can deform to allow for the complex valve leaflet biomechanics for hundreds of millions of cycles. Tissue antigenicity must be reduced prior to implantation, and can be accomplished through multiple treatments, but glutaraldehyde is the only one clinically used. Glutaraldehyde has the benefit of crosslinking collagen and devitalizing cells in the tissue, but results in a tissue possessing inadequate long-term durability [3, 59]. The lack of viable cells results in a loss of remodeling capabilities of the valve, and the ECM gradually degrades without replacement [61–66]. Two basic parameters can be changed in BHVs: material fabrication and valve design (including mounting, stent types, and suturing methods).

# 8 BHV Base Materials: Bovine Pericardium (BP) and Porcine Aortic Heart Valve Leaflets (PAVs)

"At the same time, market-driven preferences for either bovine pericardial or porcine aortic valves cultivated a pseudo sense of progress among surgeons while in fact they merely reflected the circumstances of the time rather than scientific insight. While in previous eras clinical failures due to design flaws affected the surgeon's preferences for bovine or porcine products, prion and virus concerns as well as tissue availabilities and production advantages determine today's marketing thrusts towards 'bovine' or 'porcine' valves" [5]. Still commercially available, both materials are currently implanted in patients. St. Jude Medical Inc. and Edwards Lifesciences utilize BP wrapped in proprietary methods around proprietary stents of differing composites to mimic the structure of a native aortic valve. Medtronic utilizes a line of BHVs fabricated from PAV leaflets. Figure 5 illustrates a variety of models of BHVs including transcatheter valves that can be implanted percutaneously.

Bovine pericardium is now much more predominant in the premier BHV models offered by companies. The core concept behind BHV function is mimicking the native heart valve as much as possible by forming a crosslinked animal derived tissue into the geometry of a trileaflet native heart valve around a polymeric/metal stent. This results in a robust valve with hemodynamics more like those of a native valve. This minimizes shear on the red blood cells, and anticoagulant therapy is not needed for BHV recipients. This makes BHVs more valuable than MHVs. With innovations in percutaneous valve implantation, BHVs are of great value to elderly patients. However, the tissue-based materials are susceptible to degradation and other biological insults that the MHV materials are engineered to be much more resistant to. For this reason, from a strictly mechanical standpoint, MHVs are much more structurally robust than BHVs. However, again, because of the anticoagulation therapy, there is actually *no significant difference* between the two implants when using the metrics of implant life in patients aged 50–69 [67]. Again, as illustrated in Fig. 6, it becomes a lifestyle choice when patients are outside of the bell curve demographic of 60–75 years of age.

The rest of this chapter will focus on tissue-based heart valve replacements (BHVs) because they are the preferred choice of surgeons: BHVs do not require anticoagulant therapy and offer more native-like hemodynamics than MHVs. Aldehyde treated BHVs are currently the market gold standard for the main patient segment currently receiving heart valves—elderly patients aged 65+. Figure 7 depicts a brief history of HVRs and the rise of the many designs starting in the 1960s.

BHVs are designed to be either stented or stentless. Stents are made from artificial material, either metal or plastic, and covered in a fabric sewing ring to provide a frame for the valve. Stentless BHVs allow for implantation of larger valves, yielding increased implant performance [3, 45, 59, 69]. Early stented valves used rigid metal

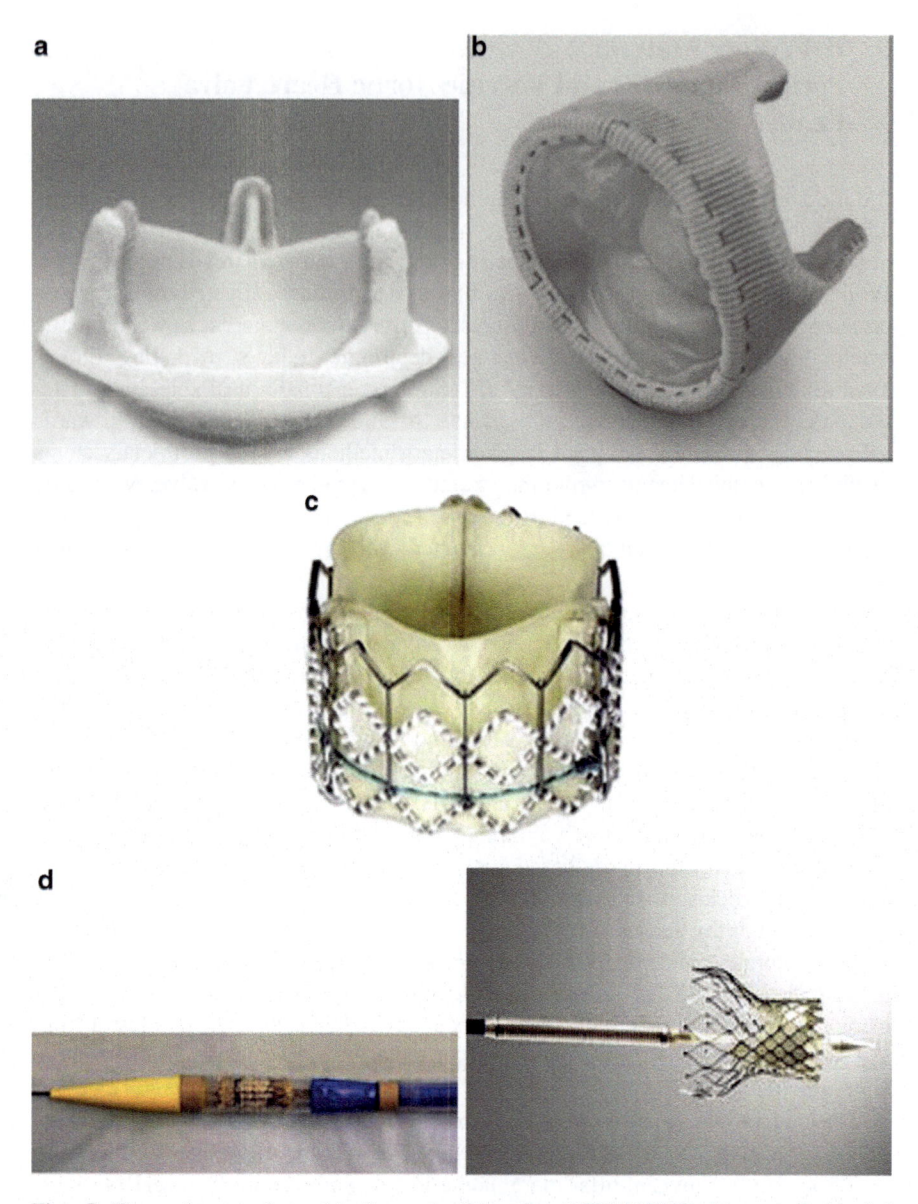

**Fig. 5** Tissue heart valves (**a**) Carpentier-Edwardss PERIMOUNT Magnas Pericardial Bioprosthesis (Stented) [45], (**b**) Edwards Prima Pluss Stentless Bioprosthesis [45], (**c**) Edwards Sapien Pericardial Percutaneous Valve [35], (**d**) CoreValve Percutaneous Valve [35]

stents with three posts. These valves exhibited a high failure rate due to tears at commissures near the posts. The concept of a flexible stent to reduce the rigidity of the frame was pursued to alleviate stress in the tissue near the posts. Initial attempts were unsuccessful at mitigating this failure and introduced further deformation

Bioprosthetic Heart Valves: From a Biomaterials Perspective

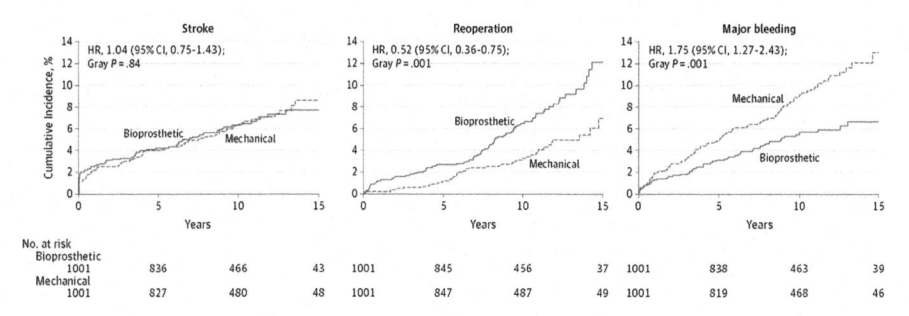

**Fig. 6** Rates of complications and adverse events of HVR requiring reoperation or intervention [67]

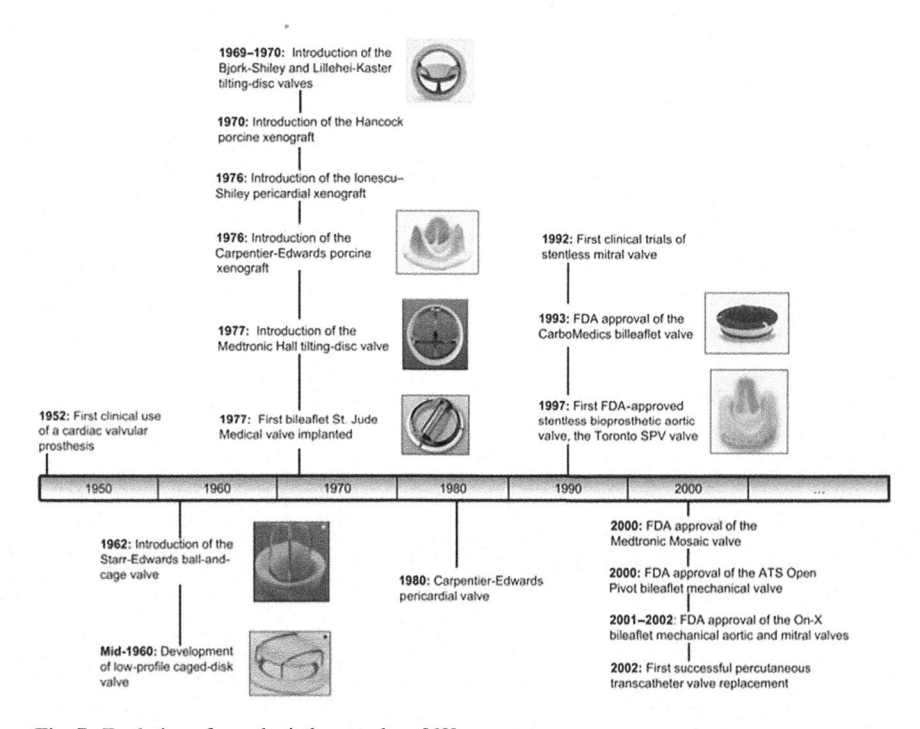

**Fig. 7** Evolution of prosthetic heart valves [68]

problems. Modern stents utilize a flexible polymer-based stent, contributing improved mechanical properties to mitigate this failure mode. Stents also increase the ease of valve implantation procedures. Expandable metal stents, similar to those used in the vasculature, are of recent interest for use with percutaneous intracatheter delivered valves [5, 6, 35, 36]. While the increased size of stentless valves allows better approximation of a patient's normal valve, their design lacks some of the benefits that stented valves present for implantation. For example, stentless valve design contains additional xenogenic aortic wall material [3, 59]. This increases the chance for calcification and pannus overgrowth in the valve, as well as inadequate implant integration [3]. Additionally, stentless valve design increases the level of difficulty of implantation and therefore limits the operation to more experienced and skilled surgeons [59].

The anatomical structure of porcine valves is highly similar to human valves, but two main differences exist. Porcine valves contain a muscle shelf to support the right coronary cusp, and also possess less fibrous continuity between the mitral and aortic valve [59, 69, 70]. This shelf is an extension of the ventricular septum, and may promote calcification and reduce valve range of motion [3, 59, 69, 70]. To eliminate this shelf in the valve, some manufacturers elect to select optimal leaflets, treat them individually, and match them for appropriate size prior to assembly [3, 59, 69]. Pericardial valves are formed from sheets of pericardium, which are cut and constructed to form a trileaflet valve [3, 45, 59]. The structure of pericardium is different from the structure of valve cusps. The pericardium contains a smooth inner layer, middle fibrosa layer, and an outer rough layer [3, 60]. The rough layer contains collagen and elastin, which must be trimmed from desired tissue to present a more uniform surface. This surface is still rough, so it becomes the inflow surface which allows for washing of the surface to prevent thrombus formation. The fibrosa contains collagen, elastin, microvessels, and nerve bundles. BP is essentially a collagen mesh that has glycosaminoglycans (GAGs) and elastin dispersed throughout the tissue composite. Though BP contains the same components (collagen, GAGs, and elastin) as PAV leaflets, BP is in stark contrast to PAV leaflet due to the composition and orientation of these extracellular matrix protein structures. BP is ~95% collagen with very little ground substances (GAGs) and interdispersed elastin throughout the mesh oriented in various directions. Pericardial-based valves allow for optimization of design for improved hemodynamics, but their vastly different structure imparts inferior biomechanics for durability [59, 69, 71].

From a manufacturing standpoint. BP is much easier to handle, comes in a sheet that can be cut in any shape desired and this is most likely the largest driving factor in it becoming the preferred design of suppliers because it drives down overhead costs and increases operational margin. Large sheets can be pre-selected according to desired thickness and mesh fiber orientation and then treated with relatively lower volumes of glutaraldehyde solution than the amount that would be required for PAV leaflets. This process can be highly automated giving BP users a distinct cost advantage at large production volumes. These sheets can then be cut into strips in a proprietary geometry such that the strips can be wrapped and stitched around stents to form the tri-valve design. It should be noted that the assembly process of all BHVs (including PAV derived BHVs) is still *done by hand* and thus is heavily protected by not only a portfolio of methods and design patents but also technical know-how in these companies' human capital. BHVs fabricated from PAV leaflets are less congruent with optimal manufacturing processes because selection requires many animals (specifically grown for heart valve materials) and is not easily automated. Only the noncoronary leaflets (one each valve) are considered as candidates for heart valve fabrication. Only the best noncoronary leaflet from a group of animals is used in the fabrication of the final BHV. Therefore, it takes multitudes of animals to produce one PAV derived BHV. These leaflets, all different in geometry, must also be individually cut and stitched together onto the stent to form the trileaflet structure. These BHVs are then stuffed and crosslinked in glutaraldehyde under a pressure head.

These differences in manufacturing technique could be producing very different products long-term, especially using a zero-set treatment technique with BP versus a pre-tension treatment technique that puts an inherent tension in the PAV leaflets [72]. Long-term data isolating these two variables in large patient groups is not available. However, BP and PAV leaflets are fundamentally the same in one respect—they are both made of the same extracellular matrix (ECM) components. The varying amounts of those ECM components and structural organization of these components are what gives rise to differing tissue composites, which will be manifested in different biomechanical responses. BP as it comes off the heart of a cow has within it a multitude of orientations varying by region. BHV manufacturers that utilize BP as the base material for BHV fabrication select BP based on alignment orientation and thickness specifications set forth within the protocol of the company. Figure 8 illustrates the range of mesh oriented collagen fibers that are within BP. Previous studies have established that these tissues are highly complex nonlinear materials that have preferred orientations [73, 74]. The biomechanical properties are dependent on the initial fixation pressure [72] and how long and what type of mechanical load/deformation is placed upon the material after fabrication [73, 75–77].

PAV leaflets were first used because they mimicked very closely the native human aortic valve. PAV and human leaflets have three layers: (1) the fibrosa, (2) the spongiosa, and (3) the ventricularis. This three-layer structure is shown in Fig. 9. The fibrosa faces the aortic side of the leaflet and contains highly circumferentially aligned type VII collagen to give the leaflet structure and strength to withstand its high cyclic pressure environment [78–80]. Facing the opposite side towards the ventricle, the ventricularis also contains collagen; but its key component is radially aligned elastin. The ventricularis functions in providing elasticity and

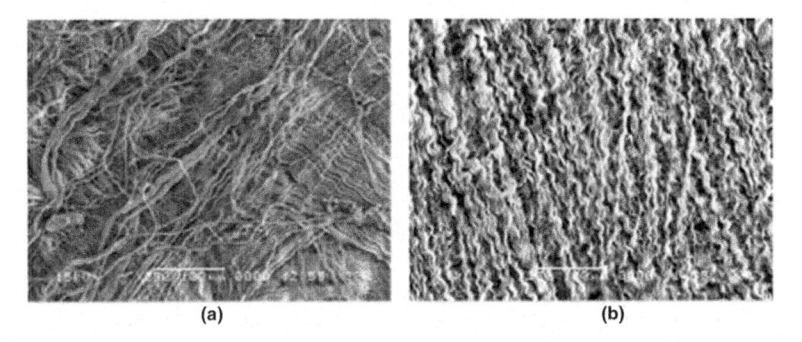

**Fig. 8** Scanning electron microscopy images of collagen in bovine pericardium. (**a**) side faces away from the heart of the pericardium (Fibrosa), (**b**) side facing the heart for the pericardium (Serosa)

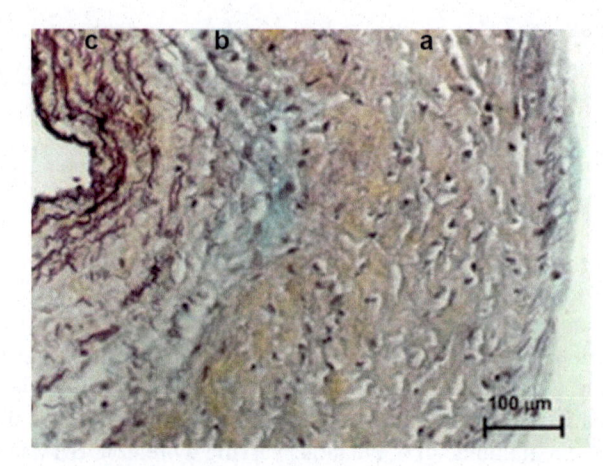

Fig. 9 Histological representation of three layers within PAV ECM. (a) Fibrosa layer with heavy circumferentially aligned collagen content illustrated in yellow. (b) Spongiosa layer comprising of GAGs, illustrated in blue, and collagen. (c) Ventricularis layer comprising collagen and radially aligned elastin illustrated in dark purple and black fibrous structures, respectively [84]

recoil to the leaflet [81, 82]. This allows the PAV to create the pressure gradients necessary for pulsatile circulation of blood after ventricular systole. In between the fibrosa and ventricularis lies the spongiosa, which primarily contains GAGs. This layer is believed to mechanically function as a shearing medium between the two other layers because the GAGs form a hydrogel-like layer [61, 83].

It has been shown that depletion of ECM proteins can cause deviations in biomechanics and could contribute to structural degradation [81–83, 85, 86]. However, standard industry practice continues to be prioritizing collagen fixation and removing potential nucleation sites for calcification. Though composed of the same ECM components, BP is a material in stark contrast to PAV leaflets in the variation in ECM components and how these components are organized. This material design elicits completely different material properties even before downstream processes such as fixation and valve design.

## 9 BHV Drawbacks

Currently marketed BHVs employ glutaraldehyde treated tissue. The resulting tissue is adequate for application, but suboptimal in long-term performance. Roughly half of these valves fail within 12–15 years [3, 8]. BHVs fail along four different paths: endocarditis, thromboembolism, structural degradation and nonstructural dysfunction, [3, 8, 15, 20, 21, 23, 59, 64, 77, 87–90]. Infective endocarditis incidence is dependent on sterility and the skill of the surgeon [18–23, 89, 91, 92]. The progression of endocarditis is similar in BHVs and native heart valves, but typically presents early on in the life of the implant. Organisms commonly infecting valves include *Staphylococcus aureus*, *Staphylococcus epidermidis*, *Streptococcus pneumoniae*, and *Aspergillus*. Early diagnosis can utilize antibiotic treatments, but disease progression can result in valve failure. The most common mode of failure associated with endocarditic prostheses is dehiscence, or the detachment of the implant from the suture line. Frequency of failure can be limited by maintaining surgical sterility. Future attempts to limit could include antibacterial coatings of sewing rings, similar to those being investigated for vascular grafts.

While BHVs have a reduced risk for thrombosis in comparison to MHVs, the risk remains, particularly early on following implantation. Although tissue-derived leaflets already possess thromboresistant material, they must go through an adaptation period following implantation. Thromboembolism risk is still high for approximately 3 months post-operation due to incomplete integration of the implant with native tissue. Consequently, anticoagulants are administered during this time period. Due to artificial materials and altered hemodynamics, stented valves have a slightly higher risk of thromboembolism than stentless valves. Nonstructural dysfunction encompasses host responses to BHVs, including pannus overgrowth, hemolysis, and paravalvular leakage. These complications, pannus overgrowth in particular, occur at similar rates across implant types. Pannus overgrowth of tissue extends over the sewing ring during integration, and can spread to affect the motion of valve cusps, eventually inhibiting their ability to open and close [93]. Hemolysis is caused by shear forces on blood cells, and typically occurs on the surface of valves. Hemodynamics performance affects hemolysis rates, and can be altered by design type. The Reynolds number is a measure of the ratio of inertial to viscous forces, and can be used as a parameter for predicting hemolysis. Since tissue valves have more natural leaflet action than MHVs, better flow profiles occur and result in decreased blood shear and hemolysis. Paravalvular leakage is typically preceded by valvular endocarditis, and consists of leakage at the interface of the valve to the patient's tissue. It allows blood to flow around the implant and to circumvent the valve when closed. This results in undesired backflow and inferior hemodynamic performance.

Structural degradation remains the primary mode of failure for BHVs, and includes calcific and noncalcific degradation [3, 8, 94]. Noncalcific structural degradation involves mechanical and material-based degradation that does not involve mineralization of the cusps. Calcific degradation involves deposition of mineralized calcium phosphate crystals in tissue to create calcific nodules. Cusp calcification typically results in leaflet hardening and stenosis. Reduction in rates of these failure modes may be accomplished through improved chemical fixation modalities. These failure modes are depicted in Fig. 10 and summarized in Fig. 11.

Roughly 75% of all porcine aortic valves experience structural degradation such as cusp tears [3]. This usually presents as valvular insufficiency. In vivo factors such as mechanical forces and enzymatic digestion initiate degradation of the extracellular matrix. Since fixed tissue loses its ability to remodel and its innate enzyme resistance, the proteins are slowly degraded, which exacerbates mechanical fatigue. Chemical treatments of BHV material cause microscopic changes to the tissue, which can increase stiffness or cause localized stresses. Mechanical failure usually

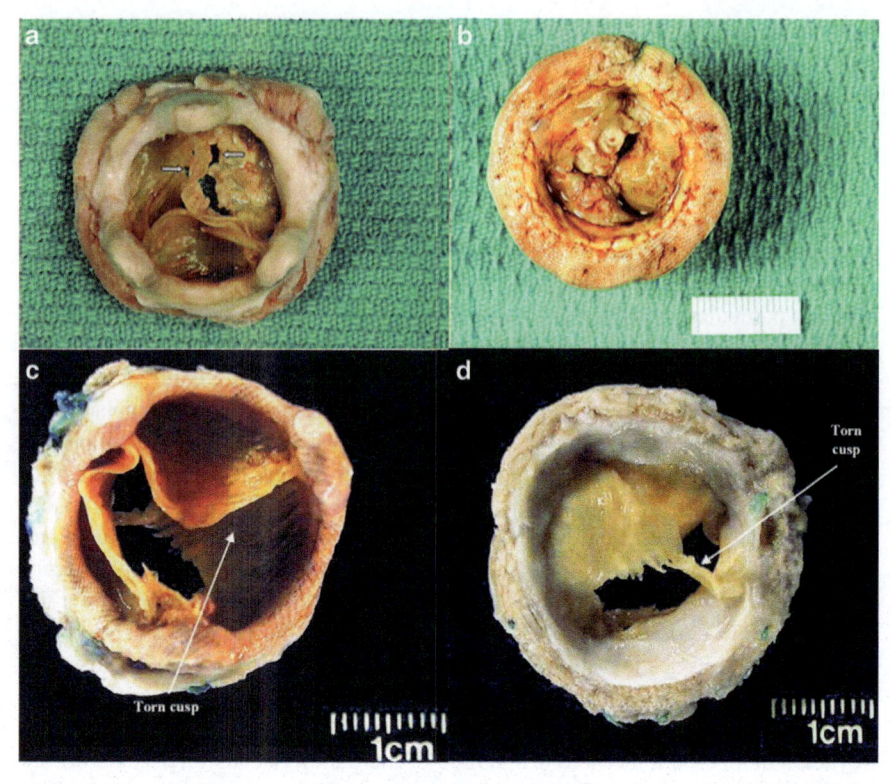

**Fig. 10** Bioprosthetic heart valve failure. (**a**) Removed porcine BHV from mitral position for 8 years. Tears and calcific nodules can be seen [8]. (**b**) Porcine BHV from mitral position, implanted for 7 years. Prosthesis shows thickening of cusps and white vegetations from infection [8]. (**c**) Aortic surface of explanted BHV showing two torn cusps with prolapse [94]. (**d**) Left ventricular outflow surface shows a torn cusp and extensive pannus formation covering portions of the cusps [94]

results in cusp tears or buckling of the tissue. Treatment with glutaraldehyde is known to cause two changes contributing to degradation: crosslinking of collagen and loss of cell viability [3]. Tissue begins deteriorating immediately following harvest, and cell death causes the release of ECM degrading proteases [61–63, 65]. The loss of extracellular matrix components, in particular collagen damage and loss of GAGs, has been observed following fixation and in explanted valves [61, 62, 65, 95–99]. Insufficient stabilization against enzymes can have drastic effects on the integrity of the ECM and durability of tissue valves in vivo [63, 100]. Glutaraldehyde crosslinking of collagen accomplishes a fair degree of extracellular matrix stabilization, but at the cost of flexibility [3]. Crosslinks and stabilization result in a locked configuration characterized by increased cusp rigidity with increased localized stresses [101–103]. Inability of glutaraldehyde to adequately preserve GAGs promotes tissue stresses and susceptibility to mechanical damage [3, 61, 64, 104, 105].

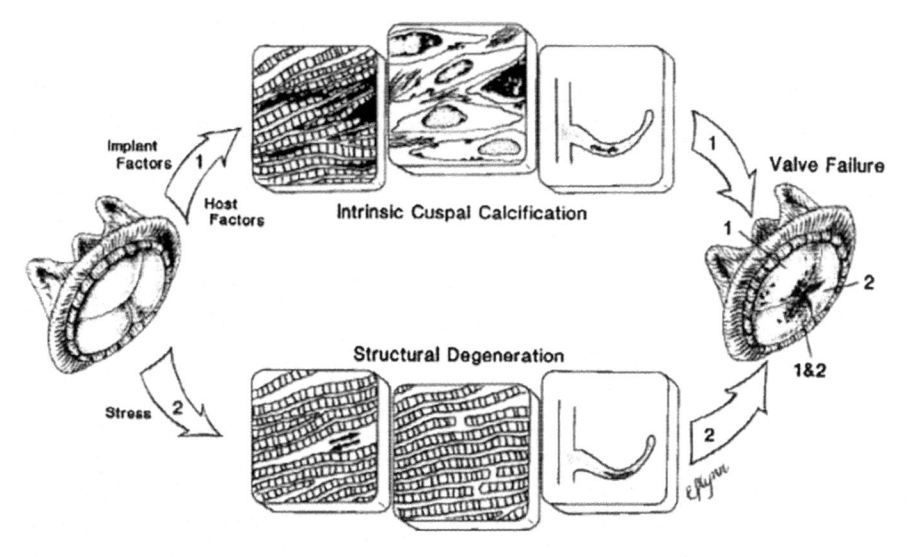

**Fig. 11** Model for structural dysfunction in bioprosthetic heart valves. Two pathways are shown, one induced by stress, resulting in structural degeneration and valve failure. The other is affected by implant and host factors, results in cuspal calcification and valve failure. Implant and host factors induce collagen-oriented deposits, seen in the ultrastructure that predominate at flexion points and commissures. The pathways are not independent, however, since structural degeneration predisposes cusps to calcification [3]

Calcification pathways are not fully understood, but are affected by various factors including tissue damage, host response, and material [3, 49]. Calcification begins with initiation or nucleation of crystals within the tissue itself. These crystals propagate, resulting in a deteriorated structure and increased stiffness. Thus, calcification has a direct effect on structural integrity and long-term durability of BHVs. Due to higher metabolic rates, calcification progresses quicker in younger patients, resulting in faster failure of BHVs than in older patients. In living tissue, cells contain ion pumps to regulate calcium levels, but following cell death this ability is lost [3, 12, 49, 65, 106, 107]. Consequently, any calcium influx cannot be removed, allowing high concentrations of calcium to build up within cells. Calcium reacts with intracellular phosphorous from organelles, nucleic acids, and membrane components, which results in nucleation and initiation of mineralization within residual cells [49]. Additional nucleation sites may also be present in the extracellular matrix. Elastin-mediated calcification occurs without chemical crosslinking of the tissue, but collagen crosslinking with glutaraldehyde creates nucleation sites [108]. Theories suggest that elastin is typically protected by glycoproteins and microfibrils, which are lost during tissue preparation or matrix metalloproteinase (MMP) activity, allowing calcification to occur. GAGs are theorized to inherently inhibit calcium through chelation or steric hindrance [49]. In vivo rat subdermal implantation models have shown that GAG loss promotes calcification, while GAG stabilization inhibits calcification [98, 103, 109, 110]. Even after 40 years of clinical use and development, structural degradation remains a problem in BHVs. Extracellular matrix loss exacerbates degradation and calcification of tissue. Additions or alterations to current chemical treatments may prove beneficial to reducing structural degradation, improving durability, and increasing the life of bioprosthetic heart valves.

## 10 Industry Practice of BHV Fabrication

Fixation methods are chemical treatments, applied to xenogenic bioprosthetic valve tissue. Utilization of these treatments accomplishes three things: sterilization of the tissue prior to implantation, rendering the tissue inert (reduce antigenicity) to reduce immune rejection, and stabilizing the extracellular matrix since remodeling is not possible. Stabilization of collagen, elastin, and GAGs in the extracellular matrix is necessary to provide mechanical integrity for long-term application. Chemicals for fixation need to create irreversible and stable crosslinking bonds between various tissue components. These bonds can be created by binding to carboxyl groups and various amino acids within the tissue to increase its stability.

Early bioprosthetic valves used formaldehyde to provide adequate stability. However, long-term durability remained a concern. With the advent of glutaraldehyde in 1969, formaldehyde treatment was replaced permanently. Glutaraldehyde remains the primary clinically used fixative due to its proven advantages. It is inexpensive and readily soluble, but its reactivity may have some harmful effects on the body. Glutaraldehyde's IUPAC name is 1,5-pentanedialdehyde, and its most stable form is a bifunctional five carbon aldehyde. In aqueous solution, glutaraldehyde can occur as a free aldehyde, monohydrate, dihydrate, hemiacetal, or any polymerized form (Fig. 12). BHVs are treated initially with low concentration glutaraldehyde for 24 h to optimize fixation. Two different reactions occur with glutaraldehyde and tissue that are important in tissue fixation: Aldol condensation and Schiff bases. Aldol condensation can occur between two aldehyde groups,

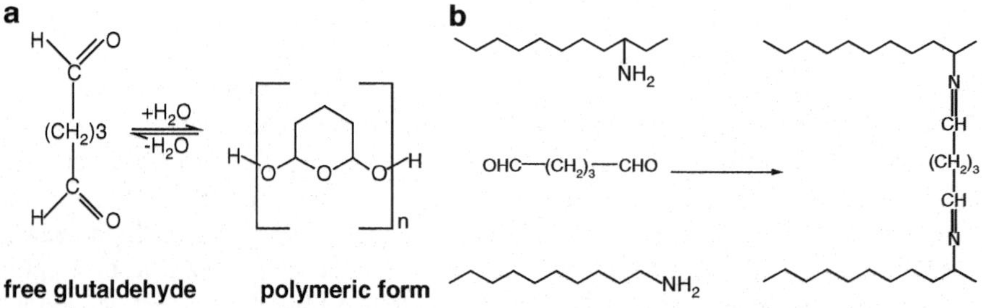

**Fig. 12** Glutaraldehyde (**a**) monomer structure free polymerizes in ring form and (**b**) schematic of reaction with collagen [111]

yielding unsaturated (unless each end is bound) polymerized glutaraldehyde. Schiff base bonds form with lysine or hydroxylysine residues in collagen, and can bind two collagen molecules together. Glutaraldehyde, which also reacts with carboxyl and amide groups in proteins, reacts most preferentially with primary amines and least preferentially with hydroxyls. Low concentrations of glutaraldehyde allow adequate penetration into the tissue for maximum fixation potential, in contrast to high concentrations which quickly create less permeable surface crosslinks.

Glutaraldehyde presents several benefits over formaldehyde crosslinking. First, formaldehyde fixation is a reversible process; the chemical can react with only one collagen molecule. Treatment with formaldehyde results in cusp stretching, deformations, and increased immune response. Although more useful than formaldehyde, glutaraldehyde depolymerizes and releases byproducts into the body. Residual glutaraldehyde exacerbates calcification, and crosslinking fails to block calcium nucleation sites. This causes the tissue to stiffen and undergo further mechanical damage. Residual glutaraldehyde, unreacted in crosslinked tissue, can also cause inflammatory or cytotoxic responses, and can promote calcification [101, 112–114]. Glutaraldehyde cytotoxicity hinders cellular ingrowth, reducing macrophage activity, and the attachment of native endothelial cells. Cytotoxicity can be reduced by thoroughly rinsing the implant in saline prior to implantation. Glutaraldehyde's pro-calcific effects can be neutralized or reduced by various anti-calcification treatments. Amino oleic acid (AOA) neutralizes glutaraldehyde by using amino groups to sequester free aldehydes, allowing endothelial growth and mitigating calcification.

Alternative fixation chemistries are constantly being investigated. Among these are epoxies, acyl azides [115], polyethylene glycol [116, 117], quercetin [118], carbodiimide [95–97, 104, 119–121], periodate [96, 109], triglycidylamine [115], and dye-mediated photooxidation [114, 122]. Researchers hope that properties of these fixation chemistries can help solve the issues experienced by glutaraldehyde treated tissue. Carbodiimide and epoxy-based chemistry have entered clinical trials. Carbodiimide treatment is further discussed in Sect. 6. Use of epoxies as a fixative yields increased pliability and a more pleasing appearance of valves. Triglycidylamine, one such polyepoxide, is being investigated for use in porcine aortic cusps, bovine pericardium, and collagen fixation. Characterization showed shrinkage temperatures similar to those of glutaraldehyde crosslinked tissues, increased collagenase resistance in vitro, improved mechanical properties, and decreased calcification levels in subdermal implants in a rat calcification model [115]. Additions to glutaraldehyde are also being investigated and used clinically. One such calcification reduction method is used in Linx™ technology (St. Jude Medical, St. Paul, MN). Ethanol preincubation reduces calcification potential in valve leaflets through the removal of phospholipids and cell debris and causing conformational changes in collagen. Ethanol has shown marked reduction in collagen-related and cell-initiated calcification, but fails to reduce elastin-mediated calcification. Additional investigations into mitigating elastin-mediated calcification could prove beneficial. In short, glutaraldehyde is clinically used to chemically crosslink tissue heart valves, but it presents its own shortcomings. These can be overcome by developing new crosslinking technology to increase stability of tissue against structural degradation and calcification. Many promising techniques exist, but proving their efficacy is a lengthy process, and after several years of investigation, failure in clinical trials remains a possibility.

## 11 Underlying Factors in Failure Modes of BHVs

BHVs start to experience failure modes around 10–15 years after implantation [7, 54]. BHVs fail due to two primary failure modes which are (1) calcification and (2) structural degradation [7]. Both these failure modes can occur independently or concurrently and can be synergistic. The tradeoffs between BHVs and MHVs are in their failure rates. Both BHV and MHV failure rates start to occur 10–15 years after implantation [56, 67]; however, BHV failure rates occur earlier in younger patients sometimes as soon as 5–7 years after implantation [7, 55, 57]. The two main modes of failure in BHVs are calcification and/or structural degeneration leading to tearing [5, 7]. These two failure modes are not mutually exclusive as both have occurred together as well as independently [76] as indicated in Fig. 12.

## 12 Mechanisms of Calcification

The major barrier to addressing calcification in BHVs is a limited understanding of the mechanisms initiating the onset of calcification. Although calcification is a passive process beginning from nucleation sites at cellular debris left in BHV materials [49, 123–126], it is currently unknown how and why these nucleation sites are manifested and how they aggregate into stenosis-inducing calcification nodules. Figure 13 depicts the predominate hypothesis on how calcification in BHVs occur.

Furthermore, there exist alternative crosslinking chemistries that do not calcify despite cellular debris being left in the implant suggesting that there are multiple pathways of calcification. Additionally, calcification can occur in collagen sponges [125]. These tissues are devoid of devitalized cells and calcification nodules were still found to have developed (Fig. 13). The five overarching processes that have demonstrated the capacity to play a role in calcification are: (1) cellular debris, particularly the phospholipid content that provides a high affinity binding site for minerals such as calcium, providing the substrate for nucleation site formation [7], (2) damaged ECM (both elastin and collagen) providing high affinity substrates for mineralization occur [103, 123, 124, 127–130], (3) glutaraldehyde itself, specifically unbound free floating aldehydes is highly toxic and may predispose tissue to calcification [125, 126, 131, 132], (4) oxidative stress [133], and (5) an active cellular response that may contribute to calcification [93]. Because there are multiple driving factors to calcification, it has been difficult to discern the exact mechanism in

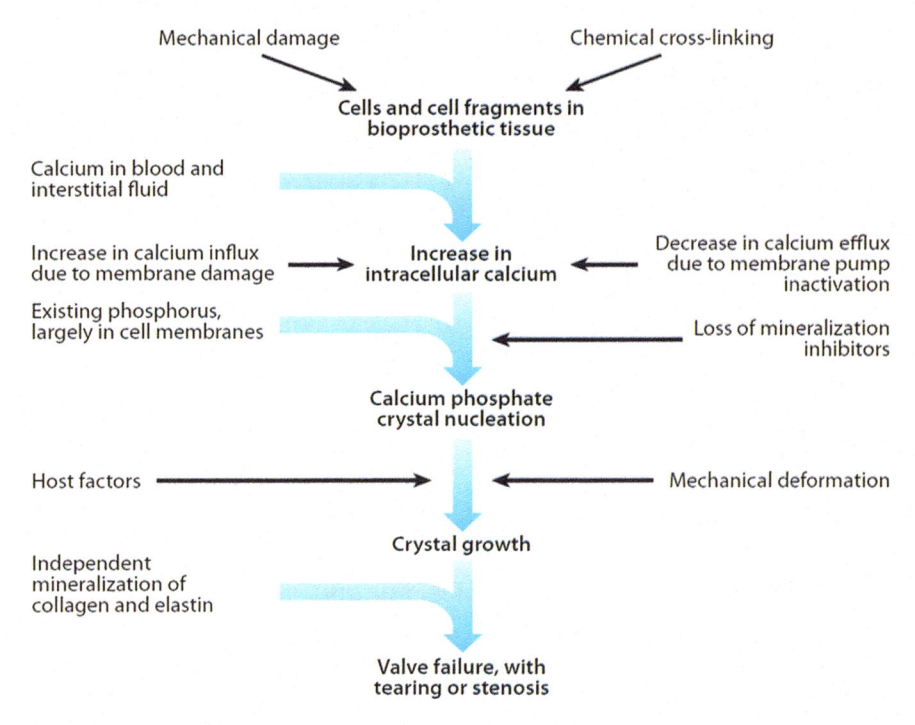

# Bioprosthetic Heart Valves: From a Biomaterials Perspective

**Fig. 13** Overarching hypothesis within the literature pertaining to mechanisms underpinning calcification for BHV failure [7]

how calcification progresses giving rise to multiple intertwined hypotheses in causes for calcification.

The literature is unclear on how the host cellular response plays into the calcification response and how it affects the biomaterial in general. Figure 14 depicts activated macrophages, phagocytic cells, and activated T-cells, which have all been found in failed BHVs explanted from compromised patients [131, 132]. TEM images showing collagen fragments in phagocytic vesicles from failed BHV explants have been reported, suggesting an active cellular response in degeneration of BHVs [134]. However, the studies do not clearly show what triggers these cells into the hostile activity towards the implant or the pathway progressions these cells are taking to produce such behavior. They do, however, show that phagocytic cells actively participate in fragmentation of ECM proteins [134, 135]. These could potentially be the sites necessary for calcification nodule formation. While the immunogenicity of the biomaterial itself may be enough to elicit such a response [135], even collagen sponges implanted subcutaneously have elicited heavy foreign body giant cell activity, phagocytic activity, and calcification [125]. Furthermore, collagen sponges implanted in nude mice (immune cell deficient mice) still presented with calcification and host cellular infiltration [136].

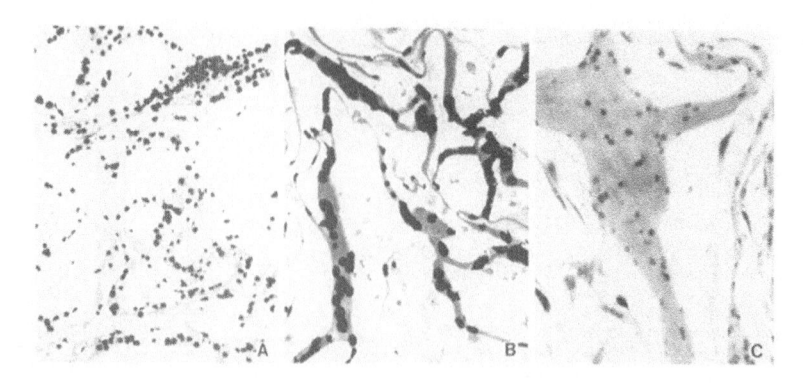

**Fig. 14** Histological analysis of explanted large-animal study sheep valves or failed human valves. TEM shows activated macrophages infiltrating between collagen fibers in ECM (A and B) as well as immune cells infiltrating into ECM of structurally failed valves (C). It has been previously shown that these cells are aggressive in destruction of foreign ECM [135, 139, 140]

## 13 Mechanisms of Structural Degradation

Regarding structural degradation, much work has been done on mechanical wear and its effects on the tissue composite. However, how alternative crosslinking chemistries and the biomechanical implications of different crosslinking networks mitigate mechanically induced structural degradation remain unexplored. Furthermore, the literature does not come to a cohesive consensus on the role of the host cellular response in regard to structural degradation. Though it is accepted that mechanical fatigue can produce many deviations in biomechanics including flexural compliance [64], permanent geometric deformation and loss of extensibility [74, 75], and large disruptions in fiber orientation [72, 73, 137, 138]; the host cellular response is capable of fragmenting ECM components [125, 135, 139, 140] which could potentially result in more structural degradation. Furthermore, it has been shown that oxidative stress can damage and weaken glutaraldehyde bonds [133].

Previous research has indicated that glutaraldehyde crosslinking itself may increase propensity of tissue calcification [125, 126]. Schiff-base crosslinking [141, 142] is shown to be reversible and susceptible to hydrolysis [142, 143]. Though the reaction can be made more irreversible by using reducing agents such as sodium tetraborate [144], such treatments are not used for clinical BHVs. The inability of GLUT crosslinking to stop degradation of the integrity of the collagen triple helix during cyclic fatigue loading was shown by shifts in amine peaks in FT-IR experiments [64]. Same studies also showed that in vitro cyclic fatigue causes change in bending stiffness in GLUT treated PAVs indicating a degradation of the ECM structure [64]. Glutaraldehyde actively crosslinks only collagen and leaves other key ECM components such as GAGs and elastin unstabilized in the overall tissue composite [141]. Therefore, these major ECM components are lost from the tissue

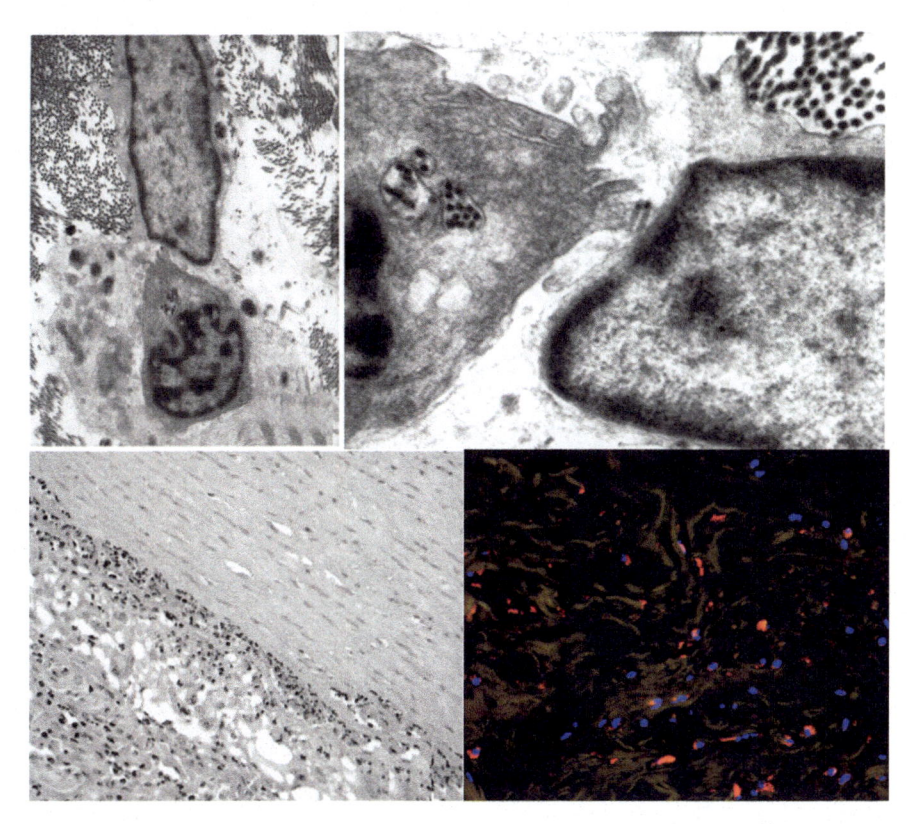

**Fig. 15** Histological analysis of explanted large animal study sheep valves or failed human valves. TEM shows activated macrophages infiltrating between collagen fibers in ECM as well as immune cells infiltrating into ECM of structurally failed valves. It has been previously shown that these cells are aggressive in destruction of foreign ECM [135, 139, 140]

due to enzymatic degradation [61, 66, 95, 104]. Slow depletion of these ECM components compounded with constant cyclic loading may lead to microtears after valve implantation. This early degeneration may then lead to calcification nodule development and/or visible tears within a few years. Figure 15 depicts that areas of loss or damage of elastin can serve as high affinity calcification sites. It has been demonstrated that depletion of elastin within soft connective tissues such as PAVs produces severely deviated biomechanical properties. Figure 16 shows that GAGs in BHVs are strongly connected with fiber-fiber and fiber-matrix interactions at low force levels and that they may be important in providing a damping mechanism to reduce leaflet flutter when the leaflet is not under high tensile stress [83]. Thus, the degradation of the essential ECM components may lead to compromised mechanical function of the leaflet; this could be a major contributor to degenerative tears of BHVs.

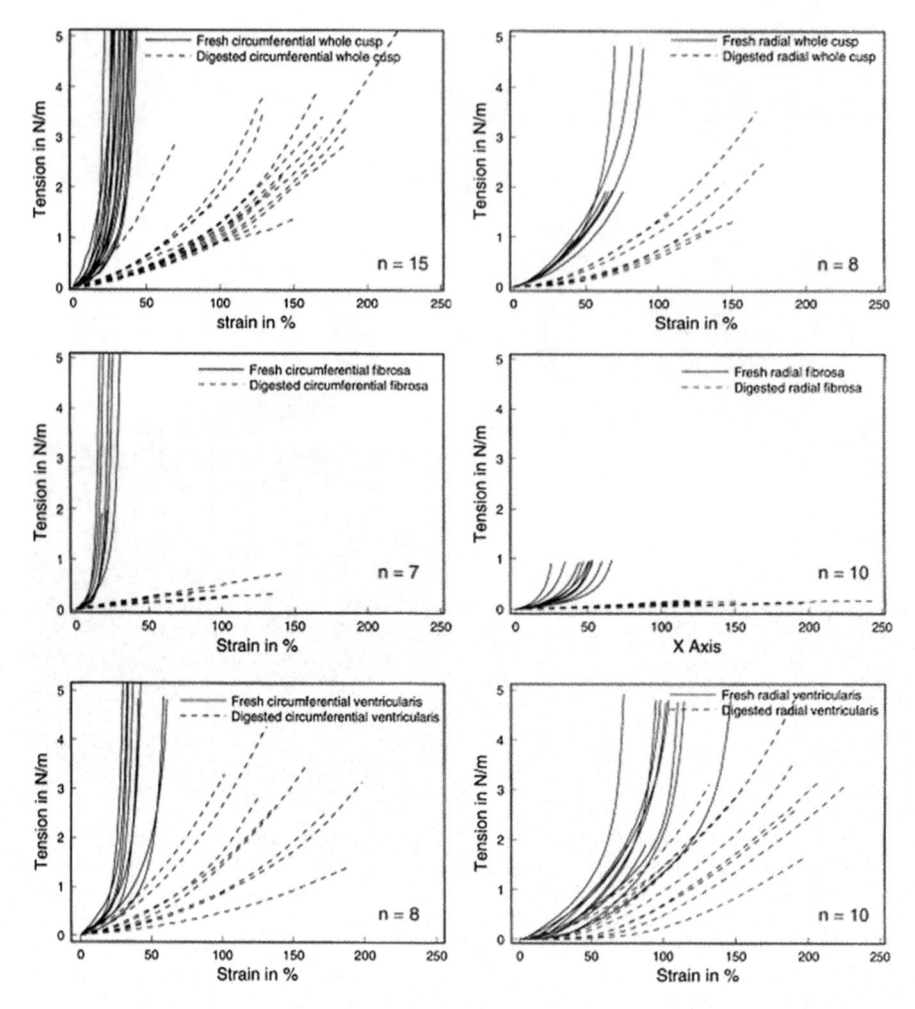

**Fig. 16** Biomechanics testing showing the alteration in mechanical behavior in elastin-digested tissues [82]

Furthermore, current treatment methods with glutaraldehyde show the structural degradation implications of using reversible chemistry in a device that requires high cyclic fatigue resistance [73, 74]. Figure 17 depicts fatigued collagen treated PAVs exhibit permanent deformations (also called permanent set), deviations in biomechanical behavior, as well as fiber orientation disruption and visible macroscopic deformations. Depicted in Fig. 18, just by the naked eye, physical sagging of the inside of the heart valve as well as stress shunting around the aortic wall can be seen [75]. Accompanying the macroscopic to microscopic structural degradation, individual collagen fibers at the protein level exhibit degradation [64, 73]. In Figure 19, FT-IR spectroscopy results of fatigued samples as compared

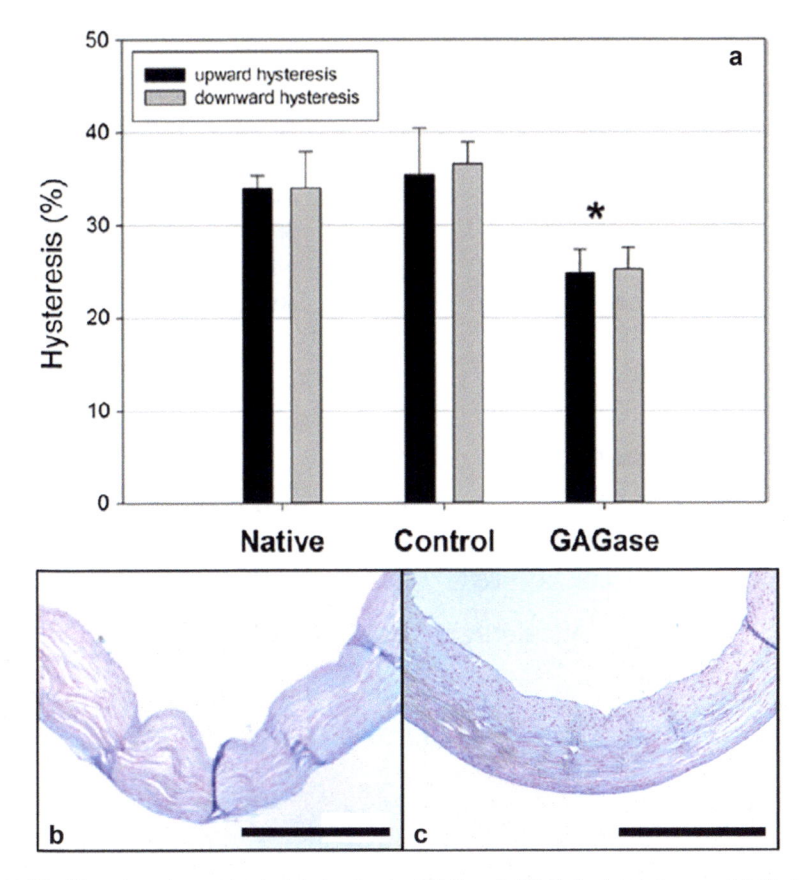

**Fig. 17** Alterations in mechanical behavior in GAG and GAG depleted tissues. (**a**) Reduced hysteresis effects suggesting that leaflets "flutter" and have lost bending stiffness. Reduction in bending stiffness and differences in low force biaxial tensile testing have been previously shown in fatigued samples that also showed a reduction in GAG content over time suggesting that GAGs play a role in maintaining fiber-fiber interaction and orientation [83]. (**b, c**) GAG depleted tissues on the left buckled more easily than tissues with GAG stabilized tissues further suggesting that GAGs play a pivotal role in maintaining an optimal bending stiffness for constant cyclic bending [104]

to unfatigued samples show a shift in peaks in the amide regions. The marked amide regions are indicative of the collagen triple helix structure in which one amide group is facing inward and two facing outward. With the disruption of collagen and "unraveling" of the triple helix, the amide group usually facing inward is now exposed to solution and shifts the peaks of the spectroscopy to a slightly altered but detectable wavelength. Therefore, it can be concluded that simply fatiguing tissues fixed with reversible chemistries such as aldehydes can cause deviations at the biochemical, microscopic, macroscopic, and gross anatomy level (Fig. 20).

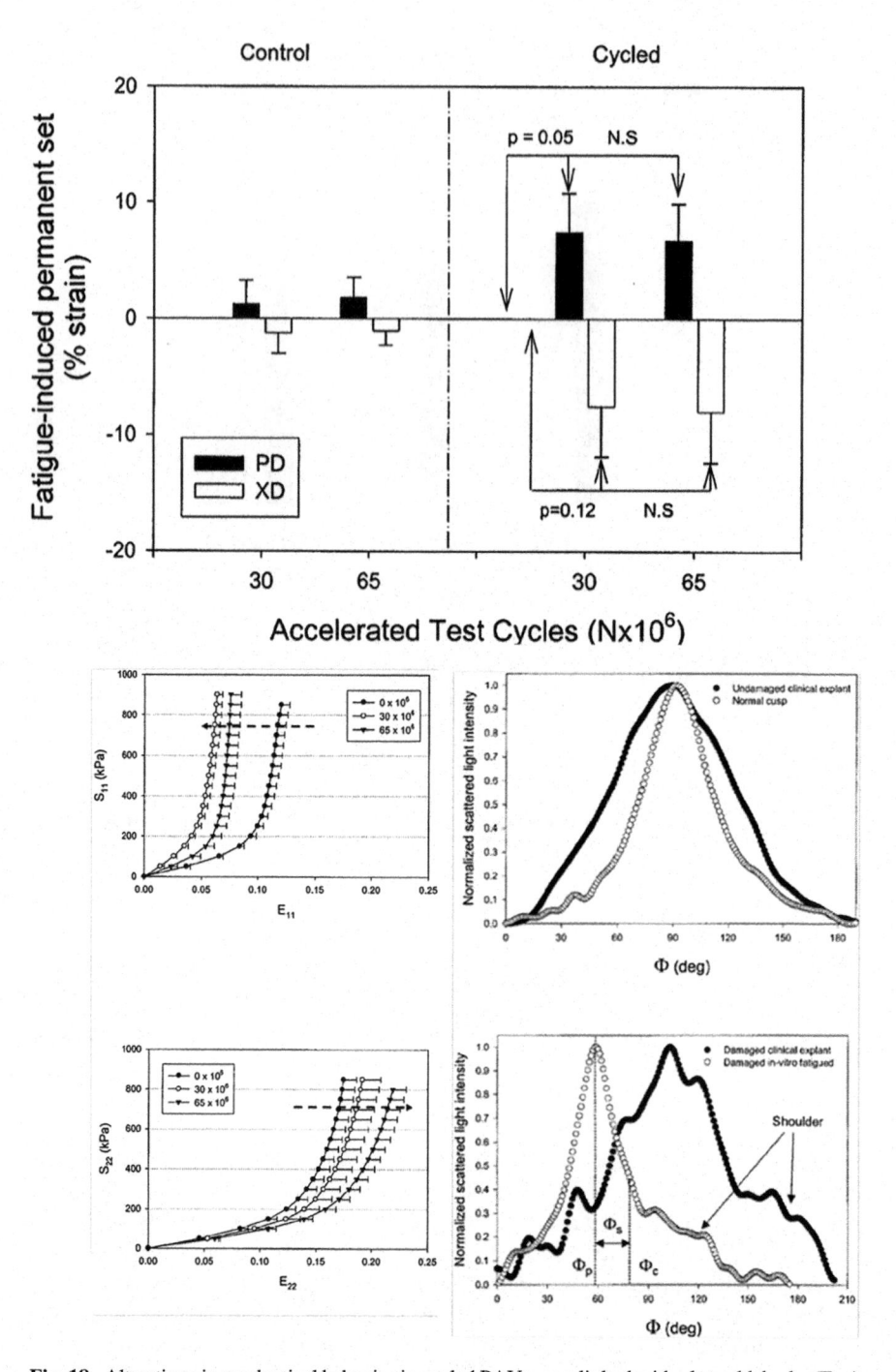

**Fig. 18** Alterations in mechanical behavior in cycled PAVs crosslinked with glutaraldehyde. (Top) Permanent geometric deformation relative to nominal geometry imparted by cyclic fatigue testing. (Bottom left) Fatigued samples show a shift in biomechanical behavior. (Bottom right) Fatigued samples show misaligned collagen fiber orientation relative to normal cusps damaged either through in vivo failure or in vitro accelerated fatigue testing [73]

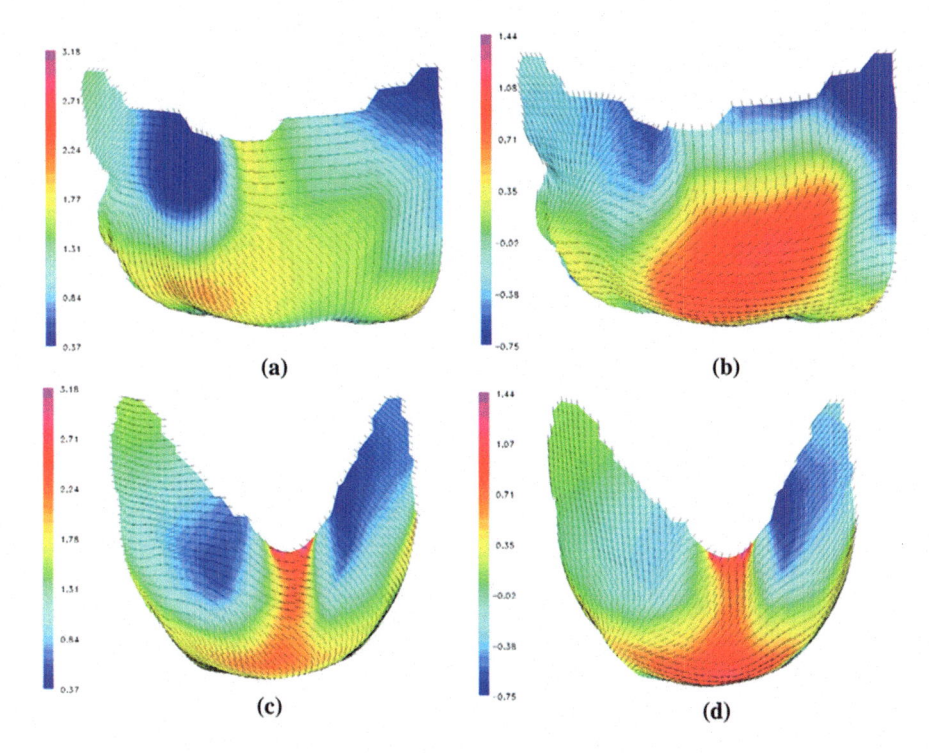

Bioprosthetic Heart Valves: From a Biomaterials Perspective 365

**Fig. 19** MRI imaging of macroscopic geometric deformations in fatigued PAVs. (**a**) Normal, unfatigued noncoronary cusp. (**b**) Fatigued noncoronary cusp with visual signs of sagging. (**c, d**) Fatigued coronary cusps showing sagging. Sagging is indicative of permanent geometric deformation and can lead to force shunting causing unpredicted stress points. These high stress points could exceed design specifications and be manifested as microtears leading to macrotears and eventually structural degradation [75]

## 14 Barriers to Innovation

The barrier to innovating the next generation of artificial heart valves that addresses a larger patient segment is in developing a biomaterial that can preserve optimal hemodynamics, has longer implant lifespan in younger patients, and does not require the use of anticoagulants. One method to accomplish this goal is in investigating alternative crosslinking methods used to chemically treat tissue biomaterials for BHVs such as porcine aortic heart valve (PAV) leaflets or bovine pericardium (BP) to produce a biomaterial that resists calcification and structural degradation. Longer lasting, more durable heart valves would minimize recurrent surgeries and improve quality of life for patients. These valve materials should (1) be biocompatible, (2) resist calcification, (3) resist progressive structural degradation, (4) improve patient quality of life, and (5) comply with best practices and processes of current heart valve producers. These metrics will be used to evaluate emerging research in BHV material design, but first, the current commercially available materials must be evaluated to show what improvements are needed.

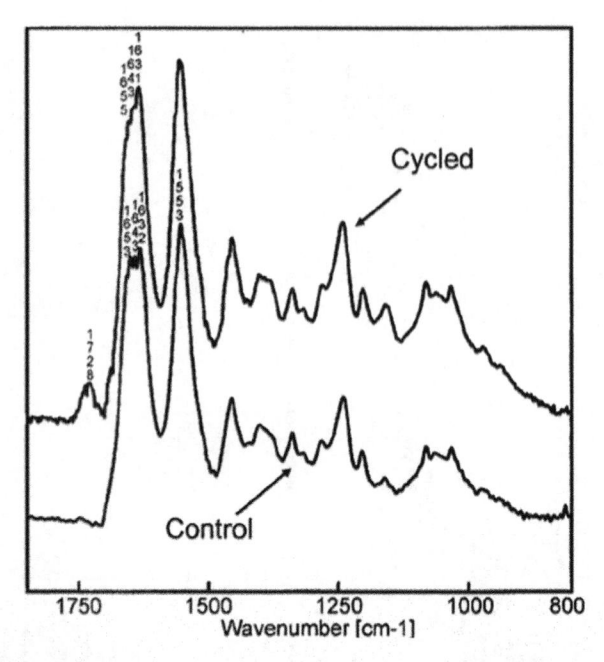

**Fig. 20** FT-IR spectroscopy investigating the integrity of the collagen triple helix of a patch of glutaraldehyde treated PAV. Note the amide shifts between the cycled and control samples in the 1500–1600 wave number spectrum. This peak pattern is indicative of the amide residues normally facing inward in the collagen triple helix that are now facing outward and exposed to solution, suggesting that the collagen triple helix is unraveling. The implication of this is that glutaraldehyde does not provide a stable enough bond to ensure the stability of the individual collagen fiber as it gets fatigued with more and more cycles. This short coming in collagen stability could play a pivotal part in structural degradation [73]

## 15 Add on Methods to Curb Shortcomings of Glutaraldehyde Crosslinking in BHVs

Three schools of thought exist in the effort to innovate biomaterials for HVR fabrication that will last longer. Either add-on technologies that can augment the core crosslinking network created by glutaraldehyde can be developed, a completely new chemistry can be used to treat BP or PAV tissues to produce a completely new biomaterial that actively resists calcification and structural degeneration, or a completely new biomaterial, synthetic or tissue engineered, can be used to fabricate new valves.

Currently, every product line of BHVs has an add-on technology involving some type of alcohol or surfactant treatment or secondary processing technique that is aimed to decrease calcification rates. For instance, St. Jude uses a proprietary alcohol or surfactant treatment in conjunction with their formaldehyde treated PAV BHVs. Edwards Lifesciences uses a treatment technique that rids the tissue of residual aldehyde and removes phospholipid content. Studies in the literature suggest that the rationale behind using surfactants and alcohols is that these chemicals rid the tissue of phospholipid content from devitalized cells that serve as high affinity binding sites for mineralization, thus avoiding calcification initiation sites (nucleation sites). Alcohol pretreatments also rid the tissue of cholesterol, alter collagen confirmation, and remove cardiolipin. However, other studies have concluded that these treatments only delay the onset of calcification in animal models. Prolonged implantation in rat subcutaneous calcification models noted a return of calcification in implanted materials in anti-calcification treatment groups. Other post glutaraldehyde crosslinking treatments are aimed at removing residual aldehyde residues that have been previously shown to decrease the propensity of the tissue to calcify. Company literature from heart valve producers such as St. Jude Medical, Medtronic, and Edwards Lifesciences also explicitly state that there are no studies conclusively demonstrating the long-term efficacy of any of these anti-calcification technologies [145–147].

To prevent structural degradation, the optimal crosslinking is a balance between using a concentrated crosslinking agent but still retaining functional compliance in the tissue. Diamine crosslinking in combination with sodium tetraborohydrate in a glutaraldehyde-based crosslinking has produced a more tightly crosslinked network and decreased risk of calcification. The diamine crosslinking reduces the glutaraldehyde bonds into more stable double bonds that are much more irreversible. However, sodium tetraborohydrate is a highly toxic chemical and the side effects of leechables has not been fully studied and cleared for use in trials. As to date, this technology has not garnered any industry interest, and furthermore innovations in the industrial market space of BHVs focuses on anti-calcification. Innovations in industry to actively prevent structural degradation (stabilize ECM proteins for better long term durability) has historically not been pursued and emphasized despite structural degradation contributing to 50% of BHV failure rates.

# 16 Alternative Crosslinking Chemistries to Innovate a Novel Biomaterial for BHV Fabrication

To date, there are no other chemistries used other than aldehyde chemistry used to fabricate BHVs for commercial use. Medtronic and Edwards Lifesciences BHVs both use glutaraldehyde [146, 147] and St. Jude Medical BHVs utilize formaldehyde [145] (Table 2).

**Table 2** Comparison of fabrication techniques in various BHV biomaterials currently on market

| Company | Base biomaterial | Crosslinker | Anti-calcification treatment |
| --- | --- | --- | --- |
| Edwards Lifesciences | BP | Glutaraldehyde | ThermaFix™ |
| Medtronic | PAV | Glutaraldehyde | AOA™ (alpha oleic acid) + Physiological Fixation™ |
| St. Jude Medical | BP | Formaldehdye | Linx™ AC |

| Structure of stabilization agents | Name | Reference |
|---|---|---|
| 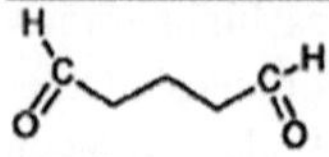 (structure) | Glutaraldehyde | Olde Damink et al., 1995; Duncan & Boughner, 1998; Langdon et al., 1999 |
| 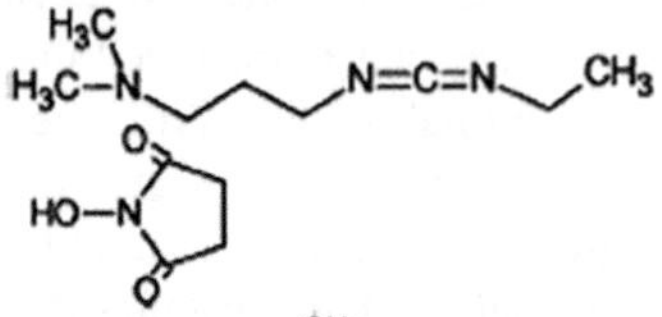 (structure) | Ethyl-aminopropyl carbodiimide (EDAC) hydroxysuccinimide (NHS) | Lee et al., 1996; Everaerts et eal., 2004; Mendoza-Novelo & Cauich-Rodríguez, 2009 |
| 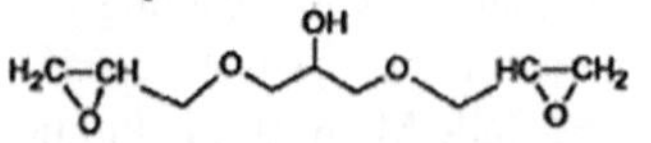 (structure) | Glycerol diglycidyl ether | Lee et al., 1994 |
| 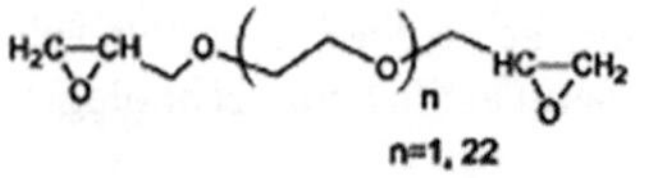 (structure) n=1, 22 | Ethylene glycol diglycidyl ether, n=1; Poly(ethylene oxide) diglycidyl ether, n=22 | Tu et al., 1993; Sung et al., 1996; Zeeman et al., 2000 |
| 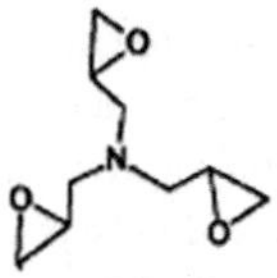 (structure) | Triglycidylamine | Conolly et al., 2005; Sack et al., 2007; Rapoport et al., 2007 |
| 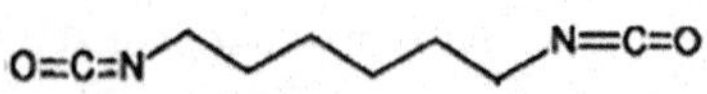 (structure) | Hexamethylene diisocyanate | Naimark et al., 1995; Olde Damink et al., 1995; Nowatzki & Tirrel, 2004 |
| 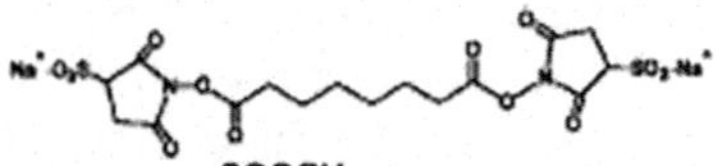 (structure) | Disuccinimidyl suberate | Pathak et al., 2001 |
| 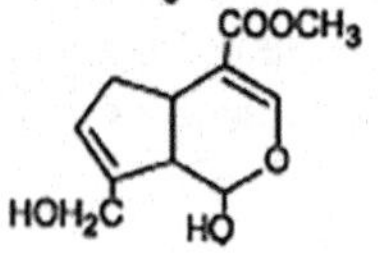 (structure) | Genipin | Sung et al. 1999; Sung et al., 2000 |
| 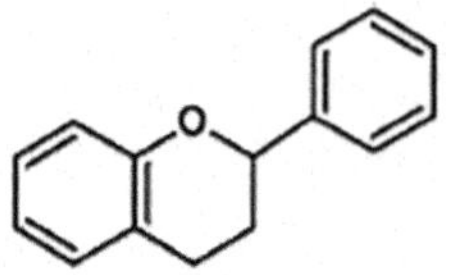 (structure) | Proanthocyanidin, Procyanidins | Han et al., 2003; Zhai et al., 2009 |
| 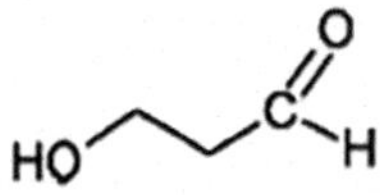 (structure) | Reuterin | Sung et al., 2002; Sung et al., 2003 |

**Fig. 21** Different stabilizing and crosslinking agents demonstrated to adequately crosslink collagen to produce biomaterials and scaffolds [141]

However, researchers in the academia are publishing a multitude of approaches to create entirely new materials. Figure 21 depicts many of the different techniques used to stabilize collagen in soft connective tissue applications such as BHV fabrication. Despite the number of methods to make a new material, all include around the same concepts, which center on using a crosslinking chemistry that is much less reversible than aldehyde chemistry and have previously been demonstrated to be resistant to calcification. The rationale behind using more irreversible chemistries is that the bonds crosslinking the ECM proteins are more stable. This is especially important when crosslinking collagen, the ECM protein that provides all the load bearing strength required of BHVs. A more stable bond should lead to a higher level of collagen stability and resistance to degradation that should lead to the resistance of permanent biomechanical alterations within the tissue composite. From this process, the BHV can be designed to minimize stress concentrations and resist stress shunting and thus microtear formation. A lower incidence of microtear formation should result in less tearing and catastrophic failure of BHVs due to structural degradation.

The art of crosslinking and the underlying fundamentals of finding the optimal chemistry for BHV fabrication lies in the crosslinking technique's ability to produce a structurally very durable, tough biomaterial, but to retain compliance even at high volumes of cyclic mechanical deformation. The crosslinking scheme must produce very stable bonds, but not produce such a rigid tissue composite that the BHV cannot function as a viscoelastic material. Alternative fixation chemistries are constantly being investigated. To name a few, fixation chemistries include epoxies, acyl azides [135], polyethylene glycol [136, 137], quercetin [138], carbodiimide [28, 139–144], periodate [95, 142], triglycidylamine [135], and dye-mediated photooxidation [104, 134]. It is hoped that properties of these fixation chemistries can help solve the issues experienced by glutaraldehyde treated tissue. The main reasoning behind using these different crosslinkers is to use the varying available binding substrates (i.e., free carboxyl and amine groups) within the ECM's soft connective tissue to form a more stable, irreversible bond between those substrates than glutaraldehyde, a bond that does not allow the tissue to calcify. Carbodiimide and epoxy-based chemistry have entered clinical trials. Use of epoxies as a fixative yields increased pliability and a more pleasing appearance of valves. Triglycidylamine is one such polyepoxide, which is being investigated for use in porcine aortic cusps, bovine pericardium, and collagen fixation. Characterization showed similar shrinkage temperatures to glutaraldehyde, increased collagenase resistance in vitro improved mechanical properties, and decreased calcification levels were seen in subdermal implants [135]. Additions to glutaraldehyde are being investigated and used clinically. One such calcification reduction method is used in Linx™ technology (St. Jude Medical, St. Paul, MN). Ethanol preincubation reduces calcification potential in valve leaflets through removal of phospholipids and cell debris and causes conformational changes in collagen. Ethanol has shown marked reduction in collagen related and cell-initiated calcification, but fails to reduce elastin-mediated calcification. Research has been done in an attempt to increase stability of glutaraldehyde bonds by reducing them to double bonds rather than single aldehyde bonds [140, 142]. Other attempts at using material such as 20% polyacrylimide gels as fillers between the free amine and carboxyl groups within the ECM have been shown to reduce calcification [148]. Additional investigations into mitigating elastin-mediated calcification could prove beneficial. Multiple, combined fixation additions to glutaraldehyde or altered fixation methods have been investigated to preserve glycosaminoglycans. Carbodiimide crosslinking, neomycin incorporation, and periodate oxidation are methods that have shown promise.

## 17 Combined Approaches to Curb Structural Degradation and Calcification

New preliminary data from the Vyavahare group has demonstrated permanent set effects in bovine pericardium (preferred material of use in industry for BHV fabrication) with glutaraldehyde-based crosslinking chemistry. However, these effects were eliminated when using carbodiimide (irreversible crosslinking chemistry that has been previously demonstrated to resist calcification) in combination with neomycin (GAGase inhibitor to enable GAG retention) and PGG, (stabilizes collagen and elastin and has implications in calcification resistance) referred to as TRI. This alternative fabrication method has been demonstrated to enable biomaterials to retain all ECM components through both in vitro enzymatic challenges and small animal in vivo studies [149]. Furthermore, this material has also been shown to resist calcification even in a long-term implant rat subcutaneous calcification model (commonly accepted model for biocompatibility and, within the space of BHVs, an accepted accelerated model to test calcification within biomaterials) [149]. This dual calcification and structural degradation resistance is suspected to be caused by the irreversible crosslinking technique protecting the ECM such that it is not susceptible to degradation and thus eliminates nucleation sites where calcification nodules to form. This also suggests that calcification and structural degradation processes in the scope of BHVs are highly intertwined and share a common underlying factor—the use of reversible chemistry such as aldehydes as a chemical fabrication method. Only by innovating around this common underlying factor can we hope to bring transformative technology to the HVR market.

This effort of using compounds previously demonstrated to stabilize individual ECM proteins and prevent calcification in combination with one another may be the new approach to engineering the next generation of BHVs that have significantly longer life span. The added value proposition of protecting all ECM proteins with an irreversible chemistry such as TRI from an outcomes standpoint is threefold.

1. More native-like biomechanical behavior, compliance and retention of more native-like biomechanical behavior and compliance after cycling or prolonged strain.
2. Significant reduction in ECM degradation by-products as well as known cytotoxic residues that may augment a negative host cellular response.
3. Calcification resistant biomaterial.

The added value proposition of innovating a new chemical fixation technique over innovating another method to fabricate heart valves from an operations standpoint has some distinct advantages.

1. Quickly scalable into industry and allows quick adoption because heart valve producers already have capital assets and infrastructure to manufacture these BHVs.

2. Predicate devices already exist, so regulatory hurdles, distribution, operator and user education, and payer models can be readily adjusted for market adoption.
3. Thus, this technology can quickly be scaled to emerging economies, where this technology is needed most.

It should be noted that other methods of creating heart valves are currently being researched and are yet not approved. Polyurethane valves, the leading candidate for synthetic-based polymer valves, lack durability. A successful polymer valve would be a more efficient manufacturing technique from a batch standpoint and would allow for better scaling as well as functional valve design for transcatheter valves.

Tissue engineering also poses a unique value proposition because a tissue engineered heart valve integrates with the human body to give an adaptable, permanent solution. However, major obstacles that have prevented realization of this concept. Decellularized heart valves were tested in the early 2000s. Investigators failed to properly decellularize the ECM, and the residual cellular components caused implant rejection immediately after implantation, causing three pediatric deaths and one emergency heart valve replacement. Thereafter, the field regressed to basic science— investigating better protocols for decellularization. Additionally, valvular interstitial cells and stem cells in general are poorly understood, especially regarding modulating a remodeling response versus a growth response. Further, decellularized valves would be problematic because many countries that are in dire need of heart valve replacements do not yet have the facilities and infrastructure that would support stem cell harvesting and growing of tissue engineered valves.

Perhaps this scattered approach to produce scientifically viable ideas that do not consider other parts of operations, supply chain, implementation, and payer models underpins the reason we have seen only marginal innovations that add onto preexisting techniques when it comes to industry adaptation of an alternative to glutaraldehyde despite all the shortcomings of the crosslinking. This phenomenon is not unique only to the vertical HVR market, but across the healthcare industry and technology spaces in general. There has not been a concept that has demonstrated compelling data from a larger picture perspective to properly derisk the amount of upfront investment required to bring a Class III medical device to market and justify the cost of adopting a different industry standard.

If we are to truly innovate for the betterment of masses and expand to all the patient segments in need of heart valve replacements, we as innovators within the field of heart valves must collaborate to solve the larger puzzle that is BHVs. We cannot continue to present the market space with tid bits and lone puzzle pieces, though pristine and compelling within their own right, of the larger picture and expect stakeholders within a market space that directly affects hundreds of millions of individuals to take a large risk on an unclear opportunity. We must start from patient clinical needs and systematically approach preventing the underlying mechanisms of action in calcification and structural degradation to consider alternative chemistries that may produce calcification resistant and structurally robust biomaterials. Furthermore, we must subject these alternative approaches to biomaterial fabrication and valve design to the harshest of objective criticism from a multidisciplinary perspective. BHV concepts groomed out of academia for industrial commercialization must be subjected to as much chemical stability analysis as biomechanical merit, from molecular merit to macro-biomechanics merit. This academic research in basic concept must also be done in parallel with and in complete alignment with clinical appointments (from surgeon, insurance, and administrative level) and business acumen (financial viability, risk assessment, strategy, target market segmentation, etc.) to adequately depict an opportunity that presents an undeniable value proposition and is investible and scalable to the market space. Only then will we have a chance at market adoption and only then are we able to fully use our skillsets to fulfill dedicating our lives to research and innovation to help people who would otherwise have no other option.

## 18 Serving a New Demographic

The current gold standard in the current heart valve market operating primarily in the developed economy space may not be able to meet all the clinical needs of the new growth sector in emerging economies. The caveat to catering to this new market segment centered in emerging economies and often resource lean regions is that this new market segment will serve a different patient segment or demographic. This means that there will be different clinical needs that will define different design criteria that will affect current paradigms in heart valve design [4, 5].

The major difference between developed economy and emerging economy patient segments in the cardiovascular market is seen if the market is segmented first by geographic location then by age. Cardiovascular disease, especially in the heart valve market, is on the rise in developed countries primarily due to chronic conditions causing health complications later in life. The average age for a heart valve recipient is over 55 years old, the median is ~65 years old. Figure 22 shows that the distribution of patients when segmented by age in the current market primarily operating in developed nations is very narrow, stark, and heavily skewed towards patients 60–69 years of age. Congenital conditions and infectious disease that damage the heart necessitating a heart valve transplant (such as rheumatic fever) in the current demographic the heart valve market serves are relatively low compared to that of the new patient segments in emerging economies.

The new patient demographic segmented by age showes a stark difference in patient distribution. The distribution curve is much flatter encompassing a larger range of ages receiving heart valve replacements. The median age receiving a heart valve transplant is 45, and the most common age receiving a heart valve replacement is ~35. A large number of the aggregate heart valve replacement recipients were younger than 35 years of age. This wide age range is due to inherent differences in the underlying conditions necessitating a heart valve replacement. In these often resource lean regions, congenital conditions and disease states often cause heart valve disease and are *no longer* negligible drivers of heart valve replacements. This is contrary to long-term chronic conditions mainly due to genetics or lifestyle leading

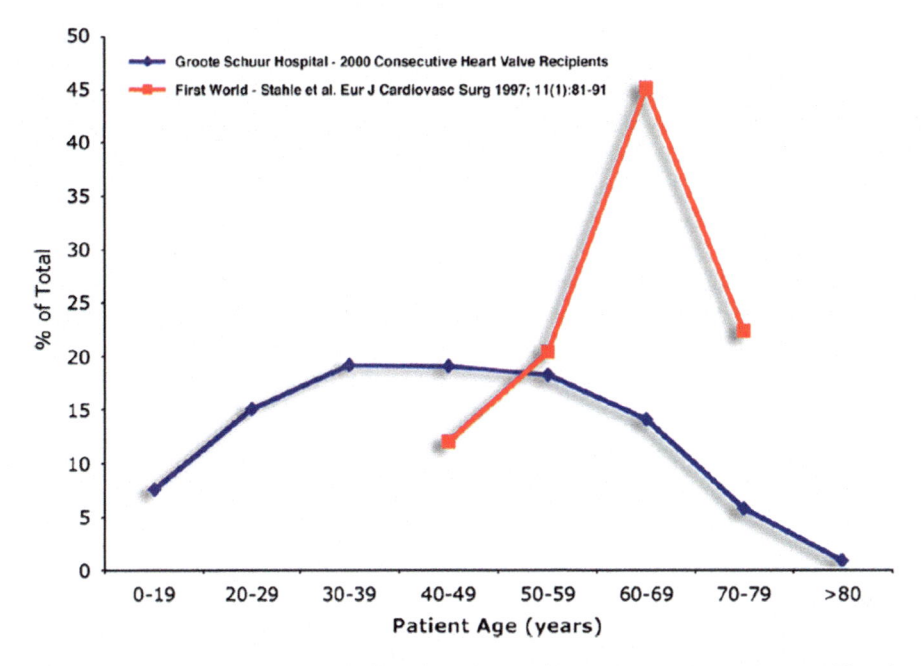

**Fig. 22** Patient population segmented by age in HVR market. Differences in age demographics of HVRs in developed economies versus developing economies. It should be noted that 85% of the healthcare market is generated in advanced economies. Only 11% of the human population resides in these regions leaving the majority of humanity to have little access to healthcare services and products

to slow but progressive degenerative conditions as seen in developed regions. This key difference in cause of need for HVRs drives a different set of demand criteria from a user and healthcare provider standpoint that HVR suppliers must address in their heart valve design.

## 19 Contrasting Patient Demographics; Contrasting Design Requirements for Heart Valves

Because of there are drastically different clinical needs between the current developed world market and the emerging economic market, there must be changes in heart valve design to fully capture market share and effectively serve the new patient demographic in emerging economies. Because the market's current patient segment served is predominantly elderly, current heart valve durability and implant lifespan is engineered to last only 10–15 years before failures start to occur [7, 57]. Strictly based on patient age demographic, the current market gold standards for heart valve replacements will not serve the emerging economy heart valve replacement market because the average recipient age is ~30 years or half the age of developed economy recipients. The value proposition offered to the patient segment of the emerging economies by the current technology available on market only marginally improves lifespan. Further, it has previously been shown that current heart valve technology fails faster and sooner as patient age decreases [55, 150]. This implies that the marginal amount of engineering that must be invested into each additional year of implant life increases as implant life increases. Therefore, the much larger patient segment distribution exhibited by emerging economy patient demographics will not be fully served by current commercially available technology, and this sector will need technological innovation to expand and capitalize on more patient segments. For a replacement valve to effectively capture market share in these new market segments in emerging economies where most of the human population resides and where there is the most clinical need, more robust biomaterials must be designed coupled with computational modeling to optimize heart valve design such that it has the lowest chances of failure modes at both the base biomaterial level and device design level.

## 20 Designing for Emerging Markets and Developed Markets

The basic question to implementing much needed technological advancement in this sector of healthcare is how we can bring value proposition to both developed markets and emerging markets *past the cannibalization of current product lines*. How can we design a valve that minimizes current market shareholder losses in adopting new technology and drive increased value in all the patient segments *and* geographical segments? By designing a heart valve with increased implant life in many patient segments and such that it leverages current heart valve supplier competencies is the quickest way to market and drives the most social impact.

## 21 Summary and Conclusion

For the 50+ years, HVRs shifted from early mechanical models such as the Bjork-Shiley valves or the ball and cage models to a diversified market of high performance designs such as the bileaflet conduit MHVs or the spectrum of proprietary BHVs in an effort to increase implant life and patient quality of life. However, there is still room for improvement. MHVs currently require anticoagulant therapy, which has many debilitating side effects for the patient and restricts its use in a younger demographic. BHVs experience failure rates at 10–15 years after implantation due to calcification and structural degradation. BHVs also fail faster in younger patients, and therefore serve only an older population. Most technological innovation has taken place in the form of transcatheter or percutaneous delivery of BHVs; however, the base concept of an HVR has not been innovated in recent decades, leading to stagnation in improved patient outcomes. To design the next generation of HVRs, we must redesign the base biomaterials these HVRs are fabricated from. Smarter, more durable, more mechanically optimal, more biocompatible, and more easily implanted designs are necessary to address the widespread clincal need of this technology. This widespread need is largely neglected in lean resource regions despite technological advances in developed regions of the world. Multidisciplinary collaboration through the different silos of science and academia as well as consideration of the needs of other stakeholders in this complex ecosystem such as industry, investors, payers, users, healthcare providers, and healthcare administrators all must accompany the distillation of science into a technology platform to be harnessed by society. This is the essence of innovation and the HVR market space is in dire need of it.

# References

1. Research and markets adds report: US implantable medical devices market report 2013–2018—reconstructive joint replacement, spinal implants, cardiovascular implants, dental implants, intraocular lens and breast implants, professional services close-up. 2013.
2. Manji RA, Menkis AH, Ekser B, Cooper DK. Porcine bioprosthetic heart valves: the next generation. Am Heart J. 2012;164(2):177–85.
3. Schoen FJ. Cardiac valves and valvular pathology: update on function, disease, repair, and replacement. Cardiovasc Pathol. 2005;14(4):189–94.
4. Yacoub MH. Establishing pediatric cardiovascular services in the developing world: a wake-up call. Circulation. 2007;116(17):1876–8.
5. Zilla P, Brink J, Human P, Bezuidenhout D. Prosthetic heart valves: catering for the few. Biomaterials. 2008;29(4):385–406.
6. Kidane AG, Burriesci G, Cornejo P, Dooley A, Sarkar S, Bonhoeffer P, Edirisinghe M, Seifalian AM. Current developments and future prospects for heart valve replacement therapy. J Biomed Mater Res B Appl Biomater. 2009;88(1):290–303.
7. Schoen FJ. Evolving concepts of cardiac valve dynamics: the continuum of development, functional structure, pathobiology, and tissue engineering. Circulation. 2008;118(18):1864–80.
8. Siddiqui RF, Abraham JR, Butany J. Bioprosthetic heart valves: modes of failure. Histopathology. 2009;55(2):135–44.
9. Writing Group Members, Lloyd-Jones D, Adams RJ, Brown TM, Carnethon M, Dai S, De Simone G, Ferguson TB, Ford E, Furie K, Gillespie C, Go A, Greenlund K, Haase N, Hailpern S, Ho PM, Howard V, Kissela B, Kittner S, Lackland D, Lisabeth L, Marelli A, McDermott MM, Meigs J, Mozaffarian D, Mussolino M, Nichol G, Roger VL, Rosamond W, Sacco R, Sorlie P, Roger VL, Thom T, Wasserthiel-Smoller S, Wong ND, Wylie-Rosett J, American Heart Association Statistics Committee, Stroke Statistics Subcommittee. Heart disease and stroke statistics—2010 update: a report from the American Heart Association. Circulation. 2010;121(7):e46–e215.
10. Cotran RS, Kumar V, Collins T. Robbins pathologic basis of disease. 6th ed. Philadelphia, PA: W.B. Saunders; 1999.
11. Akat K, Borggrefe M, Kaden JJ. Aortic valve calcification: basic science to clinical practice. Heart. 2009;95(8):616–23.

12. Weska RF, Aimoli CG, Nogueira GM, Leirner AA, Maizato MJ, Higa OZ, Polakievicz B, Pitombo RN, Beppu MM. Natural and prosthetic heart valve calcification: morphology and chemical composition characterization. Artif Organs. 2010;34(4):311–8.
13. Lee WJ, Son CW, Yoon JC, Jo HS, Son JW, Park KH, Lee SH, Shin DG, Hong GR, Park JS, Kim YJ. Massive left atrial calcification associated with mitral valve replacement. J Cardiovasc Ultrasound. 2010;18(4):151–3.
14. O'Brien KD. Pathogenesis of calcific aortic valve disease: a disease process comes of age (and a good deal more). Arterioscler Thromb Vasc Biol. 2006;26(8):1721–8.
15. Freeman RV, Otto CM. Spectrum of calcific aortic valve disease: pathogenesis, disease progression, and treatment strategies. Circulation. 2005;111(24):3316–26.
16. Piazza N, Lange R, Martucci G, Serruys PW. Patient selection for transcatheter aortic valve implantation: patient risk profile and anatomical selection criteria. Arch Cardiovasc Dis. 2012;105(3):165–73.
17. Li C, Xu S, Gotlieb AI. The response to valve injury. A paradigm to understand the pathogenesis of heart valve disease. Cardiovasc Pathol. 2011;20(3):183–90.
18. Siu SC, Silversides CK. Bicuspid aortic valve disease. J Am Coll Cardiol. 2010;55(25):2789–800.
19. Que YA, Moreillon P. Infective endocarditis. Nat Rev Cardiol. 2011;8(6):322–36.
20. El-Ahdab F, Benjamin DK Jr, Wang A, Cabell CH, Chu VH, Stryjewski ME, Corey GR, Sexton DJ, Reller LB, Fowler VG Jr. Risk of endocarditis among patients with prosthetic valves and *Staphylococcus aureus* bacteremia. Am J Med. 2005;118(3):225–9.
21. Giessel BE, Koenig CJ, Blake RL Jr. Management of bacterial endocarditis. Am Fam Physician. 2000;61(6):1725–32, 1739.
22. Griffin FM Jr, Jones G, Cobbs CC. Aortic insufficiency in bacterial endocarditis. Ann Intern Med. 1972;76(1):23–8.
23. Jegatheeswaran A, Butany J. Pathology of infectious and inflammatory diseases in prosthetic heart valves. Cardiovasc Pathol. 2006;15(5):252–5.
24. Guilherme L, Kalil J. Rheumatic fever: from innate to acquired immune response. Ann N Y Acad Sci. 2007;1107:426–33.
25. Guilherme L, Kalil J. Rheumatic fever and rheumatic heart disease: cellular mechanisms leading autoimmune reactivity and disease. J Clin Immunol. 2010;30(1):17–23.
26. Guilherme L, Ramasawmy R, Kalil J. Rheumatic fever and rheumatic heart disease: genetics and pathogenesis. Scand J Immunol. 2007;66(2–3):199–207.
27. Baasanjav S, Al-Gazali L, Hashiguchi T, Mizumoto S, Fischer B, Horn D, Seelow D, Ali BR, Aziz SA, Langer R, Saleh AA, Becker C, Nurnberg G, Cantagrel V, Gleeson JG, Gomez D, Michel JB, Stricker S, Lindner TH, Nurnberg P, Sugahara K, Mundlos S, Hoffmann K. Faulty initiation of proteoglycan synthesis causes cardiac and joint defects. Am J Hum Genet. 2011;89(1):15–27.
28. Olivier C. Rheumatic fever—is it still a problem? J Antimicrob Chemother. 2000;45(Suppl):13–21.
29. Becker AE. Acquired heart valve pathology. An update for the millennium. Herz. 1998;23(7):415–9.
30. Maganti K, Rigolin VH, Sarano ME, Bonow RO. Valvular heart disease: diagnosis and management. Mayo Clin Proc. 2010;85(5):483–500.
31. Simionescu A, Simionescu DT, Vyavahare NR. Osteogenic responses in fibroblasts activated by elastin degradation products and transforming growth factor-beta1: role of myofibroblasts in vascular calcification. Am J Pathol. 2007;171(1):116–23.
32. Liu AC, Joag VR, Gotlieb AI. The emerging role of valve interstitial cell phenotypes in regulating heart valve pathobiology. Am J Pathol. 2007;171(5):1407–18.
33. Gott VL, Alejo DE, Cameron DE. Mechanical heart valves: 50 years of evolution. Ann Thorac Surg. 2003;76(6):S2230–9.
34. Bahnson HT, Spencer FC, Busse EF, Davis FW. Cusp replacement and coronary artery perfusion in open operations on the aortic valve. Ann Surg. 1960;152(3):494–503.

35. Chiam PT, Ruiz CE. Percutaneous transcatheter aortic valve implantation: evolution of the technology. Am Heart J. 2009;157(2):229–42.
36. Fann JI, Chronos N, Rowe SJ, Michiels R, Lyons BE, Leon MB, Kaplan AV. Evolving strategies for the treatment of valvular heart disease: preclinical and clinical pathways for percutaneous aortic valve replacement. Catheter Cardiovasc Interv. 2008;71(3):434–40.
37. Ghanbari H, Viatge H, Kidane AG, Burriesci G, Tavakoli M, Seifalian AM. Polymeric heart valves: new materials, emerging hopes. Trends Biotechnol. 2009;27(6):359–67.
38. Zeltinger J, Landeen LK, Alexander HG, Kidd ID, Sibanda B. Development and characterization of tissue-engineered aortic valves. Tissue Eng. 2001;7(1):9–22.
39. Pibarot P, Dumesnil JG. Prosthetic heart valves: selection of the optimal prosthesis and long-term management. Circulation. 2009;119(7):1034–48.
40. Bachraoui K, Darghouth B, Haddad W, Saaidi I, Ben Halima A, Sdiri W, Selmi K, Makni H, Mokaddem A, Boujnah MR. [Pregnancy in patients with prosthetic heart valves]. Tunis Med. 2003;81(Suppl 8):613–6.
41. Khamooshi AJ, Kashfi F, Hoseini S, Tabatabaei MB, Javadpour H, Noohi F. Anticoagulation for prosthetic heart valves in pregnancy. Is there an answer? Asian Cardiovasc Thorac Ann. 2007;15(6):493–6.
42. Sillesen M, Hjortdal V, Vejlstrup N, Sorensen K. Pregnancy with prosthetic heart valves—30 years' nationwide experience in Denmark. Eur J Cardiothorac Surg. 2011;40(2):448–54.
43. Jonkaitiene R, Benetis R, Eidukaityte R. [Management of patients with prosthetic heart valves]. Medicina (Kaunas). 2005;41(7):553–60.
44. Vink R, Van Den Brink RB, Levi M. Management of anticoagulant therapy for patients with prosthetic heart valves or atrial fibrillation. Hematology. 2004;9(1):1–9.
45. Simionescu D. Artificial heart valves. Hoboken, NJ: Wiley; 2006.
46. Alsoufi B, Manlhiot C, McCrindle BW, Canver CC, Sallehuddin A, Al-Oufi S, Joufan M, Al-Halees Z. Aortic and mitral valve replacement in children: is there any role for biologic and bioprosthetic substitutes? Eur J Cardiothorac Surg. 2009;36(1):84–90; discussion 90.
47. Sako EY. Newer concepts in the surgical treatment of valvular heart disease. Curr Cardiol Rep. 2004;6(2):100–5.
48. Harken DE, Taylor WJ, Lefemine AA, Lunzer S, Low HB, Cohen ML, Jacobey JA. Aortic valve replacement with a caged ball valve. Am J Cardiol. 1962;9:292–9.
49. Schoen FJ, Levy RJ. Calcification of tissue heart valve substitutes: progress toward understanding and prevention. Ann Thorac Surg. 2005;79(3):1072–80.
50. Calisi P, Griffo R. [Anticoagulation in patients with heart valve prosthesis]. Monaldi Arch Chest Dis. 2002;58(2):121–7.
51. Frater RW. The development of the Starr-Edwards heart valve. Tex Heart Inst J. 1999;26(1):99.
52. Blot WJ, Ibrahim MA, Ivey TD, Acheson DE, Brookmeyer R, Weyman A, Defauw J, Smith JK, Harrison D. Twenty-five-year experience with the Bjork-Shiley convexoconcave heart valve: a continuing clinical concern. Circulation. 2005;111(21):2850–7.
53. Butany J, Collins MJ. Analysis of prosthetic cardiac devices: a guide for the practising pathologist. J Clin Pathol. 2005;58(2):113–24.
54. Bourguignon T, Bouquiaux-Stablo AL, Candolfi P, Mirza A, Loardi C, May MA, El-Khoury R, Marchand M, Aupart M. Very long-term outcomes of the Carpentier-Edwards Perimount valve in aortic position. Ann Thorac Surg. 2015;99(3):831–7.
55. Takkenberg JJ, van Herwerden LA, Eijkemans MJ, Bekkers JA, Bogers AJ. Evolution of allograft aortic valve replacement over 13 years: results of 275 procedures. Eur J Cardiothorac Surg. 2002;21(4):683–91; discussion 691.
56. Isaacs AJ, Shuhaiber J, Salemi A, Isom OW, Sedrakyan A. National trends in utilization and in-hospital outcomes of mechanical versus bioprosthetic aortic valve replacements. J Thorac Cardiovasc Surg. 2015;149(5):1262–9, e3.
57. Bourguignon T, El Khoury R, Candolfi P, Loardi C, Mirza A, Boulanger-Lothion J, Bouquiaux-Stablo-Duncan AL, Espitalier F, Marchand M, Aupart M. Very long-term

outcomes of the Carpentier-Edwards Perimount aortic valve in patients aged 60 or younger. Ann Thorac Surg. 2015;100(3):853–9.
58. Johansen P. Mechanical heart valve cavitation. Expert Rev Med Devices. 2004;1(1):95–104.
59. Vesely I. The evolution of bioprosthetic heart valve design and its impact on durability. Cardiovasc Pathol. 2003;12(5):277–86.
60. Munnelly AE, Cochrane L, Leong J, Vyavahare NR. Porcine vena cava as an alternative to bovine pericardium in bioprosthetic percutaneous heart valves. Biomaterials. 2012;33(1):1–8.
61. Lovekamp JJ, Simionescu DT, Mercuri JJ, Zubiate B, Sacks MS, Vyavahare NR. Stability and function of glycosaminoglycans in porcine bioprosthetic heart valves. Biomaterials. 2006;27(8):1507–18.
62. Simionescu DT, Lovekamp JJ, Vyavahare NR. Glycosaminoglycan-degrading enzymes in porcine aortic heart valves: implications for bioprosthetic heart valve degeneration. J Heart Valve Dis. 2003;12(2):217–25.
63. Simionescu DT, Lovekamp JJ, Vyavahare NR. Extracellular matrix degrading enzymes are active in porcine stentless aortic bioprosthetic heart valves. J Biomed Mater Res A. 2003;66(4):755–63.
64. Vyavahare N, Ogle M, Schoen FJ, Zand R, Gloeckner DC, Sacks M, Levy RJ. Mechanisms of bioprosthetic heart valve failure: fatigue causes collagen denaturation and glycosaminoglycan loss. J Biomed Mater Res. 1999;46(1):44–50.
65. Simionescu DT, Lovekamp JJ, Vyavahare NR. Degeneration of bioprosthetic heart valve cusp and wall tissues is initiated during tissue preparation: an ultrastructural study. J Heart Valve Dis. 2003;12(2):226–34.
66. Isenburg JC, Simionescu DT, Vyavahare NR. Elastin stabilization in cardiovascular implants: improved resistance to enzymatic degradation by treatment with tannic acid. Biomaterials. 2004;25(16):3293–302.
67. Chiang YP, Chikwe J, Moskowitz AJ, Itagaki S, Adams DH, Egorova NN. Survival and long-term outcomes following bioprosthetic vs mechanical aortic valve replacement in patients aged 50 to 69 years. JAMA. 2014;312(13):1323–9.
68. Dasi LP, Simon HA, Sucosky P, Yoganathan AP. Fluid mechanics of artificial heart valves. Clin Exp Pharmacol Physiol. 2009;36(2):225–37.
69. Butany J, Fayet C, Ahluwalia MS, Blit P, Ahn C, Munroe C, Israel N, Cusimano RJ, Leask RL. Biological replacement heart valves. Identification and evaluation. Cardiovasc Pathol. 2003;12(3):119–39.
70. Crick SJ, Sheppard MN, Ho SY, Gebstein L, Anderson RH. Anatomy of the pig heart: comparisons with normal human cardiac structure. J Anat. 1998;193(Pt 1):105–19.
71. Hoffmann G, Lutter G, Cremer J. Durability of bioprosthetic cardiac valves. Dtsch Arztebl Int. 2008;105(8):143–8.
72. Wells SM, Sellaro T, Sacks MS. Cyclic loading response of bioprosthetic heart valves: effects of fixation stress state on the collagen fiber architecture. Biomaterials. 2005;26(15):2611–9.
73. Sun W, Sacks M, Fulchiero G, Lovekamp J, Vyavahare N, Scott M. Response of heterograft heart valve biomaterials to moderate cyclic loading. J Biomed Mater Res A. 2004;69(4):658–69.
74. Martin C, Sun W. Modeling of long-term fatigue damage of soft tissue with stress softening and permanent set effects. Biomech Model Mechanobiol. 2013;12(4):645–55.
75. Smith DB, Sacks MS, Pattany PM, Schroeder R. High-resolution magnetic resonance imaging to characterize the geometry of fatigued porcine bioprosthetic heart valves. J Heart Valve Dis. 1997;6(4):424–32.
76. Sacks MS, Schoen FJ. Collagen fiber disruption occurs independent of calcification in clinically explanted bioprosthetic heart valves. J Biomed Mater Res. 2002;62(3):359–71.
77. Sacks MS. The biomechanical effects of fatigue on the porcine bioprosthetic heart valve. J Long-Term Eff Med Implants. 2001;11(3–4):231–47.
78. Billiar KL, Sacks MS. Biaxial mechanical properties of the natural and glutaraldehyde treated aortic valve cusp—part I: experimental results. J Biomech Eng. 2000;122(1):23–30.

79. Sacks MS, David Merryman W, Schmidt DE. On the biomechanics of heart valve function. J Biomech. 2009;42(12):1804–24.
80. Stella JA, Sacks MS. On the biaxial mechanical properties of the layers of the aortic valve leaflet. J Biomech Eng. 2007;129(5):757–66.
81. Lee TC, Midura RJ, Hascall VC, Vesely I. The effect of elastin damage on the mechanics of the aortic valve. J Biomech. 2001;34(2):203–10.
82. Vesely I. The role of elastin in aortic valve mechanics. J Biomech. 1998;31(2):115–23.
83. Eckert CE, Fan R, Mikulis B, Barron M, Carruthers CA, Friebe VM, Vyavahare NR, Sacks MS. On the biomechanical role of glycosaminoglycans in the aortic heart valve leaflet. Acta Biomater. 2013;9(1):4653–60.
84. Sierad LN, Simionescu A, Albers C, Chen J, Maivelett J, Tedder ME, Liao J, Simionescu DT. Design and testing of a pulsatile conditioning system for dynamic endothelialization of polyphenol-stabilized tissue engineered heart valves. Cardiovasc Eng Technol. 2010;1(2):138–53.
85. Scott M, Vesely I. Aortic valve cusp microstructure: the role of elastin. Ann Thorac Surg. 1995;60(2 Suppl):S391–4.
86. Grande-Allen KJ, Mako WJ, Calabro A, Shi Y, Ratliff NB, Vesely I. Loss of chondroitin 6-sulfate and hyaluronan from failed porcine bioprosthetic valves. J Biomed Mater Res A. 2003;65(2):251–9.
87. Butany J, Feng T, Luk A, Law K, Suri R, Nair V. Modes of failure in explanted mitroflow pericardial valves. Ann Thorac Surg. 2011;92(5):1621–7.
88. Vongpatanasin W, Hillis LD, Lange RA. Prosthetic heart valves. N Engl J Med. 1996;335(6):407–16.
89. Farhat F, Durand M, Delahaye F, Jegaden O. Prosthetic valve sewing-ring sealing with antibiotic and fibrin glue in infective endocarditis. A prospective clinical study. Interact Cardiovasc Thorac Surg. 2007;6(1):16–20.
90. Lepidi H, Casalta JP, Fournier PE, Habib G, Collart F, Raoult D. Quantitative histological examination of bioprosthetic heart valves. Clin Infect Dis. 2006;42(5):590–6.
91. Irving CA, Kelly D, Gould FK, O'Sullivan JJ. Successful medical treatment of bioprosthetic pulmonary valve endocarditis caused by methicillin-resistant *Staphylococcus aureus*. Pediatr Cardiol. 2010;31(4):553–5.
92. McAllister RG Jr, Samet J, Mazzoleni A, Dillon ML. Endocarditis on prosthetic mitral valves. Fatal obstruction to left ventricular inflow. Chest. 1974;66(6):682–6.
93. Zilla P, Human P, Bezuidenhout D. Bioprosthetic heart valves: the need for a quantum leap. Biotechnol Appl Biochem. 2004;40(Pt 1):57–66.
94. Beauchesne LM, Veinot JP, Higginson LA, Mesana T. Severe aortic insufficiency secondary to an aortic bioprosthesis tear. Cardiovasc Pathol. 2004;13(3):165–7.
95. Raghavan D, Simionescu DT, Vyavahare NR. Neomycin prevents enzyme-mediated glycosaminoglycan degradation in bioprosthetic heart valves. Biomaterials. 2007;28(18):2861–8.
96. Mercuri JJ, Lovekamp JJ, Simionescu DT, Vyavahare NR. Glycosaminoglycan-targeted fixation for improved bioprosthetic heart valve stabilization. Biomaterials. 2007;28(3):496–503.
97. Leong J, Munnelly A, Liberio B, Cochrane L, Vyavahare N. Neomycin and carbodiimide crosslinking as an alternative to glutaraldehyde for enhanced durability of bioprosthetic heart valves. J Biomater Appl. 2013;27(8):948–60.
98. Raghavan D, Shah SR, Vyavahare NR. Neomycin fixation followed by ethanol pretreatment leads to reduced buckling and inhibition of calcification in bioprosthetic valves. J Biomed Mater Res B Appl Biomater. 2010;92(1):168–77.
99. Friebe VM, Mikulis B, Kole S, Ruffing CS, Sacks MS, Vyavahare NR. Neomycin enhances extracellular matrix stability of glutaraldehyde crosslinked bioprosthetic heart valves. J Biomed Mater Res B Appl Biomater. 2011;99(2):217–29.
100. Purinya B, Kasyanov V, Volkolakov J, Latsis R, Tetere G. Biomechanical and structural properties of the explanted bioprosthetic valve leaflets. J Biomech. 1994;27(1):1–11.

101. Broom N, Christie GW. The structure/function relationship of fresh and glutaraldehyde-fixed aortic valve leaflets. 1st ed. New York: Yorke Medical Books; 1982.
102. Schoen FJ. Aortic valve structure-function correlations: role of elastic fibers no longer a stretch of the imagination. J Heart Valve Dis. 1997;6(1):1–6.
103. Strates B, Lian JB, Nimni ME. Calcification in cardiovascular tissues and bioprostheses. In: Nimni ME, editor. Biotechnology. Boca Raton, FL: CRC Press; 1989.
104. Shah SR, Vyavahare NR. The effect of glycosaminoglycan stabilization on tissue buckling in bioprosthetic heart valves. Biomaterials. 2008;29(11):1645–53.
105. Mirnajafi A, Raymer JM, McClure LR, Sacks MS. The flexural rigidity of the aortic valve leaflet in the commissural region. J Biomech. 2006;39(16):2966–73.
106. Erdmann E, Schwinger RH, Beuckelmann D, Bohm M. [Altered calcium homeostasis in chronic heart failure]. Z Kardiol. 1996;85(Suppl 6):123–8.
107. Li L, van Breemen C. Na(+)-Ca2+ exchange in intact endothelium of rabbit cardiac valve. Circ Res. 1995;76(3):396–404.
108. Levy RJ, Schoen FJ, Flowers WB, Staelin ST. Initiation of mineralization in bioprosthetic heart valves: studies of alkaline phosphatase activity and its inhibition by AlCl3 or FeCl3 preincubations. J Biomed Mater Res. 1991;25(8):905–35.
109. Lovekamp J, Vyavahare N. Periodate-mediated glycosaminoglycan stabilization in bioprosthetic heart valves. J Biomed Mater Res. 2001;56(4):478–86.
110. Jorge-Herrero E, Fernandez P, Gutierrez M, Castillo-Olivares JL. Study of the calcification of bovine pericardium: analysis of the implication of lipids and proteoglycans. Biomaterials. 1991;12(7):683–9.
111. Khor E. Methods for the treatment of collagenous tissues for bioprostheses. Biomaterials. 1997;18(2):95–105.
112. Chanda J, Kuribayashi R, Abe T. Prevention of calcification in glutaraldehyde-treated porcine aortic and pulmonary valves. Ann Thorac Surg. 1997;64(4):1063–6.
113. Vyavahare N, Hirsch D, Lerner E, Baskin JZ, Schoen FJ, Bianco R, Kruth HS, Zand R, Levy RJ. Prevention of bioprosthetic heart valve calcification by ethanol preincubation. Efficacy and mechanisms. Circulation. 1997;95(2):479–88.
114. Bianco RW, Phillips R, Mrachek J, Witson J. Feasibility evaluation of a new pericardial bioprosthesis with dye mediated photo-oxidized bovine pericardial tissue. J Heart Valve Dis. 1996;5(3):317–22.
115. Connolly JM, Alferiev I, Clark-Gruel JN, Eidelman N, Sacks M, Palmatory E, Kronsteiner A, Defelice S, Xu J, Ohri R, Narula N, Vyavahare N, Levy RJ. Triglycidylamine crosslinking of porcine aortic valve cusps or bovine pericardium results in improved biocompatibility, biomechanics, and calcification resistance: chemical and biological mechanisms. Am J Pathol. 2005;166(1):1–13.
116. Vasudev SC, Chandy T, Umasankar MM, Sharma CP. Inhibition of bioprosthesis calcification due to synergistic effect of Fe/Mg ions to polyethylene glycol grafted bovine pericardium. J Biomater Appl. 2001;16(2):93–107.
117. Vasudev SC, Chandy T. Polyethylene glycol-grafted bovine pericardium: a novel hybrid tissue resistant to calcification. J Mater Sci Mater Med. 1999;10(2):121–8.
118. Zhai W, Lu X, Chang J, Zhou Y, Zhang H. Quercetin-crosslinked porcine heart valve matrix: mechanical properties, stability, anticalcification and cytocompatibility. Acta Biomater. 2010;6(2):389–95.
119. Raghavan D, Starcher BC, Vyavahare NR. Neomycin binding preserves extracellular matrix in bioprosthetic heart valves during in vitro cyclic fatigue and storage. Acta Biomater. 2009;5 (4):983–92.
120. Everaerts F, Torrianni M, van Luyn M, van Wachem P, Feijen J, Hendriks M. Reduced calcification of bioprostheses, cross-linked via an improved carbodiimide based method. Biomaterials. 2004;25(24):5523–30.

121. Everaerts F, Gillissen M, Torrianni M, Zilla P, Human P, Hendriks M, Feijen J. Reduction of calcification of carbodiimide-processed heart valve tissue by prior blocking of amine groups with monoaldehydes. J Heart Valve Dis. 2006;15(2):269–77.
122. Meuris B, Phillips R, Moore MA, Flameng W. Porcine stentless bioprostheses: prevention of aortic wall calcification by dye-mediated photo-oxidation. Artif Organs. 2003;27(6):537–43.
123. Schoen FJ, Tsao JW, Levy RJ. Calcification of bovine pericardium used in cardiac valve bioprostheses. Implications for the mechanisms of bioprosthetic tissue mineralization. Am J Pathol. 1986;123(1):134–45.
124. Schoen FJ, Levy RJ, Nelson AC, Bernhard WF, Nashef A, Hawley M. Onset and progression of experimental bioprosthetic heart valve calcification. Lab Invest. 1985;52(5):523–32.
125. Levy RJ, Schoen FJ, Sherman FS, Nichols J, Hawley MA, Lund SA. Calcification of subcutaneously implanted type I collagen sponges. Effects of formaldehyde and glutaraldehyde pretreatments. Am J Pathol. 1986;122(1):71–82.
126. Golomb G, Schoen F, Smith M, Linden J, Dixon M, Levy R. The role of glutaraldehyde-induced cross-links in calcification of bovine pericardium used in cardiac valve bioprostheses. Am J Pathol. 1987;127(1):122–30.
127. Rucker RB. Calcium binding to elastin. Adv Exp Med Biol. 1974;48(0):185–209.
128. Levy RJ, Vyavahare N, Ogle M, Ashworth P, Bianco R, Schoen FJ. Inhibition of cusp and aortic wall calcification in ethanol- and aluminum-treated bioprosthetic heart valves in sheep: background, mechanisms, and synergism. J Heart Valve Dis. 2003;12(2):209–16; discussion 216.
129. Isenburg JC, Karamchandani NV, Simionescu DT, Vyavahare NR. Structural requirements for stabilization of vascular elastin by polyphenolic tannins. Biomaterials. 2006;27(19):3645–51.
130. Basalyga DM, Simionescu DT, Xiong W, Baxter BT, Starcher BC, Vyavahare NR. Elastin degradation and calcification in an abdominal aorta injury model: role of matrix metalloproteinases. Circulation. 2004;110(22):3480–7.
131. Dahm M, Lyman WD, Schwell AB, Factor SM, Frater RW. Immunogenicity of glutaraldehyde-tanned bovine pericardium. J Thorac Cardiovasc Surg. 1990;99(6):1082–90.
132. Dahm M, Husmann M, Eckhard M, Prufer D, Groh E, Oelert H. Relevance of immunologic reactions for tissue failure of bioprosthetic heart valves. Ann Thorac Surg. 1995;60(2 Suppl):S348–52.
133. Christian AJ, Lin H, Alferiev IS, Connolly JM, Ferrari G, Hazen SL, Ischiropoulos H, Levy RJ. The susceptibility of bioprosthetic heart valve leaflets to oxidation. Biomaterials. 2014;35(7):2097–102.
134. Everts V, van der Zee E, Creemers L, Beertsen W. Phagocytosis and intracellular digestion of collagen, its role in turnover and remodelling. Histochem J. 1996;28(4):229–45.
135. Human P, Zilla P. The possible role of immune responses in bioprosthetic heart valve failure. J Heart Valve Dis. 2001;10(4):460–6.
136. Levy RJ, Schoen FJ, Howard SL. Mechanism of calcification of porcine bioprosthetic aortic valve cusps: role of T-lymphocytes. Am J Cardiol. 1983;52(5):629–31.
137. Sellaro TL, Hildebrand D, Lu Q, Vyavahare N, Scott M, Sacks MS. Effects of collagen fiber orientation on the response of biologically derived soft tissue biomaterials to cyclic loading. J Biomed Mater Res A. 2007;80(1):194–205.
138. Billiar KL, Sacks MS. Biaxial mechanical properties of the native and glutaraldehyde-treated aortic valve cusp: part II—a structural constitutive model. J Biomech Eng. 2000;122(4):327–35.
139. Zilla P, Weissenstein C, Bracher M, Human P. The anticalcific effect of glutaraldehyde detoxification on bioprosthetic aortic wall tissue in the sheep model. J Card Surg. 2001;16(6):467–72.
140. Zilla P, Bezuidenhout D, Weissenstein C, van der Walt A, Human P. Diamine extension of glutaraldehyde crosslinks mitigates bioprosthetic aortic wall calcification in the sheep model. J Biomed Mater Res. 2001;56(1):56–64.

141. Mendoza-Novelo B, Cauich-Rodríguez JV. Decellularization stabilization and functionalization of collagenous tissues used as cardiovascular biomaterials. In: Biomaterials—Physics and Chemistry; 2011. p. 159–82.
142. Bezuidenhout D, Oosthuysen A, Human P, Weissenstein C, Zilla P. The effects of cross-link density and chemistry on the calcification potential of diamine-extended glutaraldehyde-fixed bioprosthetic heart-valve materials. Biotechnol Appl Biochem. 2009;54(3):133–40.
143. Deshmukh A, Deshmukh K, Nimni ME. Synthesis of aldehydes and their interactions during the in vitro aging of collagen. Biochemistry. 1971;10(12):2337–42.
144. Connolly J, Alferiev I, Kronsteiner A, Lu Z, Levy R. Ethanol inhibition of porcine bioprosthetic heart valve cusp calcification is enhanced by reduction with sodium borohydride. J Heart Valve Dis. 2004;13:487–93.
145. Trifecta Valve. https://professional.sjm.com/products/sh/tissue-valves/aortic-mitral/trifecta.
146. Mosaic Tissue Valve. https://professional.sjm.com/products/sh/tissue-valves/aortic-mitral/trifecta.
147. ThermoFix Description. http://www.edwards.com/eu/products/heartvalves/pages/thermafixdescription.aspx.
148. Oosthuysen A, Zilla PP, Human PA, Schmidt CA, Bezuidenhout D. Bioprosthetic tissue preservation by filling with a poly(acrylamide) hydrogel. Biomaterials. 2006;27(9):2123–30.
149. Tam H, Zhang W, Feaver KR, Parchment N, Sacks MS, Vyavahare N. A novel crosslinking method for improved tear resistance and biocompatibility of tissue based biomaterials. Biomaterials. 2015;66:83–91.
150. Bourguignon T, Lhommet P, El Khoury R, Candolfi P, Loardi C, Mirza A, Boulanger-Lothion J, Bouquiaux-Stablo-Duncan AL, Marchand M, Aupart M. Very long-term outcomes of the Carpentier-Edwards Perimount aortic valve in patients aged 50–65 years. Eur J Cardiothorac Surg. 2016;49(5):1462–8.

# Part III
# Computational Approaches for Heart Valve Function

# Computational Modeling of Heart Valves: Understanding and Predicting Disease

Ahmed A. Bakhaty, Ali Madani, and Mohammad R. K. Mofrad

**Abstract** Computational methods offer a robust means of studying healthy and diseased heart valves. Simulations can shed light on the elusive function of heart valves and allow us to understand the underlying causes of heart valve disease in an effort to develop effective treatments. In comparison to experimental approaches, which either may not be representative in vitro or feasible in vivo, simulations assist in obtaining detailed information on the response of valves in a cost-effective manner.

In this chapter, we take a look at how computational methods have been used to study the varying types of heart valves. In particular, we focus on the modeling of diseased valves, and the simulation of associated therapies. We aim to highlight the work done to date in computationally modeling heart valves and emphasize the need for continued development of these models.

**Keywords** Heart valve · Computational modeling · Multi-scale · Disease prediction

---

A. A. Bakhaty
Department of Civil and Environmental Engineering, University of California, Berkeley, CA, USA

Department of Electrical Engineering and Computer Science, University of California, Berkeley, CA, USA

Department of Mathematics, University of California, Berkeley, CA, USA
e-mail: abakhaty@berkeley.edu

A. Madani
Department of Applied Science and Technology and Mechanical Engineering, University of California, Berkeley, CA, USA
e-mail: madani@berkeley.edu

M. R. K. Mofrad (✉)
Department of Mechanical and Bioengineering, University of California, Berkeley, CA, USA
e-mail: mofrad@berkeley.edu

# 1 Introduction to Computational Modeling of Heart Valves

Computational methods offer a robust means of simulating the physical behavior of heart valves (HVs). They permit one to simulate in vivo conditions and probe the physical response everywhere, which is difficult—or impossible—to achieve with experimental approaches. These models can further be used in a predictive sense. For example, an anatomically accurate aortic valve (AV) model of a patient with early stages of calcific aortic stenosis (CAS) can be used to predict the progression over time of the calcification and aid the clinician in determining the most appropriate treatment course. That same model can then be used to test various treatments, such as repair, surgical replacement, or drugs.

The paradigm of using computational models to study HVs can be categorized as follows:

1. Study HVs in physiologic and pathologic states to shed light on the underlying biomechanics.
2. Develop patient-specific models to determine the efficacy of treatments, namely pre-, intra-, and postoperative modeling for predicting long-term patient outcomes.
3. Use a simulation-based methodology to design novel and robust artificial valve replacements.

Figure 1 depicts this paradigm with its components: clinical, experimental, and computational.

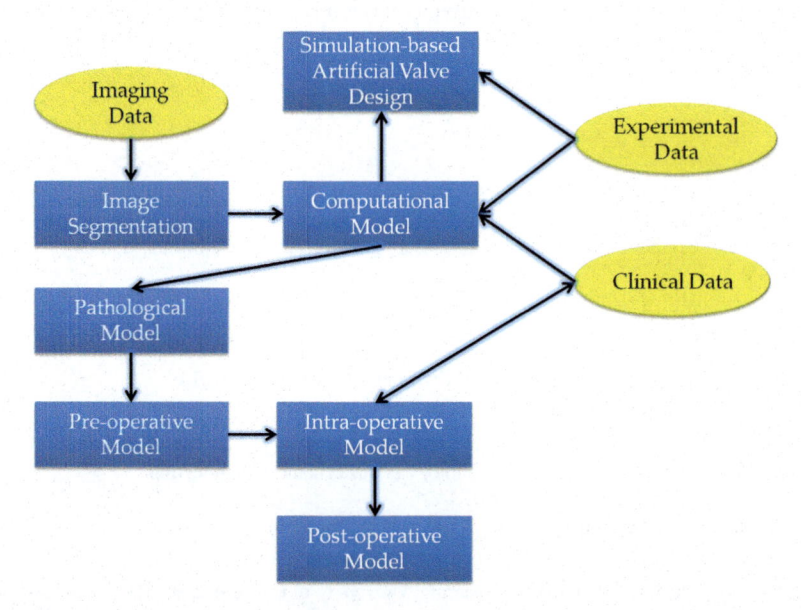

**Fig. 1** Paradigm for computational HVs. Medical imaging, laboratory experiments, and clinical data are combined to generate computational models that can be used to study pathological valves, in clinical applications, and for the design of novel artificial valves

Unfortunately, such methods come equipped with their own shortcomings. The main challenge scientists and engineers face when modeling is being able to create a model that accurately reproduces the true physical behavior of HV systems. Clearly, validation is necessary for any model. This is typically done with established experimental data, such as lab experiments or in vivo probes (e.g., echocardiography or MRI). Neumann et al. proposed a pipeline that uses microCT and in vitro data, along with machine learning and optimization to construct and validate heart valve models [1]. An approach such as this is critical for the development of sufficiently accurate models.

HVs are complex biophysical systems: they boast distinct features and biomechanics at different length scales (organ, tissue, cellular, and molecular); they exhibit highly nonlinear, inhomogeneous, and anisotropic material response; and they possess complex geometries. A detailed model that captures every aspect will be computationally intractable, with available technologies, let alone practical. On the other end of the spectrum, a highly simplified model will fail to capture important aspects of HVs. Naturally, early models were simple in scope, equipped with many simplifying assumptions that are not generally justifiable. Over time, these models have grown gradually in complexity, steadily relaxing these assumptions as advancing technology has made it feasible to do so.

In this section, we present an overview of computational modeling of heart valve systems. We first discuss the equations that govern HV biomechanics. We then go into detail into each: tissue modeling, fluid modeling, fluid-structure interaction, micromechanics, and multiscale modeling.

## 1.1 Computational Mechanics

As an organ, heart valves function as an efficient mechanism to semi-passively control blood flow through the chambers of the heart. Naturally, characterizing their behavior quantitatively amounts to analyzing the system from a mechanical point of view. More interestingly, the behavior of the valvular interstitial cells (VICs) and valvular endothelial cells (VECs) is also a mechanical process [2]. These engines play a central role in maintaining tissue homeostasis and regulating pathology [3, 4]. The cells engage in biochemical processes as a response to mechanical stimulation. Thus, we consider HVs from a biomechanical perspective across the various scales.

In mechanical systems, we aim to satisfy balance of mass,

$$\frac{d\rho}{dt} + \rho \nabla \cdot v = 0, \qquad (1)$$

where $\rho(x, t)$ is the density of particles at coordinate $x$ and time $t$, $v(x, t)$ is the velocity, and $\nabla(\bullet)$ is the divergence operator; conservation of linear and angular momenta,

$$\sum_i F_i = \frac{dp}{dt} \qquad (2)$$

$$\sum_i x \times F_i + M_i = \frac{dh}{dt}, \qquad (3)$$

where $F$ and $M$ are external forces and moments, and $p$ and $h$ are the linear and angular momenta, respectively. We also have balance of energy, given as the incremental first law of thermodynamics:

$$\delta E = \delta Q + \delta W, \qquad (4)$$

where $E$ is the total energy of the system, and $Q$ and $W$ are the heat added to and work done on the system, respectively. When considering HVs at the organ, tissue, and (to an extent) the cellular scale, it is acceptable to model the media as a continuum. Thus, the above equations reduce to the governing partial differential equations of continuum mechanics [5]. Analytical solutions exist only for very simple cases, which, in general, do not apply to the complex HV systems. Therefore, these equations are typically solved with a numerical approximation, most commonly the finite element method FEM [6]. In FEM, the geometry is discretized, and the governing equations are solved discretely at a finite number of points such that they are satisfied in an average sense over the whole domain.

Naturally, FEM and other numerical methods are approximations. Theory guarantees asymptotic convergence, and thus FEM modeling amounts to finding a balance between accuracy and computational tractability. Modeling the microstructural details (i.e., VICs and the fibrous extracellular matrix (ECM)) of HVs leads to a problem that is computationally intractable; thus, these details are often omitted. The fundamental role of this microstructure to the healthy and diseased function of HVs, however, renders models that disregard these aspects insufficient in capturing the true behavior. Methods that deal with ways to incorporate the microstructural features in a computationally tractable manner will be discussed throughout this chapter.

At the cellular scale and molecular scale, continuum mechanics becomes inappropriate to capture the discrete and stochastic nature of mechanical events occurring at that scale. Typically, we turn to molecular dynamics (MD) to solve the above equations. In MD, first principle calculations are used to satisfy the physical balance laws for a collection of particles that represent various molecules. One can see that such computations are computationally intensive, and, in general, tractable for only small systems and time scales. Extending the capabilities of these methods is a work in progress.

The remainder of this section will be an overview of how the physics (solids and fluids) of HVs are modeled and how multiple scales are considered.

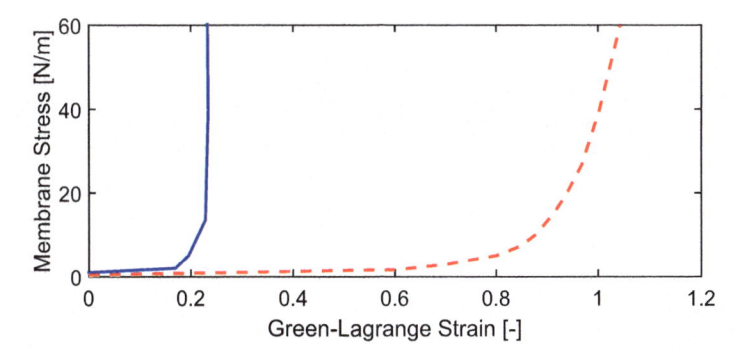

**Fig. 2** Biaxial "stress-strain" behavior of aortic valve tissue adapted from [7]. Circumferential direction is noticeably stiffer due to alignment of collagen fibers as the tissue stretches

## 1.2 Tissue Modeling

HV tissue exhibits a highly nonlinear, heterogeneous, and anisotropic response [7]. This is a manifestation of the organized network of collagen and elastin fibers that constitute the microstructure of the tissue. Figure 2 demonstrates the force-strain relationship of aortic valve tissue subjected to biaxial loading in the principal directions defined by the alignment of collagen fibers. As the valve undergoes loading due to a pressure gradient across the valve, collagen fibers are recruited and align to withstand the load. In this case, the circumferential direction is the alignment direction and is, thus, characterized by a much stiffer response.[1]

Clearly, an isotropic material model is inappropriate for capturing the behavior of HV tissue. Furthermore, a linear elastic model is similarly inadequate. Given the elasticity of biological tissue, continuum hyperelasticity is widely accepted for the modeling of HV tissue [5, 8]. For a hyperelastic material, we postulate the existence of the Helmholtz free energy $\psi$, and obtain the Piola stress $P$ as

$$P = \frac{\partial \psi}{\partial F}, \qquad (5)$$

where $F$ is the gradient of the deformation map, known as the deformation gradient. Convexity of $\psi$ is necessary to existence of a solution [9]. We typically take the Helmholtz energy as the sum of elementary energy functional:

$$\psi = \sum_i \psi_i. \qquad (6)$$

---

[1]The interesting negative stiffness behavior can be explained by the fact that samples tested are excised from tissue, and thus the fibers are detached from a larger network that acts as a unit [7].

At the minimum, these terms should typically consist of an isotropic exponential for the matrix (a la the model of Fung [10]), an anisotropic term to account for the collagen fiber alignment (a la the Holzapfel model [8]), and an initial modulus (*a la* Neo-Hookean).

A more direct treatment of fiber remodeling processes was proposed by Driessen et al. [11]. Here, collagen recruitment, alignment, and remodeling are explicitly considered by modulating the stiffness and stress in accord with the time evolution of the fiber content in $\varphi_j$ in direction $j$:

$$\frac{d\phi_j}{dt} = f(\phi_j, \lambda_j, \mu) \tag{7}$$

where $\mu$ is a rate constant, and $\lambda_j$ is the stretch in direction $j$. Given the isotropic stress $P_{iso}$, fiber $j$ Piola stress $p_j$, and Cartesian unit vectors $e_j$ in direction $j$, the Piola stress can be expressed as

$$P = P_{iso} + \sum_j \phi_j (p_j - e_j \cdot P_{iso} e_j) e_j \otimes e_j. \tag{8}$$

An interesting component usually left out of most HV simulations is contact. In FEM, the material is not aware of other material, i.e., if two geometries meet, they do not physically interact. To model the inherent phenomenon, "contact" formulations must be present in the FEM simulation.[2] Saleeb et al. [12] and Rim et al. [13] accounted for contact in HV FEM simulations and found, unsurprisingly, fundamentally different response when contact was incorporated.

## 1.3 Computational Fluid Dynamics (CFD) and Fluid Structure Interaction (FSI)

### 1.3.1 Computational Fluid Dynamics (CFD)

As an organ, heart valves consist of blood (fluid) flowing through leaflet tissue (solid). Solid modeling of valve tissue was discussed above. Herein, the fluid modeling of the blood is considered. As before, balance of mass (1), momentum (2), and energy (4) hold. Under the assumption of linear viscosity ($\mu$), this often takes the form of the Navïer-Stokes equations:

---

[2]The concept is similar to considering electrostatic repulsion in molecular dynamics simulation.

$$\rho \frac{\partial v}{\partial t} + \rho v \cdot \nabla v = -\nabla P + \mu \nabla^2 v, \quad (9a)$$

$$\nabla \cdot v = 0, \quad (9b)$$

where $P$ is the pressure. In solid mechanics, we consider a Lagrangian frame of reference, wherein we track the movement of the leaflets with respect to an initial reference configuration. In fluid mechanics, we typically use an Eulerian frame of reference, wherein we fix a control volume and temporally track fluid flow through it.

To solve these equations for complex boundary conditions and geometries, we employ a numerical scheme, such as finite differences, finite volume, finite elements, discontinuous Galerkin, and others. The reader is referred to Sotirpoulos et al. for more [14].

### 1.3.2 Suspended-Solid Fluid Models

Blood is a non-Newtonian fluid, due to the presence of suspended red blood cells, platelets, and white blood cells—which account for nearly half of the volume versus plasma. For many blood fluid mechanics simulations in literature, blood can be represented as a Newtonian fluid for vessels larger than one hundred microns. However, in many applications it might be relevant to account for solutes, and, more generally, the non-Newtonian behavior, to properly model blood flow in the heart valve and study certain pathologies. For instance, thrombosis has been linked with activation of platelets in the blood caused by nonphysiological flows from mechanical valves [15, 16]. Other suspended-solid studies include the smoothed particle hydrodynamics approach of Shahriari et al. [17], the work of Yun et al. [18] on bileaflet mechanical valves, and Pozrikidis deformable 3D cell (with membrane effects) flowing through a tube [19].

### 1.3.3 Fluid-Structure Interaction

Early heart valve models considered the solid only or just the fluid dynamics, accounting for the interaction only implicitly (e.g., as pressure loading on the solid or rigid boundaries for the fluid). Such models were not sufficient in accurately reproducing the true response of the system [20].

Computational techniques that combine the two are generally referred to as fluid-structure interaction (FSI). Three popular methods are commonly used in FSI: arbitrary Lagrangian- Eulerian (ALE) [21], immersed boundary (IB) [22], and fictitious domain (FD) [23]. In each of these methods, a solid, typically FEM, is embedded in a fluid, often FEM or finite volume model [24]. The governing equations are solved simultaneously, with kinematic conditions and equilibrium satisfied at their interface. ALE carefully treats the interface by using a body-conforming mesh that needs to be updated, a procedure that is computationally expensive. IB relaxes this requirement by having a finite difference structured fluid mesh that does not conform to the body and adjusting the equations at the interface, at the cost of losing detail at that interface. FD is similar to IB, but with a nonconforming FEM fluid mesh that is coupled with the use of Lagrange multipliers. For in-depth reviews of FSI, the reader is referred to Vigmostad et al. [25].

## 1.4 Molecular Dynamics

Although the continuum assumption can be applied to mechanics at the organ, tissue, and even (in some applications) the cellular scales, it generally does not hold as we approach molecular-scale phenomena. Molecular dynamics (MD) simulations, which are rapidly progressing due to increasingly powerful computational capabilities, must be used to analyze molecular-scale interactions. An MD simulation is a system of Newtonian equations of motion, where the Newtonian equation is special case of the Lagrangian equation of motion for mass points in a Cartesian system. The fundamental equation relating the force in terms of accelerations can be expressed as the gradient of the potential energy as follows:

$$F_i = m_i \ddot{r}_i = -\nabla U_i, \tag{10}$$

where $F_i$, $m_i$, $r_i$, and $U_i$ are the force, mass, position, and potential of particle $i$, respectively.

Therefore, to calculate the forces acting on the atoms, the potential energy of the system must be known. The nonbonded potential energy terms are traditionally considered for the pair potential, $U(r_i, r_j) = U(r_{ij})$, and neglected for higher order interactions. One of the most commonly used potentials is the Lennard-Jones potential with two parameters: $\sigma$, finite diameter at which the inter-particle potential is zero, and $\varepsilon$, the potential well depth:

$$U_{i,j} = 4\varepsilon \left[ \left(\frac{\sigma}{r}\right)^{12} - \left(\frac{\sigma}{r}\right)^{6} \right]. \tag{11}$$

In addition, there are a variety of bonding potentials that describe the intramolecular interactions. Force fields have been constructed to describe the complex interactions between atoms through the parameterization of potential functions via experimental data or ab initio and semi-empirical quantum mechanical calculations. The functional form of a force field such as AMBER [26] or CHARMM [27] can be described by the following terms:

$$U_{\text{field}} = U_{\text{bond}} + U_{\text{angle}} + U_{\text{torsion}} + U_{\text{improper}} + U_{\text{udW}} + U_{\text{elec}}, \tag{12}$$

where each term is representative of bond energies, angle, torsional effects, van der Waal's, electrostatic, etc. For a system composed of atoms with positions, potential energy, and kinetic energy (derived from atomic momenta), the classical equations of motion will be a set of coupled ordinary differential equations. Many methods exist, such as the original Verlet, velocity Verlet, and predictor-corrector algorithms, that perform the step-by-step numerical integration. With the proper selection of parameters and algorithm, MD techniques are able to accurately reproduce macroscopic properties of the system.

In addition to equilibration simulations, modelers have progressed to more investigative manipulations of biomolecules through steered molecular dynamics (SMD). Similar to atomic force microscopy (AFM) experiments or in vitro cellular mechanical events, single-molecule force spectroscopy can now be achieved in silico. Although MD simulations have not yet been systematically applied to heart valve behavior, many efforts do exist that are directly applicable. In particular, research is carried out to simulate various aspects of the behavior of collagen fiber on the molecular scale [28–31]. Madani et al. carried out a SMD simulation of mechanosensitive collagen binding bacteria that could have implications for heart valve infection [32]. Carcinoid heart disease, particularly serotonin-mediated valve disease, is a valvular pathology that has been investigated with MD. The 5-HT2B receptor was identified as the likely target leading to heart valve tissue fibrosis and has been studied at a molecular scale [33]. Of and around the heart valves, MD studies have been conducted such as on cardiac troponin, a $[Ca^{2+}]$ dependent switch for contraction in heart muscle, and the simulation of a cardiac ion-channel gating and its alteration by mutations [34, 35].

Although there have been great strides in computational capacity and parallelization techniques in MD, the computational demands of MD limit the timescale and size of simulated systems. For heart valve studies, MD is a powerful and accurate technique that can be used in conjunction with experimental methods such as NMR studies [36].

## *1.5 Multiscale Methods*

The valvular interstitial cells (VICs) play a central role in maintaining tissue homeostasis and regulating disease pathology [3, 4]. Several studies have recently been conducted to characterize the micromechanics at the cellular scale, through a continuum-based approach [37–44], in an effort to integrate them to larger computational models of HVs. It is becoming more evident that models need to capture VIC behavior to properly simulate the healthy and diseased function of HVs.

Traditional FEM does not consider the microstructure (in the case of HVs, VICs, collagen/elastin fibers, etc.) explicitly, but rather implicitly through semi-empirical stress-strain laws. One method that allows direct integration of the microstructure is computational homogenization [45]. In computational homogenization, coined $FE^2$ by Kouznetsova, stress/stiffness response is obtained by minimizing an auxiliary

FEM problem representative of the microstructure (referred to as an RVE, or representative volume element) at desired points throughout the material. Using an averaging approach, we can obtain the stress and tangent stiffness from an affine deformation field $d$ imposed on the RVE domain $\Omega$:

$$P_{\text{avg}} = \frac{1}{V_{\text{RVE}}} \int_{\Omega} \frac{\partial \psi}{\partial F} dV. \tag{13}$$

$$\mathbb{C}_{\text{avg}} = \frac{1}{V_{\text{RVE}}} \left[ \int_{\Omega} \frac{\partial^2 \psi}{\partial F^2} dV - \left( \int_{\Omega} \frac{\partial^2 \psi}{\partial F \partial d} dV \right) \left( \int_{\Omega} \frac{\partial^2 \psi}{\partial d^2} dV \right)^{-1} \left( \int_{\Omega} \frac{\partial^2 \psi}{\partial d \partial F} dV \right) \right] \tag{14}$$

Such a method allows one to probe, computationally, events occurring at the cellular scale that occur as a result of events at the macroscale (e.g., the valve as an organ). This is done in a coupled manner, i.e., events at each scale directly influence, and are influenced by, events at the other scale(s), while remaining computationally tractable. Application of this method has been validated for use in aortic valve models by Bakhaty et al. [46].

A discussion of multiscale modeling of aortic valves is presented by Weinberg et al. [47], and a more general discussion of multiscale modeling is presented by Summers et al. [48]. We return to the topic of multiscale modeling in Sect. 6 to discuss future application. In the sequels, we discuss the application of the modeling described herein to the four heart valves and artificial valve replacements.

## 2 Aortic Valve Models

Many studies of HVs, particularly computational modeling, have been focused on the aortic valve (AV), since AV disease constitutes about 67% of HV disease cases [49]. Nevertheless, much of the work done in modeling the AV can be easily extended to the three other valves, and vice versa. In this section, we will take a look at some computational studies of the aortic valve.

### 2.1 Aortic Valve Computational Models

AV tissue material properties can be calibrated to the biaxial stretch test of Billiar and Sacks [50] and bending test of Gloeckner et al. [51]. Moreover, the importance of modeling the trilayer structure of the AV is emphasized by Buchanan et al. [52]. Their study found that the AV layers act together as a unit, indicating that one may get away with using an average material description for course models, but will not capture the heterogeneous response of the AV tissue in this manner.

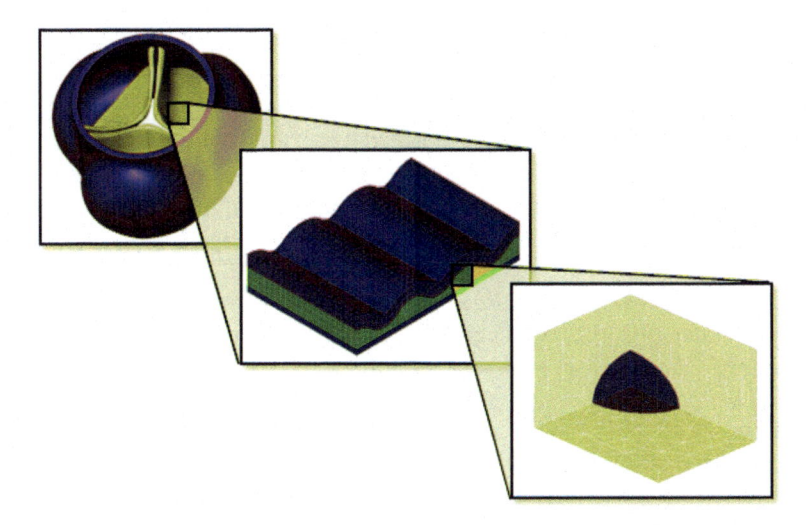

**Fig. 3** Multiscale model of the aortic valve developed by Weinberg and Mofrad [56]. Organ-scale model: FSI simulation of AV leaflets. Tissue-scale model: FEM model of AV tissue. Boundary conditions come from FSI results at critical points in the tissue. Cellular-scale model: FEM model of VIC in ECM. Boundary conditions come from tissue model

Geometrical considerations can be found in Haj-Ali et al. [53] and Thubrikar [54]. A detailed parameterization of AV geometry is provided therein for construction of idealized AV root geometry. A more detailed patient-specific modeling approach is discussed in Sect. 6.

Fluid shear stresses are important for characterizing healthy or abnormal valve behavior, and response of the endothelial cells [25]. Naturally, AV models should adopt an FSI approach. One of the earlier AV FSI models was developed by De Hart et al. [55] in 2D. The AVs trileaflet structure does not permit axisymmetry well, and thus 2D models are inadequate. Modern studies, in general, adopt 3D FSI implementations.

The importance of multiscale models is emphasized throughout this chapter. To these ends, Weinberg and Mofrad developed a model that links the organ scale and the cellular scale of the AV. [56] Illustrated in Fig. 3, an FSI model of the AV organ passes down boundary conditions to a tissue model, which in turn passes down boundary conditions to a cellular-scale model. Such a model, however, fails to capture the inherent communication between the scales, that is to say, in this model, events at the cellular scale have no repercussion on the macroscopic behavior of the AV.

For the remainder of the section, we highlight key findings of AV models used to study disease, namely, calcific aortic stenosis and bicuspid aortic valve, as well as surgical implantation of stents and valve replacements. The reader is referred to Weinberg et al. [47] and Bakhaty et al. [57] for more detail on AV computational modeling.

## 2.2 Modeling of Calcific Aortic Stenosis

Calcific aortic stenosis (CAS) is characterized by AV leaflets that do not fully open as a result of stiffening due to calcification deposits. This causes irregular blood flow that can lead to thromboembolisms and death. Stenotic behavior, and its severity, can be identified by the pressure gradient across the AV. This can be measured in vivo or computed in real time with fractional flow reserve models (see [58] and Sect. 6.2).

Weinberg and Mofrad introduced local regions of higher stiffness to model calcification deposits [59]. Although etiologically accurate, obtaining accurate distributions of CAS valves in vivo is rather difficult.

Van Loon took a different approach by altering the leaflet geometry to capture stenotic behavior [60]. The model demonstrated how small changes in the native valve geometry can lead to largely abnormal behavior. Such an approach, however, suffers from a very large parameter space that, similar to the Weinberg model, may be difficult to accurately obtain from medical images.

Both of these models share a common shortcoming: difficulty of obtaining physically realistic parameters. Maleki et al. sought to overcome this by considering the performance of the model with respect to noninvasive measurements only on average [61]. In this approach, the isotropic Fung stiffness (viz., Sect. 1.2) is modified parametrically until the simulation results match the experimental measurements.

Despite the power of these models, the progression of calcific disease remains a process driven at the cellular scale [62, 63]. One way to model this of CAS is through interstitial tissue mass growth (see Ateshian [64]) at the cellular scale of a multiscale model approach (such as that of Bakhaty and Mofrad [46]). A more complete model should account for the mechanochemical coupling.

## 2.3 Bicuspid Aortic Valve

Bicuspid aortic valve (BAV) is a congenital disease affecting 1–2% of the population and is characterized by the AV having two leaflets instead of three. This often leads to ascending aortic complications, CAS, and regurgitation [65]. Modeling of BAV is rather straightforward: replace the tricuspid geometry with a bicuspid one. The challenge lies in interpreting the model. This, in general, comes down to hemodynamic characterization (relative to the healthy trileaflet AV).

Conceptually, the bicuspid geometry will lead to abnormal fluid flows that differ from the flow patterns that the aortic root has biologically evolved to efficiently withstand. FSI models regularly predict this. Viscardi et al., for instance, found asymmetrical blood flow in the ascending aorta with BAV as opposed to the healthy trileaflet valve [66]. This supports the postulate that ascending aorta complications found in BAV patients result from aberrant fluid flow. The FSI model of Marom et al. corroborates this finding and identifies larger than normal (i.e., the trileaflet AV) fluid shear stresses and leaflet stresses [67]. Similarly, the FSI BAV model of Katayama et al. predicts excessive strains in the AV leaflets [68].

Various FSI models have demonstrated consistency in predicting abnormal mechanics (stress, strain, flow characteristics, etc.) in BAV, but this is insufficient to prove a causal relation with BAV pathology. Furthermore, such models cannot be used for patient prognosis.

Establishing a link with cellular activity may be the key to making progress towards this end.

From a multiscale standpoint, Weinberg and Mofrad found that although tissue stresses and fluid jet formation was noticeably different, when comparing BAV with the trileaflet AV, cellular-scale stresses were similar enough [69]. An implication of this is that the geometrical factors (i.e., bicuspid structure) may not be the primary cause of the progression of calcification. Rather, the phenomenon may be more related to cellular-scale events.

A different approach taken by Atkins et al. combined experimentation with a computational model of BAV to characterize cellular gene expression [70]. Shear stresses on the ascending aorta wall computed from a BAV FSI model were applied to tissue samples in the laboratory. These experiments demonstrated aortic wall remodeling and degradation. This combined simulation-experimentation multiscale methodology closes the loop that connects mathematical modeling and biology. Continued application of this approach is promising for advancing our knowledge of multiscale HV behavior.

## 2.4 Aortic Valve Surgical Modeling

Computational models offer a unique means of determining the efficacy of surgical intervention. Consider, for instance, the implantation of a stent. A parametric simulation of tissue stresses and flow characteristics can be used to predict potential tearing or leakage over the lifetime of the stent [71]. Asides from preoperative applications (see Sect. 6.5), this approach boasts a powerful tool for the design of stents.

Transcatheter aortic valve replacement (TAVR) has gained popularity in the cardiothoracic surgery community due to favorable performance [72]. These outcomes depend heavily on geometric considerations and the choice of prosthetic. Aurrichio et al. investigated these factors with patient-specific computational studies, determining the impact of geometry, effect of prostheses material, and native tissue stress distribution from various stent placements [73–75].

Currently there is no effective method for dealing with diseased AVs, and thus patients with severe cases of disease typically undergo valve replacement surgeries [49]. Sometimes valve repair surgeries are performed, but this is more common with the mitral valve (see Sect. 3.2). The methodology of simulating repair for the valves is more or less identical, except for perhaps the type of procedure appropriate for the valve in question. For instance, Labrosse et al. [76] simulated central or commissural plication [77] for AV prolapse and identified the optimal technique for restoring AV mechanical performance to normal.

## 3 Mitral Valve Models

The mitral valve (MV) has also received considerable attention from the standpoint of computational modeling, due to the persistence of mitral regurgitation. Unlike the AV, valve replacement surgeries are less common. Repair procedure is more prevalent [49]. Thus, many computational models have focused on simulating the efficacy of surgical repair. In this section, we will highlight key studies in the computational modeling of the MV.

### 3.1 Mitral Valve Computational Models

Early 3D models of the MV were developed by Kunzelmann et al. with tissue experimental data by the same group [78, 79]. This model was used to study, for instance, MV chordae tendineae repair [80, 81] and the effect of collagen concentration in the MV [82]. The latter approach used an altered stiffness of the valves to account for the biochemical variation, as opposed to a direct biochemical approach. This, of course, requires knowledge of the valve stiffness as a function of the collagen concentration from in vitro experimental data. Later models incorporated FSI with a collagen/elastin fiber recruitment and alignment tissue model (see Sects. 1.2 and 1.3) [83, 84].

The MV is distinct due to the presence of the papillary muscles and the chordae tendineae. The role of the cardiac muscles in the MV apparatus (including the leaflet) is, in fact, not passive. Incorporating muscle activity is nontrivial but it was considered by Skallerud et al. [85]. They were able to reconcile a physiological inconsistency (bulging of the MV leaflets) by incorporating parametric activation of muscle fibers during systole:

$$\sigma_f = \frac{t - t_{\text{start}}}{t_{\text{end}} - t_{\text{start}}} \sigma_{\max}, \quad t \in [t_{\text{start}}, t_{\text{end}}], \tag{15}$$

where $\sigma_f$ is the muscle fiber stress as a linear function in time (start to end of systole) of the max fiber stress. One may consider a different, biochemical approach (see for instance [86]):

$$\sigma_f = f(\lambda, [\text{Ca}^{2+}]). \tag{16}$$

Here, the calcium concentration is explicitly considered alongside mechanical factors (i.e., the stretch $\lambda$). For more on biochemical activation of muscle, see Rachev and Hayashi [87].

For a more detailed discussion of MV modeling, the reader is referred to Einstein et al. [88].

## 3.2 Mitral Valve Repair

The most common MV disease is mitral regurgitation, in which the MV does not fully close, allowing backflow of blood into the left ventricle. Severe enough cases of mitral regurgitation are often treated via surgical intervention. Unlike the AV, the MV is typically repaired as opposed to being replaced.

Computational modeling naturally lends itself as a versatile method of assessing the efficacy of various repair techniques, which depends on many factors, some of which can be modeled (for instance, suture length and placement, nature of the regurgitation, transvalvular pressure, etc.) [89]. One important factor is the age of the valve, which results in altered mechanical properties. Stiffening valve is known to be a common cause of regurgitation.

Computational modeling naturally lends itself as a versatile method of assessing the efficacy of various repair techniques, which depends on many factors, some of which can be modeled (for instance, suture length and placement, nature of the regurgitation, transvalvular pressure, etc.) [89]. One important factor is the age of the valve, which results in altered mechanical properties. Stiffening valve is known to be a common cause of regurgitation. Properly characterizing the MV properties based on age is considered by Pham et al. [90] and Lee et al. [91].

## 4 Pulmonary and Tricuspid Valve Models

The pulmonary and tricuspid valves have received far less attention than the AV and MV in the literature, presumably due to their relative rarity of disease. Nevertheless, for completeness we present a few modeling studies of these two valves.

## 4.1 Pulmonary Valve

Following the modeling paradigm introduced in Sect. 1, Tang et al. [92] pursued a study of MRI-based patient-specific modeling of pulmonary valve (PV) insertion/

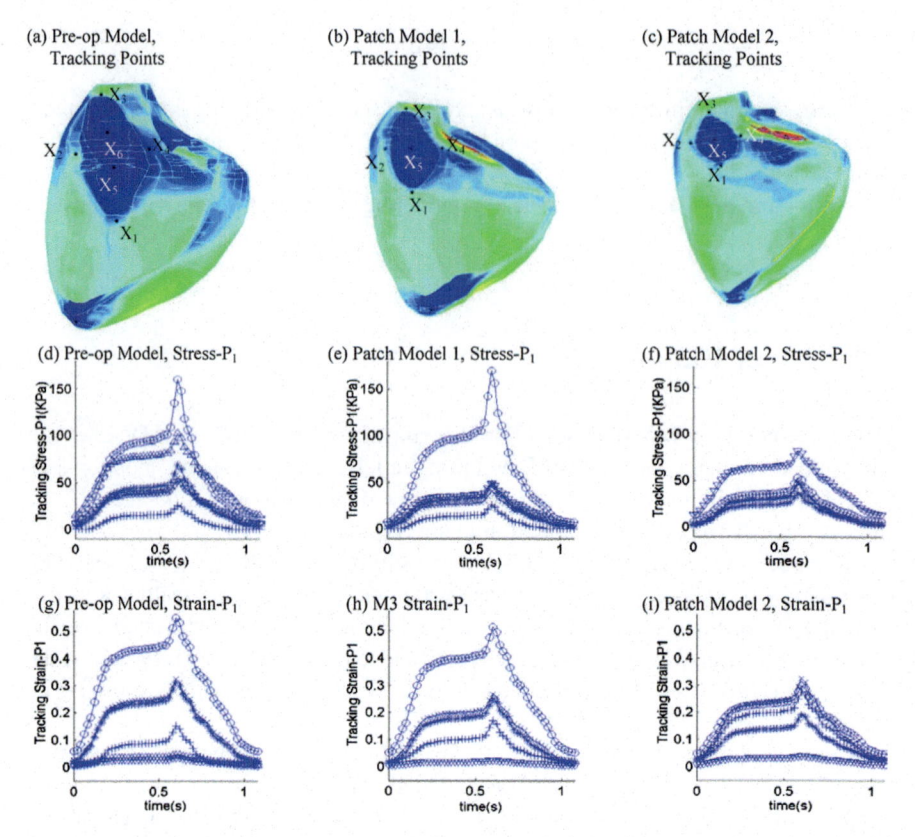

**Fig. 4** Right Ventricle (RV) volume curves and pressure-volume loops at varying states. Solid green is from image segmentation; Solid black is from preoperative simulation; Dashed blue is from simulation after percutaneous PV replacement; Dash-dotted red is from simulation after PV replacement with volume reduction. Regurgitations and curve shifts are apparent as a result of RV volume reduction. Figure originally found in Mansi et al. [93]

replacement surgery. They proposed an approach of determining the efficacy of a surgical procedure with a patient-specific computational model. They were able to develop a model of the right and left ventricles used to evaluate optimal patching and scar removal procedures by identifying lowest tissue stresses. This approach makes use of a more realistic description of the cardiac system than isolated models of HVs.

Mansi et al. followed along this paradigm of system modeling by proposing a model to simulate PV replacement surgery as the results show in Fig. 4 [93]. This model incorporates electrophysiology with the mechanics by solving the anisotropic Eikonal equation:

$$v^2\left((\nabla T_d)^T D \nabla T_d\right) = 1, \tag{17}$$

where $v$ is the local conduction velocity, $T_d$ is the electric wave depolarization time, and $D$ is the conduction anisotropy tensor.

## 4.2 Tricuspid Valve

To the authors' best knowledge, there has been almost no endeavors to computationally model the tricuspid valve (TV). This is unsurprising, given the relative scarcity of TV disease. Nevertheless, the TV is biomechanically not dissimilar from the other three valves. In particular, its structure is very similar to the other atrioventricular valve (the MV), but with three (or more [94]) leaflets. Hence, models of the MV can easily be extended to study the TV. Note, however, that the dearth of literature regarding TV tissue mechanics makes it difficult to properly model the TV. Further experimental probing is warranted for interest in modeling and understanding the TV.

# 5 Artificial Valves

When one of the four heart valves malfunctions, many patients opt to implant an artificial valve. There are three main divisions of prosthetic heart valves: mechanical, bioprosthetic, and tissue-engineered valves—with tissue-engineered valves being the newest addition to the well-established mechanical and bioprosthetic valves. Each type and subdivision will have different properties that the patient considers before implantation. Based on the patient's preference and factors such as body size, valve annular size, age, lifestyle, comorbidities, and ventricle function, a prosthesis model should be selected that provides superior hemodynamic performance that minimizes prosthesis patient mismatch (PPM). The most widely used parameter for identifying PPM is the indexed valve effective orifice area of the prosthesis divided by the patient's body surface area. In this section, we will examine the recent computational studies for various artificial heart valves that lead to a more precise quantification of valvular characteristics.

## 5.1 Artificial Valve Modeling

The ideal valve prosthetic should mimic the properties of a native heart valve in areas including, but not limited to, hemodynamics, durability, thromboresistance, and implantability. Numerical modeling can enable a precise quantification of these characteristics to ultimately influence effective valve design. Figure 5 demonstrates how a simulation allows one to determine the stresses on the valve and the native surroundings during the cardiac cycle. Consequently, the designer can make

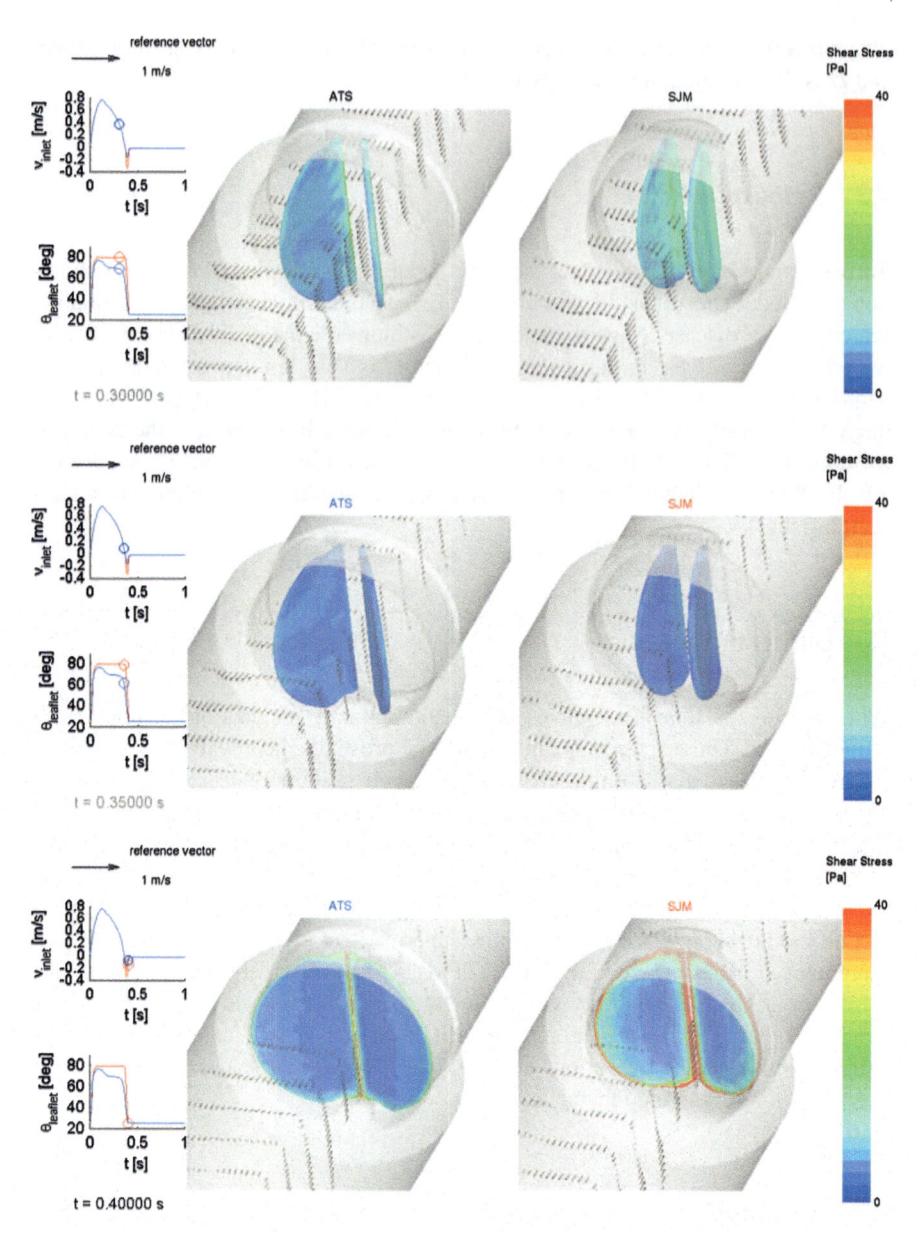

**Fig. 5** Flow and wall shear stress results for two different artificial valves at different times, as presented by Dumont et al. [95]

structural modifications or alter the prosthesis placement to insure the induced stress state is as close as possible to the native state. Furthermore, in a parametric manner, the prosthesis durability can be assessed by simulating long-term fatigue behavior.

For large artery flow simulations, blood is assumed to be an incompressible Newtonian fluid with its motion governed by the Navier-Stokes Eqs. (9a and 9b). However to fully capture the small regions of flow in and around prosthetic valve, non-Newtonian effects can become more important to capture and are included in the governing equations. The equations of motion of the valvular structure, particularly the leaflets, are governed by the balance laws of momentum. By capturing the solid, fluid, and FSI effects, advanced computational tools enable the design of new valves, even patient-specific, and the in silico performance assessment of valve prototypes without the need for extensive in vitro and animal tests.

## 5.2 Mechanical Valves

Mechanical heart valve designs have evolved significantly over the years; the most notable design is the bileaflet mechanical heart valve (BMHV), alongside the monoleaflet and caged ball valve designs. Although mechanical valves are more durable than bioprosthetic valves, the patient must consider limiting side effects. Due to high shearing of blood cells and platelets, patients with mechanical heart valves must undergo lifelong anticoagulation therapy.

Many computational models have been developed to investigate different parameters and conditions important to mechanical heart valves. Using FSI modeling techniques on three three-dimensional cases with aortic bileaflet mechanical heart valves (BMHVs), Annerel et al. examined the influence of upstream boundary conditions [1]. Borazjani et al. show the importance of aorta geometry, as opposed to the oversimplified straight aorta, on the flow physics, shear stress, and pressure drop in BMHVs [96]. To be noted, locations other than the aortic valve have been studied such as the mitral position by Choi et al. [97]. Their results indicate a mitral BMHV with a more centrally located hinge and implanted in the anatomical orientation provides the best overall performance. For further reading into the variety of numerical techniques used for simulating flow through mechanical valves, readers are encouraged to review Sotiropoulos et al. [98].

Computational models have also been used for comparison studies between differing designs of heart valves. Dumont et al. studied two commercially available BMHVs and assessed differing potential for platelet activation during regurgitation phase by their hinge mechanisms [95]. Many have sought to validate their computational models via experimental techniques such as particle image velocimetry (PIV) measurements. Guivier et al. developed an experimental model that accounts for geometric features and hydrodynamic conditions for a prosthetic valve that can later help investigate clinical issues involving the aortic valve [99]. Nobili et al. carried out a fully implicit FSI simulation of a BMHV and validated with an ultrafast cinematographic technique which proved to be a promising tool for mechanical heart valve designers and surgeons as well [100].

## 5.3 Bioprosthetic Valves

Bioprosthetic valves are tissue valves, mainly constituted from porcine or bovine pericardium tissues, which have similar geometries to native valves enabling superior hemodynamics and decrease red blood cell damage. As a result, bioprosthetic valves do not require the use of anticoagulant drugs but many fail structurally within the first 10–15 years after implantation. There are different categories of designs within bioprosthetic valves that differ by tissue type, stent, or type of replacement operation—the newest of which are transcatheter aortic valve replacement techniques.

Like the mechanical heart valves, a variety of computational models have been employed for bioprosthetic heart valves. For example, de Hart et al. developed a numerical model of a fiber-reinforced stentless aortic valve [101]. Haj-Ali et al. developed a combined computational and experimental approach for the nonlinear structural simulations of polymetric trileaflet aortic valves (PAVs) [102]. Particular events and conditions are especially relevant to capture with computational techniques. For example, even though bioprosthetic valves are known to have better hemodynamics than mechanical valves, there still might be a risk, however low, of thromboembolic events as investigated by Sirois [103]. Martin et al. developed a computational framework to evaluate noncalcific structural deterioration due to tissue leaflet fatigue [104]. And by developing a stress resultant shell model for bioprosthetic heart valves, Kim et al. correlated regions of stress concentration and large flexural deformation during opening and closing phases of the cardiac cycle with regions of calcification or mechanical damage on leaflets [105].

Lastly, the use of transcatheter heart valves is increasing each year. Numerical modeling can offer new insights such as a better understanding of the mechanism producing the radial force in transcatheter heart valves. Tzamtzis et al. simulated and compared the behavior of two transcatheter heart valves from Medtronic and Edwards Corporation [106]. Sun et al. developed computational models of a transcatheter aortic valve (TAV) to better understand the biomechanics involved with a TAV with an elliptical configuration [107].

## 5.4 Tissue-Engineered Valves

Tissue-engineered heart valves (TEHVs) have been presented as an attractive future alternative to the drawbacks of existing heart valve replacements. The main factors for TEHVs to consider are the scaffold material (whether biological or synthetic), cell source, and in vitro preconditioning settings. Computational modeling serves as a promising tool to develop a more thorough understanding of the necessary characteristics to enable a successful TEHV.

Wang et al. developed a three-dimensional aortic valve leaflet model to investigate the effect of tissue thickness and thickness variation on oxygen diffusion within the leaflet [108]. Also for tissue-engineered human heart valve leaflets, Driessen et al. performed uniaxial tensile tests to determine model parameters after various culture times [109]. They subsequently performed FEM simulations to understand the mechanical response of the leaflets to pressure loads. They specifically discovered the stresses in the leaflets induced by the pressure load increase with culture time as a result of a decrease in thickness, whereas the strains decrease, due to an increase in elastic modulus. They conclude that compared to native porcine leaflets, the mechanical response of tissue-engineered leaflets is more linear, stiffer, and less anisotropic [109].

The reader is urged to review Hasan et al. [110] for an overview of mechanical properties of human and common animal model heart valves for potential avenues of application of computational modeling.

# 6 Future Directions

## 6.1 Summary: A Discussion on the Validity of Models

The essence of computational modeling of heart valve systems is building a mathematical representation of the physics governing the system. Numerical approximations are necessary to handle the inherent complexity of such systems. Assumptions allow for tractability, but at the cost of accuracy. Improving technologies have paved the way for successive relaxations of these assumptions, but much work remains.

The notion of "accuracy" mentioned above warrants closer attention. The accuracy, or "success," of a model is tantamount to its ability to *properly* reproduce the *true* behavior of the heart valve system. Hence, the notion of validation. Any model of the heart valve must be validated against reasonable experimental data. Naturally, this requires an interdisciplinary understanding of the biological, physiological, clinical, engineering, and mathematical aspects of the problem to identify appropriate metrics of validity.

Upon validation, these models can be used to safely predict phenomena in a manner that is generally not feasible through traditional experimental techniques. This is the power of computational modeling. Naturally, the engineer or scientist performing such simulations should exercise appropriate judgment when interpreting the results and making any claims.

For the remainder of this section, we provide a brief outlook on the future of heart valve computational modeling.

## 6.2 Heart Systems

Ideally, the cardiovascular system (or some subset thereof) should be modeled. Boundary conditions of "snapshots" of the valve, with an incomplete description of its surroundings, fail to capture important interactions occurring in the cardiovascular system. Some models have considered this, but the valve in such systems becomes overly simplified. One successful implementation with clinical applications is the fractional flow reserve system of Nørgaard et al. [111]. Combining the sufficiently complex heart valve models discussed in this chapter with these larger systems is a natural extension.

## 6.3 Multiscale Considerations

The concept of coupling the scales allows for the inherent communication between the VICs and the valve organ be modeled. Models of this type have been successfully implemented of other systems and validated for AV tissue [46]. A future hope is to incorporate the nanoscale.

## 6.4 Patient-Specific Modeling

Several of the modeling studies discussed in this chapter have used geometrically accurate geometries obtained from medical images (microCT, XRAY, MRI) and reconstructed into a 3D mesh for FEM software. The goal of such studies is to create models specific to a patient in a clinical setting and use these models for diagnosis, prognosis, and for determining the best course of treatment, as well as the potential effectiveness.

## 6.5 Intraoperative Modeling

With the paradigm described above, patient-specific models can be used intraoperatively to monitor tissue and fluid stresses during surgery. Such a framework that combines noninvasive imaging with FEM to assess the MV in vivo was developed by Xu et al. [112]. Other examples include real-time surgical electromechanical model in pulmonary valve replacement surgeries [93] and intraoperative ultrasound magnetic navigation in TAVR [113].

## 6.6 Non-Physics-Based Computation

The terminology "computational modeling" in this chapter has been used in the context of the governing physics equations (Eqs. 1–4). However, numerical methods have great applicability outside of physics in biomedicine. Economic decision-making and statistical inference are two important applications of computation that have significant ramifications for clinical practice, and the biomedical field in general, from assessing the cost/risk of a surgical procedure, to determining the viability of mass-producing a bioprosthetic replacement valve. Such models offer the ability to predict the effectiveness of surgical procedures and treatment plans, and aid the clinician and patient with the decision-making process. An interesting area that has yet to be well investigated is to combine these non-physical models with the physical models, medical imaging, and databases to create a framework for pre-, intra-, and postoperative modeling.

# References

1. Neumann D, Grbic S, Mansi T. Multi-modal pipeline for comprehensive validation of mitral valve geometry and functional computational models. Comput Model. 2014;15(12):1281–312.
2. Mofrad MRK, Kamm RD. Cellular mechanotransduction: diverse perspectives from molecules to tissues. Cambridge: Cambridge University Press; 2009.
3. Taylor PM, et al. The cardiac valve interstitial cell. Int J Biochem Cell Biol. 2003;35(2):113–8.
4. Liu AC, Joag VR, Gotlieb AI. The emerging role of valve interstitial cell phenotypes in regulating heart valve pathobiology. Am J Pathol. 2007;171:1407–18.
5. Holzapfel GA. Nonlinear solid mechanics, vol. 24. Chichester: Wiley; 2000.
6. Zienkiewicz OC, et al. The finite element method, vol. 3. London: McGraw-hill; 1977.
7. Sacks MS, Yoganathan AP. Heart valve function: a biomechanical perspective. Philos Trans R Soc B. 2007;362(1484):1369–91.
8. Holzapfel GA, Gasser TC, Ogden RW. A new constitutive framework for arterial wall mechanics and a comparative study of material models. J Elas Phys Sci Solids. 2000;61:1–9.
9. Balzani D, et al. A polyconvex framework for soft biological tissues. Adjustment to experimental data. Int J Solids Struct. 2006;43(20):6052–70.
10. Sun W, Sacks MS. Finite element implementation of a generalized Fung-elastic constitutive model for planar soft tissues. Biomech Model Mechanobiol. 2005;4(2–3):190–9.
11. Driessen NJB, et al. Computational analyses of mechanically induced collagen fiber remodeling in the aortic heart valve. J Biomech Eng. 2003;125(4):549–57.
12. Saleeb AF, Kumar A, Thomas VS. The important roles of tissue anisotropy and tissue-to-tissue contact on the dynamical behavior of a symmetric tri-leaflet valve during multiple cardiac pressure cycles. Med Eng Phys. 2013;35(1):23–35.
13. Rim Y, McPherson DD, Kim H. Effect of leaflet-to-chordae contact interaction on computational mitral valve evaluation. Biomed Eng Online. 2014;13(1):31.
14. Sotiropoulos F, et al. Computational techniques for biological fluids: from blood vessel scale to blood cells. In: Image-based computational modeling of the human circulatory and pulmonary systems. New York: Springer; 2011. p. 105–55.
15. Bluestein D, Li YM, Krukenkamp IB. Free emboli formation in the wake of bileaflet mechanical heart valves and the effects of implantation techniques. J Biomech. 2002;35(12):1533–40.
16. Hellums JD, et al. Studies on the mechanisms of shear-induced platelet activation. In: Cerebral ischemia and hemorheology. New York: Springer; 1987. p. 80–9.
17. Shahriari S, et al. Evaluation of shear stress accumulation on blood components in normal and dysfunctional bileaflet mechanical heart valves using smoothed particle hydrodynamics. J Biomech. 2012;45(15):2637–44.
18. Min Yun B, et al. A numerical investigation of blood damage in the hinge area of aortic bileaflet mechanical heart valves during the leakage phase. Ann Biomed Eng. 2012;40(7):1468–85.

19. Pozrikidis C. Numerical simulation of cell motion in tube flow. Ann Biomed Eng. 2005;33(2):165–78.
20. Chandran KB. Role of computational simulations in heart valve dynamics and design of valvular prostheses. Cardiovasc Eng Technol. 2010;1(1):18–38.
21. Benson DJ. An efficient, accurate, simple ALE method for nonlinear finite element programs. Comput Methods Appl Mech Eng. 1989;72(3):305–50.
22. Peskin CS. The immersed boundary method. Acta Numerica. 2002;11:479–517.
23. Glowinski R, et al. A fictitious domain approach to the direct numerical simulation of incompressible viscous flow past moving rigid bodies: application to particulate flow. J Comput Phys. 2001;169(2):363–426.
24. LeVeque RJ. Finite volume methods for hyperbolic problems, vol. 31. Cambridge: Cambridge university press; 2002.
25. Vigmostad SC, et al. Fluid–structure interaction methods in biological flows with special emphasis on heart valve dynamics. Int J Numer Methods Biomed Eng. 2010;26(3–4):435–70.
26. Cornell WD, et al. A second generation force field for the simulation of proteins, nucleic acids, and organic molecules. J Am Chem Soc. 1995;117(19):5179–97.
27. Brooks BR, et al. CHARMM: a program for macromolecular energy, minimization, and dynamics calculations. J Comput Chem. 1983;4(2):187–217.
28. Park S, Klein TE, Pande VS. Folding and misfolding of the collagen triple helix: Markov analysis of molecular dynamics simulations. Biophys J. 2007;93(12):4108–15.
29. Sundar Raman S, et al. A molecular dynamics analysis of ion pairs formed by lysine in collagen: Implication for collagen function and stability. J Mol Struct THEOCHEM. 2008;851:299–312.
30. Sundar Raman S, et al. Role of aspartic acid in collagen structure and stability: A molecular dynamics investigation. J Phys Chem. 2006;110(41):20678–85.
31. Salsas-Escat R, Stultz CM. The molecular mechanics of collagen degradation: Implications for human disease. Exp Mech. 2009;49(1):65–77.
32. Madani A, Garakani K, Mofrad MRK. Molecular mechanics of Staphylococcus aureus adhesin, CNA, and the inhibition of bacterial adhesion by stretching collagen. PLoS One. 2017;12(6):e0179601.
33. Hutcheson JD, et al. Serotonin receptors and heart valve disease-It was meant 2B. Pharmacol Ther. 2011;132(2):146–57.
34. Varughese JF, Li Y. Molecular dynamics and docking studies on cardiac troponin C. J Biomol Struct Dyn. 2011;29(1):123–35.
35. Silva JR, et al. A multiscale model linking ion-channel molecular dynamics and electrostatics to the cardiac action potential. Proc Natl Acad Sci. 2009;106(27):11102–6.
36. DeAzevedo ER, et al. The effects of anticalcification treatments and hydration on the molecular dynamics of bovine pericardium collagen as revealed by 13C solid-state NMR. Magn Reson Chem. 2010;48(9):704–11.
37. Stella JA, et al. Tissue-to-cellular deformation coupling in cell-microintegrated elastomeric scaffolds. In: IUTAM Symposium on cellular, molecular and tissue mechanics. New York: Springer; 2010. p. 81–9.
38. David Merryman W, et al. Viscoelastic properties of the aortic valve interstitial cell. J Biomech Eng. 2009;131(4):041005.
39. Liu H, Yu S, Simmons CA. Determination of local and global elastic moduli of valve interstitial cells cultured on soft substrates. J Biomech. 2013;46(11):1967–71.
40. Zeng X, Li S. Multiscale modeling and simulation of soft adhesion and contact of stem cells. J Mech Behav Biomed Mater. 2011;4(2):180–9.
41. Unnikrishnan GU, Unnikrishnan VU, Reddy JN. Constitutive material modeling of cell: a micromechanics approach. J Biomech Eng. 2007;129(3):315–23.
42. Huang H-YS, Liao J, Sacks MS. In-situ deformation of the aortic valve interstitial cell nucleus under diastolic loading. J Biomech Eng. 2007;129:880–9.

43. Huang S, Huang H-YS. Virtual experiments of heart valve tissues. Conf Proc IEEE Eng Med Biol Soc. 2012;2012:6645–8.
44. Huang S, Huang H-YS. Virtualisation of stress distribution in heart valve tissue. Comput Methods Biomech Biomed Engin. 2014;17(15):1696–704.
45. Kouznetsova VG. Computational homogenization for the multi-scale analysis of multi-phase materials. PhD thesis. 2002.
46. Bakhaty AA, Govdindjee S and Mofrad MRK. Coupled tissue and cellular multiscale analysis of aortic valve tissue. 2016.
47. Weinberg EJ, Shahmirzadi D, Mofrad MRK. On the multiscale modeling of heart valve biomechanics in health and disease. Biomech Model Mechanobiol. 2010;9(4):373–87.
48. Ron S, Abdulla T, Schleich J-M. Progress with multiscale systems. Meas Control. 2011;44(6):180–5.
49. Alan SG, et al. Heart disease and stroke statistics–2014 update: a report from the American Heart Association. Circulation. 2014;129(3):e28–e292.
50. Billiar KL, Sacks MS. Biaxial mechanical properties of the natural and glutaraldehyde treated aortic valve cusp – part I: experimental results. J Biomech Eng. 2000;122(1):23–30.
51. Glaire Gloeckner D, Bihir KL, Sacks MS. Effects of mechanical fatigue on the bending properties of the porcine bioprosthetic heart valve. ASAIO J. 1999;45(1):59–63.
52. Buchanan RM, Sacks MS. Interlayer micromechanics of the aortic heart valve leaflet. Biomech Model Mechanobiol. 2014;13(4):813–26.
53. Haj-Ali R, et al. A general three-dimensional parametric geometry of the native aortic valve and root for biomechanical modeling. J Biomech. Sept. 2012;45(14):2392–7.
54. Thubrikar MJ. The aortic valve. Boca Raton: CRC press; 1989.
55. De Hart J, et al. A two-dimensional fluid-structure interaction model of the aortic valve. J Biomech. 2000;33:1079–88.
56. Weinberg EJ, Mofrad MRK. Transient, three-dimensional, multiscale simulations of the human aortic valve. Cardiovasc Eng. 2007;7(4):140–55.
57. Bakhaty AA, Mofrad MRK. Coupled Simulation of Heart Valves: Applications to Clinical Practice. Ann Biomed Eng. 2015;43(7):1626–39.
58. Garcia D, et al. Impairment of coronary flow reserve in aortic stenosis. J Appl Physiol. 2009;106(1):113–21.
59. Weinberg EJ, Schoen FJ, Mofrad MRK. A computational model of aging and calcification in the aortic heart valve. PLoS One. 2009;4(6):e5960.
60. Van Loon R. Towards computational modelling of aortic stenosis. Int J Numer Methods Biomed Eng. 2010;26(3–4):405–20.
61. Maleki H, et al. A metric for the stiffness of calcified aortic valves using a combined computational and experimental approach. Med Biol Eng Comput. 2014;52(1):1–8.
62. David Merryman W. Mechano-potential etiologies of aortic valve disease. J Biomech. 2010;43(1):87–92.
63. Li C, Xu S, Gotlieb AI. The progression of calcific aortic valve disease through injury, cell dysfunction, and disruptive biologic and physical force feedback loops. Cardiovasc Pathol. 2013;22(1):1–8.
64. Ateshian GA, Ricken T. Multigenerational interstitial growth of biological tissues. Biomech Model Mechanobiol. 2010;9(6):689–702.
65. Fedak PWM, et al. Clinical and pathophysiological implications of a bicuspid aortic valve. Circulation. 2002;106(8):900–4.
66. Viscardi F, et al. Comparative finite element model analysis of ascending aortic flow in bicuspid and tricuspid aortic valve. Artif Organs. 2010;34(12):1114–20.
67. Marom G, et al. Fully coupled fluid-structure interaction model of congenital bicuspid aortic valves: effect of asymmetry on hemodynamics. Med Biol Eng Comput. 2013;51(8):839–48.
68. Katayama S, et al. Bicuspid aortic valves undergo excessive strain during opening: a simulation study. J Thorac Cardiovasc Surg. June 2013;145(6):1570–6.

69. Weinberg EJ, Kaazempur Mofrad MR. A multiscale computational comparison of the bicuspid and tricuspid aortic valves in relation to calcific aortic stenosis. J Biomech. 2008;41 (16):3482–7.
70. Atkins SK, et al. Bicuspid aortic valve hemodynamics induces abnormal medial remodeling in the convexity of porcine ascending aortas. Biomech Model Mechanobiol. 2014;13 (6):1209–25.
71. Wang Q, Sirois E, Sun W. Patient-specific modeling of biomechanical interaction in transcatheter aortic valve deployment. J Biomech. 2012;45(11):1965–71.
72. Genereux P, et al. Clinical outcomes after transcatheter aortic valve replacement using valve academic research consortium definitions: a weighted meta-analysis of 3,519 patients from 16 studies. J Am Coll Cardiol. 2012;59(25):2317–26.
73. Auricchio F, et al. A computational tool to support pre-operative planning of stentless aortic valve implant. Med Eng Phys. 2011;33(10):1183–92.
74. Auricchio F, et al. Patient-specific simulation of a stentless aortic valve implant: the impact of fibres on leaflet performance. Comput Methods Biomech Biomed Engin. 2014b;17(3):277–85.
75. Auricchio F, et al. Simulation of transcatheter aortic valve implantation: a patient-specific finite element approach. Comput Methods Biomech Biomed Engin. 2014a;17(12):1347–57.
76. Labrosse MR, et al. Modeling leaflet correction techniques in aortic valve repair: A finite element study. J Biomech. 2011;44(12):2292–8.
77. Boodhwani M, et al. Repair of aortic valve cusp prolapse. Multimed Man Cardiothorac Surg. 2009;2009(702):mmcts.2008.003806.
78. Kunzelman KS, Cochran R. Stress/strain characteristics of porcine mitral valve tissue: parallel versus perpendicular collagen orientation. J Card Surg. 1992;7(1):71–8.
79. Kunzelman KS, et al. Finite element analysis of the mitral valve. J Heart Valve Dis. 1993;2 (3):326–40.
80. Kunzelman K, et al. Replacement of mitral valve posterior chordae tendineae with expanded polytetrafluoroethylene suture: a finite element study. J Card Surg. 1996;11(2):136–45.
81. Kunzelman KS, Reimink MS, Cochran RP. Flexible versus rigid ring annuloplasty for mitral valve annular dilatation: a finite element model. J Heart Valve Dis. 1998a;7(1):108–16.
82. Kunzelman KS, Quick DW, Cochran RP. Altered collagen concentration in mitral valve leaflets: biochemical and finite element analysis. Ann Thorac Surg. 1998b;66(6):S198–205.
83. Kunzelman KS, Einstein DR, Cochran RP. Fluid-structure interaction models of the mitral valve: function in normal and pathological states. Philos Trans R Soc Lond Ser B Biol Sci. 2007;362(1484):1393–406.
84. Prot V, Skallerud B. Nonlinear solid finite element analysis of mitral valves with heterogeneous leaflet layers. Comput Mech. 2008;43(3):353–68.
85. Skallerud B, Prot V, Nordrum IS. Modeling active muscle contraction in mitral valve leaflets during systole: a first approach. Biomech Model Mechanobiol. 2011;10(1):11–26.
86. Hunter PJ, Nash MP, Sands GB. Computational electromechanics of the heart. Comput Biol Heart. 1997;12:347–407.
87. Rachev A, Hayashi K. Theoretical study of the effects of vascular smooth muscle contraction on strain and stress distributions in arteries. Ann Biomed Eng. 1999;27(4):459–68.
88. Einstein DR and Del Pin F. Int. J. 2010;17(6):950–5.
89. Avanzini A. A computational procedure for prediction of structural effects of edge-to-edge repair on mitral valve. J Biomech Eng. 2008;130(3):031015.
90. Pham T, Sun W. Material properties of aged human mitral valve leaflets. J Biomed Mater Res A. 2014;102(8):2692–703.
91. Lee C-H, et al. An inverse modeling approach for stress estimation in mitral valve anterior leaflet valvuloplasty for in-vivo valvular biomaterial assessment. J Biomech. 2014;47 (9):2055–63.
92. Tang D, et al. Two-layer passive/active anisotropic FSI models with fiber orientation: MRI-based patient-specific modeling of right ventricular response to pulmonary valve insertion surgery. Mol Cell Biomech. 2007;4(3):159–76.

93. Mansi T, et al. Virtual pulmonary valve replacement interventions with a personalised cardiac electromechanical model. In: Recent Advances in the 3D Physiological Human. New York: Springer; 2009. p. 75–90.
94. Wafae N, et al. Anatomical study of the human tricuspid valve. Surg Radiol Anat. 1990;12(1):37–41.
95. Dumont K, et al. Comparison of the hemodynamic and thrombogenic performance of two bileaflet mechanical heart valves using a CFD/FSI model. J Biomech Eng. 2007;129(4):558–65.
96. Borazjani I, Ge L, Sotiropoulos F. High-resolution fluid-structure interaction simulations of flow through a bi-leaflet mechanical heart valve in an anatomic aorta. Ann Biomed Eng. 2010;38(2):326–44.
97. Choi YJ, Vedula V, Mittal R. Computational study of the dynamics of a bileaflet mechanical heart valve in the mitral position. Ann Biomed Eng. 2014;42(8):1668–80.
98. Sotiropoulos F, Borazjani I. A review of state-of-the-art numerical meth- ods for simulating flow through mechanical heart valves. Med Biol Eng Comput. 2009;47(3):245–56.
99. Guivier-Curien C, Deplano V, Bertrand E. Validation of a numerical 3-D fluid-structure interaction model for a prosthetic valve based on experimental PIV measurements. Med Eng Phys. 2009;31(8):986–93.
100. Nobili M, et al. Numerical simulation of the dynamics of a bileaflet prosthetic heart valve using a fluid-structure interaction approach. J Biomech. 2008;41(11):2539–50.
101. De Hart J, et al. A computational fluid-structure interaction analysis of a fiber- reinforced stentless aortic valve. J Biomech. 2003;36(5):699–712.
102. Haj-Ali R. Structural simulations of prosthetic tri-leaflet aortic heart valves. J Biomech. 2008;41:1510.
103. Sirois E, Sun W. Computational evaluation of platelet activation induced by a bioprosthetic heart valve. Artif Organs. 2010;35(2):157.
104. Martin C, Sun W. Simulation of long-term fatigue damage in bioprosthetic heart valves: effects of leaflet and stent elastic properties. Biomech Model Mechanobiol. 2014;13(4):759–70.
105. Kim H, et al. Dynamic simulation of bioprosthetic heart valves using a stress resultant shell model. Ann Biomed Eng. 2008;36:262–75.
106. Tzamtzis S, et al. Numerical analysis of the radial force produced by the Medtronic-CoreValve and Edwards-SAPIEN after transcatheter aortic valve implantation (TAVI). Med Eng Phys. 2013;35(1):125–30.
107. Sun W, Li K, Sirois E. Simulated elliptical bioprosthetic valve deformation: implications for asymmetric transcatheter valve deployment. J Biomech. 2010;43(16):3085–90.
108. Wang L, et al. Computational simulation of oxygen diffusion in aortic valve leaflet for tissue engineering applications. J Heart Valve Dis. 2008;17(6):700.
109. Driessen NJB, et al. Modeling the mechanics of tissue-engineered human heart valve leaflets. J Biomech. 2007;40(2):325–34.
110. Hasan A, et al. Biomechanical properties of native and tissue engineered heart valve constructs. J Biomech. 2014;47(9):1949–63.
111. Nørgaard BL, et al. Diagnostic performance of noninvasive fractional flow reserve derived from coronary computed tomography angiography in suspected coronary artery disease: the NXT trial (Analysis of Coronary Blood Flow Using CT Angiography: Next Steps). J Am Coll Cardiol. 2014;63(12):1145–55.
112. Xu C, et al. A novel approach to in vivo mitral valve stress analysis. Am J Phys Heart Circ Phys. 2010;299(6):H1790–4.
113. Luo Z, et al. Intra-operative 2-D ultrasound and dynamic 3-D aortic model registration for magnetic navigation of transcatheter aortic valve implantation. IEEE Trans Med Imaging. 2013;32(11):2152–65.

# Biomechanics and Modeling of Tissue-Engineered Heart Valves

**T. Ristori, A. J. van Kelle, F. P. T. Baaijens, and S. Loerakker**

**Abstract** Heart valve tissue engineering (HVTE) is a promising technique to overcome the limitations of currently available heart valve prostheses. However, before clinical use, still several challenges need to be overcome. The functionality of the developed replacements is determined by their biomechanical properties and, ultimately, by their collagen architecture. Unfortunately, current techniques are often not able to induce a physiological tissue remodeling, which compromises the long-term functionality. Therefore, a deeper understanding of the process of tissue remodeling is required to optimize the phenomena involved via improving the current HVTE approaches. Computational simulations can help in this process, being a valuable and versatile tool to predict and understand experimental results. This chapter first describes the similarities and differences in functionality and biomechanical properties between native and tissue-engineered heart valves. Secondly, the current status of computational models for collagen remodeling is addressed and, finally, future directions and implications for HVTE are suggested.

**Keywords** Tissue engineering · Heart valve · Remodeling · Collagen · Stress fibers · Computational · Mathematical model · Biomechanics

## Abbreviations

ECM    Extracellular matrix
HVTE   Heart valve tissue engineering

---

T. Ristori and A. J. van Kelle contributed equally to this work.

T. Ristori · A. J. van Kelle · F. P. T. Baaijens · S. Loerakker (✉)
Department of Biomedical Engineering, Eindhoven University of Technology, Eindhoven, The Netherlands

Institute for Complex Molecular Systems, Eindhoven University of Technology, Eindhoven, The Netherlands
e-mail: T.Ristori@tue.nl; A.J.V.Kelle@tue.nl; F.P.T.Baaijens@tue.nl; S.Loerakker@tue.nl

SF      Stress fiber
TE      Tissue engineering
TEHV    Tissue-engineered heart valves

# 1 Introduction

Annually, an estimated 300.000 heart valve replacements are performed worldwide [1]. Current heart valve replacements can be categorized into two groups: mechanical and bioprosthetic valves. Although both of these prostheses are lifesaving, they do exhibit a number of drawbacks. Mechanical valves have shown good structural durability, but they are prone to thromboembolic events, making lifelong treatment with anticoagulants inevitable [2]. On the other hand, while bioprosthetic valves do not require anticoagulation therapy, they do suffer from structural valve degeneration and the associated requirement for reoperation [3, 4]. Besides these type-specific drawbacks, both heart valve replacements lack the ability to grow, repair, and remodel in response to changes in their environment, which especially imposes problems for younger patients. A promising alternative for the currently available prostheses are tissue-engineered heart valves (TEHVs), as these living valves do have the intrinsic potential to grow, repair, and remodel. One can roughly distinguish two different approaches in heart valve tissue engineering (HVTE): in vitro and in situ approaches.

## 1.1 In Vitro Heart Valve Tissue Engineering

The in vitro HVTE approach is based on the classical tissue engineering (TE) paradigm, in which extracellular matrix (ECM) producing cells are seeded onto a scaffold. This cell-scaffold construct is then cultured in vitro in a bioreactor, during which various stimuli are provided to the cells to trigger them to synthesize and/or remodel their ECM. A particularly important aspect in this is the choice of scaffold. Decellularized xenografts or homografts would seem the obvious choice [5–7], as these grafts are geometrically and structurally very similar to native heart valves. However, these techniques suffer from some serious drawbacks: the use of xenografts has a zoonoses-associated risk, while the use of homografts suffers from a limited availability of donor material. Moreover, both decellularized grafts have limited recellularization potential [8]. An alternative scaffold for in vitro tissue engineering is biocompatible synthetic materials which, depending on the choice of material, offer a broad range of scaffold properties. In 1995, Shinoka et al. [9] was the first to show the potential of this classical in vitro approach, replacing a single valve leaflet with a TE construct. Since 2000, several studies performed full heart valve replacements with TEHVs, using different cell types and scaffold materials [10–13]. Unfortunately, this approach does require a period of in vitro cell culture and results in a living biological prosthesis prior to implantation, which complicates the regulatory processes. Moreover, this technique does not offer off-the-shelf capability, which makes this technique clinically less attractive.

## 1.2 In Situ Heart Valve Tissue Engineering

An alternative to the in vitro approach is the in situ HVTE approach. This technique does not rely on the classical TE paradigm, but merely uses the human body's own regenerative potential to regenerate a living, functional, heart valve. Key in this technique is the choice of scaffold, as no in vitro culture phase is used. This scaffold should be able to allow cellular infiltration and retention in vivo, while also inducing subsequent remodeling. Again, decellularized xenografts or homografts could be used as a scaffold [14–16] but these still suffer from the previously mentioned drawbacks. Decellularized in vitro tissue-engineered heart valves [8, 17] overcome this drawback, but nevertheless these valves still fail to address the issues associated with in vitro culture. The use of synthetic materials is more appealing, as the scaffold properties are well defined and can be tuned to suit the intended application. This, with the off-the-shelf availability, makes the in situ approach an attractive candidate for HVTE.

## 1.3 Limitations and Challenges

Both the classical in vitro and the in situ approach have shown promising results. However, there are still several challenges which need to be overcome. First, implanted TEHV showed leaflet retraction after in vivo remodeling, resulting in valvular insufficiency [18, 19]. Second, obtaining the native layered structure, lined with endothelial cells, has proven to be an issue [8]. Finally, current clinical studies focus mainly on the relatively short-term development of the TEHV in vivo, so the long-term tissue growth and remodeling remains yet to be elucidated. We hypothesize that any of these challenges that need to be resolved mainly arise due to a mismatch between the final structural properties of the TEHV and its native counterpart. Clearly, the starting configurations of HVTE strategies are different compared to the native situation. In order to obtain this native situation in the TEHV, inducing physiological tissue remodeling is of vital importance. Existing research recognizes that tissue remodeling of TEHVs is, besides biochemical factors, highly affected by mechanical cues [20, 21]; however, the underlying mechanisms are largely unknown. A first step towards improving TEHV structural properties is understanding the underlying mechanisms of tissue remodeling. Experimental work is a valuable tool in unraveling these underlying mechanisms; however, this tends to be very costly and time expensive. Moreover, ethical concerns naturally arise in case of animal experimentation. An attractive contribution to experimental research is the use of computational modeling. Computational models have the potential to improve the analysis of experimental observations, identify the underlying mechanisms, and predict how these mechanisms will affect the in vivo remodeling process of TEHVs with complex geometries and loading conditions.

## 1.4 Outline

In this chapter we will evaluate the history and current stance of computational modeling of tissue remodeling in TEHVs. First, an overview of the biomechanical properties of the native and tissue-engineered heart valve will be given. Next, phenomenologically motivated models are discussed (Sect. 3), followed by more complex biologically motivated models (Sects. 4 and 5). Finally, we will address the future directions of modeling heart valve tissue remodeling.

## 2 Biomechanical Properties of Native and Tissue-Engineered Heart Valves

The ultimate goal of HVTE is to create living heart valve substitutes with the same hemodynamics and biological functionality as their native counterparts. This functionality is determined by the valve's structure, and consequent biomechanical properties. Investigating the differences in structure and biomechanical properties of native valves and TEHVs is thus crucial to improve current HVTE strategies.

## 2.1 Heart Valve Structure, Function, and Biomechanics

Aortic valves are composed of three thin and half-moon-shaped leaflets attached in a crown-shape fashion to the aortic root [22]. The points of the attachment that two leaflets have in common are called commissures. When the valve is closed, the edges of the leaflets not attached to the aortic root (free edges) form a small overlap of tissue (coaptation area) that provides a safety margin for valve closure [23]. Finally, the leaflets have a body of tissue between the attachment and the free edges (belly), which bears the largest part of the pressure exerted on the leaflet [24].

The primary function of heart valves, which is to impede blood backflow, is achieved due to highly optimized biomechanical properties. Throughout the cardiac cycle, heart valves are subjected to a wide range of mechanical stimuli [25], namely shear stress during systole, bending strain in between systole and diastole, and pressure during diastole. In the context of functional mechanics, we focus our attention on the biomechanical responses to blood pressure. Native leaflets are anisotropic and highly nonlinear: they are stiffer in the circumferential direction

**SCHEME OF HEART VALVE RESPONSE TO STRETCH**

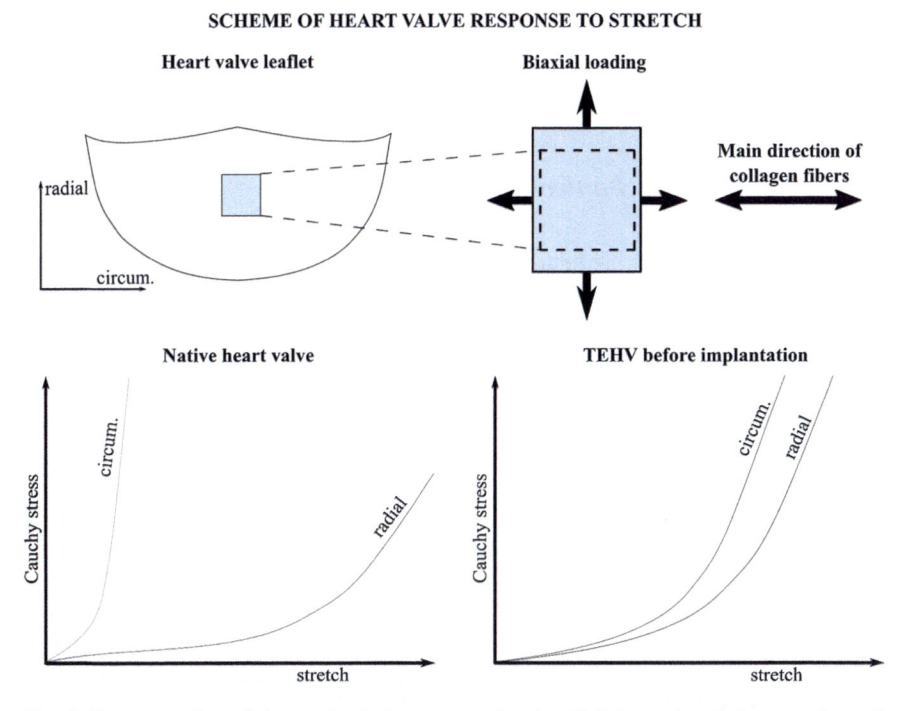

**Fig. 1** Representation of the mechanical response of native (left bottom) and tissue-engineered (right bottom) heart valves to biaxial loading (as induced by transvalvular pressure)

than the radial one, and respond to stress with a nonlinear increase of stiffness as tension increases [26, 27] (Fig. 1). Due to this biomechanical behavior, at the end of the systolic phase, the leaflets can enlarge mainly radially to close the valve, restrict circumferential strain, and then rapidly stop their deformation ensuring a proper valve closure [28]. So, there is a clear consistency between biomechanical properties and functionality.

The biomechanical behavior of aortic leaflets is determined by their components, which are mainly valvular interstitial cells, collagen, elastin, glycosaminoglycans (GAGs), and proteoglycans (PGs) [29]. The aortic leaflets have three layers, specifically the fibrosa, the spongiosa, and the ventricularis, containing different proteins: the fibrosa is mainly composed of bundles of circumferentially oriented collagen fibers [30]; the ventricularis consists of a considerable amount of elastin disposed in a dense sheet [31]; and the spongiosa comprises predominantly GAGs, PGs, and some collagen and elastin connecting the other two layers [23, 32, 33]. The biomechanical behavior of the leaflets is mostly determined by the collagen present in the fibrosa. Collagen fibers are known to be crimped in the absence of hemodynamic loading, and to withstand only tensile stress exerted along their direction. Due to their initially wavy configuration, they exert low resistance against low levels of strain while, at higher levels, they get fully extended and prevent further strain, leading to the tissue's nonlinear mechanical behavior. Similarly, the tissue anisotropy is caused by the circumferential alignment of the collagen fibers, which gives the tissue a higher stiffness in that direction.

## 2.2 TEHV Biomechanical Properties

Current HVTE strategies are focusing on replicating the biomechanical properties of native heart valves. As the collagen is a strong determinant of the biomechanical behavior of the leaflets and is the main load-bearing structure present in cardiovascular tissues, many studies have focused on creating TEHVs with a dense and organized collagen network [11, 34–41]. Mechanical tests performed on TEHVs have demonstrated that, before implantation, these constructs exhibit a biomechanical behavior which is similar to the native aortic counterpart, although the degrees of nonlinearity and anisotropy are less pronounced [10, 17, 39, 42–47] (Fig. 1). Most likely, these differences in mechanical properties are caused by insufficient collagen cross-linking [42] and collagen alignment [35] in TEHVs, characteristics that may, however, improve after implantation due to in vivo remodeling. In previous studies, TEHVs have been implanted into animal models in the pulmonary position, where they have shown good functionality in the short term. However, medium- and long-term follow-up results showed that many valves were prone to progressive leaflet thickening and/or retraction, and consequently valvular insufficiency [18, 47–49]. Concluding, the initial configuration of previous strategies did not induce physiological tissue remodeling in most cases. These experimental results highlight the necessity to increase our understanding of the underlying remodeling processes, to define a rational valve design that will induce physiological remodeling.

## 2.3 Computational Simulations of TEHVs

Computational simulations play a significant role in determining the optimal initial configurations for TEHVs. For instance, using computational modeling, Loerakker and colleagues [50, 51] demonstrated that the geometry of the leaflets can have a major influence on the mechanical behavior of the valve, and therefore also the mechanical stimuli sensed by the tissue, and that a suboptimal leaflet design can be one of the causes of in vivo leaflet retraction. In addition to knowing how the initial properties of the engineered tissues affect the mechanics of the heart valve, understanding and being able to predict the subsequent tissue remodeling process as a function of the initial valve design is essential as well. Most of the previously developed tissue remodeling algorithms focused on predicting collagen remodeling, as this is the main determinant of the mechanical behavior of most tissues. The first models that were developed to predict collagen remodeling mainly adopted a phenomenological approach to identify which mechanical parameters are most relevant for the remodeling process. An overview of these models and their similarities and differences will be given in the next section.

## 3 Mathematical Models of Collagen Remodeling

### 3.1 Collagen Architecture in Native Aortic Heart Valves

The collagen architecture in native aortic heart valve leaflets is highly organized and mainly circumferentially orientated (Fig. 2). Collagen fibers are not homogenously distributed over the leaflets, but arranged into bundles [52]. The macroscopic configuration of these bundles has often been compared to a hammock-like structure, as they seem to depart mainly from common points (the two commissure points) and branch into various bundles in between. The circumferential alignment of the bundles is evident along the free edges while, due to the branching, this orientation may be less apparent in the belly. However, microscopic examination has shown that also the collagen fibers present in this region of the leaflets are mainly oriented circumferentially [53–55].

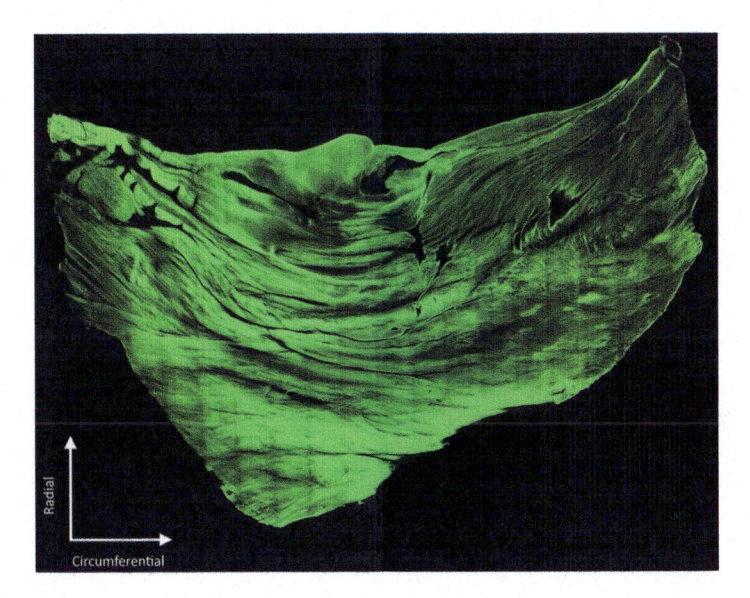

**Fig. 2** Collagen architecture in the fibrosa layer of a native human heart valve leaflet. Courtesy of Pim Oomen

## 3.2 Collagen Remodeling in Response to Mechanical Stimuli

The collagen fibers present in heart valves remodel in response to changing demands, where we define collagen remodeling as the process of realignment, degradation, and synthesis of collagen fibers. Several studies have suggested that the level of organization of the collagen architecture is proportional to the magnitude of the pressure exerted on the closed valve. For instance, pulmonary heart valves experience a lower pressure than their aortic counterpart and, although they have a similar geometry, they exhibit lower values of collagen content and alignment [53]. The same phenomenon can be observed comparing aortic valves of different ages. The aortic pressure increases with age and, consequently, the collagen structure in these heart valves adapts to these changes. Early fetal heart valves exhibit loosely arranged collagen fibers that increase in content in the first weeks of development. After birth, the circumferential alignment of the collagen fibers continues to increase with age and with the associated increase in transvalvular pressure difference, until the adult configuration is acquired [53, 56].

The influence of mechanical stimuli on collagen remodeling has been demonstrated also via in vitro experiments. For instance, it has been observed that in cell-populated collagen gels and engineered cardiovascular tissues, the collagen organization is isotropic when these constructs are biaxially constrained [57, 58], whereas in the case of uniaxial constraints, the collagen fibers orient along the constrained direction [57, 59, 60]. Additionally, Rubbens et al. [57] demonstrated that dynamic stretch increases the rate of collagen realignment and the collagen content along the stretched direction. Concluding, mechanical stimuli play a significant role in collagen remodeling, and need to be understood to successfully engineer tissues with a physiological and functional collagen network.

## 3.3 Early Computational Models

In addition to experimental work, computational models have been developed to understand the mechanisms of collagen remodeling and to predict how the final collagen organization depends on factors such as tissue geometry, mechanical properties, and external mechanical loads. Early computational studies on collagen remodeling have investigated which mechanical stimuli are the most important determinants in this process, focusing on predicting the (re)alignment of collagen fibers [61–72]. Various phenomenological hypotheses have been tested, with stress and strain often chosen as driving mechanical stimuli for this phenomenon. To validate the hypotheses for collagen remodeling, the model predictions have been compared with the collagen architecture observed in native cardiovascular tissues, such as heart valves and arteries.

## 3.4 Realignment with Principal Loading Directions

As one of the first attempts, Boerboom et al. [61], Driessen et al. [66], and Driessen et al. [63] predicted collagen remodeling considering a discrete number of fiber directions. They hypothesized that collagen fibers align towards the principal strain directions and that their amount increases with the strain magnitude (Fig. 3). Their models were able to successfully predict the mainly circumferential collagen organization present in native heart valve leaflets. Computational models with similar hypotheses, but with the stress chosen as mechanical stimulus driving the remodeling, have been used to successfully study the tissue remodeling in other tissues, such as transversely isotropic tissues and tendons [73, 74].

Conversely, the numerical algorithms resulting from these first hypotheses appeared unable to predict the collagen distribution present in native arterial walls, which is characterized by collagen fibers grouped in two distributions of which the mean orientations form a double-helical structure, with increasing pitch from inside to outside [75]. In arteries, the principal strain directions are the axial and the circumferential ones. Therefore, a numerical algorithm developed using the previous hypotheses would erroneously predict collagen fibers oriented axially or

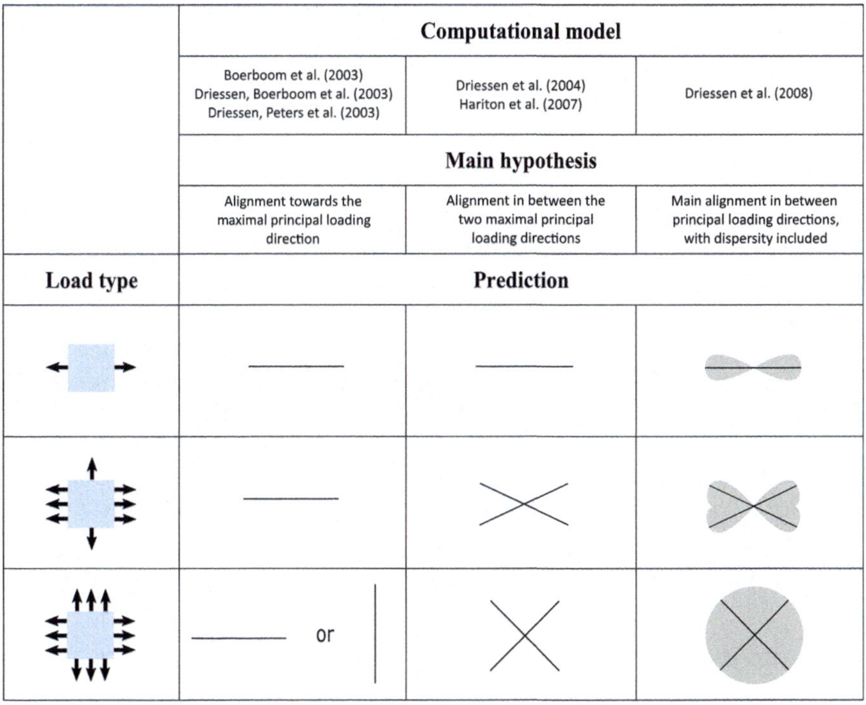

|  | Computational model | | |
|---|---|---|---|
|  | Boerboom et al. (2003) Driessen, Boerboom et al. (2003) Driessen, Peters et al. (2003) | Driessen et al. (2004) Hariton et al. (2007) | Driessen et al. (2008) |
|  | **Main hypothesis** | | |
|  | Alignment towards the maximal principal loading direction | Alignment in between the two maximal principal loading directions | Main alignment in between principal loading directions, with dispersity included |
| **Load type** | **Prediction** | | |
| → ■ → | — | — | ⋈ |
| ↕ ■ ↔ | — | ✕ | ⋈ |
| ↕↕ ■ ↔↔ | — or \| | ✕ | ⊗ |

**Fig. 3** Main hypothesis of early computational models for collagen remodeling and the predicted collagen fiber orientation under uniaxial, biaxial, or equibiaxial loading

circumferentially. Since heart valves and arteries roughly consist of the same components [76], the collagen fiber architecture present in these two cardiovascular tissues should be predictable with similar collagen remodeling laws. As a consequence, the previous hypotheses needed to be revised to successfully predict collagen remodeling in a broader range of cardiovascular tissues.

## 3.5 Realignment in Between Principal Loading Directions

Driessen et al. [64, 67] were able to predict the collagen architecture present in both native arterial walls and heart valves using a single type of remodeling law, by considering two fiber directions and hypothesizing that collagen fibers prefer to align in between the principal strain directions (Fig. 3). In addition, similar results for the collagen architecture in the arterial wall could be obtained choosing the stress as the driving mechanical stimulus [68, 69].

The preferred directions introduced in these models depend on the difference between the magnitude of the two maximal principal strains or stresses: the larger this difference, the closer the preferred direction is to the maximal principal loading direction (Fig. 3). As a result, in the case of uniaxial strain, the preferred direction and the strained direction coincide. Therefore, the predictions of this kind of mathematical models can be reformulated as follows: in the case of uniaxial strain, collagen orients towards the strained direction; in the case of biaxial strain, collagen aligns in between the principal loading directions. In the context of heart valves, since the commissure region is known to be uniaxially strained while the belly is biaxially loaded [55], these hypotheses predict a circumferential alignment of collagen close to the free edges and a more branched orientation in the belly [64], which is in accordance with experimental results [53, 55, 77].

Nevertheless, these models still have some limitations, mostly related to the use of only two fiber directions. In the belly of the leaflets, using only two fiber directions, these models predict the branching of the collagen fibers with the two fiber families diverging from the circumferential direction [64] (Fig. 3). Although the branching of collagen fibers itself is consistent with experimental observations, the complete lack of collagen fibers in the circumferential direction is not physiological [53, 55]. Further limitations can be found considering equibiaxially loaded tissues. The isotropic collagen organization observed in these tissues [78] cannot be predicted with computational models that only consider two fiber directions for the collagen network.

## 3.6 Inclusion of Fiber Dispersity

In a successive study, Driessen et al. [65] overcame the limitations of the previous models by introducing a fiber dispersity for the two collagen fiber families. Similar to the previous mathematical models, the main fiber directions were hypothesized to align in between the principal loading directions, while the fiber dispersity was hypothesized to be inversely proportional to the difference of strain or stress magnitude between the two maximal principal loading directions (Fig. 3).

The remodeling laws of this model are conceptually equivalent to the previously proposed laws [67, 68], but the use of a fiber dispersity presents significant advantages over the previous approach. For instance, this numerical algorithm is able to predict the isotropic collagen distribution in tissues subjected to equibiaxial loads. In that case the fiber dispersity is predicted to be maximal, and thus a main direction of the fibers is predicted to be absent. Moreover, Driessen et al. [65] obtained improved results for the simulation of collagen remodeling in native heart valves. The use of fiber dispersity enabled the prediction of a large amount of collagen in the circumferential direction also in the belly of the leaflets, which had not been observed in the previous computational results [64]. The main directions of the two fiber families at this location were still predicted to be in between the circumferential and radial direction. However, due to the large fiber dispersity, the two fiber families always overlapped along the circumferential direction. Consequently, for the collagen architecture as a whole, their contributions in this direction had to be summed and thus a high content of collagen in the circumferential direction could be successfully predicted (Fig. 3).

## 3.7 Main Limitations

Although the models described in this section are successful in predicting the collagen structure in various cardiovascular tissues, they are mainly phenomenological in nature and do not include biological motivations for the collagen remodeling process. The inclusion of the biological phenomena involved in collagen remodeling in computational models is, however, necessary to understand the underlying mechanisms and further enhance the predictive capacity of the models. For example, next to mechanical cues, cells are the main affecters of collagen remodeling, since they are responsible for collagen production and cell-produced enzymatic collagen degradation. Moreover, cells have shown to alter the apparent stiffness and prestress of the tissue, by pulling on their surroundings, including collagen fibers [79]. These effects need to be accounted for in computational models of tissue remodeling, as it is of major importance for obtaining accurate predictions of the mechanical state of the tissue, which in turn will drive the remodeling process in the model.

In the past years, considerable efforts have been made in developing computational models that can predict the contractile stresses exerted by cells under various conditions. The integration of such algorithms and biologically motivated models of collagen remodeling has the potential to further unravel the mechanobiological mechanisms involved in tissue remodeling. The next section will give an overview of the computational models that are able to predict the orientation of actin stress fibers, which are the major source of cellular contractility and tissue prestress.

## 4 Modeling Stress Fiber Remodeling

### 4.1 Actin Stress Fibers

Actin stress fibers (SFs) are filaments of F-actin cross-linked by α-actinin and bundled by myosin II. These acto-myosin bundles are fundamental for the interaction of cells with their surroundings, because cells are known to synthetize collagen [80] and exert contractile forces onto their surrounding material along the directions of these fibers [81]. These cellular forces play a pivotal role in the differentiation and orientation of cells, and the organization of the extracellular matrix [79, 82–84]. Furthermore, it is well established that cellular forces are among the most important contributors to tissue prestress [85]. For these reasons, to enhance the computational models for tissue remodeling, including an algorithm that is able to predict the orientation of and stress exerted by SFs, is crucial.

### 4.2 SF Remodeling

Similar to collagen fibers, SFs remodel in response to mechanical stimuli. Several studies have shown that, when cells are cultured on a substrate with anisotropic stiffness or embedded in a uniaxially constrained collagen gel, their SFs align along the stiffest or constrained direction [83, 84]. Conversely, when seeded on a stiff substrate or a biaxially constrained collagen gel that is cyclically strained in one direction, cells orient perpendicular to the strain [83, 86–88]. This phenomenon is known as strain avoidance, and depends on the frequency and amplitude of the cyclic strain, where larger frequencies or amplitudes lead to higher levels of alignment [87, 89]. Importantly, recent studies have shown that SFs do not orient perpendicular to the direction of uniaxial cyclic stretching when cells are seeded on a low stiffness substrate or embedded in a uniaxially constrained collagen gel. In these conditions, SFs prefer to align along the direction of the applied stretch [83, 90, 91]. This phenomenon seems to highlight the importance of SF stress for the organization of SFs. In fact, one possible explanation for these observations is the following: for low stiffness materials, cells cannot exert high tension perpendicular to the cyclic stretch because the material is too compliant to contrast the cellular tension, and consequently the final SF architecture do not orient in this direction; on the other hand, in the direction of the cyclic stretch, cells sense this mechanical stimulus as a mechanical resistance to their cellular forces, and thus they accordingly realign their SFs along the direction of the applied stretch.

In addition to mechanical stimuli, SFs also respond to topographical patterns, which is known as the contact guidance phenomenon. We refer to contact guidance when SFs arrange themselves along geometrical patterns such as nano/microgrooves on 2D substrates [92] or fibrous structures in 3D environments [93, 94]. The effects of strain anisotropy and contact guidance can be in competition. For instance, this competition can occur when cells are seeded on microgrooved substrates stretched along the direction of these topographical patterns. In these conditions, contact guidance would induce an alignment parallel to the direction of the microgrooves, while the cyclic strain would induce an alignment in the perpendicular direction. Prodanov et al. [95] demonstrated that the final SF configuration resulting from this competition depends on several factors, such as the microgroove size and the frequency of the cyclic strain. Similarly, in 3D, it has been observed that contact guidance has a larger influence than strain anisotropy when cells are embedded in a gel with a high density of collagen [93], or cultured on polymeric fibrous scaffolds [96].

## 4.3 Computational Models for SF Remodeling

To simulate and understand the processes of SF remodeling and cellular contraction, several computational models have been proposed. As for mathematical models for collagen remodeling, stress and strain have often been chosen as driving mechanical stimuli for SF remodeling. In addition, since it appears that strain rate influences the SF stress, also this parameter has been included in the proposed evolution laws. In particular, it has been observed that the $\alpha$-actinin and myosin present in SFs have an organization similar to muscle sarcomeres [81]. Therefore, the law of Hill [97] that describes the relationship between the strain rate and the stress for sarcomeric structures has often been used in mathematical models to predict stress exerted by SFs. For the SF turnover, two main hypotheses have been proposed, suggesting that cells rearrange their SF network to maintain either a stress or strain homeostasis. Most of the proposed models are phenomenological and have been inspired by experimental observations, but a trend towards more physically motivated models can be recognized.

## 4.4 Stress Homeostasis

Deshpande et al. [98] developed one of the first computational models for SF remodeling and cellular contraction. They based their model on the assumptions that cells avoid fast compression and seek for a stress homeostasis (Table 1). In particular, they hypothesized that: SF polymerization is isotropic and activated by a cellular signal; the SF stress can be described by a Hill-type law dependent on strain rate, in such a way that stress decreases when SFs undergo fast compression; in response to a decrease in stress, cells depolymerize the SFs along that direction. Interestingly, although this model did not include a dependency on strain, it could successfully predict the SF organization for cells under static conditions or subjected to dynamic stretch on a stiff substrate through the use of a strain rate dependency [98, 105, 106].

Table 1 Overview of hypotheses, predictions, and main limitations of the SF remodeling models in this section

| Model | Main hypothesis | SF stress dependency | SF assembly | SF disassembly | Reference state | Predictions | Main limitation |
|---|---|---|---|---|---|---|---|
| Deshpande et al. [98] | Stress homeostasis | Strain rate | Isotropic | Occurring for stress disequilibrium | Not specified | Strain avoidance dependent on strain rate | No strain dependency |
| Kaunas and Hsu [99] | Strain homeostasis | Not specified | Isotropic | Occurring for strain disequilibrium | Initial configuration | Strain avoidance dependent on stretch magnitude | No strain rate dependency |
| Hsu et al. [100] | Strain homeostasis | Not specified | Isotropic | Occurring for strain disequilibrium | Changing over time | Strain avoidance dependent on strain magnitude and rate | No SF stress specified |
| Kaunas et al. [101] | Stress and strain homeostasis | Strain | Isotropic | Occurring for stress disequilibrium | Changing over time | Strain avoidance dependent on strain magnitude and rate | Mainly phenomenological model |
| Vernerey and Farsad [102] | Stress dependent SF assembly | Strain and strain rate | Increasing with SF stress | Isotropic | Initial configuration | Strain avoidance dependent on stretch magnitude | Cell memory of stretch history |
| Obbink-Huizer et al. [90] | Stress dependent SF assembly | Strain and strain rate | Increasing with SF active stress | Isotropic | Initial configuration | Strain avoidance dependent on strain magnitude and rate | Cell memory of stretch history |
| Foucard and Vernerey [103] | Thermodynamic equilibrium | Strain and strain rate | Not specified | Not specified | Initial configuration | Strain avoidance dependent on strain magnitude and rate | Cell memory of stretch history |
| Vigliotti et al. [104] | Thermodynamic equilibrium | Strain and strain rate | Isotropic | Occurring for thermodynamic disequilibrium | Changing over time | Strain avoidance dependent on strain magnitude and rate | Many parameters |

In a successive study, this model has also been enhanced with the inclusion of differential equations for focal adhesion formation and diffusion of the intracellular signaling arising from these cellular points of attachment to the substrate [107, 108]. These enhancements have enabled the prediction of SF concentration at the cellular level [109, 110] and the simulation of contact guidance in 2D microgrooved substrates [111].

Unfortunately, due to the absence of a strain dependency, this model was unable to predict the dependency of strain avoidance on strain magnitude for constant strain rate, as was demonstrated experimentally by Faust et al. [89]. In these experiments, cells were seeded on an isotropic substrate and cyclically stretched at different strain magnitude but constant strain rate. In spite of this constant strain rate, the authors observed that the effects of strain anisotropy were larger when the strain magnitude increased.

## 4.5 Stress and Strain Homeostasis

Kaunas and Hsu [99] proposed a different hypothesis for SF remodeling, by assuming that is motivated by maintaining a strain homeostasis (Table 1). In particular, cells were hypothesized to depolymerize SFs when their strain differed from an optimal homeostatic level. The polymerization of SFs was assumed isotropic, and the total amount of SFs was kept locally constant. Since this mathematical model was dependent only on the strain experienced by SFs, it could simulate the increase of strain avoidance behavior with increasing strain magnitude, but not for increasing frequency. Therefore, Hsu et al. [100] enhanced the model by assuming that SFs are able to relax over time due to their viscoelastic characteristics, with an associated velocity of relaxation (Table 1). In this way, when the strain rate applied to the cells in the simulations was low compared to this relaxation, then SFs were able to follow the deformation without being subjected to depolymerization. Conversely, for high strain rates compared to the relaxation rate, cells preferred to orient their SFs away from that mechanical stimulus. With this enhancement, the mathematical model could successfully predict the effects that changes in strain rate have on the phenomenon of strain avoidance. However, since the magnitude of the stress exerted by the cells was not clearly specified, predicting larger stresses for stiffer materials and for directions with larger amounts of SFs was not possible.

In a successive study, Kaunas et al. [101] proposed a new model for SF turnover coupled with active cellular contraction, with the assumption that cells reorganize their SFs following the principles of both stress and strain homeostasis (Table 1). In their model, SFs depolymerized when the SF stress differed from a reference value, with this SF stress dependent on the level of strain experienced by SFs. In addition, similar to Hsu et al. [100], in this mathematical model the viscoelastic behavior of SFs was taken into account to explain the influence of strain rate on SF remodeling. Numerical simulations demonstrated that the model was indeed able to predict the final configuration of SFs for cells subjected to cyclic strain. Nevertheless, it is unclear if the model can simulate SF remodeling in cells residing in an environment that has an anisotropic stiffness.

## 4.6 SF Assembly Dependent on Strain and Strain Rate

In 2011, Vernerey and Farsad [102] developed a predictive model for SF remodeling and cellular contraction, in which they assumed that stress, strain, and strain rate affect SF polymerization instead of depolymerization (Table 1). They hypothesized that cells produce more SFs in directions where they exert more stress. Similar to the model of Kaunas et al. [101], SF stress was dependent on both the strain and the strain rate sensed by SFs. In particular, the stress was assumed to increase with the amount of strain in the SFs. The relationship between SF stress and strain rate was described by a Hill-type law as in the model of Deshpande et al. [98], with the exception that it was antisymmetric around zero. In addition, in contrast to the previous models, the model assumed that cells have a long-term memory of the stretch history, because the original configuration of the cell was used to define the strain stimulus.

With these features, the model was able to successfully predict SF remodeling occurring in cells that are statically cultured in an environment with anisotropic stiffness, and in cells that are cyclically strained. Nevertheless, the increase of strain avoidance for increased strain rate could not be captured due to the antisymmetry around zero of the Hill-law present in the model.

Due to this limitation, Obbink-Huizer et al. [90] proposed a new model for SF remodeling (Table 1). While the main hypotheses of this computational model were comparable with the model of Vernerey and Farsad [102], some modifications were introduced to the equations. For example, to avoid antisymmetry, the Hill-law was shifted in such a way that, similar to Deshpande et al. [98], SF compression caused decreased SF stress, while extension did not induce significant variations. Moreover, the SF polymerization was supposed to depend only on the stress actively exerted by the SFs, and not on the passive component of the SF stress. These adjustments were beneficial for the predictive potential of the model, which was now able to predict the SF final configuration as a result of several mechanical stimuli, such as anisotropic stiffness and cyclic strain with varying frequency and amplitude, using only a single set of parameter values.

Despite its high predictive potential, the hypotheses that are the basis of this model are mainly phenomenological and some of the consequences deriving from these assumptions are questionable. For example, it is unlikely that cells have a long-term memory of their stretch history. Therefore, to further increase the predictive potential of computational simulations, and to use the computational models as a tool to unravel the underlying mechanisms of SF remodeling process, physically motivated hypotheses should be introduced.

## 4.7 Model Based on Thermodynamics

The first promising computational models in this direction have recently been proposed by Foucard and Vernerey [103], and Vigliotti et al. [104], by introducing thermodynamical laws as driving mechanisms for SF remodeling (Table 1). In their studies, they defined chemical potentials for both unbound F-actin filaments and SFs, and they hypothesized that SF remodeling occurs when these two chemical potentials are not in equilibrium. In both models, SF stress and positive strain were assumed to stabilize SFs and thus diminish the SF chemical potential, although a different definition of strain experienced by SFs was used. While Foucard and Vernerey [103] used the usual definition of strain dependent on an original configuration such as in Vernerey and Farsad [102], Vigliotti et al. [104] introduced a new definition for the SF strain by hypothesizing that SFs change their reference configuration over time as a result of SF remodeling, with this last assumption appearing more biologically sound.

Both mathematical models exhibited excellent predictive potential both for static and dynamic conditions, and for materials with different stiffness. For instance, the model of Vigliotti et al. [104] was tested for a wide range of cyclic stretching profiles with promising results and, compared to the previous phenomenological model of Obbink-Huizer et al. [90], it showed improved predictions for different cyclic waveforms. Moreover, these models represent the first attempts to give an explanation for the biological mechanisms responsible for SF remodeling.

## 4.8 Main Challenges

The mathematical models described in this section mostly focus on the effects that mechanical cues have on SF remodeling, while only a few studies analyzed the effects that topological patterns have on SF networks. Currently the competition between strain avoidance and contact guidance has not been fully investigated through computational simulations. A first attempt has been proposed to investigate this phenomenon at the cellular level [112], but further studies could elucidate the underlying mechanisms and the factors involved.

Furthermore, in view of a development of numerical algorithms coupling both SF and collagen remodeling, the refinement of the computational models for SF remodeling should not lead to excessive computational costs. The processes of SF and collagen remodeling have different time scales, with SF remodeling occurring over minutes and collagen remodeling over days. Consequently, the prediction of the final SF architecture should be relatively fast, to enable the simulation of weeks of remodeling.

## 5 Prestress and Cell-Mediated Collagen Remodeling

The computational models described in Sect. 3 were able to successfully predict the collagen architecture in native cardiovascular structures in many cases. Yet, these phenomenological models lacked biological motivation for the proposed remodeling laws. An important part of remodeling is collagen turnover, which is mediated by cells; they deposit new collagen fibers and synthesize enzymes responsible for collagen degradation. Another important cellular aspect, besides mediating collagen remodeling, is the presence of cell-generated forces that induce collagen fiber realignment, tissue prestress and subsequent tissue compaction. Including cellular behavior into collagen remodeling algorithms can therefore greatly increase the predictive potential of the models, and help to better understand the underlying mechanisms of collagen remodeling.

### 5.1 Cell-Mediated Collagen Turnover

Cells embedded in the tissue form the bridge between mechanical stimuli and subsequent collagen turnover; in fact, cells sense mechanical loads and act accordingly to remodel the collagenous matrix. These mechanical stimuli influence three factors that are important in cell-mediated collagen turnover: collagen synthesis, cellular orientation, and cell-produced enzymatic collagen degradation.

First, new collagen fibers are synthesized (in the form of procollagen) and deposited by cells. In vitro experiments have demonstrated that fibroblasts embedded in cyclically strained tissues increase their mRNA expression of procollagen [113–115]. As a result, cells seeded on cyclically stretched scaffolds increase their collagen synthesis [34, 57].

Second, in the process of collagen synthesis, cellular orientation also plays a significant role. In fact, these new collagen fibers are primarily deposited in the direction in which cells are oriented [80]. The orientation of contractile cells is determined by the direction of SFs, which remodel in response to mechanical stimuli.

Finally, not only cell-mediated collagen synthesis and deposition are important for collagen remodeling, but also collagen degradation. Cells synthesize matrix metalloproteinases (MMPs) and tissue inhibitors of matrix metalloproteinases (TIMPs), which are the mediators of collagen degradation [116, 117]. Fibroblasts increase their expression of these MMPs and TIMPs in response to cyclic strain [118, 119]. In addition to cellular expression, strain also affects enzymatic collagen degradation. In particular, if collagen fibers get strained, they undergo a molecular change for which it has been hypothesized that it makes them less susceptible to MMP-mediated degradation [120–123]. In summary, it is apparent that mechanical stimuli influence cellular behavior which ultimately has significant consequences for collagen turnover and the resulting collagen architecture.

## 5.2 Cell-Mediated Prestress

In addition to driving the collagen turnover in response to mechanical stimuli, cells influence the biomechanical behavior of tissues and collagen realignment by applying traction forces to their surroundings. The presence of these forces has a direct effect on this biomechanical behavior, since they lead to tissue prestress, compaction and resulting collagen realignment in many soft biological tissues [124]. Rausch and Kuhl [125] attributed discrepancies between in vivo and in vitro stiffness measurements of biological tissues to the presence of prestress. Seemingly, the prestress directly influences the tissue's biomechanical behavior by changing the apparent stiffness. Furthermore, prestress indirectly affects collagen remodeling. For example, Grenier et al. [126] showed that prestressed tissues exhibit enhanced tissue formation compared to unconstrained tissues. One explanation for this observation is that the prestressed collagen fibers in the constrained samples were less susceptible for enzymatic degradation.

Cell-mediated collagen remodeling and prestress are not only complex phenomena on their own, but also highly interrelated. For this reason, including cellular behavior into computational models of collagen remodeling is imperative. In this section we will first discuss models that include cellular prestress in a phenomenological manner. These models showed the importance of including cell traction. Second, mathematical models that propose biologically motivated laws for collagen remodeling and degradation are addressed. Finally, an example of a model applied to TEHVs is described.

## 5.3 Phenomenological Models of Cell-Mediated Collagen Remodeling

One of the first models to investigate the effects of cell traction on collagen realignment was the model of Soares et al. [127]. In this phenomenological model, cell traction was added to the collagen framework proposed by Driessen et al. [65], via the inclusion of inverse isotropic volumetric growth. Although this model was relatively simple, it was able to predict the change in collagen orientation through the thickness of tissue-engineered vascular grafts. Despite its relative simplicity, the model showed the importance of including cellular behavior (in this particular case cell traction) into computational models.

Successively, Nagel and Kelly [128] studied the effects of cellular contraction on collagen realignment with a more complex theory. Rather than directly including the cell traction forces in the model, they assumed that cells, seeking for a homeostatic level of strain or stress experienced via their interaction with collagen, contract collagen fibers in a "hand-over-hand" fashion [79]. This was implemented in the model by assuming that the stress-free configuration of collagen fibers evolves towards a preferred homeostatic level. These principles were then coupled with a mathematical model for the collagen realignment along the maximum principal strain direction. Several experimental observations could be reproduced with this framework, including collagen remodeling occurring in periosteum [129], in collagen gels [58], and cruciform fibrin gels [78]. However, this model was still phenomenological, since it did not describe the underlying mechanisms driving collagen crimping and turnover, and the homeostatic collagen stress-free configuration was set a priori.

## 5.4 Biologically Motivated Models for Cell-Mediated Collagen Remodeling

Biologically motivated hypotheses for collagen synthesis and degradation were proposed by Valentín et al. [130, 131], who based their models on the framework proposed by Baek et al. [132, 133]. In particular, they assumed that collagen remodeling depends on wall shear stress induced production of vasoactive molecules and intramural stress induced production of growth factors. Newly synthesized collagen fibers were deposited at a certain preferred homeostatic stretch, while collagen degradation was implemented via a fading memory behavior; more recently produced fibers contribute more to the stress response. Finally, active smooth muscle contraction was modeled to depend on the ratio of produced constrictors and dilators. With this model, the two-step response to changes in arterial flow rate (instant contraction or dilation, followed by tissue remodeling) could be predicted, including the typical time scales at which these phenomena occur. Moreover, arterial thickening due to pressure overload could be predicted successfully. Nevertheless, these model predictions could not be obtained using one set of model parameters. The predictive power of the model has been shown by Miller et al. [134] and Khosravi et al. [135], who used this framework to identify important scaffold parameters and development of tissue-engineered vascular grafts.

Recently, Heck et al. [136] proposed a biologically motivated model in which collagen degradation is the main mediator of collagen remodeling. In this model, collagen fibers are assumed to degrade via a strain dependent mechanism for which at higher strains the collagen fibers are less susceptible to degradation [123, 137]. Furthermore, it was hypothesized that both collagen production and degradation increase with cellular deformation. Contact-guided cell traction was included by assuming that the magnitude of the cell-induced collagen fiber strain evolves towards the value at which collagen production and degradation are in equilibrium. Finally, when over-stretched, collagen fibers could rupture in a two-stage manner.

This model was able to predict a wide range of experiments. First, the inclusion of fiber rupture allowed the prediction of axially loaded bovine pericardium subjected to collagenase [138]. In addition, the experiments of Lee et al. [139] (uniaxially loaded fibroblast populated collagen gels), Hu et al. [140] (biaxially constrained collagen gels), and Foolen et al. [129] (transition stretch of periosteum) could be predicted successfully. However, despite the good predictive capacity of the model, these results could not be obtained with a single set of model parameters. Moreover, the SF remodeling is not included explicitly in the model, but is merely dictated by the collagen network. Therefore, cell-mediated remodeling is limited as collagen fiber remodeling is the real actuator of the model.

## 5.5  Inclusion of SF Remodeling

Following their phenomenological study of 2011, Soares et al. [141] mathematically described the mechanism of tissue compaction including SF contractile forces and remodeling in the computational model, by using the previously described framework of Deshpande et al. [98]. With this model, SF orientation, tissue compaction, and stresses in statically cultured tissue-engineered strips [124] could be predicted. However, the collagen remodeling process is mostly neglected in this study; the amount of collagen is assumed to correspond with the relative amount of SFs in each direction.

Successively, Loerakker et al. [142] attempted to model cell-mediated collagen remodeling using a bottom-up approach, coupling two separated mathematical models for collagen and SF remodeling. In particular, several previously proposed hypotheses were included and coupled in one common numerical algorithm. While the SF remodeling algorithm coincided with the model proposed by Obbink-Huizer et al. [90], the SF mediated tissue compaction occurred via two mechanisms: directly via SF contractile stress and via cell-mediated contraction of the collagen fibers. The latter was implemented in a similar manner as proposed by Nagel and Kelly [128], but the stress-free configuration of collagen fibers was assumed to depend on the SF stress and the presence of mechanical loads, instead of being imposed a priori. Concerning the collagen turnover, two hypotheses for collagen production and two hypotheses for collagen degradation were tested and compared. The first degradation hypothesis was the same as implemented by van Donkelaar et al. [137] and Heck et al. [136]; degradation decreases with strain and becomes constant after a certain threshold value. On the other hand, the second degradation hypothesis assumed an increased degradation after this threshold value. The production part of the model assumed either collagen deposition in the direction of the SFs, in accordance to experimental observations [80], or isotropic deposition which would correspond to the hypothesis that collagen remodeling is dictated by degradation.

This model was able to reproduce a variety of experimental outcomes [78, 124, 129] and was the first to predict a helical collagen organization in vascular grafts without a priori assuming two distinct fiber families. An advantage of this model over the previously described models is that all the simulations were performed with the same set of model parameters. Besides this versatility, this computational model also has some limitations. First of all, the model is not able to predict tissue growth or dilation. In addition, it shares all the previously described limitations of the computational model proposed by Obbink-Huizer et al. [90], since the same hypotheses were used for the SF remodeling.

## 5.6 Remodeling of TEHVs

Recently, the framework of Loerakker et al. [142] was applied to predict collagen remodeling in TEHVs [143]. In order to prevent excessive computational costs in simulating collagen and SF remodeling under dynamic loading conditions, an analytical approximation for the SF remodeling laws was used [144].

As valvular insufficiency is an important problem in HVTE, they assessed this risk for TEHVs after remodeling at the pulmonary and aortic position. In agreement with existing research, it was found that TEHVs implanted at the pulmonary position are at risk to develop valvular insufficiency, where this risk increases with the level of cellular contractility. In contrast, it was predicted that leaflet retraction and the associated insufficiency is unlikely to occur at the aortic position because in these conditions the hemodynamic loads are high enough to counterbalance the cellular traction forces. Moreover, they predicted the development of a native-like, circumferentially aligned collagen network at the aortic position.

## 5.7 Future Scope

By including cellular behavior, the predictive capacity of computational models for collagen remodeling has greatly increased. Moreover, a better understanding of the underlying mechanisms has been obtained. These models have only recently been applied to predict the remodeling of TEHVs, but have already shown their efficacy by highlighting the important determinants in the remodeling process. This shows that computational models can tremendously aid HVTE research. Therefore, developing current models even further is of vital importance. The final chapter will deal with the future perspectives and challenges in modeling the in vivo adaptation and functionality of TEHVs.

## 6 Future Directions

It is yet unclear whether current HVTE techniques are able to induce a physiological remodeling of the collagen network, which is essential for obtaining the proper tissue biomechanical properties that are strongly correlated with heart valve functionality. Computational simulations can help to improve HVTE strategies, since they are a valuable and versatile tool for increasing our understanding of and control over the process of collagen remodeling. The early computational models (Sect. 3) mainly relied on phenomenological remodeling laws. This enabled the prediction of the collagen architecture in arteries and heart valves but, on the other hand, did not allow for the investigation of the underlying biological processes driving collagen remodeling. Since cells are ultimately responsible for collagen remodeling, numerical algorithms capturing the cellular behavior need to be included in the computational models. Section 4 elucidated the development of models for the prediction of the cell orientation and contraction, through the use of numerical algorithms for SF remodeling. Finally, cell models were coupled to collagen remodeling laws (Sect. 5), which substantially improved the predictive potential of the computational simulations. Yet, there is still room for improvement of the existing computational models for TEHV remodeling. First of all, since they only capture relatively short-term remodeling, long-term prediction of tissue growth of TEHVs has not been established. In addition, the phenomenological nature of many of these models does not grant a mechanistic understanding of the biological processes involved. This final section will deal with the future directions of modeling growth and remodeling of TEHVs, and their implications for the improvement of current HVTE techniques.

## 6.1 Growth

Current models mainly focus on short-term tissue remodeling and, in general, they rely on the hypothesis of conservation of mass, which impair predictions of long-term volumetric growth. As TEHV are aimed at being life-lasting valve replacements, modeling and optimizing this long-term tissue growth is crucial.

Several research groups have included growth into their numerical algorithms for collagen remodeling in cardiovascular tissues. Humphrey and Rajagopal [145] implemented growth using the concept of an evolving stress-free tissue configuration. Successively, the group of Kuhl implemented volumetric growth, assuming stress- or stretch-driven growth along the principal loading directions [146]. This approach was successfully applied to arteries (stress-driven [147]), skeletal muscle (stretch-driven [148]), skin (stretch-driven [149]), and heart muscle (stretch- and stress-driven [150]). In particular, the latter demonstrated that assuming stress- or stretch-driven growth correspond to the prediction of different physiological pathologies. This shows that the choice of the driving stimulus determines to a great extent the model predictions. Therefore, identifying the driving stimulus of TEHV growth and remodeling is of vital importance to accurately predict long-term tissue development. Oomen et al. [53] made a first attempt, by investigating the driving stimulus of human heart valve growth. In this study, geometric and mechanical characterization, followed by finite element simulations of the loaded configuration of native heart valves of increasing age in diastole, revealed increasing tissue stress with age, while stretch differences were relatively small between the aortic and pulmonary valve as well as valves of different ages. Consequently, these results suggest that heart valve growth is a stretch mediated process.

## 6.2 Agent-Based Models

In addition to tissue growth, obtaining a mechanistic understanding of tissue growth and remodeling is essential. A promising approach towards this end is the use of agent-based computational models, where both the behavior of individual agents and the interaction between them determines the outcome of computational simulations. Obviously, single cells can potentially form the basis of such models for TEHVs.

Recently, Werfel et al. [151] have used an agent-based model to predict cancer development due to physical changes in the surroundings of the cell. In their study, they modeled cells on a 2D substrate as adhesive spheres, in which the tension experienced by the cells determines their fate; in particular, cells under extension had a high probability to grow and proliferate, while compression induced apoptosis. The computational simulations suggested that geometric changes and increased variability in phenotype may be a mechanism for tissue disorganization and subsequent non-mutagenic cancer development. Interestingly, in this model both cell-cell interaction and growth were incorporated, and this could be translated to models for growth and remodeling of TEHVs.

In addition, an agent-based mathematical model predicting scar development after a myocardial infarction has been proposed by Rouillard and Holmes [152]. In this study, a previous agent-based model [153] was coupled to a macroscopic model for tissue mechanics. The fibroblasts in the agent-based part were assumed to deposit and remodel the collagen in response to chemical, structural, and mechanical stimuli. The model was able to predict infarct healing of a coronary ligation rat model [154]. This was the first agent-based model specifically developed for the prediction of tissue remodeling, and it showed the potential of this multi-scale approach.

The advantage of using agent-based models is that important cellular processes such as cell migration can be modeled specifically, which would not be possible using a continuum approach. Moreover, the use of single agents allows the inclusion of stochastic factors into computational models. Not only do these agent-based models help to gain a mechanistic understanding of the underlying processes of growth and remodeling, but they also have the potential to tackle other multi-scale problems, such as disease modeling.

## 6.3 How Far Should We Go?

In general, increasing the complexity of mathematical models can lead to a greater predictive potential of the computational simulations. However, a disadvantage of increasing the complexity is that increasing the number of differential equations related to the tissue remodeling laws generally corresponds to an increase in model parameters. Consequently, the estimation of these parameter values and the interpretation of the simulation results become more difficult, and often a single set of model parameters cannot be identified. Albert Einstein once said:

It can scarcely be denied that the supreme goal of all theory is to make the irreducible basic elements as simple and as few as possible...

Reflecting on this quote, we strongly believe that a model should have a certain degree of complexity to capture the behavior one aims to describe accurately; however, all additional complexity will unnecessarily increase the model uncertainty and should therefore be avoided.

## 6.4 Implications for TEHVs

Computational simulations have already demonstrated their utility in improving HVTE techniques [51, 143] and in understanding complex experimental results [143]. Their use is extremely promising for the optimization of the in vivo remodeling response of TEHVs, through the prediction of the influence of several factors involved, such as scaffold design, cell type, or transvalvular pressure magnitude. However, further improvements are still necessary.

While the enhancement of current computational models continues, this procedure constantly needs validation. In this last process, in vitro experimental results backing the model predictions are fundamental and allow for a comparison between computational and experimental results. For this reason, many current mathematical models have shown their predictive capacity by reproducing multiple in vitro experimental outcomes. While experimental data are already available to analyze and model short-term tissue remodeling, there is a lack of experimental results for the development of computational models for long-term growth. To this aim, controlled experiments to identify the driving stimulus of TEHV growth are needed. Only with the information provided by these experiments computational models for tissue growth will be able to be coupled with biologically motivated hypotheses that are fundamental to understanding, predicting, and optimizing the long-term behavior of implanted TEHVs.

Concluding, computational models and experiments can mutually benefit from one another, and the synergy of combined numerical-experimental approaches will continue to help the HVTE research to obtain a fully functional, life-lasting heart valve substitute.

## References

1. Yacoub MH, Takkenberg JJM. Will heart valve tissue engineering change the world? Nat Clin Pract Cardiovasc Med. 2005;2(2):60–1. https://doi.org/10.1038/ncpcardio0112.
2. Zilla P, Brink J, Human P, Bezuidenhout D. Prosthetic heart valves: catering for the few. Biomaterials. 2008;29(4):385–406. https://doi.org/10.1016/j.biomaterials.2007.09.033.
3. Hammermeister K, Sethi GK, Henderson WG, Grover FL, Oprian C, Rahimtoola SH. Outcomes 15 years after valve replacement with a mechanical versus a bioprosthetic

valve: final report of the Veterans Affairs randomized trial. J Am Coll Cardiol. 2000;36 (4):1152–8. https://doi.org/10.1016/S0735-1097(00)00834-2.
4. Oxenham H, Bloomfield P, Wheatley DJ. Twelve-year comparison of a Bjork-Shiley mechanical heart valve with porcine bioprostheses. N Engl J Med. 2003;324(9):573–9. https://doi.org/10.1056/NEJM199102283240901.
5. Curtil A, Pegg DE, Wilson A. Repopulation of freeze-dried porcine valves with human fibroblasts and endothelial cells. J Heart Valve Dis. 1997;6(3):296–306.
6. Knight RL, Booth C, Wilcox HE, Fisher J, Ingham E. Tissue engineering of cardiac valves: re-seeding of acellular porcine aortic valve matrices with human mesenchymal progenitor cells. J Heart Valve Dis. 2005;14(6):806–13.
7. Schenke-Layland K. Complete dynamic repopulation of decellularized heart valves by application of defined physical signals—an in vitro study. Cardiovasc Res. 2003;60(3):497–509. https://doi.org/10.1016/j.cardiores.2003.09.002.
8. Weber B, Dijkman PE, Scherman J, Sanders B, Emmert MY, Grünenfelder J, et al. Off-the-shelf human decellularized tissue-engineered heart valves in a non-human primate model. Biomaterials. 2013;34(30):7269–80. https://doi.org/10.1016/j.biomaterials.2013.04.059.
9. Shinoka T, Breuer CK, Tanel RE, Zund G, Miura T, Ma PX, et al. Tissue engineering heart valves: valve leaflet replacement study in a lamb model. Ann Thorac Surg. 1995;60(95): S513–6. https://doi.org/10.1016/0003-4975(95)00733-4.
10. Hoerstrup SP, Sodian R, Daebritz S, Wang J, Bacha EA, Martin DP, et al. Functional living trileaflet heart valves grown in vitro. Circulation. 2000;102(19 Suppl 3):III44–9. https://doi.org/10.1161/01.CIR.102.suppl_3.III-44.
11. Mol A, Driessen NJB, Rutten MCM, Hoerstrup SP, Bouten CVC, Baaijens FPT. Tissue engineering of human heart valve leaflets: a novel bioreactor for a strain-based conditioning approach. Ann Biomed Eng. 2005;33(12):1778–88. https://doi.org/10.1007/s10439-005-8025-4.
12. Schmidt D, Mol A, Breymann C, Achermann J, Odermatt B, Go M, et al. Living autologous heart valves engineered from human prenatally harvested progenitors. Circulation. 2006;114:125–32. https://doi.org/10.1161/CIRCULATIONAHA.105.001040.
13. Sutherland FWH, Perry TE, Yu Y, Sherwood MC, Rabkin E, Masuda Y, et al. From stem cells to viable autologous semilunar heart valve. Circulation. 2005;111:2783–91. https://doi.org/10.1161/CIRCULATIONAHA.104.498378.
14. Dohmen PM, da Costa F, Holinski S, Lopes SV, Yoshi S, Reichert LH, et al. Is there a possibility for a glutaraldehyde-free porcine heart valve to grow? Eur Surg Res. 2006;38 (1):54–61. https://doi.org/10.1159/000091597.
15. Goldstein S, Clarke DR, Walsh SP, Black KS, O'Brien MF. Transpecies heart valve transplant: advanced studies of a bioengineered xeno-autograft. Ann Thorac Surg. 2000;70 (6):1962–9. https://doi.org/10.1016/S0003-4975(00)01812-9.
16. Leyh RG, Wilhelmi M, Rebe P, Fischer S, Kofidis T, Haverich A, et al. In vivo repopulation of xenogeneic and allogeneic acellular valve matrix conduits in the pulmonary circulation. Ann Thorac Surg. 2003;75(5):1457–63; discussion 1463.
17. Dijkman PE, Driessen-Mol A, Frese L, Hoerstrup SP, Baaijens FPT. Decellularized homologous tissue-engineered heart valves as off-the-shelf alternatives to xeno- and homografts. Biomaterials. 2012;33(18):4545–54. https://doi.org/10.1016/j.biomaterials.2012.03.015.
18. Driessen-Mol A, Emmert MY, Dijkman PE, Frese L, Sanders B, Weber B, et al. Transcatheter implantation of homologous "off-the-shelf" tissue-engineered heart valves with self-repair capacity: long-term functionality and rapid in vivo remodeling in sheep. J Am Coll Cardiol. 2014;63(13):1320–9. https://doi.org/10.1016/j.jacc.2013.09.082.
19. Flanagan TC, Cornelissen C, Koch S, Tschoeke B, Sachweh JS, Schmitz-Rode T, et al. The in vitro development of autologous fibrin-based tissue-engineered heart valves through optimised dynamic conditioning. Biomaterials. 2007;28(23):3388–97. https://doi.org/10.1016/j.biomaterials.2007.04.012.

20. Mol A, Smits AIPM, Bouten CVC, Baaijens FPT. Tissue engineering of heart valves: advances and current challenges. Expert Rev Med Devices. 2009;6(3):259–75. https://doi.org/10.1586/erd.09.12.
21. Sacks MS, Schoen FJ, Mayer JE. Bioengineering challenges for heart valve tissue engineering. Annu Rev Biomed Eng. 2009;11(1):289–313. https://doi.org/10.1146/annurev-bioeng-061008-124903.
22. Anderson RH. The surgical anatomy of the aortic root. Multimed Man Cardiothorac Surg. 2007;102:1–8. https://doi.org/10.1510/mmcts.2006.002527.
23. Thubrikar MJ. The aortic valve. Boca Raton, FL: CRC Press; 1990. 232 p
24. Sutton JP, Ho SY, Anderson RH. The forgotten interleaflet triangles: a review of the surgical anatomy of the aortic valve. Ann Thorac Surg. 1995;59(2):419–27.
25. Balachandran K, Sucosky P, Yoganathan AP. Hemodynamics and mechanobiology of aortic valve inflammation and calcification. Int J Inflam. 2011;2011:263870. https://doi.org/10.4061/2011/263870.
26. Billiar KL, Sacks MS. Biaxial mechanical properties of the natural and glutaraldehyde treated aortic valve cusp—part I: experimental results. J Biomech Eng. 2000;122(1):23–30. https://doi.org/10.1115/1.429624.
27. Mavrilas D, Missirlis Y. An approach to the optimization of preparation of bioprosthetic heart valves. J Biomech. 1991;24(5):331–9.
28. Sacks MS, Yoganathan AP. Heart valve function: a biomechanical perspective. Philos Trans R Soc Lond B Biol Sci. 2007;362(1484):1369–91. https://doi.org/10.1098/rstb.2007.2122.
29. Schoen FJ. Evolving concepts of cardiac valve dynamics: the continuum of development, functional structure, pathobiology, and tissue engineering. Circulation. 2008;118 (18):1864–80. https://doi.org/10.1161/CIRCULATIONAHA.108.805911.
30. Ho SY. Structure and anatomy of the aortic root. Eur J Echocardiogr. 2009;10(1):i3–10. https://doi.org/10.1093/ejechocard/jen243.
31. Scott MJ, Vesely I. Morphology of porcine aortic valve cusp elastin. J Heart Valve Dis. 1996;5 (5):464–71.
32. Buchanan RM, Sacks MS. Interlayer micromechanics of the aortic heart valve leaflet. Biomech Model Mechanobiol. 2014;13(4):813–26. https://doi.org/10.1007/s10237-013-0536-6.
33. Hasan A, Ragaert K, Swieszkowski W, Selimović Š, Paul A, Camci-Unal G, et al. Biomechanical properties of native and tissue engineered heart valve constructs. J Biomech. 2014;47:1949–63. https://doi.org/10.1016/j.jbiomech.2013.09.023.
34. Boerboom RA, Rubbens MP, Driessen NJB, Bouten CVC, Baaijens FPT. Effect of strain magnitude on the tissue properties of engineered cardiovascular constructs. Ann Biomed Eng. 2008;36(2):244–53. https://doi.org/10.1007/s10439-007-9413-8.
35. Cox MAJ, Kortsmit J, Driessen NJB, Bouten CVC, Baaijens FPT. Tissue-engineered heart valves develop native-like collagen fiber architecture. Tissue Eng Part A. 2010;16 (5):1527–37.
36. Engelmayr GC, Papworth GD, Watkins SC, Mayer JE, Sacks MS. Guidance of engineered tissue collagen orientation by large-scale scaffold microstructures. J Biomech. 2006;39 (10):1819–31. https://doi.org/10.1016/j.jbiomech.2005.05.020.
37. Engelmayr GC, Rabkin E, Sutherland FWH, Schoen FJ, Mayer JE, Sacks MS. The independent role of cyclic flexure in the early in vitro development of an engineered heart valve tissue. Biomaterials. 2005;26(2):175–87. https://doi.org/10.1016/j.biomaterials.2004.02.035.
38. Engelmayr GC, Sales VL, Mayer JE, Sacks MS. Cyclic flexure and laminar flow synergistically accelerate mesenchymal stem cell-mediated engineered tissue formation: implications for engineered heart valve tissues. Biomaterials. 2006;27(36):6083–95. https://doi.org/10.1016/j.biomaterials.2006.07.045.
39. Neidert MR, Tranquillo RT. Tissue-engineered valves with commissural alignment. Tissue Eng. 2006;12(4):891–903. https://doi.org/10.1089/ten.2006.12.891.
40. Ramaswamy S, Boronyak SM, Le T, Holmes A, Sotiropoulos F, Sacks MS. A novel bioreactor for mechanobiological studies of engineered heart valve tissue formation under pulmonary

arterial physiological flow conditions. J Biomech Eng. 2014;136:1–14. https://doi.org/10.1115/1.4028815.
41. Ramaswamy S, Gottlieb D, Engelmayr GC, Aikawa E, Schmidt DE, Gaitan-Leon DM, et al. The role of organ level conditioning on the promotion of engineered heart valve tissue development in-vitro using mesenchymal stem cells. Biomaterials. 2010;31(6):1114–25. https://doi.org/10.1016/j.biomaterials.2009.10.019.
42. Balguid A, Rubbens MP, Mol A, Bank RA, Bogers AJJC, van Kats JP, et al. The role of collagen cross-links in biomechanical behavior of human aortic heart valve leaflets—relevance for tissue engineering. Tissue Eng. 2007;13(7):1501–11. https://doi.org/10.1089/ten.2006.0279.
43. Driessen NJB, Mol A, Bouten CVC, Baaijens FPT. Modeling the mechanics of tissue-engineered human heart valve leaflets. J Biomech. 2007;40(2):325–34. https://doi.org/10.1016/j.jbiomech.2006.01.009.
44. Mol A, Rutten MCM, Driessen NJB, Bouten CVC, Zünd G, Baaijens FPT, et al. Autologous human tissue-engineered heart valves: prospects for systemic application. Circulation. 2006;114(Suppl. 1):152–9. https://doi.org/10.1161/CIRCULATIONAHA.105.001123.
45. Schmidt D, Dijkman PE, Driessen-Mol A, Stenger R, Mariani C, Puolakka A, et al. Minimally-invasive implantation of living tissue engineered heart valves: a comprehensive approach from autologous vascular cells to stem cells. J Am Coll Cardiol. 2010;56(6):510–20. https://doi.org/10.1016/j.jacc.2010.04.024.
46. Sodian R, Hoerstrup SP, Sperling JS, Daebritz S, Martin DP, Moran AM, et al. Early in vivo experience with tissue-engineered trileaflet heart valves. Circulation. 2000;102:III22–9.
47. Syedain ZH, Lahti MT, Johnson SL, Robinson PS, Ruth GR, Bianco RW, et al. Implantation of a tissue-engineered heart valve from human fibroblasts exhibiting short term function in the sheep pulmonary artery. Cardiovasc Eng Technol. 2011;2(2):101–12. https://doi.org/10.1007/s13239-011-0039-5.
48. Flanagan TC, Sachweh JS, Frese J, Schnöring H, Gronloh N, Koch S, et al. In vivo remodeling and structural characterization of fibrin-based tissue-engineered heart valves in the adult sheep model. Tissue Eng Part A. 2009;15(10):2965–76. https://doi.org/10.1089/ten.tea.2009.0018.
49. Gottlieb D, Kunal T, Emani S, Aikawa E, Brown DW, Powell AJ, et al. In vivo monitoring of function of autologous engineered pulmonary valve. J Thorac Cardiovasc Surg. 2010;139(3):723–31. https://doi.org/10.1016/j.jtcvs.2009.11.006.
50. Loerakker S, Argento G, Oomens CWJ, Baaijens FPT. Effects of valve geometry and tissue anisotropy on the radial stretch and coaptation area of tissue-engineered heart valves. J Biomech. 2013;46(11):1792–800. https://doi.org/10.1016/j.jbiomech.2013.05.015.
51. Sanders B, Loerakker S, Fioretta ES, Bax DJP, Driessen-Mol A, Hoerstrup SP, et al. Improved geometry of decellularized tissue engineered heart valves to prevent leaflet retraction. Ann Biomed Eng. 2016;44:1061–71. https://doi.org/10.1007/s10439-015-1386-4.
52. Rock CA, Han L, Doehring TC. Complex collagen fiber and membrane morphologies of the whole porcine aortic valve. PLoS One. 2014;9(1):e86087. https://doi.org/10.1371/journal.pone.0086087.
53. Oomen PJA, Loerakker S, van Geemen D, Neggers J, Goumans MTH, van den Bogaerdt AJ, et al. Age-dependent changes of stress and strain in the human heart valve and their relation with collagen remodeling. Acta Biomater. 2016;29:161–9. https://doi.org/10.1016/j.actbio.2015.10.044.
54. Sacks MS, Smith DB. A small angle light scattering device for planar connective tissue microstructural analysis. Ann Biomed Eng. 1997;25(4):678–89.
55. Sacks MS, Smith DB, Hiester ED. The aortic valve microstructure: effects of transvalvular pressure. J Biomed Mater Res. 1998;41(1):131–41. https://doi.org/10.1002/(SICI)1097-4636(199807)41:1<131::AID-JBM16>3.0.CO;2-Q.
56. Aikawa E. Human semilunar cardiac valve remodeling by activated cells from fetus to adult: implications for postnatal adaptation, pathology, and tissue engineering. Circulation. 2006;113(10):1344–52. https://doi.org/10.1161/CIRCULATIONAHA.105.591768.

57. Rubbens MP, Driessen-Mol A, Boerboom RA, Koppert MMJ, van Assen HC, TerHaar Romeny BM, et al. Quantification of the temporal evolution of collagen orientation in mechanically conditioned engineered cardiovascular tissues. Ann Biomed Eng. 2009;37 (7):1263–72. https://doi.org/10.1007/s10439-009-9698-x.
58. Thomopoulos S, Fomovsky GM, Holmes JW. The development of structural and mechanical anisotropy in fibroblast populated collagen gels. J Biomech Eng. 2005;127(5):742–50. https://doi.org/10.1115/1.1992525.
59. Costa KD, Lee EJ, Holmes JW. Creating alignment and anisotropy in engineered heart tissue: role of boundary conditions in a model three-dimensional culture system. Tissue Eng. 2003;9 (4):567–77. https://doi.org/10.1089/107632703768247278.
60. Kostyuk O, Brown RA. Novel spectroscopic technique for in situ monitoring of collagen fibril alignment in gels. Biophys J. 2004;87(1):648–55. https://doi.org/10.1529/biophysj.103.038976.
61. Boerboom RA, Driessen NJB, Bouten CVC, Huyghe JM, Baaijens FPT. Finite element model of mechanically induced collagen fiber synthesis and degradation in the aortic valve. Ann Biomed Eng. 2003;31(9):1040–53. https://doi.org/10.1114/1.1603749.
62. Creane A, Maher E, Sultan S, Hynes N, Kelly DJ, Lally C. Prediction of fibre architecture and adaptation in diseased carotid bifurcations. Biomech Model Mechanobiol. 2011;10 (6):831–43. https://doi.org/10.1007/s10237-010-0277-8.
63. Driessen NJB, Boerboom RA, Huyghe JM, Bouten CVC, Baaijens FPT. Computational analyses of mechanically induced collagen fiber remodeling in the aortic heart valve. J Biomech Eng. 2003;125(4):549–57. https://doi.org/10.1115/1.1590361.
64. Driessen NJB, Bouten CVC, Baaijens FPT. Improved prediction of the collagen fiber architecture in the aortic heart valve. J Biomech Eng. 2005;127(2):329. https://doi.org/10.1115/1.1865187.
65. Driessen NJB, Cox MAJ, Bouten CVC, Baaijens FPT. Remodelling of the angular collagen fiber distribution in cardiovascular tissues. Biomech Model Mechanobiol. 2008;7(2):93–103. https://doi.org/10.1007/s10237-007-0078-x.
66. Driessen NJB, Peters GWM, Huyghe JM, Bouten CVC, Baaijens FPT. Remodelling of continuously distributed collagen fibres in soft connective tissues. J Biomech. 2003;36 (8):1151–8. https://doi.org/10.1016/S0021-9290(03)00082-4.
67. Driessen NJB, Wilson W, Bouten CVC, Baaijens FPT. A computational model for collagen fibre remodelling in the arterial wall. J Theor Biol. 2004;226(1):53–64. https://doi.org/10.1016/j.jtbi.2003.08.004.
68. Hariton I, DeBotton G, Gasser TC, Holzapfel G a. Stress-driven collagen fiber remodeling in arterial walls. Biomech Model Mechanobiol. 2007;6(3):163–75. https://doi.org/10.1007/s10237-006-0049-7.
69. Kuhl E, Holzapfel GA. A continuum model for remodeling in living structures. J Mater Sci. 2007;42(21):8811–23. https://doi.org/10.1007/s10853-007-1917-y.
70. Menzel A, Harrysson M, Ristinmaa M. Towards an orientation-distribution-based multi-scale approach for remodelling biological tissues. Comput Methods Biomech Biomed Eng. 2008;11 (5):505–24. https://doi.org/10.1080/10255840701771776.
71. Menzel A, Waffenschmidt T. A microsphere-based remodelling formulation for anisotropic biological tissues. Philos Trans A Math Phys Eng Sci. 2009;367(1902):3499–523. https://doi.org/10.1098/rsta.2009.0103.
72. Sáez P, Peña E, Doblaré M, Martinez MÁ. An anisotropic microsphere-based approach for fiber orientation adaptation in soft tissue. IEEE Trans Biomed Eng. 2011;58(12 Part 2):3500–3. https://doi.org/10.1109/TBME.2011.2166154.
73. Kuhl E, Garikipati K, Arruda EM, Grosh K. Remodeling of biological tissue: mechanically induced reorientation of a transversely isotropic chain network. J Mech Phys Solids. 2005;53 (7):1552–73. https://doi.org/10.1016/j.jmps.2005.03.002.

74. Menzel A. Modelling of anisotropic growth in biological tissues: a new approach and computational aspects. Biomech Model Mechanobiol. 2005;3(3):147–71. https://doi.org/10.1007/s10237-004-0047-6.
75. Schriefl AJ, Reinisch AJ, Sankaran S, Pierce DM, Holzapfel GA. Quantitative assessment of collagen fibre orientations from two-dimensional images of soft biological tissues. J R Soc Interface. 2012;9:3081–93. https://doi.org/10.1098/rsif.2012.0339.
76. Holzapfel GA, Gasser TC, Ogden RW. A new constitutive framework for arterial wall mechanics and a comperative study of material models. J Elast. 2000;61:1–48.
77. Sauren AAHJ. The mechanical behaviour of the aortic valve. PhD Thesis Eindhoven: Technische Hogeschool Eindhoven. 1981. https://doi.org/10.6100/IR94978.
78. Sander EA, Barocas VH, Tranquillo RT. Initial fiber alignment pattern alters extracellular matrix synthesis in fibroblast-populated fibrin gel cruciforms and correlates with predicted tension. Ann Biomed Eng. 2011;39(2):714–29. https://doi.org/10.1007/s10439-010-0192-2.
79. Meshel AS, Wei Q, Adelstein RS, Sheetz MP. Basic mechanism of three-dimensional collagen fibre transport by fibroblasts. Nat Cell Biol. 2005;7(2):157–64. https://doi.org/10.1038/ncb1216.
80. Wang JH-C, Jia F, Gilbert TW, Woo SL-Y. Cell orientation determines the alignment of cell-produced collagenous matrix. J Biomech. 2003;36(1):97–102. https://doi.org/10.1016/S0021-9290(02)00233-6.
81. Burridge K, Wittchen ES. The tension mounts: stress fibers as force-generating mechanotransducers. J Cell Biol. 2013;200(1):9–19. https://doi.org/10.1083/jcb.201210090.
82. Engler AJ, Sen S, Sweeney HL, Discher DE. Matrix elasticity directs stem cell lineage specification. Cell. 2006;126(4):677–89. https://doi.org/10.1016/j.cell.2006.06.044.
83. Foolen J, Deshpande VS, Kanters FMW, Baaijens FPT. The influence of matrix integrity on stress-fiber remodeling in 3D. Biomaterials. 2012;33(30):7508–18. https://doi.org/10.1016/j.biomaterials.2012.06.103.
84. Ghibaudo M, Saez A, Trichet L, Xayaphoummine A, Browaeys J, Silberzan P, et al. Traction forces and rigidity sensing regulate cell functions. Soft Matter. 2008;4(9):1836. https://doi.org/10.1039/b804103b.
85. Van Vlimmeren MAA, Driessen-Mol A, Oomens CWJ, Baaijens FPT. Passive and active contributions to generated force and retraction in heart valve tissue engineering. Biomech Model Mechanobiol. 2012;11(7):1015–27. https://doi.org/10.1007/s10237-011-0370-7.
86. Kaunas R, Usami S, Chien S. Regulation of stretch-induced JNK activation by stress fiber orientation. Cell Signal. 2006;18(11):1924–31. https://doi.org/10.1016/j.cellsig.2006.02.008.
87. Tondon A, Hsu HJ, Kaunas R. Dependence of cyclic stretch-induced stress fiber reorientation on stretch waveform. J Biomech. 2012;45(5):728–35. https://doi.org/10.1016/j.jbiomech.2011.11.012.
88. Wang JH. Substrate deformation determines actin cytoskeleton reorganization: a mathematical modeling and experimental study. J Theor Biol. 2000;202(1):33–41. https://doi.org/10.1006/jtbi.1999.1035.
89. Faust U, Hampe N, Rubner W, Kirchgeßner N, Safran S, Hoffmann B, et al. Cyclic stress at mHz frequencies aligns fibroblasts in direction of zero strain. PLoS One. 2011;6(12):e28963. https://doi.org/10.1371/journal.pone.0028963.
90. Obbink-Huizer C, Oomens CWJ, Loerakker S, Foolen J, Bouten CVC, Baaijens FPT. Computational model predicts cell orientation in response to a range of mechanical stimuli. Biomech Model Mechanobiol. 2014;13(1):227–36. https://doi.org/10.1007/s10237-013-0501-4.
91. Tondon A, Kaunas R. The direction of stretch-induced cell and stress fiber orientation depends on collagen matrix stress. PLoS One. 2014;9(2):e89592. https://doi.org/10.1371/journal.pone.0089592.
92. Lamers E, Frank Walboomers X, Domanski M, te Riet J, van Delft FCMJM, Luttge R, et al. The influence of nanoscale grooved substrates on osteoblast behavior and extracellular matrix

deposition. Biomaterials. 2010;31(12):3307–16. https://doi.org/10.1016/j.biomaterials.2010. 01.034.
93. Foolen J, Janssen-van den Broek MWJT, Baaijens FPT. Synergy between Rho signaling and matrix density in cyclic stretch-induced stress fiber organization. Acta Biomater. 2014;10 (5):1876–85. https://doi.org/10.1016/j.actbio.2013.12.001.
94. De Jonge N, Kanters FMW, Baaijens FPT, Bouten CVC. Strain-induced collagen organization at the micro-level in fibrin-based engineered tissue constructs. Ann Biomed Eng. 2013;41 (4):763–74. https://doi.org/10.1007/s10439-012-0704-3.
95. Prodanov L, te Riet J, Lamers E, Domanski M, Luttge R, van Loon JJWA, et al. The interaction between nanoscale surface features and mechanical loading and its effect on osteoblast-like cells behavior. Biomaterials. 2010;31(30):7758–65. https://doi.org/10.1016/j. biomaterials.2010.06.050.
96. Niklason LE, Yeh AT, Calle EA, Bai Y, Valentín A, Humphrey JD. Enabling tools for engineering collagenous tissues integrating bioreactors, intravital imaging, and biomechanical modeling. Proc Natl Acad Sci U S A. 2010;107(8):3335–9. https://doi.org/10.1073/pnas. 0907813106.
97. Hill AV. The heat of shortening and the dynamic constants of muscle. Proc R Soc Lond B Biol Sci. 1938;126:136–95.
98. Deshpande VS, McMeeking RM, Evans AG. A bio-chemo-mechanical model for cell contractility. Proc Natl Acad Sci USA. 2006;103(38):14015–20.
99. Kaunas R, Hsu H-J. A kinematic model of stretch-induced stress fiber turnover and reorientation. J Theor Biol. 2009;257(2):320–30. https://doi.org/10.1016/j.jtbi.2008.11.024.
100. Hsu H-J, Lee C-F, Kaunas R. A dynamic stochastic model of frequency-dependent stress fiber alignment induced by cyclic stretch. PLoS One. 2009;4(3):e4853. https://doi.org/10.1371/journal.pone.0004853.
101. Kaunas R, Hsu HJ, Deguchi S. Sarcomeric model of stretch-induced stress fiber reorganization. Cell Health Cytoskelet. 2011;3(1):13–22. https://doi.org/10.2147/CHC.S14984.
102. Vernerey FJ, Farsad M. A constrained mixture approach to mechano-sensing and force generation in contractile cells. J Mech Behav Biomed Mater. 2011;4(8):1683–99. https://doi.org/10.1016/j.jmbbm.2011.05.022.
103. Foucard L, Vernerey FJ. A thermodynamical model for stress-fiber organization in contractile cells. Appl Phys Lett. 2012;100(1):13702–137024. https://doi.org/10.1063/1.3673551.
104. Vigliotti A, Ronan W, Baaijens FPT, Deshpande VS. A thermodynamically motivated model for stress-fiber reorganization. Biomech Model Mechanobiol. 2016;15:761–89. https://doi.org/10.1007/s10237-015-0722-9.
105. Deshpande VS, McMeeking RM, Evans AG. A model for the contractility of the cytoskeleton including the effects of stress-fibre formation and dissociation. Proc R Soc A Math Phys Eng Sci. 2007;463(2079):787–815. https://doi.org/10.1098/rspa.2006.1793.
106. Wei Z, Deshpande VS, McMeeking RM, Evans AG. Analysis and interpretation of stress fiber organization in cells subject to cyclic stretch. J Biomech Eng. 2008;130(3):031009–1. https://doi.org/10.1115/1.2907745.
107. Deshpande VS, Mrksich M, McMeeking RM, Evans AG. A bio-mechanical model for coupling cell contractility with focal adhesion formation. J Mech Phys Solids. 2008;56 (4):1484–510. https://doi.org/10.1016/j.jmps.2007.08.006.
108. Pathak A, McMeeking RM, Evans AG, Deshpande VS. An analysis of the cooperative mechano-sensitive feedback between intracellular signaling, focal adhesion development, and stress fiber contractility. J Appl Mech. 2011;78(4):041001. https://doi.org/10.1115/1. 4003705.
109. Pathak A, Deshpande VS, McMeeking RM, Evans AG. The simulation of stress fibre and focal adhesion development in cells on patterned substrates. J R Soc Interface. 2008;5(22):507–24. https://doi.org/10.1098/rsif.2007.1182.

110. Ronan W, Deshpande VS, McMeeking RM, McGarry JP. Cellular contractility and substrate elasticity: a numerical investigation of the actin cytoskeleton and cell adhesion. Biomech Model Mechanobiol. 2014;13(2):417–35. https://doi.org/10.1007/s10237-013-0506-z.
111. Vigliotti A, Mcmeeking RM, Deshpande VS. Simulation of the cytoskeletal response of cells on grooved or patterned substrates. Interface. 2015;12:20141320. https://doi.org/10.1098/rsif.2014.1320.
112. Ristori T, Vigliotti A, Baaijens FPT, Loerakker S, Deshpande VS. Prediction of cell alignment on cyclically strained grooved substrates. Biophys J. 2016;111(10):2274–85. https://doi.org/10.1016/j.bpj.2016.09.052.
113. Breen EC. Mechanical strain increases type I collagen expression in pulmonary fibroblasts in vitro. J Appl Physiol. 2000;88:203–9.
114. Butt R, Bishop JE. Mechanical load enhances the stimulatory effect of PDGF on pulmonary artery fibroblast procollagen synthesis. Chest. 1998;114(1):25S. https://doi.org/10.1378/chest.114.1.
115. Yang G, Crawford RC, Wang JH-C. Proliferation and collagen production of human patellar tendon fibroblasts in response to cyclic uniaxial stretching in serum-free conditions. J Biomech. 2004;37(10):1543–50. https://doi.org/10.1016/j.jbiomech.2004.01.005.
116. Visse R. Matrix metalloproteinases and tissue inhibitors of metalloproteinases: structure, function, and biochemistry. Circ Res. 2003;92(8):827–39. https://doi.org/10.1161/01.RES.0000070112.80711.3D.
117. Wojtowicz-Praga SM, Dickson RB, Hawkins MJ. Matrix metalloproteinase inhibitors. Invest New Drugs. 1997;15(1):61–75.
118. Shelton L, Rada JS. Effects of cyclic mechanical stretch on extracellular matrix synthesis by human scleral fibroblasts. Exp Eye Res. 2007;84(2):314–22. https://doi.org/10.1016/j.exer.2006.10.004.
119. Yang G, Im H-J, Wang JH-C. Repetitive mechanical stretching modulates IL-1β induced COX-2, MMP-1 expression, and PGE2 production in human patellar tendon fibroblasts. Gene. 2005;363:166–72. https://doi.org/10.1016/j.gene.2005.08.006.
120. Bhole AP, Flynn BP, Liles M, Saeidi N, Dimarzio CA, Ruberti JW. Mechanical strain enhances survivability of collagen micronetworks in the presence of collagenase: implications for load-bearing matrix growth and stability. Philos Trans R Soc A Math Phys Eng Sci. 2009;367(1902):3339–62. https://doi.org/10.1098/rsta.2009.0093.
121. Huang C, Yannas IV. Mechanochemical studies of enzymatic degradation of insoluble collagen fibers. J Biomed Mater Res. 1977;11(1):137–54. https://doi.org/10.1002/jbm.820110113.
122. Ruberti JW, Hallab NJ. Strain-controlled enzymatic cleavage of collagen in loaded matrix. Biochem Biophys Res Commun. 2005;336(2):483–9. https://doi.org/10.1016/j.bbrc.2005.08.128.
123. Wyatt KE-K, Bourne JW, Torzilli PA. Deformation-dependent enzyme mechanokinetic cleavage of type I collagen. J Biomech Eng. 2009;131(5):051004. https://doi.org/10.1115/1.3078177.
124. van Vlimmeren MAA, Driessen-Mol A, Oomens CWJ, Baaijens FPT. Model system to quantify stress generation, compaction, and retraction in engineered heart valve tissue. Tissue Eng Part C Methods. 2011;17(10):983–91. https://doi.org/10.1089/ten.tec.2011.0070.
125. Rausch MK, Kuhl E. On the effect of prestrain and residual stress in thin biological membranes. J Mech Phys Solids. 2013;61(9):1955–69. https://doi.org/10.1016/j.jmps.2013.04.005.
126. Grenier G, Rémy-Zolghadri M, Larouche D, Gauvin R, Baker K, Bergeron F, et al. Tissue reorganization in response to mechanical load increases functionality. Tissue Eng. 2005;11(1–2):90–100. https://doi.org/10.1089/ten.2005.11.90.
127. Soares ALF, Stekelenburg M, Baaijens FPT. Remodeling of the collagen fiber architecture due to compaction in small vessels under tissue engineered conditions. J Biomech Eng. 2011;133(7):071002. https://doi.org/10.1115/1.4003870.

128. Nagel T, Kelly DJ. Remodelling of collagen fibre transition stretch and angular distribution in soft biological tissues and cell-seeded hydrogels. Biomech Model Mechanobiol. 2012;11 (3–4):325–39. https://doi.org/10.1007/s10237-011-0313-3.
129. Foolen J, van Donkelaar CC, Soekhradj-Soechit S, Ito K. European Society of Biomechanics S.M. Perren Award 2010: an adaptation mechanism for fibrous tissue to sustained shortening. J Biomech. 2010;43(16):3168–76. https://doi.org/10.1016/j.jbiomech.2010.07.040.
130. Valentín A, Cardamone L, Baek S, Humphrey JD. Complementary vasoactivity and matrix remodelling in arterial adaptations to altered flow and pressure. J R Soc Interface. 2009;6 (32):293–306. https://doi.org/10.1098/rsif.2008.0254.
131. Valentín A, Humphrey JD, Holzapfel GA. A finite element-based constrained mixture implementation for arterial growth, remodeling, and adaptation: theory and numerical verification. Int J Numer Method Biomed Eng. 2013;29(8):822–49. https://doi.org/10.1002/cnm.2555.
132. Baek S, Rajagopal KR, Humphrey JD. A theoretical model of enlarging intracranial fusiform aneurysms. J Biomech Eng. 2006;128(1):142. https://doi.org/10.1115/1.2132374.
133. Baek S, Valentín A, Humphrey JD. Biochemomechanics of cerebral vasospasm and its resolution: II. Constitutive relations and model simulations. Ann Biomed Eng. 2007;35 (9):1498–509. https://doi.org/10.1007/s10439-007-9322-x.
134. Miller KS, Khosravi R, Breuer CK, Humphrey JD. A hypothesis-driven parametric study of effects of polymeric scaffold properties on tissue engineered neovessel formation. Acta Biomater. 2015;11:283–94. https://doi.org/10.1016/j.actbio.2014.09.046.
135. Khosravi R, Miller KS, Best CA, Shih YC, Lee Y-U, Yi T, et al. Biomechanical diversity despite mechanobiological stability in tissue engineered vascular grafts two years post-implantation. Tissue Eng Part A. 2015;21:1529–38. https://doi.org/10.1089/ten.tea.2014.0524.
136. Heck TAM, Wilson W, Foolen J, Cilingir AC, Ito K, van Donkelaar CC. A tissue adaptation model based on strain-dependent collagen degradation and contact-guided cell traction. J Biomech. 2015;48(5):823–31. https://doi.org/10.1016/j.jbiomech.2014.12.023.
137. van Donkelaar CC, Heck TAM, Wilson W, Foolen J, Ito K. Versatility of a collagen adaptation model that includes strain-dependent degeneration and cell traction. Vol. 1A: Abdominal aortic aneurysms; active and reactive soft matter; atherosclerosis; biofluid mechanics; education; biotransport phenomena; bone, joint and spine mechanics; brain injury; cardiac mechanics; cardiovascular devices, fluids and imaging, C. ASME; 2013. p. V01AT02A003. https://doi.org/10.1115/SBC2013-14214.
138. Ellsmere JC, Khanna RA, Lee JM. Mechanical loading of bovine pericardium accelerates enzymatic degradation. Biomaterials. 1999;20(v):1143–50. https://doi.org/10.1016/S0142-9612(99)00013-7.
139. Lee EJ, Holmes JW, Costa KD. Remodeling of engineered tissue anisotropy in response to altered loading conditions. Ann Biomed Eng. 2008;36(8):1322–34. https://doi.org/10.1007/s10439-008-9509-9.
140. Hu J-J, Humphrey JD, Yeh AT. Characterization of engineered tissue development under biaxial stretch using nonlinear optical microscopy. Tissue Eng Part A. 2009;15(7):1553–64. https://doi.org/10.1089/ten.tea.2008.0287.
141. Soares ALF, Oomens CWJ, Baaijens FPT. A computational model to describe the collagen orientation in statically cultured engineered tissues. Comput Methods Biomech Biomed Engin. 2014;17:251–62. https://doi.org/10.1080/10255842.2012.680192.
142. Loerakker S, Obbink-Huizer C, Baaijens FPT. A physically motivated constitutive model for cell-mediated compaction and collagen remodeling in soft tissues. Biomech Model Mechanobiol. 2014;13(5):985–1001. https://doi.org/10.1007/s10237-013-0549-1.
143. Loerakker S, Ristori T, Baaijens FPT. A computational analysis of cell-mediated compaction and collagen remodeling in tissue-engineered heart valves. J Mech Behav Biomed Mater. 2016;58:173–87. https://doi.org/10.1016/j.jmbbm.2015.10.001.

144. Ristori T, Obbink-Huizer C, Oomens CWJ, Baaijens FPT, Loerakker S. Efficient computational simulation of actin stress fiber remodeling. Comput Methods Biomech Biomed Engin. 2016;19(12):1347–58. https://doi.org/10.1080/10255842.2016.1140748.
145. Humphrey JD, Rajagopal KR. A constrained mixture model for growth and remodeling of soft tissues. Math Model Methods Appl Sci. 2002;12(03):407–30. https://doi.org/10.1142/S0218202502001714.
146. Kuhl E. Growing matter: a review of growth in living systems. J Mech Behav Biomed Mater. 2014;29:529–43. https://doi.org/10.1016/j.jmbbm.2013.10.009.
147. Kuhl E, Maas R, Himpel G, Menzel A. Computational modeling of arterial wall growth. Biomech Model Mechanobiol. 2007;6(5):321–31. https://doi.org/10.1007/s10237-006-0062-x.
148. Zöllner AM, Abilez OJ, Böl M, Kuhl E. Stretching skeletal muscle: chronic muscle lengthening through sarcomerogenesis. PLoS One. 2012;7(10):e45661. https://doi.org/10.1371/journal.pone.0045661.
149. Zöllner AM, Buganza Tepole A, Gosain AK, Kuhl E. Growing skin: tissue expansion in pediatric forehead reconstruction. Biomech Model Mechanobiol. 2012;11(6):855–67. https://doi.org/10.1007/s10237-011-0357-4.
150. Göktepe S, Acharya SNS, Wong J, Kuhl E. Computational modeling of passive myocardium. Int J Numer Method Biomed Eng. 2011;27(1):1–12. https://doi.org/10.1002/cnm.1402.
151. Werfel J, Krause S, Bischof AG, Mannix RJ, Tobin H, Bar-Yam Y, et al. How changes in extracellular matrix mechanics and gene expression variability might combine to drive cancer progression. PLoS One. 2013;8(10):e76122. https://doi.org/10.1371/journal.pone.0076122.
152. Rouillard AD, Holmes JW. Coupled agent-based and finite-element models for predicting scar structure following myocardial infarction. Prog Biophys Mol Biol. 2014;115(2–3):235–43. https://doi.org/10.1016/j.pbiomolbio.2014.06.010.
153. Rouillard AD, Holmes JW. Mechanical regulation of fibroblast migration and collagen remodelling in healing myocardial infarcts. J Physiol. 2012;590(Pt 18):4585–602. https://doi.org/10.1113/jphysiol.2012.229484.
154. Fomovsky GM, Holmes JW. Evolution of scar structure, mechanics, and ventricular function after myocardial infarction in the rat. Am J Physiol Heart Circ Physiol. 2010;298(1):H221–8. https://doi.org/10.1152/ajpheart.00495.2009.

# Fluid–Structure Interaction Analysis of Bioprosthetic Heart Valves: the Application of a Computationally-Efficient Tissue Constitutive Model

Rana Zakerzadeh, Michael C. H. Wu, Will Zhang, Ming-Chen Hsu, and Michael S. Sacks

**Abstract** This paper builds on a recently developed computationally tractable material model merged with an immersogeometric fluid–structure interaction methodology for bioprosthetic heart valve modeling and simulation. Our main objective is to enable improved application of the use of exogenous crosslinked tissues in prosthesis design through computational methods by utilizing physically realistic constitutive models. To enhance constitutive modeling, valve leaflets are modeled with a computationally efficient phenomenological constitutive relation stemmed from a full structural model to explore the influence of incorporating a high-fidelity material model for the leaflets. We call this phenomenological version as the effective model. This effective model constitutive form is incorporated in the context of the isogeometric analysis to develop an efficient fluid–structure interaction method for thin shell structure of the leaflet tissues. The implementation is supported by representative simulations showing the applicability and usefulness of our effective material model in heart valve simulation framework.

**Keywords** Constitutive model · Fluid–structure interaction · Immersogeometric analysis · Isogeometric analysis · Heart valve

---

R. Zakerzadeh · W. Zhang
James T. Willerson Center for Cardiovascular Modeling and Simulation, The Oden Institute and the Department of Biomedical Engineering, The University of Texas at Austin, Austin, TX, USA

M. C. H. Wu · M.-C. Hsu
Department of Mechanical Engineering, Iowa State University, Ames, IA, USA

M. S. Sacks (✉)
The Oden Institute and the Department of Biomedical Engineering, The University of Texas at Austin, Austin, TX, USA
e-mail: msacks@ices.utexas.edu

© Springer Nature Switzerland AG 2018
M. S. Sacks, J. Liao (eds.), *Advances in Heart Valve Biomechanics*,
https://doi.org/10.1007/978-3-030-01993-8_17

# 1 Introduction

Bioprosthetic heart valves (BHVs) are the most popular replacement heart valve owing to their natural hemodynamic characteristics, high resistance to thrombosis, and good medium-term durability. However, durability of BHVs is still the major limitation of the current technology. Structural valve deterioration may occur which necessitates replacement particularly in younger patients and the replacement surgery has higher risk of mortality. Therefore, the heterograft tissue needs to be refined and modified to improve the life span of BHV. Otherwise durability in the range of 10–15 years continues to hamper BHVs.

Almost all BHV designs use exogenous crosslinked (EXL) collagenous soft tissues obtained from bovine pericardium to manufacture leaflets. Exogenous crosslinking of soft collagenous tissue is a common method for biomaterial development and medical therapies. To enable improved application of the use of EXL tissues in computational methods and prosthetic design, the development of physically realistic constitutive models is clearly required. Recent advances in computational methods to enhance the physical realism of BHV simulations are reviewed in [29]. These computational techniques utilize the advanced numerical methods for the simulation of heart valve function to assess the impact of selecting different constitutive models in the organ-level simulation of a BHV. Computational simulations have shown the impact of different choices of constitutive models to simulate valve function and have observed that the stress and strain distribution in the leaflets was severely impacted by the material modeling choice, as well as the effect of permanent set as the mechanism responsible for the geometry change of BHV leaflets in the long-term cyclic loading. This information should serve not only to infer reliable and dependable BHV function, but also to establish guidelines and insight for the design of future prosthetic valves by analyzing the influence of design, hemodynamics, and tissue mechanics.

BHV leaflet mechanical properties have been demonstrated to be anisotropic. Fiber bundles have a preferred direction of alignment and some degree of dispersion. Wu et al. [27] recently explored the effects of anisotropy on BHV leaflet strain patterns using an immersogeometric fluid–structure interaction (FSI) analysis. Considering that collagen fibers is the main load-bearing structure in the tissue, the results confirmed that the anisotropy of the BHV leaflets is important for its overall function. However, this work was primarily concerned with phenomenologically modeling of the anisotropy and nonlinearity with a limited number of parameters and the proposed Lee–Sacks model cannot account for interactions between fibers or inelastic changes to the material's local stress-free configuration [30]. Moreover, since phenomenological models do not take into account the underlying mechanisms, this category of material models can only match the response of the leaflet tissue in a limited range, where experimental data were acquired for parameter estimation, but cannot predict how the mechanical response will differ outside of that range. However, they have been used in computational simulations to predict the tissue response under unpredictable ranges of deformation. Therefore, high-level simulations still require more complex material models.

The first rigorous full structural-based model which explicitly incorporating various features of the collagen fiber architecture for EXL soft tissues is developed by Sacks et al. [21]. This novel structural model accounts for three contributors to the mechanical response of soft tissue: collagen fibers, EXL matrix, and the interactions between fiber–fiber and fiber–matrix components. The authors concluded that fiber-ensemble interactions played a large role in stress development and often dominating the total tissue response. However, although structural models proposed in Sacks et al. [21] can accurately reproduce the soft tissue response, these models require a multiple integrals over the collagen fiber architecture to accurately compute the strain energy of fiber interactions. This can lead to increase in computational cost, and significant loss of numerical precision when the number of quadrature points is not sufficient. To reproduce the characteristics of this structural model in a computationally efficient way, a simple phenomenological form of this high-fidelity structural constitutive model is described in Zhang et al. [31]. This computationally efficient version, which they referred to it as the effective material model, is able to fully reproduce the response of complex structural model for the entire range of deformations and the material parameters are identified by developing a systematic parameter estimation approach. This effective material model has some advantages over other soft tissue models in literature, which are explained in details in [31]. Effective model can be implemented in place of structural-based material models in numerical simulations, but receive its parameters from the structural-based model through a proper parameter estimation techniques.

Here in this work, we want to explore the significance of using this computationally efficient effective model to BHV modeling and explain why this constitutive model is a suitable choice for heart valve simulation. To answer this question, we demonstrate how this model is formulated within the Kirchhoff–Love shell formulation framework of [15] and we simulate the response of EXL tissue to the cyclic loading using our previously developed immersogeometric FSI method. We demonstrate that this effective model [31] can be used for different tissue types, at the same time, "effective" for finite element simulation of heart valve problems. The material model's strain-energy functional contains an exponential form which takes care of fiber and interaction parts; and the matrix part can be modeled using modified Yeoh model or a simple neo-Hookean form. We show the derivatives of invariants for the effective model in curvilinear coordinates to calculate the stress and elasticity tensors. Following the approach presented in [27], we define the fibers in the physical space and then we transform them to the computational coordinates. Our implementation has been validated using biaxial simulations.

The relative simplicity of the effective material model is suitable for computationally expensive applications such as FSI analysis of the heart valve. Moreover, we have a single functional form that covers all the tissue materials and wide range of deformations. We have shown this by using the model for both crosslinked and native heart valve tissues and comparing them. This is not a trivial feature, as other widely used soft tissue models are not able to match wide range of material properties (e.g., [6, 16]) and also this characteristic is important as the aortic valve leaflet material properties change as a result of biological processes. Moreover, this model allows us to separate the effect of crosslinking from the tissue matrix component and identify the interaction effects. To the best of our knowledge, no previous study to date has focused on investigating how the crosslinking behavior of BHV leaflet influences various characteristics of the valve dynamics. Although based on the results in Sacks et al. [21] we know that the EXL effect is important in stress development, still at this point we don't know how much this effect matters in BHV dynamics. We explore the effect of EXL on BHVs dynamics and strain distribution by separating the effect of exogenous crosslinking from the full structural-based model. This crosslinking behavior takes into account the bending rigidity of EXL fibers, fiber–fiber interaction, and fiber–matrix interaction. To optimally utilize heterograft biomaterials in replacement of heart valves, an improved understanding of the effects of chemical modification on their mechanical behavior is clearly a critical step. Investigations of this type may provide useful clinical information for increases in fatigue resistance resulting from crosslink augmentation.

This manuscript is organized as follows: Sect. 2 reviews the hyperelastic Kirchhoff–Love thin shell formulation of Kiendl et al. [15]. It also provides an overview of the computationally efficient material model that we use for the effective response of planar soft tissues and explains how to implement this constitutive model within the curvilinear shell formulation framework. We verify our implementation by reproducing analytical results following the biaxial test in [23] for BHV leaflets. In Sect. 3 we summarize the immersogeometric FSI simulation setup and propose numerical simulations to explain the importance of our material model tool in heart valve simulation. We discuss the numerical results in Sect. 4. Finally, conclusions and future directions are presented in Sect. 5.

## 2 Modeling Framework

### 2.1 Thin Shell Formulations for the BHV Leaflets

BHV leaflets are modeled as hyperelastic thin shells and discretized isogeometrically, as in [15] and [9]. We assume that this thin shell structure is represented mathematically by its mid-surface. Further, we assume this surface to be piecewise $C^1$-continuous and apply the Kirchhoff–Love shell formulation and isogeometric discretization studied by Kiendl et al. [13, 14] and Kiendl [12]. The spatial coordinates of the shell mid-surface in the reference and current configurations are given by $\mathbf{X}(\xi_1, \xi_2)$ and $\mathbf{x}(\xi_1, \xi_2)$, respectively, parameterized by $\xi_1$ and $\xi_2$.

The Kirchhoff–Love hypothesis of normal cross sections implies that a point $\mathbf{x}$ in the shell continuum can be described by a point $\mathbf{r}$ on the midsurface and a vector $\mathbf{a}_3$ normal to the midsurface:

$$\mathbf{x}(\xi^1, \xi^2, \xi^3) = \mathbf{r}(\xi^1, \xi^2) + \xi^3 \, \mathbf{a}_3(\xi^1, \xi^2), \quad (1)$$

where $\xi^1$, $\xi^2$ are midsurface coordinates, $\xi^3 \in [-h_{\text{th}}/2, h_{\text{th}}/2]$ is the thickness coordinate, and $h_{\text{th}}$ is the shell thickness. Covariant base vectors and metric coefficients are defined by $\mathbf{g}_i = \mathbf{x},_i$ and $g_{ij} = \mathbf{g}_i \cdot \mathbf{g}_j$, respectively, where $(\cdot),_i = \partial(\cdot)/\partial \xi^i$ indicates partial differentiation. We adopt the convention that Latin indices take on values $\{1, 2, 3\}$ while Greek indices take on values $\{1, 2\}$. Contravariant base vectors $\mathbf{g}^i$ are defined by the property $\mathbf{g}^i \cdot \mathbf{g}_j = \delta^i_j$. Contravariant metric coefficients can be obtained by matrix inversion $[g^{ij}] = [g_{ij}]^{-1}$. For the shell model, only in-plane components of $g_{ij}$ and linear terms in $\xi^3$ are considered:

$$g_{\alpha\beta} = a_{\alpha\beta} - 2\, \xi^3 b_{\alpha\beta}, \tag{2}$$

where $a_{\alpha\beta} = \mathbf{a}_\alpha \cdot \mathbf{a}_\beta$, $b_{\alpha\beta} = \mathbf{a}_{\alpha,\beta} \cdot \mathbf{a}_3$, $\mathbf{a}_\alpha = \mathbf{r},_\alpha$, and $\mathbf{a}_3 = \frac{\mathbf{a}_1 \times \mathbf{a}_2}{\|\mathbf{a}_1 \times \mathbf{a}_2\|}$. The definition holds for both deformed and undeformed configurations. Variables of the latter are indicated by $(\mathring{\cdot})$, e.g., $\mathring{\mathbf{x}}$, $\mathring{\mathbf{g}}_{,i}$, $\mathring{g}_{ij}$, etc. The Jacobian determinant of the structure's motion is $J = \sqrt{|g_{ij}|/|\mathring{g}_{ij}|}$ and the in-plane Jacobian determinant is $\hat{J} = \sqrt{|g_{\alpha\beta}|/|\mathring{g}_{\alpha\beta}|}$.

The weak form of the shell structural formulation is

$$\int_{\Gamma_0} \mathbf{w} \cdot \rho h_{th} \frac{\partial^2 \mathbf{y}}{\partial t^2}\bigg|_\mathbf{X} d\Gamma + \int_{\Gamma_0} \int_{-\frac{h_{th}}{2}}^{\frac{h_{th}}{2}} \delta \mathbf{E} : \mathbf{S} \, d\xi^3 d\Gamma \\ - \int_{\Gamma_0} \mathbf{w} \cdot \rho h_{th} \mathbf{f} \, d\Gamma - \int_{\Gamma_t} \mathbf{w} \cdot \mathbf{h}^{net} \, d\Gamma = 0, \tag{3}$$

where $\mathbf{y}$ is the midsurface displacement, the derivative $\partial(\cdot)/\partial t|_\mathbf{X}$ holds material coordinates $\mathbf{X}$ fixed, $\rho$ is the density, $\mathbf{S}$ is the second Piola–Kirchhoff stress, $\delta \mathbf{E}$ is the variation of the Green–Lagrange strain corresponding to displacement variation $\mathbf{w}$, $\mathbf{f}$ is a prescribed body force, $\mathbf{h}^{net}$ is the total traction from the two sides of the shell, and $\Gamma_0$ and $\Gamma_t$ are the shell midsurfaces in the reference and deformed configurations. The Green–Lagrange strain is $\mathbf{E} = \frac{1}{2}(\mathbf{C} - \mathbf{I})$, where $\mathbf{C}$ is the right Cauchy–Green deformation tensor and $\mathbf{I}$ is the identity tensor. For Kirchhoff–Love shell kinematics, only in-plane strain components are computed, using $E_{\alpha\beta} = \frac{1}{2}(g_{\alpha\beta} - \mathring{g}_{\alpha\beta})$. The second Piola–Kirchhoff stress tensor is obtained from a hyperelastic strain-energy density functional $\psi$:

$$\mathbf{S} = \frac{\partial \psi}{\partial \mathbf{E}} = 2\frac{\partial \psi}{\partial \mathbf{C}}. \tag{4}$$

Linearizing Eq. (4) yields the tangent material tensor

$$\mathbb{C} = \frac{\partial \mathbf{S}}{\partial \mathbf{E}} = 4\frac{\partial^2 \psi}{\partial \mathbf{C}^2}. \tag{5}$$

In this work, we assume the material to be incompressible; the elastic strain-energy functional form for the effective model $\psi_{eff}$ is augmented by a constraint term enforcing $J = 1$, via a Lagrange multiplier $p$:

$$\psi = \psi_{eff} - p(J - 1). \tag{6}$$

For shell analysis, one can use the plane stress condition, $S^{33} = 0$, to analytically determine the Lagrangian multiplier $p$ [15, Eq. (46)]. Furthermore, one can express $E_{33}$ in terms of $E_{\alpha\beta}$ [15, Eq. (38)] and derive an in-plane material tangent tensor $\hat{\mathbb{C}}$ such that $dS^{\alpha\beta} = \hat{\mathbb{C}}^{\alpha\beta\gamma\delta} dE_{\gamma\delta}$, where dS and dE are total differentials of **S** and **E**:

$$S^{\alpha\beta} = 2\frac{\partial \psi_{eff}}{\partial C_{\alpha\beta}} - 2\frac{\partial \psi_{eff}}{\partial C_{33}} J_o^{-2} g^{\alpha\beta}, \tag{7}$$

$$\begin{aligned}\hat{\mathbb{C}}^{\alpha\beta\gamma\delta} &= 4\frac{\partial^2 \psi_{eff}}{\partial C_{\alpha\beta} \partial C_{\gamma\delta}} + 4\frac{\partial^2 \psi_{eff}}{\partial C_{33}^2} J_o^{-4} g^{\alpha\beta} g^{\gamma\delta} \\ &\quad - 4\frac{\partial^2 \psi_{eff}}{\partial C_{33} \partial C_{\alpha\beta}} J_o^{-2} g^{\gamma\delta} - 4\frac{\partial^2 \psi_{eff}}{\partial C_{33} \partial C_{\gamma\delta}} J_o^{-2} g^{\alpha\beta} \\ &\quad + 2\frac{\partial \psi_{eff}}{\partial C_{33}} J_o^{-2} (2g^{\alpha\beta} g^{\gamma\delta} + g^{\alpha\gamma} g^{\beta\delta} + g^{\alpha\delta} g^{\beta\gamma}).\end{aligned} \tag{8}$$

With Eqs. (7) and (8), arbitrary 3D constitutive models providing $\frac{\partial \psi_{eff}}{\partial C_{ij}}$ and $\frac{\partial^2 \psi_{eff}}{\partial C_{ij} \partial C_{kl}}$ can be directly used for shell analysis. The whole shell formulation is completely described in terms of the shell midsurface deformation. For more details of the shell formulation, see [15].

## 2.2 Leaflet Tissue Material Model

Recently, Zhang et al. [31] developed a constitutive model form for the effective response of soft tissues. This effective material model is applicable to a wide range of soft tissue responses, while being as computationally efficient as most common effective models, such as Holzapfel-Gasser-Ogden [7], generalized Rivlin model [19], May-Newmann model [17], or the generalized Fung model [5] which are popular for numerical simulations. The effective material model proposed by Zhang et al. along with optimal loading paths is able to fully replicate the response of complex meso-scale structural constitutive models for the entire range of deformations. It has been shown to be very effective at capturing the mechanical response of the high-fidelity structural model presented in [21] for EXL collagenous tissue, while provides significant decrease in its computational cost and is relatively easy to implement. We model the constitutive behavior of BHV leaflets using this model which we refer to as the "effective model" form in this paper. Here we first provide an overview of the structural model and then we summarize its effective version.

To briefly summarize, the aforementioned structural model is composed of three components: collagen, $\psi_C$, matrix, $\psi_M$, and interactions, $\psi_I$.

$$\psi = \psi_M + \psi_C + \psi_I \tag{9}$$

The matrix term, $\psi_M$, can be modeled as a modified version of the Yeoh model or neo-Hookean form. As the modified Yeoh model, it is composed of two terms both a function of the invariant $I_1 = \mathrm{tr}(\mathbf{C})$,

$$\psi_M = \frac{\eta_M}{2}\left(\frac{1}{a}(I_1 - 3)^a + \frac{r}{b}(I_1 - 3)^b\right), \tag{10}$$

with $1 < a < b, ab < 2, 0 \leq r$.

This model contains four parameters: $\eta_M$ is the modulus parameter corresponding to the same parameter in the neo-Hookean model, $a$, $b$, and $r$ are the shape parameters, where $a$ and $b$ control the shape of the two terms, while $r$ is the weight between the two terms. In general, we find $a$, $b$, and $r$ to be very consistent between specimens, where $a \approx 1$, $b \approx 1.87$, and $r \approx 15$ can be treated as constants.

$$\psi_M = \frac{\mu_a}{2a}(I_1 - 3)^a + \frac{\mu_b}{2b}(I_1 - 3)^b, \tag{11}$$

with $1 < a < b, a.b < 2$.

In the case of neo-Hookean model for the extracellular matrix phase we have

$$+\psi_M = \frac{c_{NH}}{2}(I_1 - 3). \tag{12}$$

Here $c_{NH}$ represents matrix modulus. Modified Yeoh model produces a response that is more linear in second Piola–Kirchhoff stress and stretch in comparison to the neo-Hookean model.

The collagen contribution is

$$\psi_C = \phi_c \eta_C \int_\theta \Gamma_1(\theta) \int_1^{\lambda_\theta} D_1(\lambda_s)\left(\frac{\lambda_\theta}{\lambda_s} - 1\right)^2 \mathrm{d}\lambda_s \mathrm{d}\theta. \tag{13}$$

In Eq. (13), $\phi_c$ is the mass fraction of the collagen fibers and $D_1$ and $\Gamma_1$ are the collagen fiber recruitment distribution function and orientation distribution function, respectively. $\lambda_\theta = \sqrt{n_\theta \cdot \mathbf{C} n_\theta}$ is the stretch of the collagen fiber ensemble referenced to the referential configuration of the tissue and $\lambda_s$ is the slack stretch required to straighten the collagen fiber crimp. $\lambda_t = \lambda_\theta/\lambda_s$ is the stretch of the collagen fibers after it's straightened, which is also the true fiber stretch. $\eta_C$ is defined as the modules of collagen and terms.

Similarly, the response of the interaction term is given by:

$$\psi_I = \frac{\eta_I}{2} \int_\alpha \int_\beta \Gamma(\alpha)\Gamma(\beta) \int_1^{\lambda_\alpha} \int_1^{\lambda_\beta} D(x_\alpha)D(x_\beta)\left(\frac{\lambda_\alpha \lambda_\beta}{x_\alpha x_\beta} - 1\right)^2 dx_\alpha\, dx_\beta\, d\alpha\, d\beta \qquad (14)$$

$\eta_I$ accounts for interactions modulus and $\alpha$ and $\beta$ are directions in the tissue that two fiber ensembles are oriented along them, and $\lambda_\alpha$ and $\lambda_\beta$ denote stretches of the collagen fibers in these directions. For details of this model, readers are referred to [21].

The second Piola–Kirchhoff stress of all three components of the full model (Eq. (9)), using $\mathbf{S} = 2\frac{\partial \psi}{\partial \mathbf{C}}$, is given by

$$\begin{aligned}
\mathbf{S} = \mathbf{S}_C + \mathbf{S}_I + \mathbf{S}_M = &\ \phi_C \eta_C \int_\theta \Gamma_1(\theta) \left\{ \int_1^{\lambda_\theta} \frac{D_1(x)}{x}\left(\frac{1}{x} - \frac{1}{\lambda_\theta}\right) dx \right\} \mathbf{n}_\theta \otimes \mathbf{n}_\theta d\theta \\
&+ \phi_C \eta_I \int_\alpha \int_\beta \Gamma_1(\alpha)\Gamma_1(\beta) \\
&\times \Bigg[\left\{ \int_1^{\lambda_\alpha}\int_1^{\lambda_\beta} \frac{2\lambda_\beta D_1(x_\alpha)D_1(x_\beta)}{x_\alpha x_\beta}\left(\frac{\lambda_\alpha \lambda_\beta}{x_\alpha x_\beta} - 1\right) dx_\alpha\, dx_\beta \\
&\qquad + \int_1^{\lambda_\beta} D_1(x_\beta)\left(\frac{\lambda_\beta}{x_\beta} - 1\right)^2 dx_\beta \right\} \frac{\mathbf{n}_\alpha \otimes \mathbf{n}_\alpha}{\lambda_\alpha} \\
&+ \left\{ \int_1^{\lambda_\alpha}\int_1^{\lambda_\alpha} \frac{2\lambda_\beta D_1(x_\alpha)D_1(x_\beta)}{x_\alpha x_\beta}\left(\frac{\lambda_\alpha \lambda_\beta}{x_\alpha x_\beta} - 1\right) dx_\alpha\, dx_\beta \\
&\qquad + \int_1^{\lambda_\alpha} D_1(x_\alpha)\left(\frac{\lambda_\alpha}{x_\alpha} - 1\right)^2 dx_\alpha \right\} \frac{\mathbf{n}_\beta \otimes \mathbf{n}_\beta}{\lambda_\beta} \Bigg] d\alpha\, d\beta \\
&+ \phi_M \eta_M \left[ \left((I_1 - 3)^{\alpha-1} + r(I_1 - 3)^{\beta-1}\right)\left(\mathbf{I} - C_{33}\mathbf{C}^{-1}\right) \right]
\end{aligned} \qquad (15)$$

where $\phi_C$ and $\phi_M$ are the mass fraction of collagen and matrix, respectively.

Using the structural model (Eq. (15)), the mechanical response of bovine pericardium leaflets under a wider range of deformations is explored in and the effective material model is introduced [31]. The model functional form uses a transversely isotropic exponential type term to model the network of collagen fibers and the interaction effects. The response of soft tissues in the low stress region and to the shearing is mainly due to the matrix, especially before collagen recruitment. To account for the extracellular matrix, modified Yeoh model previously introduced in [21] or neo-Hookean model similar to the approach in [27] is perfectly good to simulate the response of tissues at low stress. The final form of the effective model for heart valve tissue replacement is given by

$$\psi_{eff} = c_0(e^Q - 1) + \psi_M$$
$$Q = b_1 E_M^2 + b_2 E_P^2 + b_4 E_M E_P + b_5 E_M^4 + b_6 E_P^4 + b_9 E_M E_P^3 \quad (16)$$
$$+ b_{10} E_\phi^4 + b_{11} E_M^2 E_\phi^2 + b_{12} E_P^2 E_\phi^2,$$

where $c_0, b_1, \ldots b_{12}$ are material parameters. Given a preferred fiber direction $\vec{M}$ as a unit vector defining the collagen fiber direction in the reference configuration, and its orthogonal direction $\vec{P}$, we define the Green strains relative to this basis as follows:

$$E_M = \vec{M} \cdot \mathbf{E} \ \vec{M} = \frac{1}{2} M^i M^j g_{ij} - \frac{1}{2}, \quad (17)$$

$$E_P = \vec{P} \cdot \mathbf{E} \ \vec{P} = \frac{1}{2} P^i P^j g_{ij} - \frac{1}{2}, \quad (18)$$

$$E_\phi = \vec{M} \cdot \mathbf{E} \ \vec{P} = \frac{1}{2} M^i P^j g_{ij}, \quad (19)$$

$$E_M = \vec{M} \cdot \mathbf{E} \ \vec{M}, \quad (20)$$

$$E_P = \vec{P} \cdot \mathbf{E} \ \vec{P}, \quad (21)$$

$$E_\phi = \vec{M} \cdot \mathbf{E} \ \vec{P}, \quad (22)$$

$E_M, E_P, E_\phi$ are Green–Lagrange strain with respect to the material axis. These will be used in place of the invariants in the effective material mode so that the model parameters are invariant with respect to rigid body motion and changes in the reference coordinate system. The second Piola–Kirchhoff stress is given by

$$\mathbf{S} = \frac{\partial \psi}{\partial E_M} \vec{M} \otimes \vec{M}^T + \frac{\partial \psi}{\partial E_S} \vec{s} \otimes \vec{s}^T + \frac{1}{2} \frac{\partial \psi}{\partial E_\phi} (\vec{M} \otimes \vec{P} + \vec{P} \otimes \vec{M}). \quad (23)$$

The elasticity tensor is given by the second derivative of the strain energy function with respect to the right Cauchy strain. Using the chain rules, fully expanding all terms, and enforcing symmetry of partial derivatives, the generalized form for the elasticity tensor is obtained. Equations (7) and (8) require the following derivatives of $\psi_{eff}(E_M, E_S, E_\phi)$:

$$\frac{\partial \Psi_{eff}(E_M, E_S, E_\phi)}{\partial C_{ij}} = \frac{\partial \Psi_{eff}}{\partial E_M}\frac{\partial E_M}{\partial C_{ij}} + \frac{\partial \Psi_{eff}}{\partial E_S}\frac{\partial E_S}{\partial C_{ij}} + \frac{\partial \Psi_{eff}}{\partial E_\phi}\frac{\partial E_\phi}{\partial C_{ij}},$$

$$\frac{\partial^2 \Psi_{eff}(E_M, E_S, E_\phi)}{\partial C_{ij}\partial C_{kl}} = \frac{\partial^2 \Psi_{eff}}{\partial E_M \partial E_M}\frac{\partial E_M}{\partial C_{ij}}\frac{\partial E_M}{\partial C_{kl}} + \frac{\partial^2 \Psi_{eff}}{\partial E_S \partial E_S}\frac{\partial E_S}{\partial C_{ij}}\frac{\partial E_S}{\partial C_{kl}}$$

$$+ \frac{\partial^2 \Psi_{eff}}{\partial E_\phi \partial E_\phi}\frac{\partial E_\phi}{\partial C_{ij}}\frac{\partial E_\phi}{\partial C_{kl}} + \frac{\partial^2 \Psi_{eff}}{\partial E_M \partial E_S}\left(\frac{\partial E_M}{\partial C_{ij}}\frac{\partial E_S}{\partial C_{kl}}\right.$$

$$\left.+ \frac{\partial E_S}{\partial C_{ij}}\frac{\partial E_M}{\partial C_{kl}}\right) \quad (24)$$

$$+ \frac{\partial^2 \Psi_{eff}}{\partial E_M \partial E_\phi}\left(\frac{\partial E_M}{\partial C_{ij}}\frac{\partial E_\phi}{\partial C_{kl}} + \frac{\partial E_\phi}{\partial C_{ij}}\frac{\partial E_M}{\partial C_{kl}}\right)$$

$$+ \frac{\partial^2 \Psi_{eff}}{\partial E_S \partial E_\phi}\left(\frac{\partial E_S}{\partial C_{ij}}\frac{\partial E_\phi}{\partial C_{kl}} + \frac{\partial E_\phi}{\partial C_{ij}}\frac{\partial E_S}{\partial C_{kl}}\right),$$

We present the following derivatives of strain invariants in curvilinear coordinate system for the effective model form with Green–Lagrange strain:

$$\frac{\partial E_M}{\partial C} = \frac{1}{2}\vec{M}\otimes\vec{M} = \frac{1}{2}M^i M^j, \quad (25a)$$

$$\frac{\partial E_S}{\partial C} = \frac{1}{2}\vec{P}\otimes\vec{P} = \frac{1}{2}P^i P^j, \quad (25b)$$

$$\frac{\partial E_\phi}{\partial C} = \frac{1}{4}(\vec{M}\otimes\vec{P} + \vec{P}\otimes\vec{M}) = \frac{1}{4}(M^i P^j + P^i M^j) \quad (25c)$$

Moreover, based on the structure of the bovine pericardium used in BHVs we assume that collagen fibers lie primarily in the plane of the tissue (i.e., $M^3 = 0$ in the curvilinear coordinate system of Sect. 2) and our nine response functions are

$$\frac{\partial \Psi_{eff}}{\partial E_M}, \frac{\partial \Psi_{eff}}{\partial E_S}, \frac{\partial \Psi_{eff}}{\partial E_\phi}, \frac{\partial^2 \Psi_{eff}}{\partial E_M^2}, \frac{\partial^2 \Psi_{eff}}{\partial E_S^2}, \frac{\partial^2 \Psi_{eff}}{\partial E_\phi^2}, \frac{\partial^2 \Psi_{eff}}{\partial E_M \partial E_S},$$

$$\frac{\partial^2 \Psi_{eff}}{\partial E_M \partial E_\phi}, \frac{\partial^2 \Psi_{eff}}{\partial E_S \partial E_\phi}. \quad (26)$$

The matrix response can be estimated from the low stress region where collagen does not contribute any stress. A modified Yeoh model as presented in [21] can be utilized. This form was found to fit the low-stress data of crosslinked BHV quite well. Therefore, for the matrix contribution we have

$$\frac{\partial \psi_M(I_1)}{\partial C_{ij}} = \frac{\partial \psi_M}{\partial I_1} \frac{\partial I_1}{\partial C_{ij}}, \tag{27}$$

$$\frac{\partial^2 \psi_M(I_1)}{\partial C_{ij} \partial C_{kl}} = \frac{\partial^2 \psi_M}{\partial I_1 \partial I_1} \frac{\partial I_1}{\partial C_{ij}} \frac{\partial I_1}{\partial C_{kl}}, \tag{28}$$

full expressions for the terms appearing in (27) and (28) are:

$$I_1 = \text{tr}\mathbf{C} = g_{\alpha\beta} \overset{\circ}{g}{}^{\alpha\beta} + C_{33}, \tag{29}$$

$$\frac{\partial \psi_M}{\partial I_1} = \frac{\eta_M}{2}(I_1 - 3)^{a-1} + \frac{\eta_M r}{2}(I_1 - 3)^{b-1}, \tag{30}$$

$$\frac{\partial^2 \psi_M}{\partial I_1 \partial I_1} = \frac{\eta_M}{2}(a-1)(I_1 - 3)^{a-2} + \frac{\eta_M r}{2}(b-1)(I_1 - 3)^{b-2}, \tag{31}$$

$$\frac{\partial I_1}{\partial C_{ij}} = \overset{\circ}{g}{}^{ij}, \tag{32}$$

## 2.3 Implementation Verification of the Material Model

To ensure that the stress update and tangent stiffness were properly implemented, stress-controlled single element biaxial tests were performed. We adopt the static testing setup of [20] similar to the validation procedure presented in [27]. Briefly, node stress of the single-element was prescribed as boundary condition to control the element deformation. The nominal stress imposed on each side of the specimen varies from 0 to 100 MPa. The Cartesian material coordinates $X_1$ and $X_2$ are oriented at a 45° angle relative to the Cartesian specimen coordinates $X'_1$ and $X'_2$. The fiber direction **M** is aligned with $X_1$. We simulate the equibiaxial tests using one cubic B-spline element. The material parameters reproduced using the effective model for an EXL bovine pericardium tissue sample [21] are used. The resulting finite element updated strains were used as input to the constitutive model of Eq. (7) to calculate the theoretical updated stresses. The theoretical stresses were then compared with the finite element stresses to validate the correctness of the model implementation. The result demonstrates exact agreement between the theoretical results and numerical simulation. It indicated that the material model was successfully incorporated into our isogeometric analysis (IGA). The left panel in Fig. 1 shows the stress-controlled equibiaxial testing setup. Arrows on the edges represent nominal stresses. The stress-strain curve shown in the right panel demonstrates that the stretches in the circumferential direction were small compared to the radial stretches, due to the difference in extensibility between both directions. The lines are analytically evaluation of the constitutive equation for the selected values of parameters characterizing the model. The dots represent the effective model response for values of radial and circumferential stresses plotted against strain.

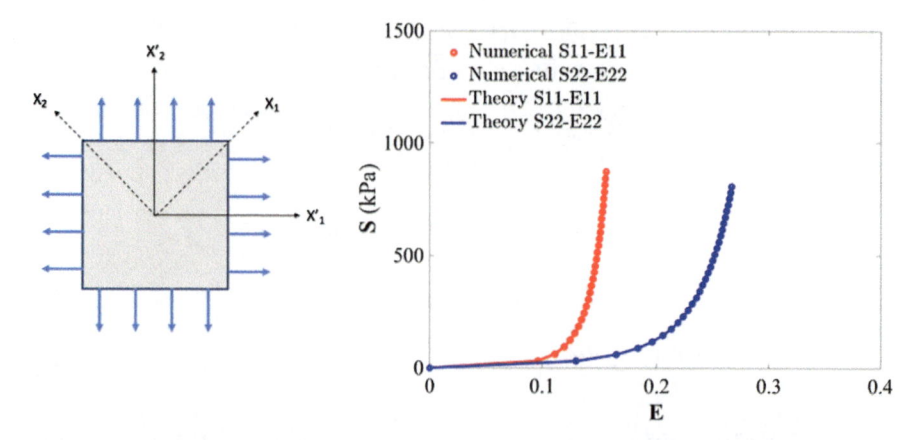

**Fig. 1** Verification and validation of the correct implementation of the material model. Predicted Green strains from an equibiaxial protocol from a single element compared with the theoretical result, demonstrating exact agreement. Line plots are theoretical results and dot plots are numerical results

## 3 Numerical Simulations

In this section we discuss some computational simulations that elaborate the usefulness of the effective model form for heart valve biomechanics simulations. The first test case is designed to show that this model form is suitable to represent different level and direction of anisotropy, and therefore can be used for different tissue types. The second test case shows the effect of interaction term in heart valve simulation under physiological loading condition. The effect of fiber–ensemble interaction in computational simulation of heart valve has not been studied before due to the difficulties to separate the effect of EXL from the material model but we have the tools to perform this investigation. Indeed, one advantage of the full structural model [21] and therefore the effective model is that the various contributions to the total stress can be separated to reveal the EXL effect by examining the effect of it on the individual stress component under the physiological loading of the heart valve. In particular, we perform a qualitative study of the EXL effects on FSI simulation of heart valve by comparing the dynamics of a control and cross-linked specimens. For the purpose of numerical simulations of this paper, we use a neo-Hookean term to model the extracellular matrix and a combination of exponential type terms to model the network of collagen fibers and the interaction effects, according to Eq. (16). The neo-Hookean model is simpler than the modified Yeoh model and is described by fewer material parameters. Before explaining the test cases, we provide a short summary of our computational setup.

## 3.1 Setup of the Immersogeometric FSI Simulation

In this section we summarize the main components of the immersogeometric framework for FSI, as it applies to the simulation of BHVs. For mathematical and implementation details, the reader is referred to [8, 9, 11]. For the simulation of BHV with anisotropic leaflets, the immersogeometric FSI techniques developed in [10, 11] and [8, 9] are utilized. We consider an aortic configuration and model the artery as a 16 cm long elastic cylindrical tube with a three-lobed dilation near the BHV region modeling the aortic sinus (Fig. 2). The cylindrical portion of the artery has an inside diameter of 2.6 cm and a wall thickness of 0.15 cm. The blood flow in the deforming artery is governed by the Navier–Stokes equations of incompressible flow posed on a moving domain. The domain motion is handled using the Arbitrary Lagrangian–Eulerian (ALE) formulation [4] and follows the motion of the deformable arterial wall, which is governed by equations of large-deformation elastodynamics written in the Lagrangian frame [2]. The discretization of the Navier–Stokes equations makes use of a combination of trivariate quadratic NURBS and variational multiscale formulation [1]. The discretization of the solid arterial wall also makes use of NURBS-based IGA. The discretization between blood flow and arterial wall is assumed to be conforming. BHV leaflets are modeled as rotation-free hyperelastic Kirchhoff–Love shells presented in Sect. 2.1 and discretized by T-splines using Bézier elements [22]. The mesh resolutions are shown in Fig. 2 and are identical to those reported in [9]. The BHV is immersed into the moving blood-flow domain; the immersed FSI problem is formulated using an augmented Lagrangian approach and solved using a semi-implicit time integration procedure [10, 11]. Contact between leaflets is handled by a penalty-based approach and imposed at quadrature points of the shell structure [11]. A combination of the quasi-direct and block-iterative FSI coupling strategies [24] is adopted for solving the coupled FSI problem [8].

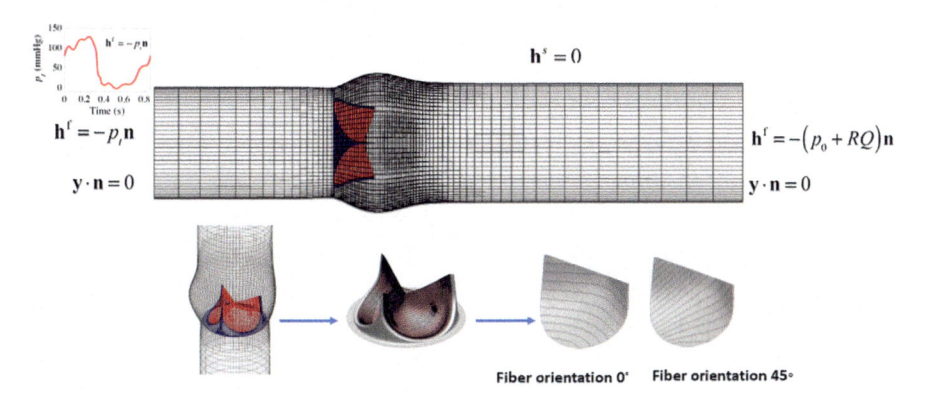

**Fig. 2** Computational domain for the FSI simulations of a trileaflet aortic valve. Top: View of the arterial wall, lumen, valve leaflets, and rigid stent. Bottom: T-Spline surface and fiber orientations used for simulations

The arterial wall is modeled as a neo-Hookean material with dilatational penalty [2], where the shear and bulk moduli of the model are selected to produce a Young's modulus of $1.0 \times 10^7$ dyn/cm$^2$ and Poisson's ratio of 0.45 in the small-strain limit. The density of the arterial wall is set to 1.0 g/cm$^3$. Mass-proportional damping is added to model the interaction of the artery with surrounding tissue and interstitial fluid, with the damping coefficient set to $1.0 \times 10^4$ s$^{-1}$. The density and dynamic viscosity of blood flow are set to 1.0 g/cm$^3$ and $3.0 \times 10^{-2}$ g/(cm s), respectively. The BHV leaflet density and thickness are set to 1.0 g/cm$^3$ and 0.0386 cm, respectively. The choice of these values is based on the discussions in [9] and the references therein.

The inflow and outflow boundaries are located at the ventricular and aortic ends of the channel shown in Fig. 2. The designations of inflow and outflow are based on the prevailing flow direction during systole. We apply a physiologically realistic left ventricular pressure waveform (Fig. 2) as a traction boundary condition at the inflow. The applied pressure signal is periodic with a period of 0.86 s. The traction $-(p_0 + RQ)\mathbf{n}$ is applied at the outflow for the resistance boundary condition [26], where $p_0$ is a constant physiological pressure level, $\mathbf{n}$ is the outward-facing normal of the fluid domain, $R > 0$ is a resistance coefficient, and $Q$ is the volumetric flow rate through the outflow. In the present computation, we set $p_0 = 80$ mmHg and $R = 70$ (dyn s)/cm$^5$. These values ensure a realistic transvalvular pressure difference of 80 mmHg across a closed valve while permitting a reasonable flow rate during systole.

The inlet and outlet cross sections are free to slide in their tangential planes and deform radially, but constrained not to move in the orthogonal directions [3]. The outer wall of the artery has a zero-traction boundary condition. The metal frame and suture ring of the BHV are assumed to be rigid and fixed in space; homogeneous Dirichlet conditions are applied to any control point of the solid portion of the artery mesh whose corresponding basis function's support intersects the stationary metal frame and suture ring. The edges of the BHV leaflets are clamped to the rigid frame. The time step size for the FSI simulation is $\Delta t = 1.0 \times 10^{-4}$ s.

Each FSI simulation takes about 48 h to compute a full cardiac cycle using 144 processor cores on Lonestar 5.

## 3.2 Different Levels and Directions of Anisotropy Using Effective Model

In this test case we simulate the coupling of the BHV and the surrounding blood flow under physiological conditions. The BHV leaflets are modeled using various materials with different degree of anisotropy and various direction of anisotropy, demonstrating the effectiveness of the proposed techniques for wide range of material properties and deformations in practical computations. In particular, to study the effect of material anisotropy due to the level of anisotropy and fiber orientation, we perform FSI simulations of two cases. We perform FSI simulation of a valve using the native porcine tissue for the leaflets and a valve of the same geometry modeled

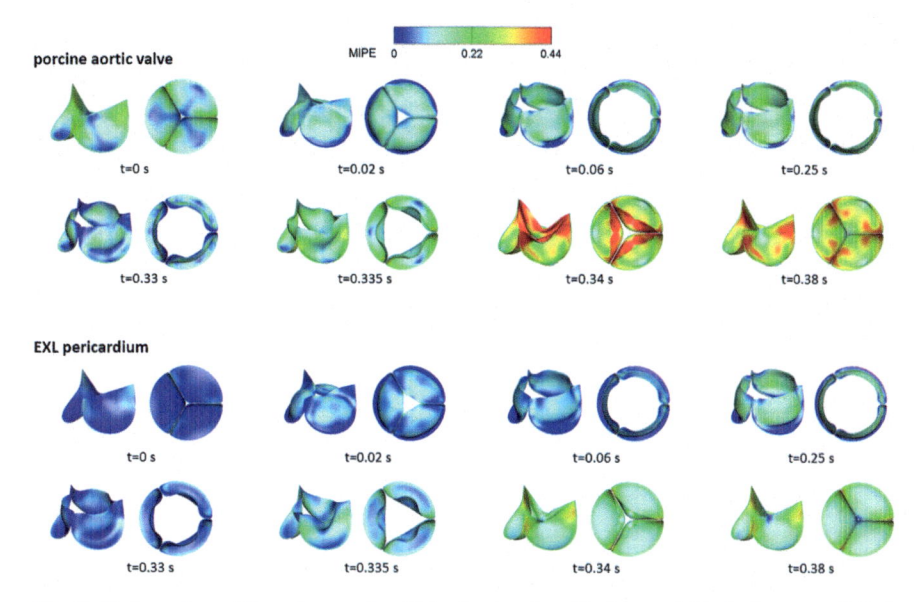

**Fig. 3** Deformation of the valve results obtained using the effective model both for a crosslinked BP and a valve of the same geometry modeled using native porcine parameters, fiber direction is zero for native valve and 45 for BHV. The color represents maximum in-plane principal strain

using crosslinked bovine pericardium tissue, and we compare the results. Bovine pericardium has mild anisotropy and native porcine tissue has more extreme anisotropy.

Starting from the structural model (Eq. (15)), we reproduced the mechanical response of bovine pericardium and porcine aortic valve leaflets using our single form model, i.e. effective model, to determine the rest of the parameters of the effective form. Details of the parameter estimation procedure have been previously presented in [31]. The coefficients $c_{NH}$ of the neo-Hookean model are assumed to be $1.0 \times 10^6$ dyn/cm$^2$ for crosslinked bovine pericardium and $1.0 \times 10^5$ dyn/cm$^2$ for the native porcine tissue simulation. Fiber direction of $45°$ for crosslinked pericardium sample and $0°$ for native porcine sample has been assumed. The result is shown in Figs. 3 and 4.

Figure 3 illustrates the deformations and strain distributions of the valve in both models throughout a period of the FSI simulation. Figure 4 depicts the corresponding flow velocity fields in the fluid domain obtained for both cases. We observe that there is no that much change in overall flow field between two tissue type cases, however the valve kinematics is very different. Several important qualitative differences between the valve deformations in two models are observed. Firstly, the peak strain in native tissue is much larger and regions of concentrated strains are observed. In the crosslinked pericardium strains are more evenly distributed through the leaflets in a more uniform manner. The results show that the deformations of the native tissue during the whole cycle are markedly different from those computed using crosslinked tissue properties, due to the fact that

**porcine aortic valve**

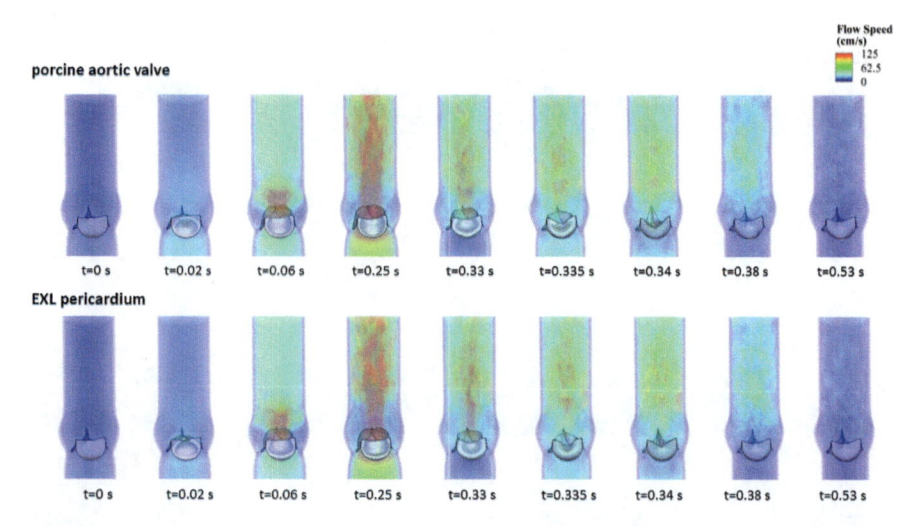

t=0 s  t=0.02 s  t=0.06 s  t=0.25 s  t=0.33 s  t=0.335 s  t=0.34 s  t=0.38 s  t=0.53 s

**EXL pericardium**

t=0 s  t=0.02 s  t=0.06 s  t=0.25 s  t=0.33 s  t=0.335 s  t=0.34 s  t=0.38 s  t=0.53 s

**Fig. 4** Velocity field at several points during a cardiac cycle obtained using the effective model both for a crosslinked BP and a valve of the same geometry modeled using native porcine parameters

chemical modification stiffened the pericardium tissue. This difference is more noticeable during diastolic pressure. In particular, in native case diastolic state exhibits much greater sagging of the belly region. The strain field at time $t = 0:34$ s for native valve is also interesting in that the strain near the commissure points is significantly higher than it is at the same time point for the other model (and also when the valve is fully closed). This is due to the effect of the fluid hammer striking the valve as it initially closes. At the fully closed stage, asymmetrical fiber orientation in the $45°$ case produces an asymmetrical strain distribution as shown in Fig. 3.

The simulated valves in both models open and close at roughly the same time; we define the fully opened stage at $t = 0.25$ s and the fully closed stage at $t = 0.53$ s. The valve opens with the rising left ventricular pressure at the beginning stage of systole (0.0–0.06 s) and stays fully open near the peak systole (0.25 s), allowing sufficient blood flow to enter the ascending aorta. A very quick valve closure, which minimizes the reverse flow into the left ventricle, is observed at the beginning of diastole (0.34–0.38 s). Finally, the valve properly seals and the flow reaches a near-hydrostatic state (0.53–0.78 s).

## 3.3 A Qualitative Study of the Crosslinking Effects on FSI Simulation of BHV

In this test case we explore the effects of the crosslinking behavior of the leaflet tissues on the valve dynamics and its strain distribution. Crosslinking induces both intramolecular and intermolecular crosslinks within the collagen fibers, leading to enhanced interfiber crosslinking (i.e., interaction). Related works have revealed micromechanical interactions due to EXL formation, but the effect of these interactions on the heart valve dynamics under physiological condition remains largely unknown.

We perform FSI simulations to model valve dynamics for a native pericardium and a chemically modified pericardium, by using different material parameters to delineate the mechanical effects of exogenous cross-links and corresponding interaction effects. For this purpose, we separate the effect of EXLs on the fibers and matrix from the full structural model and obtain the corresponding material parameters of the effective model functional form, so that the interaction effects could be separately identified. In particular, we fit the effective model to the synthetic data obtained from our structural-based model when contributions from all the terms are included (matrix, fiber, interaction). This gives us one set of parameters for the simulation which represents the "crosslinked bovine pericardium" case. Next, we create another set of synthetic data using structural-based model when we shut down the interaction contribution term, and we fit the effective model to that. This gives us another set of parameters which represents "native bovine pericardium" case. We perform simulation using both sets and compare the results. Collagen fiber modulus, recruitment distribution function, and orientation distribution function are kept the same between the both specimens. 0° fiber orientation is assumed. We also identify this EXL effect for various extracellular matrix modulus. The coefficients $c_{NH}$ of the neo-Hookean model set to be $1.0 \times 10^6$ dyn/cm$^2$ for the first set of simulations. We also explored this effect for much less stiffer matrix, when we have $c_{NH} = 1.0 \times 10^5$ dyn/cm$^2$.

The deformation of the leaflet cross-section of both native and crosslinked case for several points of a cardiac cycle is presented in Fig. 5 for simulations with different matrix modulus. The chosen time points are consistent with Fig. 3. For the case with smaller matrix modulus, the deformation and strain distribution of the leaflets at a few selected time points in the cardiac cycle, that are a selected time point during opening phase of the leaflet, and when the valve is fully opened and is fully closed (after reaching a periodic solution) is illustrated in Fig. 6.

Based on the results shown in Figs. 5 and 6, the overall behavior of the crosslinked case is very close to that of the native case, especially for the case of using physiological matrix stiffness $c_{NH} = 1.0 \times 10^6$ dyn/cm$^2$. This is expected, given that the force resulted from crosslinking contribution is too small to affect valve deformation at peak flow and at the end of the cycle. In fact the main loading at peak flow when the valve is fully opened and at the end of the diastole when the valve is fully closed is caused by matrix effect and tensile stresses, respectively.

We observed that native tissue undergoes slightly larger peak strains. This is due to its less stiffness as the chemical modification stiffens bovine pericardium. We further noted that the interaction produced substantial contribution in the valve deformation due to bending effects, when the valve transitions from the opening position to the closed state. In particular, the displacement of cross-sections of the leaflet in the fully closed and full-opened stages is almost the same for both native

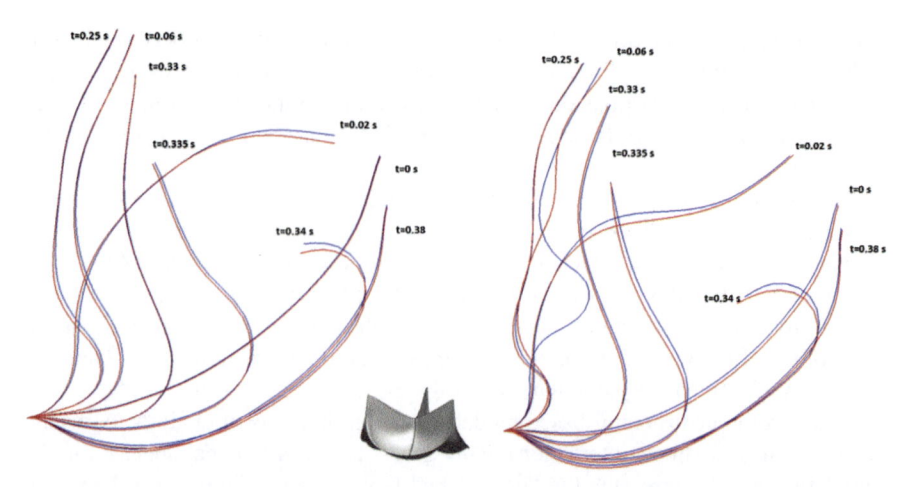

**Fig. 5** Cross-sections of the time dependent leaflet profile, comparison between native BP (blue line) and crosslinked BP (red line) for different matrix modulus. Left plot obtained by using modulus equal to $1.0 \times 10^6$ dyn/cm$^2$ and right plot obtained using modulus $1.0 \times 10^5$ dyn/cm$^2$. Results show that ignoring the effect of EXL causes small difference in valve deformation

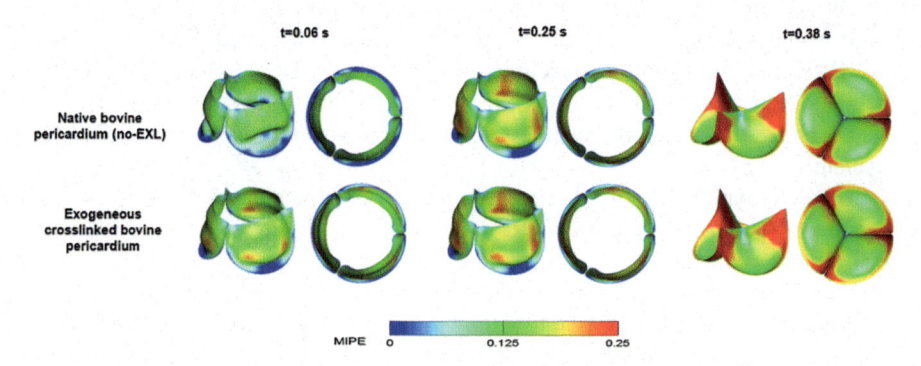

**Fig. 6** Comparison of valve deformation at a selected time point during valve opening, fully opened, and fully closed state between native bovine pericardium and crosslinked bovine pericardium for the right panel plot in Fig. 5 using matrix modulus $1.0 \times 10^5$ dyn/cm$^2$. The color represents maximum in-plane principal strain

and crosslinked tissues, and for different values of matrix stiffness (Fig. 5). The interactions actually produced the largest contributions for the case of softer matrix coefficient $c_{NH} = 1.0 \times 10^5$ dyn/cm$^2$, for the snapshots at time 2 and 6 ms when a portion of the leaflet near belly region undergoes larger strains, while the deformations of the attachment edge are quite similar at all times. When the valve was fully closed the collagen fiber contribution dominates the response and therefore the effect of interaction term is not noticeable. Similar procedure happens when the valve is fully opened and the isotropic matrix contribution dominates the deformation. We also observed that there are no significant differences in the fluid flow between both cases.

## 4 Discussion

A multi-component model is presented in this paper for simulation of the interaction between the blood flow and the valve with realistic representation of the material properties of the leaflets. The material modeling tool incorporated in our framework can be used for large range of material properties and is able to capture the effect of crosslinking behavior and the resulting interaction effect. In this section we discuss the results of our FSI simulation, which underscore the effectiveness of our effective material model in simulation of heart valve.

To investigate the effect of alteration of the tissue mechanical properties, we simulated the deformation of the valve for both native porcine tissue and crosslinked pericardium using the single form of the effective material model. We showed that the effective material model is able to fully reproduce the response of BHV dynamics for various range of deformation. It is not a trivial result as most other constitutive models cannot fit all tissue responses and they are more specialized. We observed that the crosslinked bovine pericardium tissue has more uniform elastic properties and strain distribution. Also the native porcine valve is under significantly higher strain especially near the commissure regions of the valve (see Fig. 3). These observations are mainly due to the differences in mechanical properties in vivo and the measured mechanical response obtained from experiments. In the case of porcine aortic valve, which is extremely anisotropic with very high compliance in the radial direction of the leaflets, residual strain has also a significant impact on its apparent stiffness and therefore the deformation of the leaflets.

We also investigated the contribution of the fiber interaction term and crosslinking behavior in BHV simulation. This is based on the fact that chemical modification induces both intramolecular and intermolecular crosslinks within the collagen fibers, leading to alteration of mechanical properties. There is no measurement of the effect of interaction term available at this time. Hence using this computational framework we separate the effects of EXLs on the fiber and matrix, so that the matrix, collagen, and interaction effects could be clearly identified. We estimated how it affects the valve dynamics throughout the cardiac cycle by eliminating the interaction contribution from the full model and obtaining the resulting material parameters. We noted that when we consider interaction term, the effective model fits the experimental data slightly better comparing to when we don't have EXL term. For the native and crosslinked pericardium, the values of total $R^2$ are 0.96 and 0.97, respectively. By looking at the results (Fig. 5) for valve deformation in a cardiac cycle, an increase in strain peak value is observed in native case which is due to the elimination of interaction related stiffness. The bending stiffness of chemically treated tissue can be up to four times greater than the fresh tissue according to Vesley and Boughner [25]. The increased bending stiffness of the treated tissue is responsible for increased stresses that may ultimately lead to the leaflet tearing and calcification. Furthermore we observed that the difference in valve dynamics is much more pronounced when the valve transitions from the closed stage to the opened stage, as the bending effects are dominating valve response during the valve opening phase (right panel, Fig. 5). During opening, fibers contribute very little to the bending resistance. Bending stiffness of the native tissue is likely responsible for the buckling and localized kinking during the valve opening.

The heart valves operate under a complex cyclic tensile–shear–flexural loading environment. The structure of the aortic valve ensures the high tensile strength for resisting the high transvalvular pressures and the low flexural stiffness as required for normal opening of the valve. Flexure is a major mode of deformation for BHV leaflet during valve operation that induces more complex deformation patterns within the tissue compared to tensile loads; with one portion of tissue in tension while the remaining portion in compression. In other words, flexural deformation is a sensitive method of evaluation of mechanical properties in the low strain range.

Results of this study indicate that the flexural properties of bovine pericardium are dominated by crosslinking behavior. Because crosslinking affects the matrix modulus, it affects the bending modulus and affects the flexural properties. The effective tissue stiffness in the crosslinked tissue is higher than the native tissue. In other words, the glutaraldehyde treated tissue flexural properties is different. This has been also studied earlier by Mirnajafi et al. [18]. We further note that the contribution of the interaction is sensitive to extracellular matrix modulus. More precisely, increasing matrix modulus decreases the effect of interaction terms due to decreasing the bending behavior. The EXL effect is intensified when we have smaller matrix modulus, as it escalates the bending effect. Therefore, increase in the matrix modulus can act as a way to compensate for the EXL effect./AQPlease check the sentence, " In other words, the glutaraldehyde" for clarity.

After chemical fixation the entire extracellular matrix is highly bonded, inducing an increase in tissue stiffness in the low strain range. Further, a slightly stiffer belly region may accommodate smoother multidirectional flexure of the leaflet during valve opening, avoiding wrinkling of the opened leaflet and hence better performance of the valve. However, the bending stiffness due to crosslinking appears to be only slightly affected when matrix modulus is high (left panel, Fig. 5). The displacement of the leaflet cross-section remained largely unchanged following chemical fixation in diastole, when the dominant mode of deformation is tensile stresses, which ensures proper leaflet coaption at the fully closed stage.

Our results suggest that crosslinking may act useful to avoid buckling during valve opening phase. However, if crosslinking increases stiffness too much, this might not be desired as it increases the peak stress within the leaflet and moreover the valve may not be opened properly. It would be optimal for design purpose to produce materials that lead to reduced bending stresses during valve function, yet provide sufficient tensile strength. These findings can be used to guide the development of novel chemical treatment methods that seek to optimize biomechanical properties of prosthetic valve. From a valve design perspective minimizing valve opening forces and transvalvular pressure gradients also requires more knowledge about the heterograft flexural behavior. Flexural deformation modes thus provide a highly sensitive and straightforward means to explore the subtle effects of chemical modification on the mechanics of soft tissues.

## 5 Conclusions

To enhance the realism of the BHV simulations for the design purpose, choosing an accurate material model is imperative as the material properties affect the mechanical behavior of the valve and may have an impact on the stress distribution. The focus of this paper is to provide a framework which enables us to use more physically realistic soft tissue material model in heart valve simulations. We started from the microstructural material model presented in [21] and a phenomenological material model form presented in [31] that covers the interaction between the tissues collagen fibers and its ground matrix as well as the interaction between different fibers. We use this computationally tractable material model to facilitate the integration of the model into our FSI solver. We showed that this effective material model is suitable for the simulation of valve dynamics using different range of leaflet physical properties. Moreover, we observed that crosslinking affects bending stiffness and bending behavior.

Applications of this approach include the utilization in the design of novel chemical treatments to produce specific mechanical responses and the study of fatigue damage in bioprosthetic heart valve biomaterials. Also, since our model capture both fluid flow and valve deformation for large range of material properties, it can be extended to propose the optimized replacement valve that works very similar to the native valve. In particular, we study the interaction between the blood flow and the patient-specific structure of the aortic root and heart valve's leaflets to obtain the geometry of the leaflets which results in best valve performance based on our design criteria similar to the approach presented in [28]. We can also study the effect of age related leaflet structural changes. To capture this process, the model presented in this paper needs to be coupled with an optimization algorithm and prestress. This will be implemented in our future work. These results can provide guidelines for designing leaflet tissues to improve valve durability.

**Acknowledgements** This work was supported by the National Heart, Lung, and Blood Institute of the National Institutes of Health under award number R01HL129077. The author Rana Zakerzadeh is partially supported by ICES Postdoctoral Fellowship. We thank the Texas Advanced Computing Center (TACC) at the University of Texas at Austin for providing HPC resources that have contributed to the research results reported in this paper.

## References

1. Bazilevs Y, Calo VM, Cottrel JA, Hughes TJR, Reali A, Scovazzi G. Variational multiscale residual-based turbulence modeling for large eddy simulation of incompressible flows. Comput Methods Appl Mech Eng. 2007;197:173–201.
2. Bazilevs Y, Calo VM, Hughes TJR, Zhang Y. Isogeometric fluid–structure interaction: theory, algorithms, and computations. Comput Mech. 2008;43:3–37.

3. Bazilevs Y, Hsu M-C, Zhang Y, Wang W, Kvamsdal T, Hentschel S, Isaksen J. Computational fluid–structure interaction: methods and application to cerebral aneurysms. Biomech Model Mechanobiol. 2010;9:481–98.
4. Donea J, Giuliani S, Halleux JP. An arbitrary Lagrangian-Eulerian finite element method for transient dynamics fluid–structure interactions. Comput Methods Appl Mech Eng. 1982;33:689–723.
5. Fung Y-C. Biomechanics. New York: Springer; 1993.
6. Gilmanov A, Stolarski H, Sotiropoulos F. Non-linear rotation-free shell finite-element models for aortic heart valves. J Biomech. 2017;50:56–62.
7. Holzapfel GA, Gasser TC. A new constitutive framework for arterial wall mechanics and a comparative study of material models. J Elast. 2000;61:1–48.
8. Hsu M-C, Kamensky D, Bazilevs Y, Sacks MS, Hughes TJR. Fluid–structure interaction analysis of bioprosthetic heart valves: significance of arterial wall deformation. Comput Mech. 2014;54:1055–71.
9. Hsu M-C, Kamensky D, Xu F, Kiendl J, Wang C, Wu MCH, Mineroff J, Reali A, Bazilevs Y, Sacks MS. Dynamic and fluid–structure interaction simulations of bioprosthetic heart valves using parametric design with T-splines and Fung-type material models. Comput Mech. 2015;55:1211–25.
10. Kamensky D, Evans JA, Hsu M-C. Stability and conservation properties of collocated constraints in immersogeometric fluid-thin structure interaction analysis. Commun Comput Phys. 2015;18:1147–80.
11. Kamensky D, Hsu M-C, Schillinger D, Evans JA, Aggarwal A, Bazilevs Y, Sacks MS, Hughes TJR. An immersogeometric variational framework for fluid–structure interaction: application to bioprosthetic heart valves. Comput Methods Appl Mech Eng. 2015;284:1005–53.
12. Kiendl J. Isogeometric analysis and shape optimal design of shell structures. PhD thesis, Lehrstuhl für Statik, Technische Universität München; 2011.
13. Kiendl J, Bletzinger K-U, Linhard J, Wüchner R. Isogeometric shell analysis with Kirchhoff–Love elements. Comput Methods Appl Mech Eng. 2009;198:3902–14.
14. Kiendl J, Bazilevs Y, Hsu MC, Wüchner R, Bletzinger K-U. The bending strip method for isogeometric analysis of Kirchhoff–Love shell structures comprised of multiple patches. Comput Methods Appl Mech Eng. 2010;199:2403–16.
15. Kiendl J, Hsu M-C, Wu MCH, Reali A. Isogeometric Kirchhoff–Love shell formulations for general hyperelastic materials. Comput Methods Appl Mech Eng. 2015;291:280–303.
16. Kim H, Lu J, Sacks MS, Chandran KB. Dynamic simulation pericardial bioprosthetic heart valve function. J Biomech Eng. 2006;128(5):717–24.
17. May-Newman K, Yin FCP. A constitutive law for mitral valve tissue. J Biomech Eng. 1998;120 (1):38.
18. Mirnajafi A, Raymer J, Scott MJ, Sacks MS. The effects of collagen fiber orientation on the flexural properties of pericardial heterograft biomaterials. Biomaterials 2005;26(7):795–804.
19. Rivlin RS, Saunders DW. Large elastic deformations of isotropic materials. VII. Experiments on the deformation of rubber. Philos Trans R Soc A Math Phys Eng Sci. 1951;243 (865):251–88.
20. Sacks MS, Sun W. Multiaxial mechanical behavior of biological materials. Annu Rev Biomed Eng. 2003;5(1):251–84.
21. Sacks MS, Zhang W, Wognum S. A novel fibre-ensemble level constitutive model for exogenous cross-linked collagenous tissues. Interface Focus. 2015;6(1):20150090.
22. Scott MA, Borden MJ, Verhoosel CV, Sederberg TW, Hughes TJR. Isogeometric finite element data structures based on Bézier extraction of T-splines. Int J Numer Methods Eng. 2011;88:126–56.
23. Sun W, Sacks MS. Finite element implementation of a generalized Fung-elastic constitutive model for planar soft tissues. Biomech Model Mechanobiol. 2005;4:190–9.

24. Tezduyar TE, Sathe S, Stein K. Solution techniques for the fully-discretized equations in computation of fluid–structure interactions with the space–time formulations. Comput Methods Appl Mech Eng. 2006;195:5743–53.
25. Vesely I, Boughner D. Analysis of the bending behaviour of porcine xenograft leaflets and of natural aortic valve material: bending stiffness, neutral axis and shear measurements. J Biomech. 1989;22(6–7):655–71.
26. Vignon-Clementel IE, Figueroa CA, Jansen KE, Taylor CA. Outflow boundary conditions for three-dimensional finite element modeling of blood flow and pressure in arteries. Comput Methods Appl Mech Eng. 2006;195:3776–96.
27. Wu MCH, Zakerzadeh R, Kamensky D, Kiendl J, Sacks MS, Hsu M-C. An anisotropic constitutive model for immersogeometric fluidstructure interaction analysis of bioprosthetic heart valves. J Biomech. 2018;74:23–31.
28. Xu F, Morganti S, Zakerzadeh R, Kamensky D, Auricchio F, Reali A, Hughes TJR, Sacks MS, Hsu M-C. A framework for designing patient-specific bioprosthetic heart valves using immersogeometric fluid–structure interaction analysis. Int J Numer Methods Biomed Eng. 2018;34(4):e2938. https://doi.org/10.1002/cnm.2938.
29. Zakerzadeh R, Hsu M-C, Sacks MS. Computational methods for the aortic heart valve and its replacements. Expert Rev Med Devices. 2017;14(11):849–66.
30. Zhang W, Sacks MS. Modeling the response of exogenously crosslinked tissue to cyclic loading: the effects of permanent set. J Mech Behav Biomed Mater. 2017;75:336–50.
31. Zhang W, Zakerzadeh R, Zhang W, Sacks MS. A computationally efficient material model for the effective response of planar soft tissues. J Mech Behav Biomed Mater.

# Towards Patient-Specific Mitral Valve Surgical Simulations

Amir H. Khalighi, Bruno V. Rego, Andrew Drach, Robert C. Gorman, Joseph H. Gorman, and Michael S. Sacks

**Abstract** Ischemic mitral regurgitation (IMR) occurs when a mitral valve (MV) is rendered incompetent by left ventricular (LV) remodeling induced by a myocardial infarction (MI). Hemodynamically significant, IMR affects at least 300,000 Americans. This important clinical problem is expected to grow substantially during the next 20 years as the population ages. MV repair with undersized ring annuloplasty has been the preferred treatment for IMR. However, 1/3 of all patients treated this way develop significant recurrent IMR within 6 months. Using real-time 3D echocardiography (rt-3DE) image analysis software, it has been demonstrated that IMR in humans is etiologically heterogeneous. In one subset of patients the predominant cause of IMR is annular dilatation and flattening; in the remaining patients, leaflet tethering is the dominant pathology. It has been demonstrated that recurrent IMR after ring annuloplasty occurs most commonly when leaflet tethering is the primary cause of IMR. There is now agreement that adjunctive procedures are required to treat IMR caused by leaflet tethering. However, there is no consensus regarding the best procedure. Multicenter registries and randomized trials would be necessary to prove which procedure is superior. Given the number of proposed procedures and the complexity and duration of such studies, it is highly unlikely that IMR procedure optimization will be achieved by prospective clinical trials. Novel computational approaches directed towards optimized surgical repair procedures can substantially reduce the need for such trial-and-error approaches. We thus present a state-of-the-art means to produce patient-specific MV computational models, which can directly

---

A. H. Khalighi · B. V. Rego · A. Drach
James T. Willerson Center for Cardiovascular Modeling and Simulation, Institute for Computational Engineering and Sciences, Department of Biomedical Engineering, The University of Texas at Austin, Austin, TX, USA

R. C. Gorman · J. H. Gorman
Gorman Cardiovascular Research Group, Smilow Center for Translational Research, Perelman School of Medicine, University of Pennsylvania, Philadelphia, PA, USA

M. S. Sacks (✉)
The Oden Institute and the Department of Biomedical Engineering, The University of Texas at Austin, Austin, TX, USA
e-mail: msacks@ices.utexas.edu

utilize rt-3DE imaging data that can be used develop quantitatively optimized devices and procedures for MV repair.

**Keywords** Mitral valve · Ischemic mitral regurgitation · Finite element modeling · Image-based modeling · Patient-specific model · Chordae tendineae · Surgical repair

## 1 Introduction

IMR occurs when a structurally normal mitral valve (MV) is rendered incompetent as the result of LV remodeling induced by a MI [1]. IMR increases mortality even when mild, with a strongly graded relationship between severity and reduced survival [2, 3]. IMR is present to some degree in over 50% of patients with reduced LV function undergoing coronary artery bypass grafting (CABG) [4, 5] and affects at least 300,000 Americans [4]. The magnitude of the clinical problem is significant and is expected to grow substantially over the next two decades as the population ages [2, 4–6]. MV repair with undersized ring annuloplasty has been the preferred treatment strategy for IMR [7–9]. However, the overall persistence and recurrence rate of moderate or severe IMR within 6–12 months of surgery has been consistently reported to affect a third of the treated patients [10–12]. The high incidence of recurrence after repair may explain the difficulty in demonstrating a survival benefit of annuloplasty for IMR over coronary revascularization alone [12–14]. The negative clinical implications of recurrent IMR were confirmed by the recent randomized multicenter study conducted by the Cardiothoracic Surgical Trials Network (CTSN), which evaluated the relative benefits and risks of repair versus replacement in patients with IMR [12]. The CTSN study demonstrated no significant difference in LV volume or survival at 12 months between repair and replacement groups. However, the recurrence of moderate or severe IMR after 12 months was 33% in the repair group [12]. Subgroup analysis demonstrated that repair patients with recurrent IMR had no reduction in LV volume while repair patients without recurrence experienced LV volume reduction that was superior to patients having valve replacement [12]. These results suggest that a patient-specific approach to treatment guided by preoperative imaging-based risk stratification that is predictive of recurrent IMR would be useful for optimizing surgical results. Our group has created such a tool [15]. Our long-standing laboratory work and that of others [16–19] along with clinical studies [14, 20–22] have helped to elucidate that IMR is associated with annular dilatation and annular flattening as well as leaflet tethering. More recent work using our state-of-the-art real-time 3D echocardiography (rt-3DE) image analysis software has demonstrated that the pathogenesis of IMR in humans is etiologically heterogeneous [20, 23–25]. That is, patients have varying combinations of the pathogenic anatomical abnormalities. In some patients the major contributor to the pathogenesis of IMR is annular dilatation and flattening. While in others annular dilatation is mild to moderate and leaflet tethering is the major pathology driving the IMR.

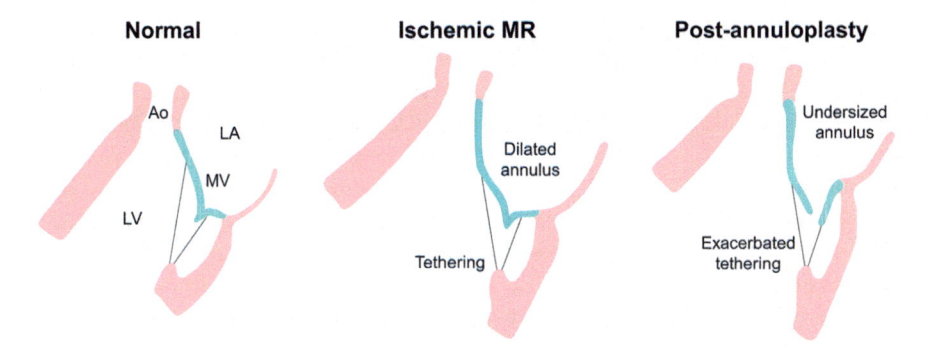

**Fig. 1** IMR is caused by a variable combination of annular dilatation and leaflet tethering. Undersized annuloplasty reduces annular size but exacerbates leaflet tethering. Patients with severe leaflet tethering are at greater risk for recurrent IMR after annuloplasty placement

Undersized ring annuloplasty, while effective in reducing annular area, does little to improve and may actually exacerbate leaflet tethering by displacing the posterior annulus anteriorly, which leads to increased posterior leaflet tethering (Fig. 1) [26]. Therefore, it is likely that the relative contributions of annular dilatation and leaflet tethering to IMR in individual patients determine the clinical response to annuloplasty. Patients with annular dilatation and relatively limited tethering are likely to respond best to ring annuloplasty, while those with severe tethering are more likely to develop recurrent IMR after annuloplasty.

In a recent study, our group demonstrated that real-time 3D echocardiography combined with custom image analysis and valve modeling algorithms provided a powerful tool for quantifying the complex geometry of the entire MV and more importantly is able to effectively predict the risk of recurrent IMR after undersized ring annuloplasty. The study demonstrated that, in fact, patients with leaflet tethering as the predominant cause of IMR responded poorly to ring annuloplasty [15]. It is in this patient population that improvements in MV repair techniques are required. We have made significant advances in our understanding of IMR pathogenesis, valve function/anatomy, leaflet structure and micromechanical properties, and especially in the development of imaging and computational models. These advances have put us in the position to develop computational tools to design procedures and devices for the now well-defined and clinically identifiable patient population that is poorly served by repair with simple annuloplasty. The development of those computational tools, devices, and procedures is the goal of our ongoing MV research endeavors [27–30].

## 2 Image-Guided 3D MV Analysis

Since its invention by Carpentier over 40 years ago, the objective of MV repair has been to reestablish normal leaflet geometry and coaptation [27–33]. Our rt-3DE imaging work and the computational MV modeling has supported Carpentier's early intuition that IMR repair durability is strongly correlated with post-repair valve geometry. We have shown that saddle-shaped annuloplasty reduces valve stress by increasing both leaflet curvature [34, 35] and leaflet coaptation [25] when compared with flat annuloplasty. These studies have also demonstrated that IMR repairs using leaflet augmentation strategies also reduce valve stress [36] and increase leaflet coaptation [37]. While these approaches have improved IMR repair and annuloplasty ring design by putting both on a more quantitative footing, the associated computational tools have lacked the systematic methodology to fully predict post-repair MV geometry based on pre-repair geometry and repair technique. Through clinical research, two points have become clear regarding the treatment of IMR:

1. Successful repair produces better results than replacement.
2. Simple undersized ring annuloplasty is ineffective in at least 1/3 of patients [12, 15].

There is agreement that adjunctive procedures are required for patients with IMR caused by a predominance of leaflet tethering; however, there is no consensus regarding what the best procedure is [38]. Although there is significant clinical interest in developing these procedures the reported data are from small, single-center retrospective or observational studies [39–44]. Multicenter registries and randomized trials would be necessary to prove which procedure is superior. Given the number of proposed procedures and the complexity and duration of such comparative studies, it is highly unlikely that IMR procedure optimization could be effectively carried out this way. As a result, the systematic simulation-based strategy we have designed using our comprehensive computational pipeline offers a unique opportunity for solving this difficult and important clinical problem. Proposed adjunctive procedures fall into two major categories: intra-ventricular papillary muscle/chordal interventions and leaflet augmentation procedures. While the papillary muscle procedures have their putative benefits [39, 40, 42, 43], leaflet augmentation is a technically more straightforward and reproducible way to alleviate tethering [36, 37, 43–47].

## 3 A Comprehensive Pipeline for Multi-Resolution Modeling of the Mitral Valve

Multiple studies have demonstrated that the pathological geometries unique to each patient can affect the durability of MV repairs [12, 15, 34]. While computational modeling of the MV is a promising approach to improve the surgical outcomes, its complex anatomy precludes use of simple geometries. Moreover, the lack of the complete in vivo geometric information presents significant challenges in the development of patient-specific computational models. To address this issue, we have developed a novel pipeline for building attribute-rich computational models of the

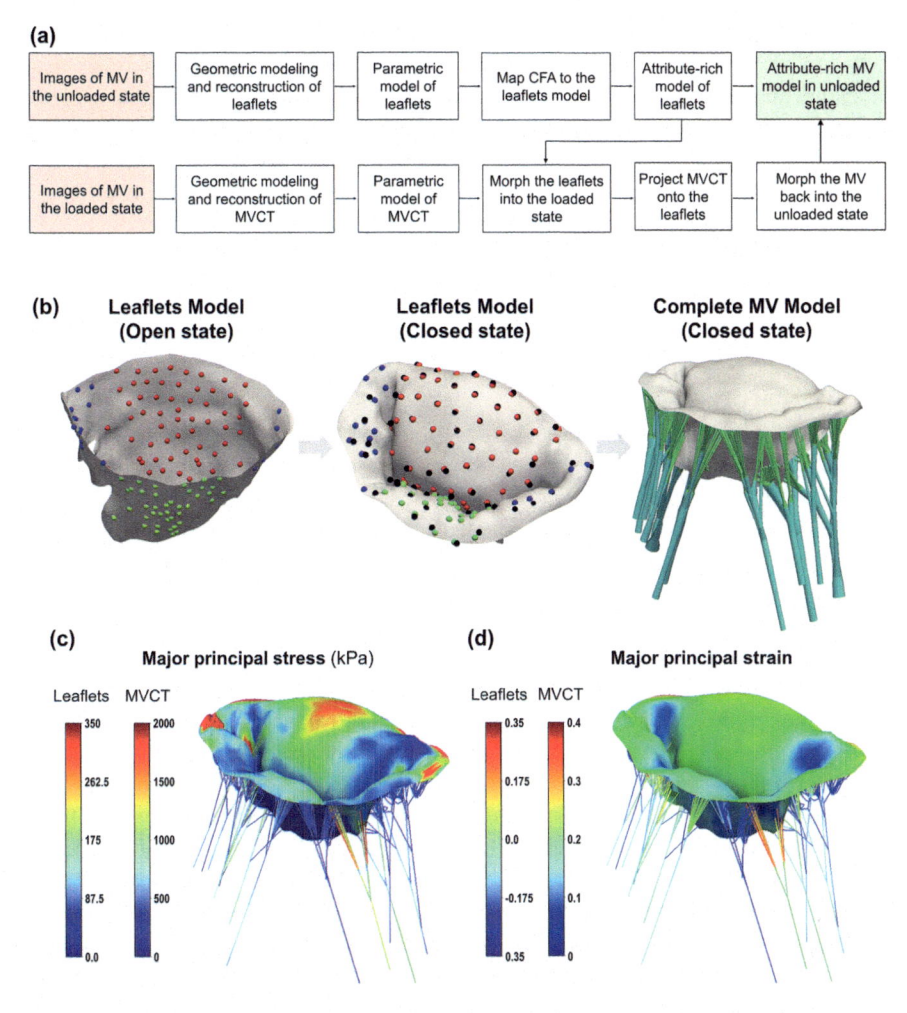

**Fig. 2** (a) Multi-resolution MV model pipeline, based on in vitro images. (b) Our pipeline relies on high-resolution geometric models of the MV leaflets and chordae tendineae from the fully open and fully closed states, respectively, to develop complete MV models with an unparalleled level of geometric accuracy. (c) and (d) demonstrate the results for predicting the stress and strain response of the MV at the end-systolic time point of the cardiac cycle

MV of varying fidelity from in vitro imaging data (Fig. 2). The approach combines high-resolution micro-computed tomography (micro-CT) geometric information from loaded and unloaded states to achieve a high level of anatomic detail, followed by mapping and parametric embedding of tissue attributes to build high-resolution, attribute-rich computational models. Moreover, subsequent lower resolution models were then developed and evaluated by comparing the displacements and surface strains to those extracted from the imaging data. Based on our novel framework to build micron-resolution MV models with local fiber architectural infomartion incorporated, we then established required levels of fidelity for building predictive MV models in the diseased and post-repaired states, demonstrating that a model with a feature and mesh size of ~1 mm was sufficient to predict overall MV shape, stress, and strain distributions with high accuracy. Further, as an extension of the approach described in Fan and Sacks [48], we developed the following total fiber recruitment tensor $\chi$ to provide insight into the deformations and structural response of leaflet tissue microstructure under the mechanical loading, defined as

$$\chi(\mathbf{E}) = \int_{-\pi/2}^{\pi/2} \Gamma(\theta) \left[ \int_0^{E_{ens}(\theta)} \mathbf{D}(\mathbf{x})\, d\mathbf{x} \right] \mathbf{n}(\theta) \bigotimes \mathbf{n}(\theta)\, d\theta \qquad (1)$$

where $\Gamma$ is the fiber orientation distribution function, $\mathbf{D}$ is the fiber ensemble recruitment function, and $E_{ens}$ is the fiber ensemble strain. This is an important indicator of the tissue-level response of the leaflets under different loading conditions. The magnitude of $\chi$ is a normalized index of local structural changes in the leaflet tissue, which varies between zero (slack tissue with full structural reserve) and 1.0 (fully recruited tissue with depleted structural reserve). The choice of the tensorial formulation instead of the single scalar index provides a deeper insight into the directional tissue response.

In addition to the simulations of in vitro states of the MV, we also performed a series of simulations mimicking various pathological and surgically intervened in vivo conditions. The goal of this exploratory study was to estimate the effect of different annulus shapes on the coaptation behavior of the MV. We relied on reconstruction of extant images obtained from live Dorset sheep using rt-3DE to acquire mitral annulur geometries and then averaged three annulus reconstructions to build a representative in vivo annulus. The average annulus was then mapped onto the in vitro MV model (Fig. 2) via arc-length parameterization. Next, the morphed annulus was deformed into additional states: (1) flattened; (2) isotropically enlarged by 20%; (3) isotropically reduced by 20%; (4) anisotropically stretched by 20% in the short-axis direction (SL); and (5) anisotropically stretched by 20% in the long-axis direction (CW) (Fig. 3a). Simulations of the in vitro simulated dilated and repaired states demonstrated a high degree of predictive accuracy [49]. The simulated dilated states, which are based on realistic in vivo estimates of post-IMR human MV, demonstrated that dilation alone (i.e., without changes in papillary muscle position) induced large regional increases in leaflet stress. More interestingly, the structural reserve (as indicated by $\chi$) in the commissure regions increased dramatically. This result underscores why IMR can put such demands on the MV leaflets, leading to deleterious changes in leaflet function and further demonstrates why study of the underlying structural adaptions provides insight into the etiology of the treatment of MV disease].

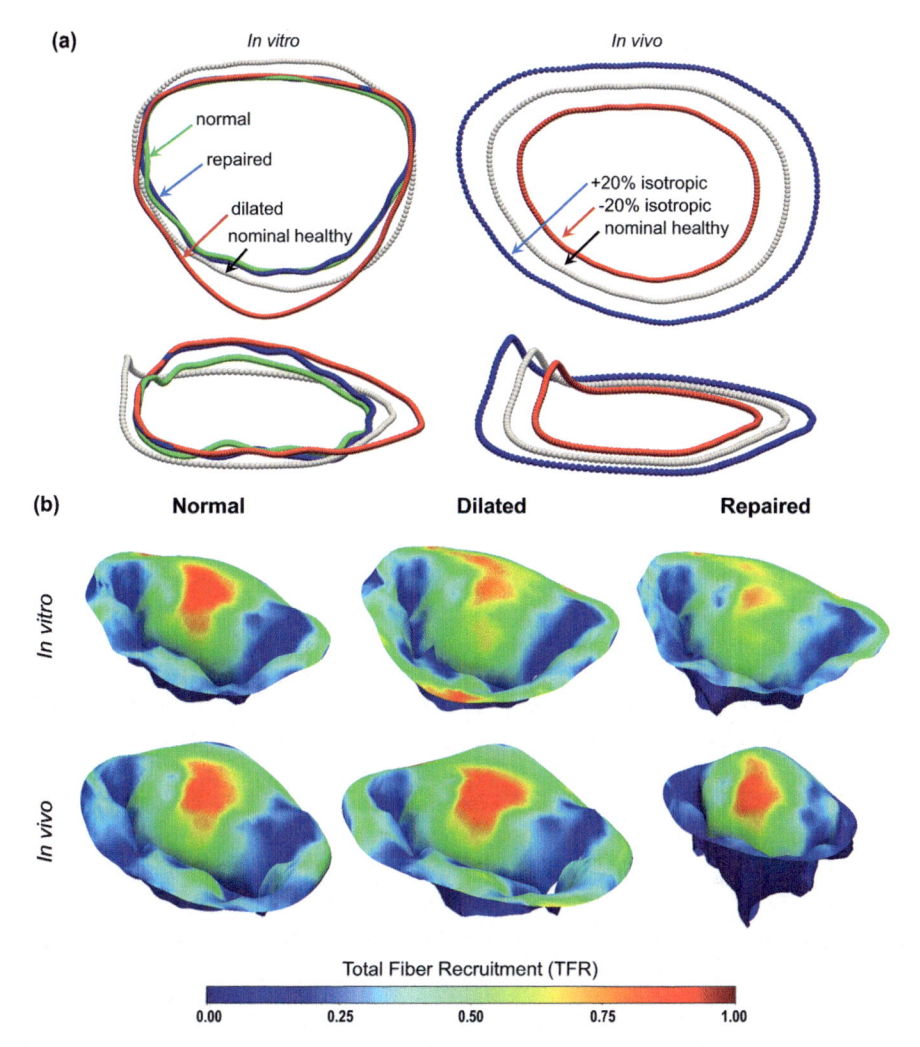

Towards Patient-Specific Mitral Valve Surgical Simulations

**Fig. 3** (**a**) Illustrations of both in vitro and in vivo annular shapes that were enforced as displacement boundary conditions thoughout our MV closure simulations to replicate the effects of MV annulus deformations due to disease (IMR) and surgical treatment (annuloplasty repair) (**b**) Elevated of $\chi$ were observed in the center of the anterior leaflet the diseased and repaired states

## 4 Exploitation of Image-Based Biomechanical Modeling

While the above findings demonstrate the value and utility of detailed, high quality MV models, our ultimate aim is patient-specific surgical planning and understanding of the underlying processes. Such approaches are possible due to recent advances in quantitative rt-3DE MV imaging, which have allowed us to produce images of IMR valves throughout the cardiac cycle. This data will serve as input for our newly developed approaches, described in the following sections that directly incorporate MV-specific geometries and a functionally equivalent subvalvular apparatus (effectively the equivalent MVCT structure) to develop patient-specific models, as described below.

## 4.1 rt-3DE In Vivo Data Acquisition and Segmentation

Following established methods [34, 50], electrocardiographically gated full-volume images were acquired from normal anesthetized sheep (randomized to sex) using a iE33 platform (Philips Medical Systems, Andover, MA) with a 2–7 MHz transesophageal matrix-array transducer over four consecutive cardiac cycles. From each subject's data series, rt-3DE images of the MV in the fully opened, mid-systole, and fully closed state were selected for analysis. These 3D TEE images were exported in Cartesian format (224 × 208 × 208 voxels), with an approximate isotropic resolution of 0.6–0.8 mm. Next, from each rt-3DE image, following [25, 51], the plane of the MV orifice was rotated into a short-axis view. The geometric center of the MV orifice was then be translated to the intersection of two long-axis planes corresponding to the intercommissural and septolateral axes of the MV orifice. A rotational template consisting of 18 long-axis cross-sectional planes separated by 10° increments was superimposed on the 3D image. Two annular points intersecting each of the 18 long-axis rotational planes were then identified and marked interactively (Fig. 4). The anterior and posterior commissures were defined as annular points at the junction between the anterior and posterior leaflets (middle of commissural region) and interactively identified. Finally, the two leaflets were traced separately in parallel long-axis cross-sections, 1 mm apart and sufficient to encompass the entire MV from commissure to commissure (Fig. 4).

## 4.2 In Vivo Leaflet Geometric Model Development Pipeline

Starting from the segmented fully open rt-3DE image of each MV, we constructed MV-specific meshed geometries of the entire leaflet surface, including both the annular and leaflet free-edge boundaries (Fig. 4). First, the interactively traced cross-sections of the leaflet mid-surface were all parameterized using shape-preserving piecewise cubic interpolating polynomial curves, and re-discretized with an equal number of spline segments of uniform arc length [52]. It is straightforward from this representation to obtain separate preliminary meshes of the anterior and posterior leaflets, using a 2D Delaunay triangulation scheme (Fig. 4). To join the two leaflets in the commissural regions, circumferential cross-sections of the entire leaflet surface were generated using a cubic spline interpolation of corresponding parametric (i.e., relative arc length) locations on each trace curve, with periodic tangent and curvature boundary conditions enforced to preserve

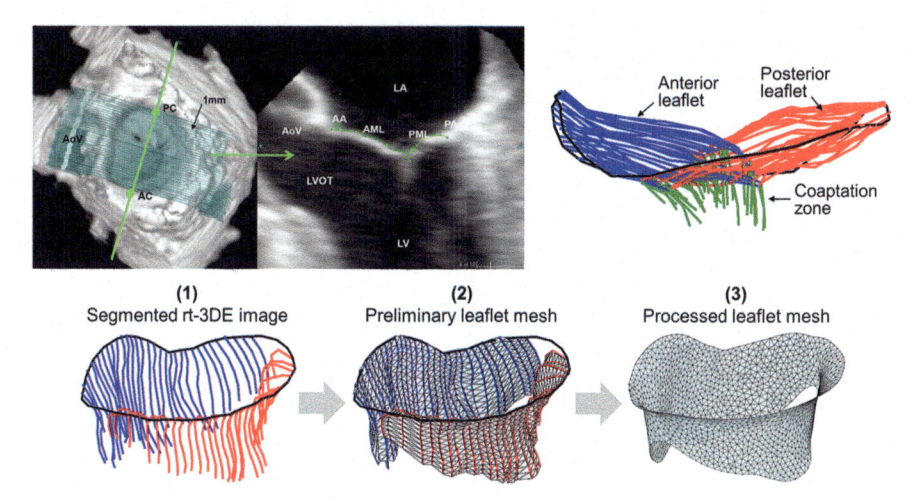

**Fig. 4** Images (rt-3DE) of a representative ovine MV showing the centering and sectioning planes used for tracing the MV leaflet structure. The segmented images were then converted to computational meshes by interpolating the leaflet curve traces and building a triangular surface representation of the leaflet geometry

surface smoothness. This procedure yields a fully parameterized spline representation of the leaflet surface. Lastly, we discretized the surface uniformly using a Poisson-disk sampling method, tuning the sampling density to arrive at a 1 mm mesh resolution. The end product of this pipeline is a *complete MV-specific leaflet geometry* for each specimen (Fig. 4).

## 4.3 Development of the MV FE Model

The above pipeline directly generated accurate 3D models of the MV leaflets from rt-3DE images. However, the subvalvular apparatus, which is composed primarily of the MV chordae tendineae (MVCT), cannot be accurately imaged in vivo. To address this limitation, we developed a novel approach to generate a *MV-specific functionally equivalent subvalvular apparatus*, as described in the following [53]. We first started with converting the leaflet geometry into a working MV FE model. To model the leaflets' mechanical behavior, we applied a nonlinear transversely isotropic material model for the leaflets, which we have extensively demonstrated to be sufficient for obtaining accurate organ-level simulation [49, 54–60]. To implement our models within a computational framework, we used the commercial FE software package ABAQUS 6.14.1 (SIMULIA, Dassault Systèmes, Providence, RI, USA) that was configured to use a nonlinear quasi-static solver with direct explicit time intgration and automatic time stepping. In terms of computational elements, the leaflet geometry discretized using first-order isoparametric triangular elements and assigned shell element type (ABAQUS type S3). This approach was based on our extensive studies on the native MV using high-resolution, in vitro-derived models (Fig. 2) [49, 54, 60–63]. Further, the MVCT geometry was discretized using the first-order isoparametric line elements and assigned uniaxial truss element type (ABAQUS type T3D3) with element-wise cross-sectional areas. The convergence criteria for simulations of valve closure was set to 0.01 in relative displacements and 0.0001 in relative residuals.

The final step in the development of the computational model of the MV is calibration of the complete geometric model of MV. This is necessary to adjust the generic tissue parameters to accurately match the observed behavior of the specific MV. This process involves simultaneous calibration of MVCT (to match the overall shape of the MV in the fully coapted state) and leaflets (to match tenting volume of the leaflets). First, the fully coapted model of MV was morphed into the unloaded (open) state using the previously established full field deformation map for leaflets. Due to lack of the deformation map for the MVCT, the morphing step was performed using a stiff material model with all elements subjected to the gravity load to emulate the conditions of the imaged state. These two conditions allowed us to effectively recover the open-state shape of the imaged MV without changing the length of individual branches of the MVCT. In the next step, we performed a closing simulation of the MV to determine the effective pre-strain in MVCT. The leaflets were subjected to the uniform pressure loading and the MVCT were modeled using the same "stiff" material model. This simulation effectively recovered the closed state of the MV from which we started the calibration step. However, at this stage, each of the MVCT elements is now tensioned with realistic stresses. We then recovered the corresponding strain levels by inverting the actual material model of MVCT. Lastly, to produce fully calibrated models of the full MV, we populated the open-state model with these pre-strain values and re-run the closing simulation using the realistic MVCT material model to verify the accuracy of the calibration step [49].

## 4.4 Development of a Functionally Equivalent Subvalvular Apparatus

To develop the geometrically synthetised funcationally equivalent models of the MV chordal structure, we started with performing the a series of sensitivity studies by starting from the most anatomically accurate model, then simplified the geometry ina systematic approach. Based on our unique previous work on the MVCT anatomy [64], we identified the main geometric attributes which render the CT network rather complex to be the following: (1) pointwise cross-sectional area fields, (2) branching patterns, (3) chordae-to-leaflet insertion locations, and (4) chordae-to-papillary muscle attachment locations. Thus, in our sensitivity studies we perturbed the chordal geometry by simplifying the native structure along the above characteristic attributes (Fig. 5). Simulation results, based on high-resolution in vitro reconstructions [65],

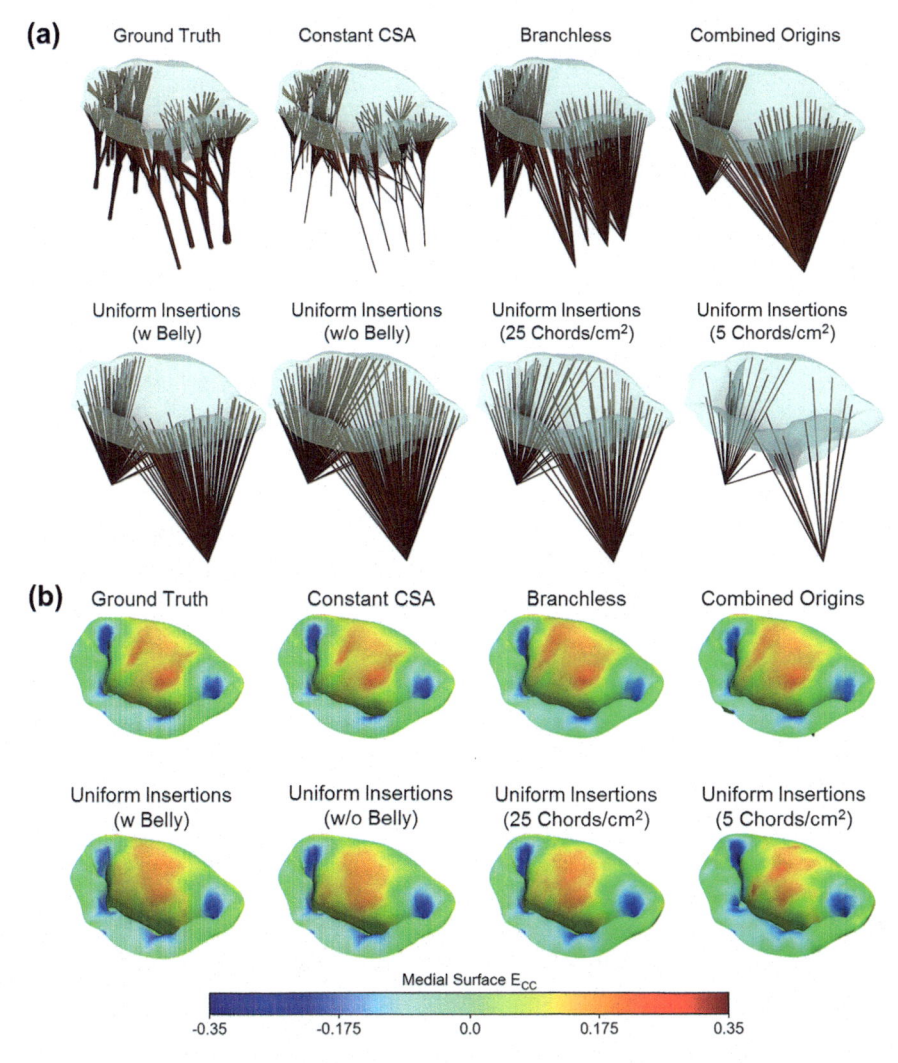

**Fig. 5** (a) Geometric representations of the native (top left) and various MVCT resolution models utilized to represent the effective subvalvular apparatus (i.e., MVCT network). (b) FE simulations using these MVCT network integrated into the MV leaflet model clearly demonstrate that a functionally equivalent MVCT network can be developed

have indicated that several levels of MVCT density and detail can produce accurate results for the normal MV (Fig. 5). Specifically, starting with the native valve MVCT geometry, a series of progressive simplifications of the MVCT were undertaken, including common cross-sectional area, removal of all branching, use of two common origins, and a gradual density reduction. The resulting simulations using the same leaflet geometry clearly demonstrate nearly equivalent results (the circumferential strain field shown in Fig. 5b). The optimal arrangement appeared to be using a

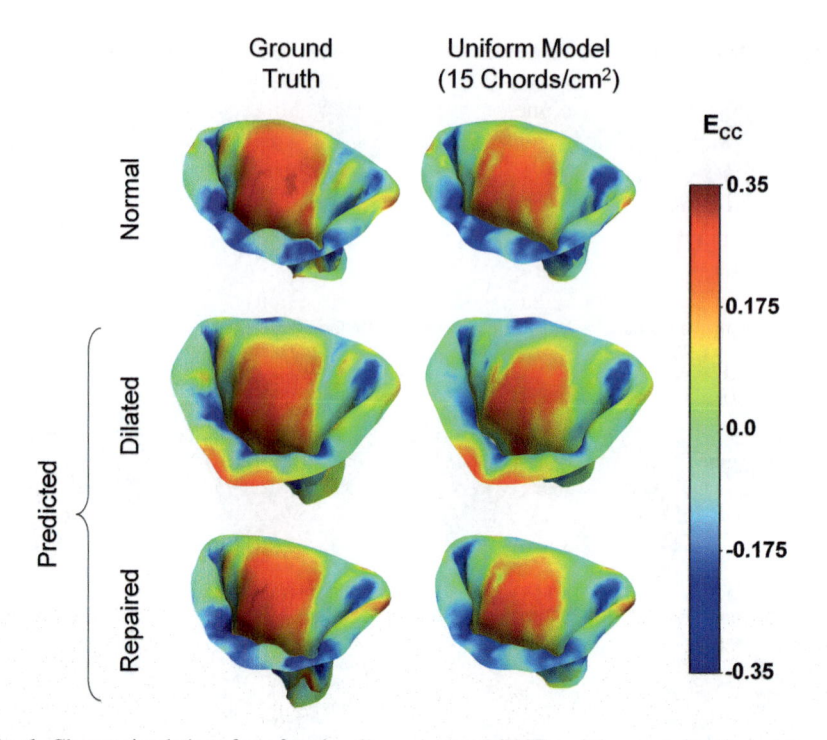

**Fig. 6** Closure simulations for a functionally equivalent MVCT and the ground truth models

chordal density of $15 \pm 2$ insertions/cm² with uniform density and common origin for each papillary muscle group.

Once developed, the simulated MVCT approach's predictive capabilities were evaluated. Note that these evaluations were based entirely on the developed "normal" model with no further modifications. The MV model was tested against two states, dilated and repaired, taken from the same in vitro measurements [49]. Results indicated that, when compared to the fully reconstructed MV, the simulated chordae model was able to predict both the geometric shape and strain distributions in both states (Fig. 6). Thus, reproduction of detailed MVCT network is not required for accurate *organ-level* simulations [63]. Most importantly, this compelling result indicates that *a functionally equivalent subvalvular apparatus can be developed on a valve-specific basis, based on available in vivo leaflet and papillary muscle tip geometry.*

## 5 Future Directions

The results of these study preceding aims suggest that determining the optimal repair scenario in silico. We have shown that there is much more to optimization of an MV repair than basic stress and strain analysis [61]. The various methods that can be used for MV repair can be divided into two main categories: the shape and size of the AP ring, and augmentation patch geometry. While at first glance these appear to be fairly basic parameters, the large number of parameter combinations, combined with complexity of the MV geometric and material nonlinearities, preclude an easy, straightforward determination of the optimal repair strategy. Moreover, there is an inherent question of what constitutes an optimal repair. We have shown that the "structural reserve" of the collagen fiber network (as quantified by the $\chi$ tensor) provides an important additional functional metric. To address this complex task, an MV-specific optimized repair can be defined as that scenario that combines leaflet stress minimization, homogenization of the stress field, maximization of leaflet tissue structural reserve, and coaptation [30]. The next step is clearly to apply the completed pipeline (Fig. 2) to examine its translation to human studies using existing pre- and postsurgical imaging data from our previous work [15].

**Acknowledgments** This work was supported by National Heart, Lung, and Blood Institute of the National Institutes of Health under grant no. R01-HL119297, the National Science Foundation grant no. DGE-1610403, and the American Heart Association grant no. 18PRE34030258.

# References

1. Atluri P, Panlilio CM, Liao GP, Suarez EE, McCormick RC, Hiesinger W, Cohen JE, Smith MJ, Patel AB, Feng W, Woo YJ. Transmyocardial revascularization to enhance myocardial vasculogenesis and hemodynamic function. J Thorac Cardiovasc Surg. 2008;135(2):283–91, 291.e1; discussion 91. https://doi.org/10.1016/j.jtcvs.2007.09.043. Epub 2008/02/05.
2. Grigioni F, Enriquez-Sarano M, Zehr KJ, Bailey KR, Tajik AJ. Ischemic mitral regurgitation: long-term outcome and prognostic implications with quantitative Doppler assessment. Circulation. 2001;103(13):1759–64. Epub 2001/04/03.
3. Lamas GA, Mitchell GF, Flaker GC, Smith SC Jr, Gersh BJ, Basta L, Moye L, Braunwald E, Pfeffer MA. Clinical significance of mitral regurgitation after acute myocardial infarction. Survival and Ventricular Enlargement Investigators. Circulation. 1997;96(3):827–33. Epub 1997/08/05.
4. Trichon BH, Glower DD, Shaw LK, Cabell CH, Anstrom KJ, Felker GM, O'Connor CM. Survival after coronary revascularization, with and without mitral valve surgery, in patients with ischemic mitral regurgitation. Circulation. 2003;108(Suppl 1):II103–10. https://doi.org/10.1161/01.cir.0000087656.10829.df. Epub 2003/09/13.
5. Borger MA, Alam A, Murphy PM, Doenst T, David TE. Chronic ischemic mitral regurgitation: repair, replace or rethink? Ann Thorac Surg. 2006;81(3):1153–61. https://doi.org/10.1016/j.athoracsur.2005.08.080. Epub 2006/02/21.
6. Trichon BH, Felker GM, Shaw LK, Cabell CH, O'Connor CM. Relation of frequency and severity of mitral regurgitation to survival among patients with left ventricular systolic dysfunction and heart failure. Am J Cardiol. 2003;91(5):538–43. Epub 2003/03/05.
7. Gillinov AM, Wierup PN, Blackstone EH, Bishay ES, Cosgrove DM, White J, Lytle BW, McCarthy PM. Is repair preferable to replacement for ischemic mitral regurgitation? J Thorac Cardiovasc Surg. 2001;122(6):1125–41.
8. Grossi EA, Goldberg JD, LaPietra A, Ye X, Zakow P, Sussman M, Delianides J, Culliford AT, Esposito RA, Ribakove GH, Galloway AC, Colvin SB. Ischemic mitral valve reconstruction and replacement: comparison of long-term survival and complications. J Thorac Cardiovasc Surg. 2001;122(6):1107–24. https://doi.org/10.1067/mtc.2001.116945. Epub 2001/12/01.

9. Nishimura RA, Otto CM, Bonow RO, Carabello BA, Erwin JP 3rd, Guyton RA, O'Gara PT, Ruiz CE, Skubas NJ, Sorajja P, Sundt TM 3rd, Thomas JD, American College of Cardiology/American Heart Association Task Force on Practice Guidelines. 2014 AHA/ACC guideline for the management of patients with valvular heart disease: executive summary: a report of the American College of Cardiology/American Heart Association Task Force on Practice Guidelines. J Am Coll Cardiol. 2014;63(22):2438–88. https://doi.org/10.1016/j.jacc.2014.02.537. Epub 2014/03/08.
10. McGee EC, Gillinov AM, Blackstone EH, Rajeswaran J, Cohen G, Najam F, Shiota T, Sabik JF, Lytle BW, McCarthy PM, Cosgrove DM. Recurrent mitral regurgitation after annuloplasty for functional ischemic mitral regurgitation. J Thorac Cardiovasc Surg. 2004;128(6):916–24. https://doi.org/10.1016/j.jtcvs.2004.07.037.
11. Hung J, Papakostas L, Tahta SA, Hardy BG, Bollen BA, Duran CM, Levine RA. Mechanism of recurrent ischemic mitral regurgitation after annuloplasty: continued LV remodeling as a moving target. Circulation. 2004;110(11 Suppl 1):II85–90. https://doi.org/10.1161/01.CIR.0000138192.65015.45. Epub 2004/09/15.
12. Acker MA, Parides MK, Perrault LP, Moskowitz AJ, Gelijns AC, Voisine P, Smith PK, Hung JW, Blackstone EH, Puskas JD. Mitral-valve repair versus replacement for severe ischemic mitral regurgitation. N Engl J Med. 2014;370(1):23–32.
13. Mihaljevic T, Lam BK, Rajeswaran J, Takagaki M, Lauer MS, Gillinov AM, Blackstone EH, Lytle BW. Impact of mitral valve annuloplasty combined with revascularization in patients with functional ischemic mitral regurgitation. J Am Coll Cardiol. 2007;49(22):2191–201. https://doi.org/10.1016/j.jacc.2007.02.043. Epub 2007/06/05.
14. Kron IL, Hung J, Overbey JR, Bouchard D, Gelijns AC, Moskowitz AJ, Voisine P, O'Gara PT, Argenziano M, Michler RE, Gillinov M, Puskas JD, Gammie JS, Mack MJ, Smith PK, Sai-Sudhakar C, Gardner TJ, Ailawadi G, Zeng X, O'Sullivan K, Parides MK, Swayze R, Thourani V, Rose EA, Perrault LP, Acker MA, CTSN Investigators. Predicting recurrent mitral regurgitation after mitral valve repair for severe ischemic mitral regurgitation. J Thorac Cardiovasc Surg. 2015;149(3):752–61.e1. https://doi.org/10.1016/j.jtcvs.2014.10.120. Epub 2014/12/17.
15. Bouma W, Lai EK, Levack MM, Shang EK, Pouch AM, Eperjesi TJ, Plappert TJ, Yushkevich PA, Mariani MA, Khabbaz KR, Gleason TG, Mahmood F, Acker MA, Woo YJ, Cheung AT, Jackson BM, Gorman JH 3rd, Gorman RC. Preoperative three-dimensional valve analysis predicts recurrent ischemic mitral regurgitation after mitral annuloplasty. Ann Thorac Surg. 2016;101(2):567–75. https://doi.org/10.1016/j.athoracsur.2015.09.076.
16. Gorman RC, McCaughan JS, Ratcliffe MB, Gupta KB, Streicher JT, Ferrari VA, St John-Sutton MG, Bogen DK, Edmunds LH Jr. Pathogenesis of acute ischemic mitral regurgitation in three dimensions. J Thorac Cardiovasc Surg. 1995;109(4):684–93.
17. Gorman JH 3rd, Gorman RC, Jackson BM, Enomoto Y, St John-Sutton MG, Edmunds LH Jr. Annuloplasty ring selection for chronic ischemic mitral regurgitation: lessons from the ovine model. Ann Thorac Surg. 2003;76(5):1556–63. https://doi.org/10.1016/S0003-4975(03)00891-9. pii: S0003497503008919. Epub 2003/11/07.
18. Tibayan FA, Rodriguez F, Langer F, Zasio MK, Bailey L, Liang D, Daughters GT, Ingels NB Jr, Miller DC. Annular remodeling in chronic ischemic mitral regurgitation: ring selection implications. Ann Thorac Surg. 2003;76(5):1549–54; discussion 54–5. https://doi.org/10.1016/S0003-4975(03)00880-4. pii: S0003497503008804. Epub 2003/11/07.
19. Minakawa M, Robb JD, Morital M, Koomalsinghl KJ, Vergnat M, Gillespie MJ, Gorman JH 3rd, Gorman RC. A model of ischemic mitral regurgitation in pigs with three-dimensional echocardiographic assessment. J Heart Valve Dis. 2014;23(6):713–20. Epub 2015/03/21.
20. Jassar AS, Vergnat M, Jackson BM, McGarvey JR, Cheung AT, Ferrari G, Woo YJ, Acker MA, Gorman RC, Gorman JH. Regional annular geometry in patients with mitral regurgitation: implications for annuloplasty ring selection. Ann Thorac Surg. 2014;97(1):64–70.
21. Otsuji Y, Handschumacher MD, Schwammenthal E, Jiang L, Song JK, Guerrero JL, Vlahakes GJ, Levine RA. Insights from three-dimensional echocardiography into the mechanism of functional mitral regurgitation: direct in vivo demonstration of altered leaflet tethering geometry. Circulation. 1997;96(6):1999–2008.

22. Srichai MB, Grimm RA, Stillman AE, Gillinov AM, Rodriguez LL, Lieber ML, Lara A, Weaver JA, McCarthy PM, White RD. Ischemic mitral regurgitation: impact of the left ventricle and mitral valve in patients with left ventricular systolic dysfunction. Ann Thorac Surg. 2005;80 (1):170–8. https://doi.org/10.1016/j.athoracsur.2005.01.068. Epub 2005/06/25.
23. Gorman JH 3rd, Jackson BM, Enomoto Y, Gorman RC. The effect of regional ischemia on mitral valve annular saddle shape. Ann Thorac Surg. 2004;77(2):544–8.
24. Ryan LP, Jackson BM, Parish LM, Plappert TJ, St John-Sutton MG, Gorman JH 3rd, Gorman RC. Regional and global patterns of annular remodeling in ischemic mitral regurgitation. Ann Thorac Surg. 2007;84(2):553–9. https://doi.org/10.1016/j.athoracsur.2007.04.016. Epub 2007/07/24.
25. Vergnat M, Jassar AS, Jackson BM, Ryan LP, Eperjesi TJ, Pouch AM, Weiss SJ, Cheung AT, Acker MA, Gorman JH 3rd, Gorman RC. Ischemic mitral regurgitation: a quantitative three-dimensional echocardiographic analysis. Ann Thorac Surg. 2011;91(1):157–64. https://doi.org/10.1016/j.athoracsur.2010.09.078. Epub 2010/12/22.
26. Kuwahara E, Otsuji Y, Iguro Y, Ueno T, Zhu F, Mizukami N, Kubota K, Nakashiki K, Yuasa T, Yu B, Uemura T, Takasaki K, Miyata M, Hamasaki S, Kisanuki A, Levine RA, Sakata R, Tei C. Mechanism of recurrent/persistent ischemic/functional mitral regurgitation in the chronic phase after surgical annuloplasty: importance of augmented posterior leaflet tethering. Circulation. 2006;114(1 Suppl):I529–34. https://doi.org/10.1161/CIRCULATIONAHA.105.000729. Epub 2006/07/06.
27. Khang A, Buchanan RM, Ayoub S, Rego BV, Lee CH, Ferrari G, Anseth KS, Sacks MS. Mechanobiology of the heart valve interstitial cell: Simulation, experiment, and discovery. In: Verbruggen SW, editor. Mechanobiology in health and disease, vol. 2018. London: Elsevier; 2018. p. 249–83.
28. Sacks MS, Khalighi A, Rego B, Ayoub S, Drach A. On the need for multi-scale geometric modelling of the mitral heart valve. Healthc Technol Lett. 2017;4(5):150.
29. Rego BV, Ayoub S, Khalighi AH, Drach A, Gorman JH, Gorman RC, Sacks MS. Alterations in mechanical properties and in vivo geometry of the mitral valve following myocardial infarction. Summer biomechanics, bioengineering and biotransport conference, Tucson, AZ, USA; 2017.
30. Khalighi AH, Drach A, Sacks MS. Patient-specific mitral valve annuloplasty repair: The optimal ring design for treating ischemic mitral regurgitation. Summer biomechanics, bioengineering and biotransport conference, Tucson, AZ, USA; 2017.
31. Carpentier A. [Reconstructive valvuloplasty. A new technique of mitral valvuloplasty]. La Presse medicale. 1969;77(7):251–3.
32. Carpentier A, Deloche A, Dauptain J, Soyer R, Blondeau P, Piwnica A, Dubost C, McGoon DC. A new reconstructive operation for correction of mitral and tricuspid insufficiency. J Thorac Cardiovasc Surg. 1971;61(1):1–13.
33. Carpentier A. Cardiac valve surgery—the "French correction". J Thorac Cardiovasc Surg. 1983;86(3):323–37.
34. Ryan LP, Jackson BM, Eperjesi TJ, Plappert TJ, St John-Sutton M, Gorman RC, Gorman JH 3rd. A methodology for assessing human mitral leaflet curvature using real-time 3-dimensional echocardiography. J Thorac Cardiovasc Surg. 2008;136(3):726–34. https://doi.org/10.1016/j.jtcvs.2008.02.073. Epub 2008/09/23.
35. Ryan LP, Jackson BM, Hamamoto H, Eperjesi TJ, Plappert TJ, St John-Sutton M, Gorman RC, Gorman JH 3rd. The influence of annuloplasty ring geometry on mitral leaflet curvature. Ann Thorac Surg. 2008;86(3):749–60; discussion 60. https://doi.org/10.1016/j.athoracsur.2008.03.079. pii: S0003-4975(08)00726-1. Epub 2008/08/30.
36. Robb JD, Minakawa M, Koomalsingh KJ, Shuto T, Jassar AS, Ratcliffe SJ, Gorman RC, Gorman JH 3rd. Posterior leaflet augmentation improves leaflet tethering in repair of ischemic mitral regurgitation. Eur J Cardiothorac Surg. 2011;40:1501–7. https://doi.org/10.1016/j.ejcts.2011.02.079. Epub 2011/05/07.
37. Jassar AS, Minakawa M, Shuto T, Robb JD, Koomalsingh KJ, Levack MM, Vergnat M, Eperjesi TJ, Jackson BM, Gorman JH III. Posterior leaflet augmentation in ischemic mitral regurgitation increases leaflet coaptation and mobility. Ann Thorac Surg. 2012;94(5):1438–45.

38. Onorati F, Rubino AS, Marturano D, Pasceri E, Santarpino G, Zinzi S, Mascaro G, Renzulli A. Midterm clinical and echocardiographic results and predictors of mitral regurgitation recurrence following restrictive annuloplasty for ischemic cardiomyopathy. J Thorac Cardiovasc Surg. 2009;138(3):654–62. https://doi.org/10.1016/j.jtcvs.2009.01.020. Epub 2009/08/25.
39. Kron IL, Green GR, Cope JT. Surgical relocation of the posterior papillary muscle in chronic ischemic mitral regurgitation. Ann Thorac Surg. 2002;74(2):600–1. Epub 2002/08/14.
40. Masuyama S, Marui A, Shimamoto T, Nonaka M, Tsukiji M, Watanabe N, Ikeda T, Yoshida K, Komeda M. Chordal translocation for ischemic mitral regurgitation may ameliorate tethering of the posterior and anterior mitral leaflets. J Thorac Cardiovasc Surg. 2008;136(4):868–75. https://doi.org/10.1016/j.jtcvs.2008.06.034. Epub 2008/10/29.
41. Arai H, Itoh F, Someya T, Oi K, Tamura K, Tanaka H. New surgical procedure for ischemic/functional mitral regurgitation: mitral complex remodeling. Ann Thorac Surg. 2008;85(5):1820–2. https://doi.org/10.1016/j.athoracsur.2007.11.073. Epub 2008/04/30.
42. Fayad G, Marechaux S, Modine T, Azzaoui R, Larrue B, Ennezat PV, Bekhti H, Decoene C, Deklunder G, Le Tourneau T, Warembourg H. Chordal cutting VIA aortotomy in ischemic mitral regurgitation: surgical and echocardiographic study. J Card Surg. 2008;23(1):52–7. https://doi.org/10.1111/j.1540-8191.2007.00503.x. Epub 2008/02/23.
43. Kincaid EH, Riley RD, Hines MH, Hammon JW, Kon ND. Anterior leaflet augmentation for ischemic mitral regurgitation. Ann Thorac Surg. 2004;78(2):564–8; discussion 8. https://doi.org/10.1016/j.athoracsur.2004.02.040. Epub 2004/07/28.
44. de Varennes B, Chaturvedi R, Sidhu S, Cote AV, Shan WL, Goyer C, Hatzakorzian R, Buithieu J, Sniderman A. Initial results of posterior leaflet extension for severe type IIIb ischemic mitral regurgitation. Circulation. 2009;119(21):2837–43. https://doi.org/10.1161/CIRCULATIONAHA.108.831412. Epub 2009/05/20.
45. Dobre M, Koul B, Rojer A. Anatomic and physiologic correction of the restricted posterior mitral leaflet motion in chronic ischemic mitral regurgitation. J Thorac Cardiovasc Surg. 2000;120(2):409–11. https://doi.org/10.1067/mtc.2000.106521. Epub 2000/08/05.
46. Rendon F, Aramendi JI, Rodrigo D, Baraldi C, Martinez P. Patch enlargement of the posterior mitral leaflet in ischemic regurgitation. Asian Cardiovasc Thorac Ann. 2002;10(3):248–50. https://doi.org/10.1177/021849230201000313. Epub 2002/09/06.
47. Rabbah J-P, Saikrishnan N, Yoganathan AP. A novel left heart simulator for the multi-modality characterization of native mitral valve geometry and fluid mechanics. Ann Biomed Eng. 2013;41(2):305–15. https://doi.org/10.1007/s10439-012-0651-z.
48. Fan R, Sacks MS. Simulation of planar soft tissues using a structural constitutive model: finite element implementation and validation. J Biomech. 2014;47:2043–54.
49. Drach A, Khalighi AH, Sacks MS. A comprehensive pipeline for multi-resolution modeling of the mitral valve: validation, computational efficiency, and predictive capability. Int J Numer Methods Biomed Eng. 2017;34:e2921. https://doi.org/10.1002/cnm.2921. Epub 2017/08/05.
50. Pouch AM, Jackson BM, Yushkevich PA, Gorman JH 3rd, Gorman RC. 4D-transesophageal echocardiography and emerging imaging modalities for guiding mitral valve repair. Ann Cardiothorac Surg. 2015;4(5):461–2. https://doi.org/10.3978/j.issn.2225-319X.2015.02.01.
51. Jassar AS, Brinster CJ, Vergnat M, Robb JD, Eperjesi TJ, Pouch AM, Cheung AT, Weiss SJ, Acker MA, Gorman JH 3rd, Gorman RC, Jackson BM. Quantitative mitral valve modeling using real-time three-dimensional echocardiography: technique and repeatability. Ann Thorac Surg. 2011;91(1):165–71. https://doi.org/10.1016/j.athoracsur.2010.10.034. Epub 2010/12/22.
52. Rego BV, Khalighi AH, Drach A, Lai EK, Pouch AM, Gorman RC, Gorman JH 3rd, Sacks MS. A non-invasive method for the determination of in vivo mitral valve leaflet strains. Int J Numer Meth Biomed Eng. https://doi.org/10.1002/cnm.3142; e3142.
53. Khalighi AH, Rego BV, Drach A, Gorman RC, Gorman JH 3rd, Sacks MS. Development of a functionally equivalent model of the mitral valve chordae tendineae through topology optimization. Ann Biomed Eng. https://doi.org/10.1007/s10439-018-02122-y.
54. Aggarwal A, Aguilar VS, Lee C-H, Ferrari G, Gorman JH, Gorman RC, Sacks MS. Patient-specific modeling of heart valves: from image to simulation. In: Ourselin S, Rueckert D, Smith N, editors. Functional imaging and modeling of the heart. London: Springer; 2013. p. 141–9.

55. Lee C-H, Oomen PA, Rabbah J, Yoganathan A, Gorman R, Gorman J III, Amini R, Sacks M. A high-fidelity and micro-anatomically accurate 3D finite element model for simulations of functional mitral valve. In: Ourselin S, Rueckert D, Smith N, editors. Functional imaging and modeling of the heart. Berlin: Springer; 2013. p. 416–24.
56. Lee CH, Amini R, Gorman RC, Gorman JH 3rd, Sacks MS. An inverse modeling approach for stress estimation in mitral valve anterior leaflet valvuloplasty for in-vivo valvular biomaterial assessment. J Biomech. 2014;47(9):2055–63. https://doi.org/10.1016/j.jbiomech.2013.10.058.
57. Lee C-H, Amini R, Sakamoto Y, Carruthers CA, Aggarwal A, Gorman RC, Gorman JH III, Sacks MS. Mitral valves: a computational framework. In: Multiscale modeling in biomechanics and mechanobiology. London: Springer; 2015. p. 223–55.
58. Rego BV, Wells SM, Lee CH, Sacks MS. Mitral valve leaflet remodelling during pregnancy: insights into cell-mediated recovery of tissue homeostasis. J R Soc Interface. 2016;13 (125):20160709. https://doi.org/10.1098/rsif.2016.0709.
59. Lee CH, Zhang W, Feaver K, Gorman RC, Gorman JH 3rd, Sacks MS. On the in vivo function of the mitral heart valve leaflet: insights into tissue-interstitial cell biomechanical coupling. Biomech Model Mechanobiol. 2017;16:1613–32. https://doi.org/10.1007/s10237-017-0908-4.
60. Lee CH, Rabbah JP, Yoganathan AP, Gorman RC, Gorman JH 3rd, Sacks MS. On the effects of leaflet microstructure and constitutive model on the closing behavior of the mitral valve. Biomech Model Mechanobiol. 2015;14(6):1281–302. https://doi.org/10.1007/s10237-015-0674-0.
61. Khalighi AH, Drach A, ter Huurne FM, Lee C-H, Bloodworth C, Pierce EL, Jensen MO, Yoganathan AP, Sacks MS. A comprehensive framework for the characterization of the complete mitral valve geometry for the development of a population-averaged model. In: Functional imaging and modeling of the heart. Cham: Springer; 2015. p. 164–71.
62. Lee CH, Zhang W, Liao J, Carruthers CA, Sacks JI, Sacks MS. On the presence of affine fibril and fiber kinematics in the mitral valve anterior leaflet. Biophys J. 2015;108(8):2074–87. https://doi.org/10.1016/j.bpj.2015.03.019.
63. Khalighi AH, Drach A, Bloodworth CH 4th, Pierce EL, Yoganathan AP, Gorman RC, Gorman JH 3rd, Sacks MS. Mitral valve chordae tendineae: topological and geometrical characterization. Ann Biomed Eng. 2016;45:378–93. https://doi.org/10.1007/s10439-016-1775-3.
64. Bloodworth CH 4th, Pierce EL, Easley TF, Drach A, Khalighi AH, Toma M, Jensen MO, Sacks MS, Yoganathan AP. Ex vivo methods for informing computational models of the mitral valve. Ann Biomed Eng. 2016;45:496–507. https://doi.org/10.1007/s10439-016-1734-z.
65. Khalighi AH, Drach A, Gorman RC, Gorman JH 3rd, Sacks MS. Multi-resolution geometric modeling of the mitral heart valve leaflets. Biomech Model Mechanobiol. 2018;17:351–66. https://doi.org/10.1007/s10237-017-0965-8. Epub 2017/10/07.

Printed by Printforce, the Netherlands